Introduction to Linear Algebra and Differential Equations

JOHN W. DETTMAN

Professor of Mathematics
Oakland University
Rochester, Michigan

Dover Publications, Inc.
New York

This Dover edition, first published in 1986, is an unabridged, corrected republication of the work originally published by the McGraw-Hill Book Company, New York, 1974.

Library of Congress Cataloging-in-Publication Data

Dettman, John W. (John Warren)
Introduction to linear algebra and differential equations.

Originally published: New York : McGraw-Hill, 1974.
Includes bibliographical references and index.
1. Algebras, Linear. 2. Differential equations. I. Title.
QA184.D47 1986 512'.5 86-6311
ISBN-13: 978-0-486-65191-0
ISBN-10: 0-486-65191-6

Manufactured in the United States by LSC Communications
65191611 2017
www.doverpublications.com

CONTENTS

PREFACE

Since 1965, when the Committee on the Undergraduate Program in Mathematics (CUPM) of the Mathematical Association of America recommended that linear algebra be taught as part of the introductory calculus sequence, it has become quite common to find substantial amounts of linear algebra in the calculus curriculum of all students at the sophomore level. This is a natural development because it is now pretty well conceded that not only is linear algebra indispensable to the mathematics major, but that it is that part of algebra which is most useful in the application of mathematical analysis to other areas, e.g., linear programming, systems analysis, statistics, numerical analysis, combinatorics, and mathematical physics. Even in nonlinear analysis, linear algebra is essential because of the commonly used technique of dealing with the nonlinear phenomenon as a perturbation of the linear case. We also find linear algebra prerequisite to areas of mathematics such as multivariable analysis, complex variables, integration theory, functional analysis, vector and tensor analysis, ordinary and partial differential equations, integral equations, and probability. So much for the case for linear algebra.

The other two general topics usually found in the sophomore program are multivariable calculus and differential equations. In fact, modern calculus texts have generally included (in the second volume) large portions of linear algebra and multivariable calculus, and, to a more limited extent, differential equations. I have written this book to show that it makes good sense to package linear algebra with an introductory course in differential equations. On the other hand, the linear algebra included here (vectors, matrices, vector spaces, linear transformations, and characteristic value problems) is an essential prerequisite for multivariable calculus. Hence, this volume could become the text for the first half of the sophomore year, followed by any one of a number of good multivariable calculus books which either include linear algebra or depend on it. The prerequisite for this material is a one-year introductory calculus course with some mention of partial derivatives.

I have tried throughout this book to progress from familiar ideas to the more difficult and abstract. Hence, two-dimensional vectors are introduced after a study of complex numbers, matrices with linear equations, vector spaces after two- and three-dimensional euclidean vectors, linear transformations after matrices, higher order linear differential equations after first order linear equations, etc. Systems of differential equations are left to the end after the student has gained some experience with scalar equations. Geometric ideas are kept in the forefront while treating algebraic concepts, and applications are brought in as often as possible to illustrate the theory. There are worked-out examples in every section and numerous exercises to reinforce or extend the material of the text. Numerical methods are introduced in Chap. 5 in connection with first order equations. The starred sections at the end of each chapter are not an essential part of the book. In fact none of the unstarred sections depend on them. They are included in the book because (1) they are related to or extend the basic material and (2) I wanted to include some advanced topics to challenge and stimulate the more ambitious student to further study. These starred sections include a variety of mathematical topics such as:

1 Analytic functions of a complex variable.
2 Power series.
3 Existence and uniqueness theory for algebraic equations.
4 Hilbert spaces.
5 Jordan forms.
6 Picard iteration.
7 Green's functions.
8 Integral equations.
9 Weierstrass approximation theorem.
10 Bernstein polynomials.
11 Lerch's theorem.

12 Power series solution of differential equations.
13 Existence and uniqueness theory for systems of differential equations.
14 Grönwald's inequality.

The book can be used in a variety of different courses. The ideal situation would be a two-quarter course with linear algebra for the first and differential equations for the second. For a semester course with more emphasis on linear algebra, Chaps. 1–6 would give a fairly good introduction to linear differential equations with applications to engineering (damped harmonic oscillator). If one wished less emphasis on linear algebra and more on differential equations, Chap. 4 could be skipped since the characteristic value problem is not used in an essential way until Chap. 9. Chapters 7, 8, and 9 are independent so that a variety of topics could be introduced after Chap. 6, depending on the interests of the class. For a class with a good background in complex variables and linear algebraic equations, Chaps. 1 and 2 could be skipped.

About half of the book was written during the academic year 1970–71 while I was a Senior Research Fellow at the University of Glasgow. I want to thank Professors Ian Sneddon and Robert Rankin for allowing me to use the facilities of the University. I also wish to thank Mr. Alexander McDonald and Mr. Iain Bain, students at the University of Glasgow, who checked the exercises and made many helpful suggestions. The first six chapters have been used in a course at Oakland University. I am indebted to these students for their patience in studying from a set of notes which were far from polished. Finally, I want to thank my family for putting up with my lack of attentiveness while I was in the process of preparing this manuscript.

JOHN W. DETTMAN

1

COMPLEX NUMBERS

1.1 INTRODUCTION

There are several reasons for beginning this book with a chapter on complex numbers. (1) Many students do not feel confident in calculating with complex numbers, even though this is a topic which should be carefully covered in the high school curriculum. (2) The complex numbers represent a very elementary example of a vector space. We shall, in fact, use the complex numbers to introduce the two-dimensional euclidean vectors. (3) Even if we were to attempt to avoid vector spaces over the complex numbers by using only real scalar multipliers, we would eventually have to deal with complex characteristic values and characteristic vectors. (4) The most efficient way to deal with the solution of linear differential equations with constant coefficients is through the exponential function of a complex variable.

We shall first define the algebra of complex numbers and then the geometry of the complex plane. This will lead us in a natural way to a treatment of two-dimensional euclidean vectors. Next we shall introduce complex-valued functions, both of a single real variable and of a single complex variable. This will be followed by a careful treatment of the exponential function. The last section (which is starred) is intended for the more ambitious students. It discusses power series as a function of a complex variable. Here we shall justify the properties of the exponential function and lay the groundwork for the study of analytic functions of a complex variable.

1.2 THE ALGEBRA OF COMPLEX NUMBERS

We shall represent complex numbers in the form $z = x + iy$, where x and y are real numbers. As a matter of notation we say that x is the *real part* of z [$x = \text{Re}(z)$] and y is the *imaginary part* of z [$y = \text{Im}(z)$]. We say that two complex numbers are equal if and only if their real parts are equal and their imaginary parts are equal. We could say that $i = \sqrt{-1}$ except that for a person who has experience only with real numbers, there is no number which when squared gives -1 (if a is real, $a^2 \geq 0$). It is better simply to say that i is a complex number and then define its powers; i, $i^2 = -1$, $i^3 = -i$, $i^4 = 1$, etc. We can now define *addition* and *multiplication* of complex numbers in a natural way:

$$z_1 + z_2 = (x_1 + iy_1) + (x_2 + iy_2) = (x_1 + x_2) + i(y_1 + y_2)$$

$$\begin{aligned} z_1 z_2 &= (x_1 + iy_1)(x_2 + iy_2) = x_1 x_2 + i^2 y_1 y_2 + i x_1 y_2 + i y_1 x_2 \\ &= (x_1 x_2 - y_1 y_2) + i(x_1 y_2 + y_1 x_2) \end{aligned}$$

With these definitions it is easy to show that addition and multiplication are both associative and commutative operations, that is,

$$\begin{aligned} (z_1 + z_2) + z_3 &= z_1 + (z_2 + z_3) \\ z_1 + z_2 &= z_2 + z_1 \\ z_1(z_2 z_3) &= (z_1 z_2) z_3 \\ z_1 z_2 &= z_2 z_1 \end{aligned}$$

If a is a real number, we can represent it as a complex number as follows: $a = a + i0$. Hence we see that the real numbers are contained in the complex numbers. This statement would have little meaning, however, unless the algebraic operations of the real numbers were preserved within the context of the complex numbers. As a starter we have

$$\begin{aligned} a + b &= (a + i0) + (b + i0) = (a + b) + i0 \\ ab &= (a + i0)(b + i0) = ab + i0 \end{aligned}$$

We can, of course, verify the consistency of the other operations as they are defined for the complex numbers.

Let a be real and let $z = x + iy$. Then $az = (a + i0)(x + iy) = ax + iay$. In other words, multiplication of a complex number by a real number a is accomplished by multiplying both real and imaginary parts by the real number a. With this in mind, we define the *negative* of a complex number z by

$$-z = (-1)z = (-x) + i(-y)$$

The *zero* of the complex numbers is $0 + i0 = 0$, and we have the obvious property

$$\begin{aligned} z + (-z) &= x + iy + (-x) + i(-y) \\ &= 0 + i0 \\ &= 0 \end{aligned}$$

We can now state an obvious theorem, which we put together for a reason which will become clear later.

Theorem 1.2.1

(i) For all complex numbers z_1 and z_2, $z_1 + z_2 = z_2 + z_1$.

(ii) For all complex numbers z_1, z_2, and z_3,

$$z_1 + (z_2 + z_3) = (z_1 + z_2) + z_3$$

(iii) For all complex numbers z, $z + 0 = z$.

(iv) For each complex number z there exists a negative $-z$ such that $z + (-z) = 0$.

We define *subtraction* in terms of addition of the negative, that is,

$$\begin{aligned} z_1 - z_2 = z_1 + (-z_2) &= (x_1 + iy_1) + (-x_2) + i(-y_2) \\ &= (x_1 - x_2) + i(y_1 - y_2) \end{aligned}$$

Suppose $z = x + iy$, and we look for a reciprocal complex number $z^{-1} = u + iv$ such that $zz^{-1} = 1$. Then

$$(x + iy)(u + iv) = (xu - yv) + i(xv + yu) = 1 + i0$$

Then $xu - yv = 1$ and $xv + yu = 0$. These equations have a unique solution if and only if $x^2 + y^2 \neq 0$. The solution is

$$u = \frac{x}{x^2 + y^2} \qquad v = \frac{-y}{x^2 + y^2}$$

Therefore, we see that every complex number z, *except zero*, has a unique reciprocal,

$$z^{-1} = \frac{x}{x^2 + y^2} - i\frac{y}{x^2 + y^2}$$

We can now define *division* by any complex number, other than zero, in terms of multiplication by the reciprocal; that is, if $z_2 \neq 0$,

$$\frac{z_1}{z_2} = z_1 z_2^{-1} = (x_1 + iy_1)\left(\frac{x_2}{x_2^2 + y_2^2} - i\frac{y_2}{x_2^2 + y_2^2}\right)$$

$$= \frac{x_1 x_2 + y_1 y_2}{x_2^2 + y_2^2} + i\frac{x_2 y_1 - x_1 y_2}{x_2^2 + y_2^2}$$

As a mnemonic, note that $(x_2 + iy_2)(x_2 - iy_2) = x_2^2 + y_2^2$ and hence

$$\frac{z_1}{z_2} = \frac{x_1 + iy_1}{x_2 + iy_2}\frac{x_2 - iy_2}{x_2 - iy_2} = \frac{x_1 x_2 + y_1 y_2 + i(x_2 y_1 - x_1 y_2)}{x_2^2 + y_2^2}$$

EXAMPLE 1.2.1 Let $z_1 = 2 + 3i$ and $z_2 = -1 + 4i$. Then $z_1 + z_2 = (2 - 1) + (3 + 4)i = 1 + 7i$, $z_1 - z_2 = (2 + 1) + (3 - 4)i = 3 - i$, $z_1 z_2 = (-2 - 12) + i(8 - 3) = -14 + 5i$, and

$$\frac{z_1}{z_2} = \frac{2 + 3i}{-1 + 4i}\frac{-1 - 4i}{-1 - 4i} = \frac{-2 + 12 - 3i - 8i}{1 + 16} = \frac{10}{17} - \frac{11}{17}i$$

The reader should recall the important *distributive law* from his study of the real numbers. The same property holds for complex numbers; that is,

$$z_1(z_2 + z_3) = z_1 z_2 + z_1 z_3$$

The proof will be left to the reader.

We summarize what we have said so far in the following omnibus theorem (the reader will be asked for some of the proofs in the exercises).

Theorem 1.2.2 The operations of addition, multiplication, subtraction, and division (except by zero) are defined for complex numbers. As far as these operations are concerned, the complex numbers behave like real numbers.† The real numbers are contained in the complex numbers, and the above operations are consistent with the previously defined operations for the real numbers.

There is one property of the real numbers which does not carry over to the complex numbers. The complex numbers are not ordered as the reals are.

† In algebra we say that both the real and complex numbers are *algebraic fields*. The reals are a subfield of the complex numbers.

Recall that for real numbers 1 and -1 cannot both be positive. Also if $a \neq 0$, then a^2 is positive. If the complex numbers had the same properties of order, both $1^2 = 1$ and $i^2 = -1$ would be positive. Therefore, we shall not try to order the complex numbers and/or write inequalities between complex numbers.

We conclude this section with two special definitions which are important in the study of complex numbers. The first is *absolute value*, which we denote with the symbol $|z|$. The absolute value is defined for every complex number $z = x + iy$ as

$$|z| = \sqrt{x^2 + y^2}$$

It is easy to show that $|z| \geq 0$ and $|z| = 0$ if and only if $z = 0$.

The other is the *conjugate*, denoted by $\bar{z}$. The conjugate of $z = x + iy$ is defined as

$$\bar{z} = x - iy$$

The proof of the following theorem will be left for the reader.

Theorem 1.2.3

(i) $\overline{z_1 + z_2} = \bar{z}_1 + \bar{z}_2$.
(ii) $\overline{z_1 - z_2} = \bar{z}_1 - \bar{z}_2$.
(iii) $\overline{z_1 z_2} = \bar{z}_1 \bar{z}_2$.
(iv) $\overline{z_1/z_2} = \bar{z}_1/\bar{z}_2$.
(v) $z\bar{z} = |z|^2$.

EXERCISES 1.2

1 Let $z_1 = 2 + i$ and $z_2 = -3 + 5i$. Compute $z_1 + z_2$, $z_1 - z_2$, $z_1 z_2$, z_1/z_2, $\bar{z}_1$, $\bar{z}_2$, $|z_1|$, and $|z_2|$.

2 Let $z_1 = -1 + 3i$ and $z_2 = 2 - 4i$. Compute $z_1 + z_2$, $z_1 - z_2$, $z_1 z_2$, z_1/z_2, $\bar{z}_1$, $\bar{z}_2$, $|z_1|$, and $|z_2|$.

3 Prove that addition of complex numbers is associative and commutative.

4 Prove that multiplication of complex numbers is associative and commutative.

5 Prove the distributive law.

6 Show that subtraction and division of real numbers is consistent within the context of complex numbers.

7 Show that the equations $xu - yv = 1$ and $xv + yu = 0$ have a unique solution for u and v if and only if $x^2 + y^2 \neq 0$.

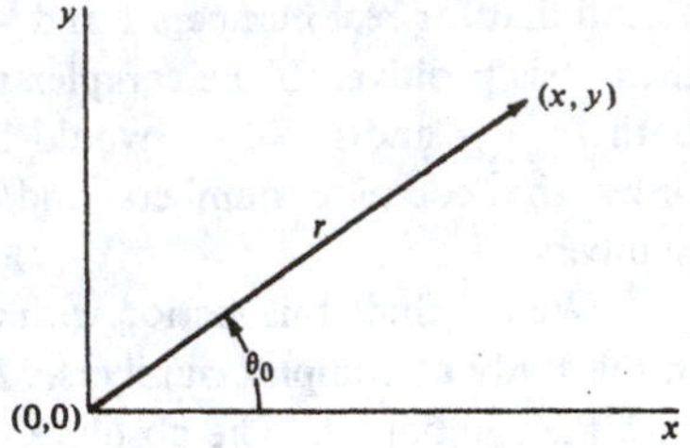

FIGURE 1

8 Let a and b be real and z_1 and z_2 be complex. Prove the following:
(a) $a(z_1 + z_2) = az_1 + az_2$ (b) $(a + b)z_1 = az_1 + bz_1$
(c) $a(bz_1) = (ab)z_1$ (d) $1z_1 = z_1$

9 Show that $|z| \geq 0$ and $|z| = 0$ if and only if $z = 0$.

10 Prove Theorem 1.2.3.

11 Show that

$$\operatorname{Re}(z) = \tfrac{1}{2}(z + \bar{z}) \qquad \operatorname{Im}(z) = \frac{1}{2i}(z - \bar{z})$$

12 Show that the definition of absolute value for real numbers is consistent with that for complex numbers.

13 Let $z = x + iy$ and $w = u + iv$. Prove that $(xu + yv)^2 \leq |z|^2|w|^2$ and hence that $|xu + yv| \leq |z|\,|w|$.

14 Use the result of Exercise 13 to show that $|z + w| \leq |z| + |w|$.

15 Use the result of Exercise 14 to show that

$$|z_1 + z_2 + \cdots + z_n| \leq |z_1| + |z_2| + \cdots + |z_n|$$

16 Show that $|z - w| \geq \big||z| - |w|\big|$.

1.3 THE GEOMETRY OF COMPLEX NUMBERS

It will be very useful to give the complex numbers a geometric interpretation. This will be done by associating the complex number $z = x + iy$ with the point (x,y) in the euclidean plane (see Fig. 1). It is customary to draw an arrow from the origin (0,0) to the point (x,y). For each complex number $z = x + iy$ there is a unique point (x,y) in the plane and (except for $z = 0$) a unique arrow from the origin to the point (x,y).

There is also a polar-coordinate representation of the complex numbers. Let r equal the length of the arrow and θ_0 be the minimum angle measured

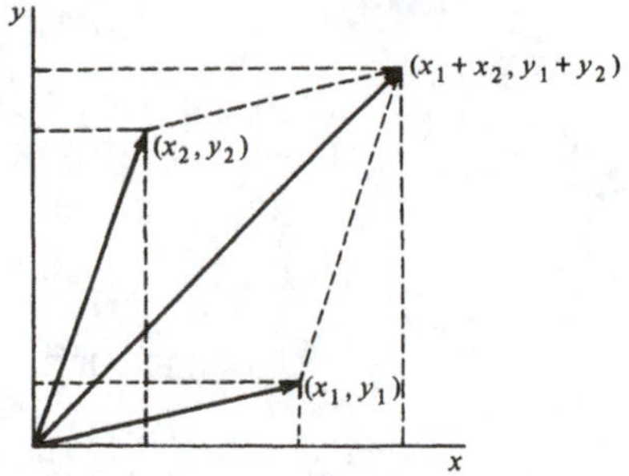

FIGURE 2

from the positive x axis to the arrow in the counterclockwise direction. Then

$$r = \sqrt{x^2 + y^2} = |z|$$

$$\theta_0 = \tan^{-1} \frac{y}{x}$$

where θ_0 is that value of the inverse tangent such that $0 \le \theta_0 < 2\pi$, where $\cos \theta_0 = x/|z|$ and $\sin \theta_0 = y/|z|$. Then

$$z = x + iy = r(\cos \theta_0 + i \sin \theta_0)$$

According to the convention we have adopted, r and θ_0 are uniquely defined except for $z = 0$ (θ_0 is not defined for $z = 0$). However, we note that

$$\cos \theta_0 + i \sin \theta_0 = \cos (\theta_0 \pm 2k\pi) + i \sin (\theta_0 \pm 2k\pi)$$

for any positive integer k. Therefore, we shall let $\theta = \theta_0 + 2k\pi$, $k = 0, 1, 2, 3, \ldots$, and then $z = r(\cos \theta + i \sin \theta)$, and we call θ the argument of z ($\theta = \arg z$), realizing full well that $\arg z$ is defined only to within multiples of 2π.

The algebraic operations on complex numbers can now be interpreted geometrically. Let us begin with addition. Let $z_1 = x_1 + iy_1$ and $z_2 = x_2 + iy_2$. Then $z_1 + z_2 = (x_1 + x_2) + i(y_1 + y_2)$. Referring to Fig. 2, we see that the arrow which corresponds to the sum $z_1 + z_2$ is along the diagonal of a parallelogram formed with the sides corresponding to z_1 and z_2. Thus the rule to form the sum of two complex numbers geometrically is as follows: construct the parallelogram formed by the two arrows corresponding to the complex numbers z_1 and z_2; then the sum $z_1 + z_2$ corresponds to the arrow from the origin along the diagonal to the opposite vertex. If the arrows lie along the same line, obvious modifications in the rule need to be made.

The difference between two complex numbers, $z_1 - z_2$, can be formed

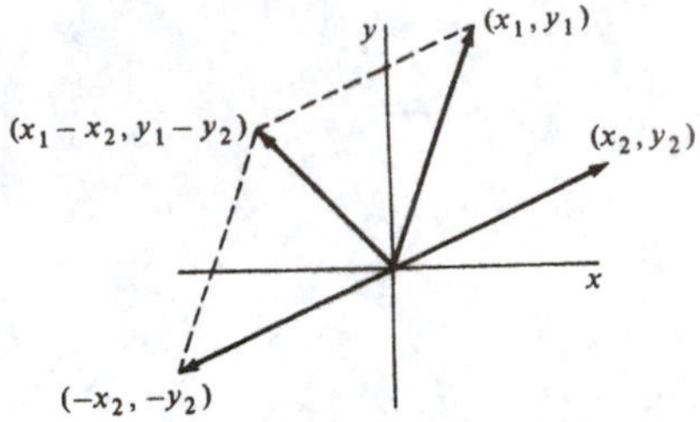

FIGURE 3

geometrically by constructing the diagonal of the parallelogram formed by z_1 and $-z_2$ (see Fig. 3).

To interpret the product of two complex numbers geometrically we use the polar-coordinate representation. Let $z_1 = r_1(\cos\theta_1 + i\sin\theta_1)$ and $z_2 = r_2(\cos\theta_2 + i\sin\theta_2)$. Then

$$\begin{aligned} z_1 z_2 &= r_1 r_2(\cos\theta_1 + i\sin\theta_1)(\cos\theta_2 + i\sin\theta_2) \\ &= r_1 r_2[(\cos\theta_1 \cos\theta_2 - \sin\theta_1 \sin\theta_2) \\ &\qquad + i(\sin\theta_1 \cos\theta_2 + \cos\theta_1 \sin\theta_2)] \\ &= r_1 r_2[\cos(\theta_1 + \theta_2) + i\sin(\theta_1 + \theta_2)] \end{aligned}$$

Figure 4 shows the interpretation of this result geometrically. This result also gives us an important theorem.

Theorem 1.3.1 For all complex numbers z_1 and z_2, $|z_1 z_2| = |z_1|\,|z_2|$. For all nonzero complex numbers z_1 and z_2, $\arg z_1 z_2 = \arg z_1 + \arg z_2$.

The quotient of two complex numbers can be similarly interpreted. Let $z_2 \neq 0$ and $z_1/z_2 = z_3$. Then $z_1 = z_2 z_3$ and $|z_1| = |z_2|\,|z_3|$, $\arg z_1 = \arg z_2 + \arg z_3$. Since $|z_2| \neq 0$, $|z_3| = |z_1|/|z_2|$; and if $z_1 \neq 0$, $z_3 \neq 0$, then $\arg z_3 = \arg z_1 - \arg z_2$.

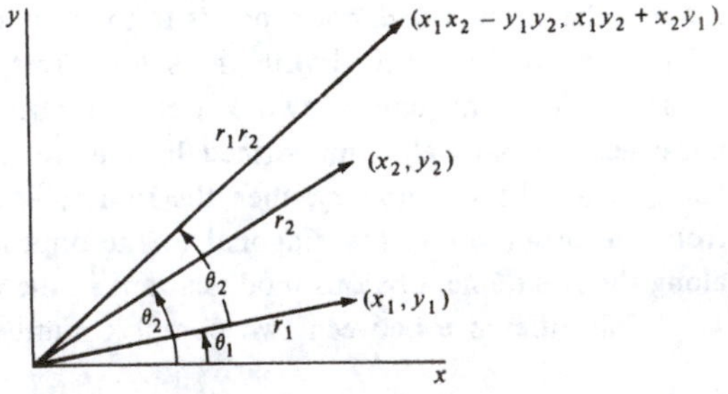

FIGURE 4

This proves the following theorem:

Theorem 1.3.2 For all complex numbers z_1 and $z_2 \neq 0$, $|z_1/z_2| = |z_1|/|z_2|$. For all nonzero complex numbers z_1 and z_2, $\arg(z_1/z_2) = \arg z_1 - \arg z_2$.

Powers of a complex number z have the following simple interpretation. Let $z = r(\cos\theta + i\sin\theta)$; then $z^2 = r^2(\cos 2\theta + i\sin 2\theta)$, and by induction $z^n = r^n(\cos n\theta + i\sin n\theta)$ for all positive integers n. For all $z \neq 0$ we define $z^0 = 1$, and of course $z^{-1} = r^{-1}[\cos(-\theta) + i\sin(-\theta)]$. Then for all positive integers m, $z^{-m} = r^{-m}[\cos(-m\theta) + i\sin(-m\theta)]$. Therefore, we have for all integers n and all $z \neq 0$

$$z^n = r^n(\cos n\theta + i\sin n\theta)$$

Having looked at powers, we can study roots of complex numbers. We wish to solve the equation $z^n = c$, where n is a positive integer and c is a complex number. If $c = 0$, clearly $z = 0$, so let us consider only $c \neq 0$. Let $|c| = \rho$ and $\arg c = \phi$, keeping in mind that ϕ is multiple-valued. Then

$$z^n = r^n(\cos n\theta + i\sin n\theta) = \rho(\cos\phi + i\sin\phi)$$

and $r^n = \rho$, $n\theta = \phi$. Let $\phi = \phi_0 + 2k\pi$, where ϕ_0 is the smallest nonnegative argument of c. Then $\theta = (\phi_0 + 2k\pi)/n$ and $r = \rho^{1/n}$, where k is any integer. However, not all values of k will produce distinct complex roots z. Suppose $k = 0, 1, 2, \ldots, n - 1$. Then the angles

$$\frac{\phi_0}{n}, \frac{\phi_0 + 2\pi}{n}, \frac{\phi_0 + 4\pi}{n}, \ldots, \frac{\phi_0 + (2n-2)\pi}{n}$$

are all distinct angles. However, if we let $k = n, n + 1, n + 2, \ldots, 2n - 1$, we obtain the angles

$$\frac{\phi_0}{n} + 2\pi, \frac{\phi_0 + 2\pi}{n} + 2\pi, \frac{\phi_0 + 4\pi}{n} + 2\pi, \ldots, \frac{\phi_0 + (2n-2)\pi}{n} + 2\pi$$

which differ by 2π from the angles obtained above and therefore do not produce new solutions. Similarly for other values of k we shall obtain roots included for $k = 0, 1, 2, \ldots, n - 1$. We have proved the following theorem.

Theorem 1.3.3 For $c = \rho(\cos\phi_0 + i\sin\phi_0)$, $\rho \neq 0$, the equation $z^n = c$, n a positive integer, has precisely n distinct solutions

$$z = \rho^{1/n}\left(\cos\frac{\phi_0 + 2k\pi}{n} + i\sin\frac{\phi_0 + 2k\pi}{n}\right)$$

$k = 0, 1, 2, \ldots, n - 1$. These solutions are all the distinct nth roots of c.

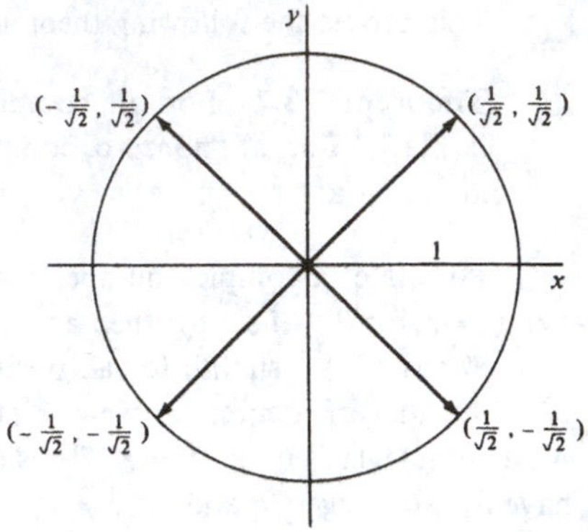

FIGURE 5

EXAMPLE 1.3.1 Find all solutions of the equation $z^4 + 1 = 0$. Since

$$z^4 = -1 = \cos(\pi + 2k\pi) + i \sin(\pi + 2k\pi)$$

according to Theorem 1.3.3, the only distinct solutions are

$$\cos\frac{\pi}{4} + i\sin\frac{\pi}{4} = \frac{1}{\sqrt{2}} + \frac{i}{\sqrt{2}}$$

$$\cos\frac{3\pi}{4} + i\sin\frac{3\pi}{4} = -\frac{1}{\sqrt{2}} + \frac{i}{\sqrt{2}}$$

$$\cos\frac{5\pi}{4} + i\sin\frac{5\pi}{4} = -\frac{1}{\sqrt{2}} - \frac{i}{\sqrt{2}}$$

$$\cos\frac{7\pi}{4} + i\sin\frac{7\pi}{4} = \frac{1}{\sqrt{2}} - \frac{i}{\sqrt{2}}$$

These roots can be plotted on a unit circle separated by an angle of $2\pi/4 = \pi/2$ (see Fig. 5).

EXAMPLE 1.3.2 Find all solutions of the equation $z^2 + 2z + 2 = 0$. This is a quadratic equation with real coefficients. However, we write the variable as z to emphasize that the roots could be complex. By completing the square, we can write $z^2 + 2z + 1 = (z + 1)^2 = -1$. Then taking the square root of -1, we get $z + 1 = \cos(\pi/2) + i \sin(\pi/2) = i$ and $z + 1 = \cos(3\pi/2) + i \sin(3\pi/2) = -i$. Therefore, the only two solutions are $z = -1 + i$ and $z = -1 - i$. Note that if we had written the equation as

$az^2 + bz + c = 0$, $a = 1$, $b = 2$, $c = 2$ and applied the *quadratic formula*

$$z = \frac{-b \pm \sqrt{b^2 - 4ac}}{2a}$$

we would have got the same result by saying $\pm\sqrt{-1} = \pm i$. The reader will be asked to verify the quadratic formula in the exercises in cases where a, b, and c are real *or* complex.

We conclude this section by proving some important inequalities which also have a geometrical interpretation. We begin with the *Cauchy-Schwarz inequality*

$$|x_1x_2 + y_1y_2| \leq |z_1|\,|z_2|$$

where $z_1 = x_1 + iy_1$ and $z_2 = x_2 + iy_2$. Consider the squared version

$$(x_1x_2 + y_1y_2)^2 = x_1^2x_2^2 + 2x_1x_2y_1y_2 + y_1^2y_2^2$$
$$(x_1x_2 + y_1y_2)^2 \leq |z_1|^2|z_2|^2 = (x_1^2 + y_1^2)(x_2^2 + y_2^2)$$
$$(x_1x_2 + y_1y_2)^2 \leq x_1^2x_2^2 + y_1^2y_2^2 + x_1^2y_2^2 + y_1^2x_2^2$$

This inequality will be true if and only if

$$2x_1x_2y_1y_2 \leq x_1^2y_2^2 + y_1^2x_2^2$$

But this is obvious from $(x_1y_2 - x_2y_1)^2 \geq 0$. This proves the Cauchy-Schwarz inequality.

We have the following geometrical interpretation of the Cauchy-Schwarz inequality. Let $\theta_1 = \arg z_1$ and $\theta_2 = \arg z_2$. Then $x_1 = |z_1| \cos\theta_1$, $y_1 = |z_1| \sin\theta_1$, $x_2 = |z_2| \cos\theta_2$, $y_2 = |z_2| \sin\theta_2$, and

$$\begin{aligned} x_1x_2 + y_1y_2 &= |z_1|\,|z_2|(\cos\theta_1\cos\theta_2 + \sin\theta_1\sin\theta_2) \\ &= |z_1|\,|z_2|\cos(\theta_1 - \theta_2) \end{aligned}$$

and hence the inequality merely expresses the fact that $|\cos(\theta_1 - \theta_2)| \leq 1$.

Next we consider the *triangle inequality*

$$|z_1 + z_2| \leq |z_1| + |z_2|$$

Again we consider the squared version

$$\begin{aligned} |z_1 + z_2|^2 &= (x_1 + x_2)^2 + (y_1 + y_2)^2 \\ &= x_1^2 + y_1^2 + x_2^2 + y_2^2 + 2x_1x_2 + 2y_1y_2 \\ |z_1 + z_2|^2 &\leq |z_1|^2 + |z_2|^2 + 2|x_1x_2 + y_1y_2| \\ &\leq |z_1|^2 + |z_2|^2 + 2|z_1|\,|z_2| = (|z_1| + |z_2|)^2 \end{aligned}$$

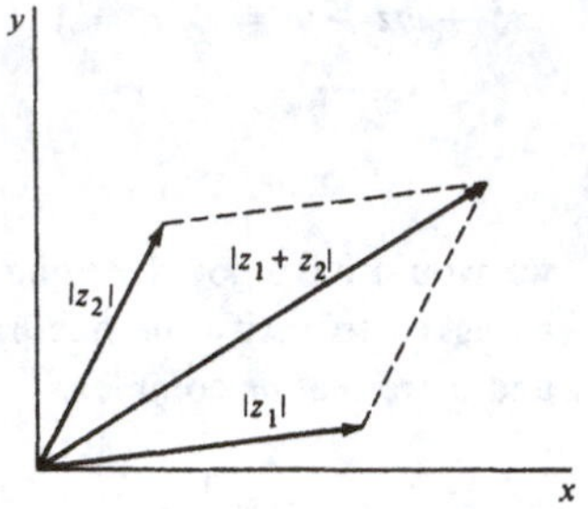

FIGURE 6

making use of the Cauchy-Schwarz inequality. The triangle inequality follows by taking the positive square root of both sides. The geometrical interpretation is simply that *the length of one side of a triangle is less than the sum of the lengths of the other two sides* (see Fig. 6).

Finally, we prove the following very useful inequality:

$$|z_1 - z_2| \geq \big||z_1| - |z_2|\big|$$

Consider $|z_1| = |z_1 - z_2 + z_2| \leq |z_1 - z_2| + |z_2|$ and $|z_2| = |z_2 - z_1 + z_1| \leq |z_1 - z_2| + |z_1|$. Therefore, $|z_1 - z_2| \geq |z_1| - |z_2|$ and $|z_1 - z_2| \geq |z_2| - |z_1|$. Since both inequalities hold, the strongest statement that can be made is

$$|z_1 - z_2| \geq \max(|z_1| - |z_2|, |z_2| - |z_1|) = \big||z_1| - |z_2|\big|$$

EXERCISES 1.3

1 Draw arrows corresponding to $z_1 = -1 + i$, $z_2 = \sqrt{3} + i$, $z_1 + z_2$, $z_1 - z_2$, z_1z_2, and z_1/z_2. For each of these arrows compute the length and the least positive argument.

2 Draw arrows corresponding to $z_1 = 1 + i$, $z_2 = 1 - \sqrt{3}\,i$, $z_1 + z_2$, $z_1 - z_2$, z_1z_2, and z_1/z_2. For each of these arrows compute the length and the least positive argument.

3 Give a geometrical interpretation of what happens to $z \neq 0$ when multiplied by $\cos\alpha + i\sin\alpha$.

4 Give a geometrical interpretation of what happens to $z \neq 0$ when divided by $\cos\alpha + i\sin\alpha$.

5 Give a geometrical interpretation of what happens to $z \neq 0$ when multiplied by -1.

6 Give a geometrical interpretation of what happens to $z \neq 0$ under the operation of conjugation.

7 Give a geometrical interpretation of what happens to $z \neq 0$ when one takes its reciprocal. Distinguish between cases $|z| < 1$, $|z| = 1$, and $|z| > 1$.

8 How many distinct powers of $\cos \alpha\pi + i \sin \alpha\pi$ are there if α is rational? Irrational? *Hint:* If α is rational, assume $\alpha = p/q$, where p and q have no common divisors other than 1.

9 Find all solutions of $z^3 + 8 = 0$.

10 Find all solutions of $z^2 + 2(1 + i)z + 2i = 0$.

11 Show that the quadratic formula is valid for solving the quadratic equation $az^2 + bz + c = 0$ when a, b, and c are complex.

12 Find the nth roots of unity; that is, find all solutions of $z^n = 1$. If w is an nth root of unity not equal to 1, show that $1 + w + w^2 + \cdots + w^{n-1} = 0$.

13 Show that the Cauchy-Schwarz inequality is an equality if and only if $z_1\bar{z}_2 = 0$ or $z_2 = \alpha z_1$, α real.

14 Show that the triangle inequality is an equality if and only if $z_1\bar{z}_2 = 0$ or $z_2 = \alpha z_1$, α a nonnegative real number.

15 Show that $|z_1 - z_2|$ is the euclidean distance between the points $z_1 = x_1 + iy_1$ and $z_2 = x_2 + iy_2$. If $d(z_1,z_2) = |z_1 - z_2|$, show that:

(*a*) $d(z_1,z_2) = d(z_2,z_1)$ (*b*) $d(z_1,z_2) \geq 0$

(*c*) $d(z_1,z_2) = 0$ if and only if $z_1 = z_2$

(*d*) $d(z_1,z_2) \leq d(z_1,z_3) + d(z_2,z_3)$, where z_3 is any other point.

16 Describe the set of points z in the plane which satisfy $|z - z_0| = r$, where z_0 is a fixed point and r is a positive constant.

17 Describe the set of points z in the plane which satisfy $|z - z_1| = |z - z_2|$, where z_1 and z_2 are distinct fixed points.

18 Describe the set of points z in the plane which satisfy $|z - z_1| \leq |z - z_2|$, where z_1 and z_2 are distinct fixed points.

19 Describe the set of points z in the plane which satisfy $|z - z_1| \leq 2|z - z_2|$, where z_1 and z_2 are distinct fixed points.

1.4 TWO-DIMENSIONAL VECTORS

In this section we shall lean heavily on the geometrical interpretation of complex numbers to introduce the system of two-dimensional euclidean vectors. The algebraic properties of these vectors will be those based on the operation of addition and multiplication by real numbers (scalars). For the moment we shall completely ignore the operations of multiplication and division of complex numbers. These operations will have no meaning for the system of vectors we are about to describe.

We shall say that a two-dimensional euclidean vector (from now on we shall say simply *vector*) is defined by a pair of real numbers (x,y), and we shall write $\mathbf{v} = (x,y)$. Two vectors $\mathbf{v}_1 = (x_1,y_1)$ and $\mathbf{v}_2 = (x_2,y_2)$ are *equal* if and

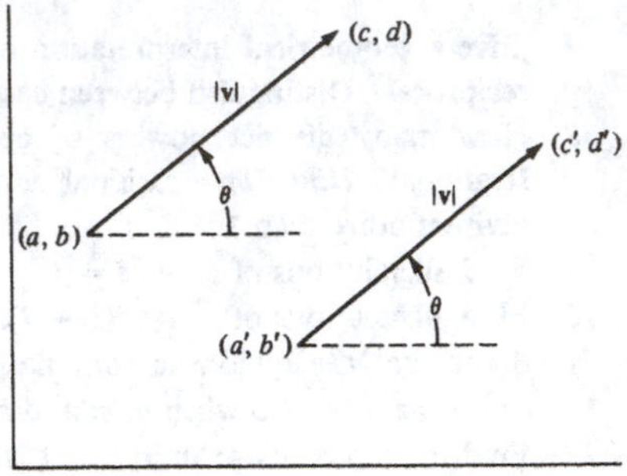

FIGURE 7

only if $x_1 = x_2$ and $y_1 = y_2$. We define *addition* of two vectors $\mathbf{v}_1 = (x_1,y_1)$ and $\mathbf{v}_2 = (x_2,y_2)$ by $\mathbf{v}_1 + \mathbf{v}_2 = (x_1 + x_2, y_1 + y_2)$. We see that the result is a vector and the operation is associative and commutative. We define the zero vector as $\mathbf{0} = (0,0)$, and we have immediately that $\mathbf{v} + \mathbf{0} = (x,y) + (0,0) = (x,y) = \mathbf{v}$ for all vectors $\mathbf{v}$. The negative of a vector $\mathbf{v}$ is $-\mathbf{v} = (-x,-y)$, and the following is obviously true: $\mathbf{v} + (-\mathbf{v}) = \mathbf{0}$ for all vectors $\mathbf{v}$.†

We define the operation of multiplication of vector $\mathbf{v} = (x,y)$ by a real scalar a as follows: $a\mathbf{v} = (ax,ay)$. The result is a vector, and it is easy to verify that the operation has the following properties:‡

1 $a(\mathbf{v}_1 + \mathbf{v}_2) = a\mathbf{v}_1 + a\mathbf{v}_2$.
2 $(a + b)\mathbf{v} = a\mathbf{v} + b\mathbf{v}$.
3 $a(b\mathbf{v}) = (ab)\mathbf{v}$.
4 $1\mathbf{v} = \mathbf{v}$.

The geometrical interpretation of vectors will be just a little different from that for complex numbers for a reason which will become clearer as we proceed. Consider a two-dimensional euclidean plane with two points (a,b) and (c,d) (see Fig. 7). Let $x = c - a$ and $y = d - b$. A geometrical interpretation of the vector $\mathbf{v} = (x,y)$ is the arrow drawn from (a,b) to (c,d). We think of this vector as having *length* $|\mathbf{v}| = \sqrt{x^2 + y^2} = \sqrt{(c - a)^2 + (d - b)^2}$ and *direction* (if $|\mathbf{v}| \neq 0$) specified by the least nonnegative angle θ such that $x = |\mathbf{v}| \cos \theta$ and $y = |\mathbf{v}| \sin \theta$. There is a difficulty in this geometrical interpretation, however. Consider another pair of points (a',b') and (c',d') such that $c - a = c' - a'$ and $d - b = d' - b'$. According to our definition of equality of vectors, the vector $(c' - a', d' - b')$ is equal to the vector $(c - a, d - b)$. In fact, it is easy to see that both vectors have the same length and direction.

† Compare these statements with Theorem 1.2.1 for complex numbers.
‡ Compare with Exercise 1.2.8.

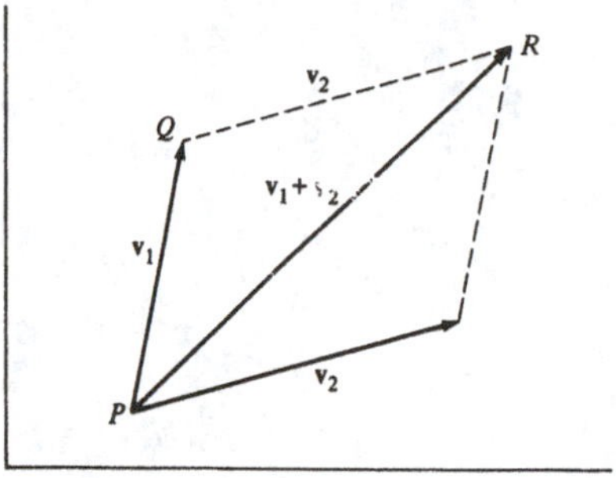

FIGURE 8

This forces us to take a broader geometrical interpretation of vectors. We shall say that a vector $(x,y) \neq (0,0)$ can be interpreted geometrically by *any* arrow which has length $|\mathbf{v}| = \sqrt{x^2 + y^2}$ and direction determined by the least nonnegative angle θ satisfying $x = |\mathbf{v}| \cos \theta$ and $y = |\mathbf{v}| \sin \theta$. The zero vector (0,0) has no direction and therefore has no comparable interpretation.

The geometrical interpretation of vector addition can now be made as follows. Consider vectors $\mathbf{v}_1 = (x_1,y_1)$ and $\mathbf{v}_2 = (x_2,y_2)$. Then $\mathbf{v}_1 + \mathbf{v}_2 = (x_1 + x_2, y_1 + y_2)$. See Fig. 8 for a geometrical interpretation of this result. The rule can be stated as follows. Place $\mathbf{v}_1$ in the plane from a point P to a point Q so that $\mathbf{v}_1$ has the proper magnitude and direction. Place $\mathbf{v}_2$ from point Q to point R so that $\mathbf{v}_2$ has the proper magnitude and direction. Then the vector $\mathbf{v}_1 + \mathbf{v}_2$ is the vector from point P to point R. If P and R coincide, $\mathbf{v}_1 + \mathbf{v}_2 = \mathbf{0}$.

An immediate corollary follows from this rule of vector addition and the triangle inequality:

$$|\mathbf{v}_1 + \mathbf{v}_2| \leq |\mathbf{v}_1| + |\mathbf{v}_2|$$

Next let us give a geometrical interpretation of multiplication of a vector by a scalar. Let a be a scalar and $\mathbf{v} = (x,y)$ a vector. Then $a\mathbf{v} = (ax,ay)$ and

$$|a\mathbf{v}| = \sqrt{a^2x^2 + a^2y^2} = \sqrt{a^2}\,\sqrt{x^2 + y^2} = |a|\,|\mathbf{v}|$$

since $\sqrt{a^2} = |a|$. Therefore, multiplication by a modifies the length of $\mathbf{v}$ if $|a| \neq 1$. If $|a| < 1$, the vector is shortened, and if $|a| > 1$, the vector is lengthened. If a is positive, ax and ay are the same sign as x and y and hence the direction of $\mathbf{v}$ is not changed. However, if a is negative, ax and ay are of the opposite sign from x and y and, in this case, $a\mathbf{v}$ has the opposite direction from $\mathbf{v}$. See Fig. 9 for a summary of the various cases. Notice that $-\mathbf{v} = -1\mathbf{v}$ has the same length as $\mathbf{v}$ but the opposite direction. Using this, we have the following

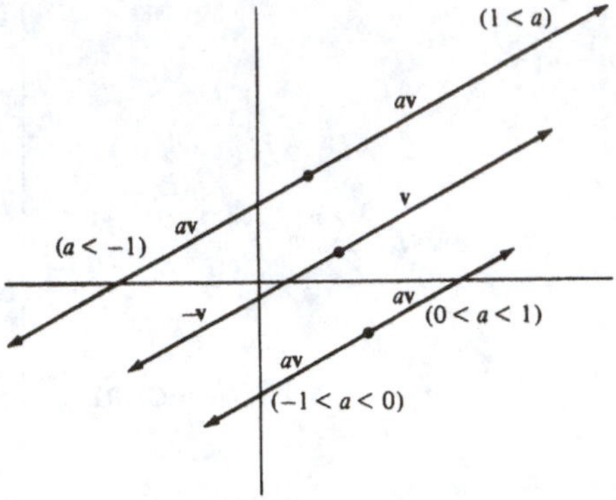

FIGURE 9

interpretation of vector subtraction $\mathbf{v}_1 - \mathbf{v}_2 = \mathbf{v}_1 + (-\mathbf{v}_2)$ (see Fig. 10). Alternatively, $\mathbf{v}_1 - \mathbf{v}_2$ is that vector which when added to $\mathbf{v}_2$ gives $\mathbf{v}_1$ (see triangle *PQR* in Fig. 10).

There is another very useful operation between vectors known as scalar product (not to be confused with multiplication by a scalar). Consider Fig. 11.

$$\mathbf{v}_1 = (x_1,y_1) = (|\mathbf{v}_1| \cos \theta_1, |\mathbf{v}_1| \sin \theta_1)$$
$$\mathbf{v}_2 = (x_2,y_2) = (|\mathbf{v}_2| \cos \theta_2, |\mathbf{v}_2| \sin \theta_2)$$

Then

$$\begin{aligned} x_1x_2 + y_1y_2 &= |\mathbf{v}_1|\,|\mathbf{v}_2|(\cos \theta_1 \cos \theta_2 + \sin \theta_1 \sin \theta_2) \\ &= |\mathbf{v}_1|\,|\mathbf{v}_2| \cos (\theta_2 - \theta_1) \end{aligned}$$

This operation, denoted by $\mathbf{v}_1 \cdot \mathbf{v}_2$, is called the *scalar product*, and the result, as we have already seen, is a scalar quantity given by the product of the lengths of the two vectors times the cosine of the angle between the vectors. If either or both of the vectors are the zero vector, then $\mathbf{v}_1 \cdot \mathbf{v}_2 = 0$.

The reader should verify the following obvious properties of the scalar product:

1 $\mathbf{v}_1 \cdot \mathbf{v}_2 = \mathbf{v}_2 \cdot \mathbf{v}_1$.
2 $\mathbf{v}_1 \cdot (\mathbf{v}_2 + \mathbf{v}_3) = (\mathbf{v}_1 \cdot \mathbf{v}_2) + (\mathbf{v}_1 \cdot \mathbf{v}_3)$.
3 $a\mathbf{v}_1 \cdot \mathbf{v}_2 = a(\mathbf{v}_1 \cdot \mathbf{v}_2)$.
4 $\mathbf{v} \cdot \mathbf{v} = |\mathbf{v}|^2$.
5 $|\mathbf{v}_1 \cdot \mathbf{v}_2| \le |\mathbf{v}_1|\,|\mathbf{v}_2|$.
6 If $|\mathbf{v}_1| \ne 0$ and $|\mathbf{v}_2| \ne 0$, then $\mathbf{v}_1 \cdot \mathbf{v}_2 = 0$ if and only if $\mathbf{v}_1$ and $\mathbf{v}_2$ are perpendicular.†

† If $|\mathbf{v}_1| \ne 0$ and $|\mathbf{v}_2| \ne 0$ and $\mathbf{v}_1 \cdot \mathbf{v}_2 = 0$, we say that $\mathbf{v}_1$ and $\mathbf{v}_2$ are *orthogonal*.

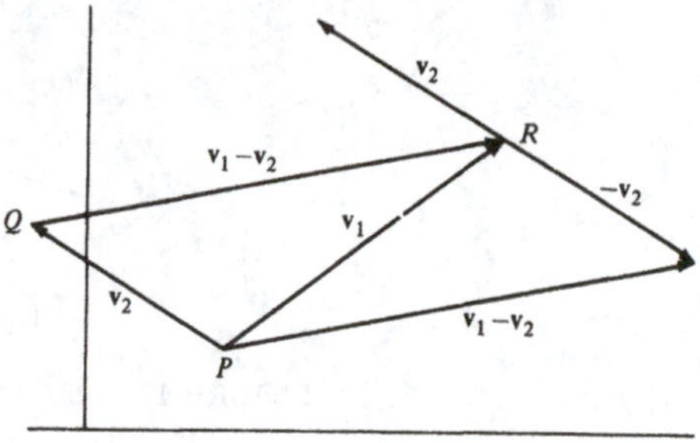

FIGURE 10

EXAMPLE 1.4.1 Find the equation of a straight line passing through the point (x_0,y_0) in the direction of the vector (a,b). Let (x,y) be the coordinates of a point on the line. Then the vector from (0,0) to (x,y) is the vector (x,y). This vector is the sum of the vector (x_0,y_0) and a multiple of (a,b) (see Fig. 12). Hence the equation of the line is $(x,y) = (x_0,y_0) + t(a,b)$. We usually refer to t as a parameter and to this representation as a *parametric representation* of the line. The parameter t clearly runs between $-\infty$ and ∞.

EXAMPLE 1.4.2 What geometrical figure is represented parametrically by $(x,y) = (x_0,y_0) + (r\cos\theta, r\sin\theta)$, where $r > 0$ is constant and the parameter θ runs between 0 and 2π? In this case, $(x - x_0, y - y_0) = (r\cos\theta, r\sin\theta)$ and $|(x - x_0, y - y_0)| = r$. The figure is therefore a circle with center at (x_0,y_0) and radius r (see Fig. 13).

The two examples illustrate the usefulness of the concept of *vector-valued functions*. Suppose for each value of t in some set of real numbers D, called the *domain* of the function, a vector $\mathbf{v}(t)$ is unambiguously defined; then we say

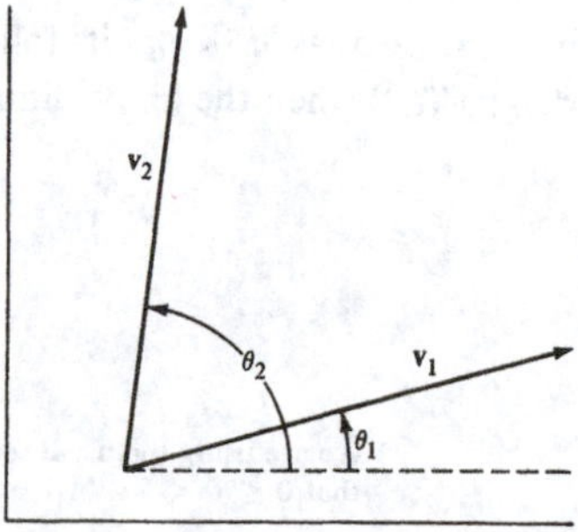

FIGURE 11

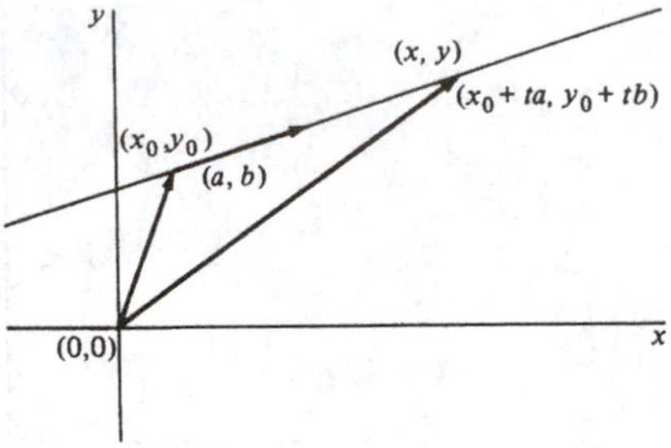

FIGURE 12

that **v** is a vector-valued function of t; t is called the *independent variable*, and **v** is called the *dependent variable*. The collection of all values of $\mathbf{v}(t)$ taken on for t in the domain is called the *range* of the function.

In Example 1.4.1, if $\mathbf{v}_0 = (x_0,y_0)$ and $\mathbf{u} = (a,b)$, then we can write $(x,y) = \mathbf{v}(t) = \mathbf{v}_0 + t\mathbf{u}$, where $-\infty < t < \infty$. Then **v** is a vector-valued function of t. The domain is the set of all real t, and the range is the set of all vectors from the origin to points on the line through (x_0,y_0) in the direction of **u**.

In Example 1.4.2, if $\mathbf{v}_0 = (x_0,y_0)$ and $0 \leq \theta < 2\pi$, then $(x,y) = \mathbf{v}(\theta) = \mathbf{v}_0 + (r\cos\theta, r\sin\theta)$. The domain† is $\{\theta \mid 0 \leq \theta < 2\pi\}$, and the range is the set of vectors from the origin to all points on the circle with center at (x_0,y_0) and radius r.

The concept of *derivative* of a vector-valued function is very easy to define. Suppose for some t_0 and some $\delta > 0$, all t satisfying $t_0 - \delta < t < t_0 + \delta$ are in the domain of $\mathbf{v}(t)$ and there is a vector $\mathbf{v}'(t_0)$ such that

$$\lim_{t \to t_0} \left| \frac{\mathbf{v}(t) - \mathbf{v}(t_0)}{t - t_0} - \mathbf{v}'(t_0) \right| = 0$$

then $\mathbf{v}'(t_0)$ is the derivative of $\mathbf{v}(t)$ at t_0. Since the length of the vector

$$\frac{\mathbf{v}(t) - \mathbf{v}(t_0)}{t - t_0} - \mathbf{v}'(t_0)$$

goes to zero as $t \to t_0$, it follows that if $\mathbf{v}(t) = (x(t),y(t))$ and $\mathbf{v}'(t_0) = (x'(t_0),y'(t_0))$, then the above limit is zero if and only if

$$\lim_{t \to t_0} \left[\frac{x(t) - x(t_0)}{t - t_0} - x'(t_0) \right] = 0$$

$$\lim_{t \to t_0} \left[\frac{y(t) - y(t_0)}{t - t_0} - y'(t_0) \right] = 0$$

† We are using the usual set notation: $\{\theta \mid 0 \leq \theta < 2\pi\}$ is read "the set of all θ such that $0 \leq \theta < 2\pi$."

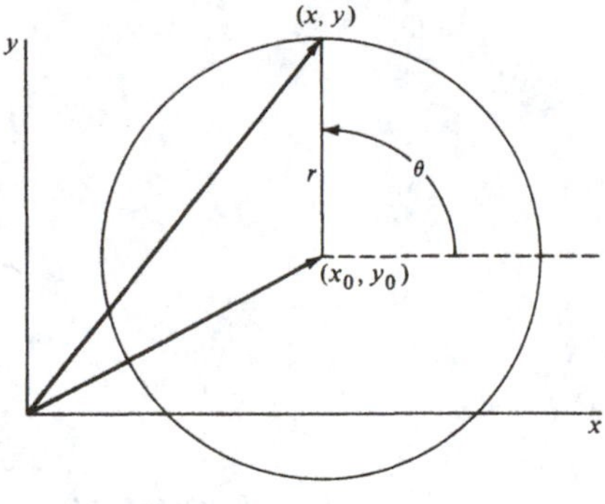

FIGURE 13

which imply that $x'(t_0)$ and $y'(t_0)$ are, respectively, the derivatives of $x(t)$ and $y(t)$ at t_0. We have therefore proved the following theorem.

Theorem 1.4.1 The vector-valued function $\mathbf{v}(t) = (x(t), y(t))$ has the derivative $\mathbf{v}'(t_0) = (x'(t_0), y'(t_0))$ at t_0 if and only if $x(t)$ and $y(t)$ have the derivatives $x'(t_0)$ and $y'(t_0)$ at t_0.

In Example 1.4.1,

$$\frac{\mathbf{v}(t) - \mathbf{v}(t_0)}{t - t_0} = (a,b)$$

Therefore,

$$\lim_{t \to t_0} \left[\frac{\mathbf{v}(t) - \mathbf{v}(t_0)}{t - t_0} - (a,b) \right] = 0$$

and hence $\mathbf{v}'(t) = (a,b)$. Notice that the derivative (a,b) is tangent to the line. This will also be the case if we take the derivative in Example 1.4.2. Using Theorem 1.4.1, we have

$$\mathbf{v}'(\theta) = (-r \sin \theta, r \cos \theta)$$

and $[\mathbf{v} - (x_0, y_0)] \cdot \mathbf{v}' = 0$, which shows by property 6 for the scalar product that $\mathbf{v}'$ is perpendicular to the vector drawn from the center of the circle to the point where $\mathbf{v}'(\theta)$ is calculated. Therefore, $\mathbf{v}'$ is tangent to the circle. This result is true in general. In fact, it is easy to see from Fig. 14 why this should be the case.

If $\mathbf{v}(t) = (x(t), y(t))$ is a vector from the origin to a point on the curve C with a tangent line L at $\mathbf{v}(t_0) = (x_0, y_0)$, then it is clear that the direction of

$$\frac{\mathbf{v}(t) - \mathbf{v}(t_0)}{t - t_0}$$

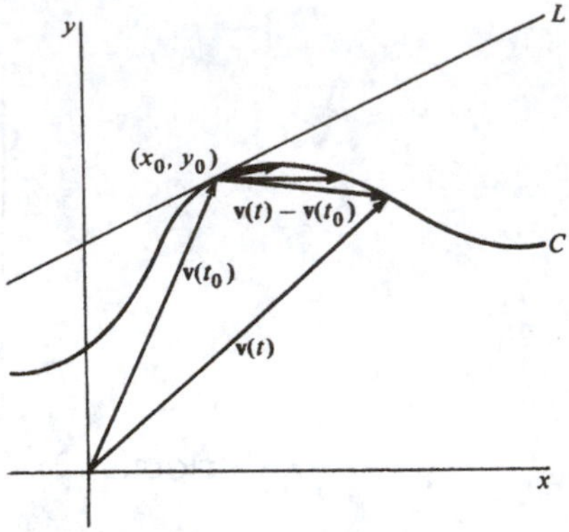

FIGURE 14

approaches the direction of the tangent line as $t \to t_0$, provided $\mathbf{v}'(t_0)$ exists and is not the zero vector. We shall, in fact, use this limit to define the tangent to the curve.

Let $\mathbf{v}(t)$ be the vector from the origin to a point on the curve C. The point moves along the curve as the parameter t varies. If $\mathbf{v}(t)$ has a nonzero derivative $\mathbf{v}'(t_0)$ at $\mathbf{v}(t_0)$, then C has a tangent line at $\mathbf{v}(t_0)$ and a tangent vector $\mathbf{v}'(t_0)$ and the equation of the tangent line is

$$(x,y) = \mathbf{v}(t_0) + (t - t_0)\,\mathbf{v}'(t_0) \qquad -\infty < t < \infty$$

EXERCISES 1.4

1 Let $\mathbf{v}_1 = (1,-2)$ and $\mathbf{v}_2 = (-3,5)$. Find and draw sketches of $\mathbf{v}_1 + \mathbf{v}_2$, $\mathbf{v}_1 - \mathbf{v}_2$, $2\mathbf{v}_1 + \mathbf{v}_2$, and $\frac{1}{2}(\mathbf{v}_2 - \mathbf{v}_1)$.

2 A man is walking due east at 2 miles per hour and the wind seems to be coming from the north. He speeds up to 4 miles per hour and the wind seems to be from the northeast. What is the wind speed, and from what direction is it coming?

3 An airplane is 200 miles due west of its destination. The wind is out of the northeast at 50 miles per hour. What should be the airplane's heading and airspeed in order for it to reach its destination in 1 hour?

4 Let $\mathbf{v} = (1,-\sqrt{3})$. Find $|\mathbf{v}|$ and the least nonnegative angle θ such that $x = |\mathbf{v}| \cos \theta$, $y = |\mathbf{v}| \sin \theta$.

5 Find the vector equation of the straight line passing through the points $(-1,2)$ and $(3,4)$. Find a vector perpendicular to this line.

6 Find the vector equation of the circle with center at $(1,3)$ passing through the point $(4,7)$. Find the equation of the tangent line to this circle at $(4,7)$.

7 A curve is given by $(x,y) = (3t^2 - t,\ t^3 - 2t^2)$, $0 \le t \le 4$. Find a tangent vector to this curve at the point $(10,0)$.

8 Verify the six properties of the scalar product listed in this section.

9 Assuming that $\mathbf{v}(t)$, $\mathbf{v}_1(t)$, $\mathbf{v}_2(t)$ are differentiable vector functions and $a(t)$ is a differentiable scalar function, show that

(*a*) $[\mathbf{v}_1(t) + \mathbf{v}_2(t)]' = \mathbf{v}_1'(t) + \mathbf{v}_2'(t)$

(*b*) $[a(t)\mathbf{v}(t)]' = a'(t)\mathbf{v}(t) + a(t)\mathbf{v}'(t)$

(*c*) $[\mathbf{v}_1(t) \cdot \mathbf{v}_2(t)]' = [\mathbf{v}_1(t) \cdot \mathbf{v}_2'(t)] + [\mathbf{v}_1'(t) \cdot \mathbf{v}_2(t)]$

(*d*) $[|\mathbf{v}(t)|^2]' = 2[\mathbf{v}(t) \cdot \mathbf{v}'(t)]$

10 Using part (*d*) of Example 9, prove that if $\mathbf{v}(t)$ has constant nonzero length and $\mathbf{v}'(t) \neq \mathbf{0}$, then $\mathbf{v}(t)$ is orthogonal to $\mathbf{v}'(t)$.

11 Let $\mathbf{v}(t) = (x(t),y(t))$ represent a curve parametrically. If $\mathbf{v}'(t) \neq \mathbf{0}$, then $\mathbf{T}(t) = \mathbf{v}'(t)/|\mathbf{v}'(t)|$ is a unit tangent vector $[|\mathbf{T}| = 1]$. Show that $\mathbf{T}'$ is normal to the curve, that is, $\mathbf{T} \cdot \mathbf{T}' = 0$.

12 If a physical particle moves along a curve given parametrically by $\mathbf{r}(t) = (x(t),y(t))$, where t is time, then $\mathbf{v}(t) = \mathbf{r}'(t)$ is called the *velocity*, $s(t) = |\mathbf{v}(t)|$ is called the *speed*, and $\mathbf{a}(t) = \mathbf{v}'(t)$ is called the *acceleration*. If the speed is never zero, show that $\mathbf{a}(t) = s'(t)\mathbf{T} + s(t)|\mathbf{T}'|\mathbf{n}$, where $\mathbf{T}$ is the unit tangent and $\mathbf{n}$ is a unit normal.

1.5 FUNCTIONS OF A COMPLEX VARIABLE

We now return to our study of complex numbers to consider functions of a complex variable. We do not need an extensive treatment of this subject, concentrating on the things we shall need for our study of differential equations. However, the reader should be aware that there is a vast literature on the subject.†

Suppose that for each complex number z in some set D (*domain* of the function) of the complex plane there is assigned a complex number $f(z)$; then we say that we have a complex-valued function f of the complex variable z defined in D. The set of values $f(z)$ is called the *range* of f. Let $z = x + iy$ and $f(z) = u + iv$, where x, y, u, v are all real. Then clearly $u(x,y)$ and $v(x,y)$ are real-valued functions of two real variables x and y defined for z in D.

EXAMPLE 1.5.1 Let $f(z) = z^2$ and D be the entire complex plane. Then $f(z) = u(x,y) + iv(x,y) = x^2 - y^2 + i(2xy)$. Particular values are, for example, $f(1 + i) = 2i$, $f(-i) = -1$.

EXAMPLE 1.5.2 Let $f(z) = \sqrt{z} = |z|^{1/2}[\cos(\frac{1}{2}\arg z) + i\sin(\frac{1}{2}\arg z)]$, $0 \leq \arg z < 2\pi$, $f(0) = 0$. This function is defined for all z in the complex

†See, for example, J. W. Dettman, "Applied Complex Variables," Macmillan, New York, 1965 (rpt. Dover, New York, 1984).

plane. For each $z \neq 0$ there are two distinct square roots. The function in this example defines one of these square roots. To describe the other square root we could define $g(z) = -f(z)$.

There are two concepts of derivative of a function of a complex variable which we shall introduce, *derivative along a curve* and *derivative at a point*. These two notions of derivative are closely related, and as these relations are pointed out, we shall see that our definitions are quite consistent.

Suppose that the domain of f contains a curve C parameterized as follows: $z(t) = x(t) + iy(t)$, $a \leq t \leq b$. Then

$$\begin{aligned} f(z(t)) &= u(x(t),y(t)) + iv(x(t),y(t)) \\ &= U(t) + iV(t) \end{aligned}$$

so the real and imaginary parts of f are defined along C as functions of the real parameter t. If U and V are differentiable as functions of t, then the derivative of f along C is defined by

$$\frac{df}{dt} = U'(t) + iV'(t)$$

If $x'(t)$ and $y'(t)$ exist and the partial derivatives $\partial u/\partial x$, $\partial u/\partial y$, $\partial v/\partial x$, and $\partial v/\partial y$ are continuous as functions of x and y at points on C, then by the chain rule

$$\frac{df}{dt} = \frac{\partial u}{\partial x}x'(t) + \frac{\partial u}{\partial y}y'(t) + i\frac{\partial v}{\partial x}x'(t) + i\frac{\partial v}{\partial y}y'(t)$$

Suppose that for some $\delta > 0$ all z satisfying $|z - z_0| < \delta$ are in the domain of f. Further, suppose that for all z satisfying this inequality $\partial u/\partial x$, $\partial u/\partial y$, $\partial v/\partial x$, and $\partial v/\partial y$ are continuous and the Cauchy-Riemann equations $\partial u/\partial x = \partial v/\partial y$ and $\partial u/\partial y = -(\partial v/\partial x)$ are satisfied at z_0. Then we say that f is differentiable† at z_0 and the derivative is

$$\begin{aligned} f'(z_0) &= \frac{\partial u}{\partial x} + i\frac{\partial v}{\partial x} \\ &= \frac{\partial v}{\partial y} - i\frac{\partial u}{\partial y} \end{aligned}$$

where the partial derivatives are evaluated at $z_0 = x_0 + iy_0$. This seems like a rather arbitrary definition of derivative, but we shall show that it is consistent with our definition of derivative along a curve.

† If f is differentiable for all z satisfying $|z - z_0| < \varepsilon$, for some $\varepsilon > 0$, then we say that f is *analytic* at z_0.

Let f be differentiable at $z_0 = x(t_0) + iy(t_0)$ on a curve C, where the derivative along C at t_0 exists. Then at t_0

$$\begin{aligned}\frac{df}{dt} &= \left(\frac{\partial u}{\partial x} + i\frac{\partial v}{\partial x}\right)[x'(t_0) + iy'(t_0)] \\ &= \left(\frac{\partial v}{\partial y} - i\frac{\partial u}{\partial y}\right)[x'(t_0) + iy'(t_0)] \\ &= f'(z_0)z'(t_0)\end{aligned}$$

where $z'(t_0) = x'(t_0) + iy'(t_0)$. Hence, we see that our definition of derivative at a point is a natural one in that it leads to a natural *chain-rule* result for the derivative of a function along a curve C. Also the value of the derivative at a point does not depend on the definition of any particular curve passing through the point.†

EXAMPLE 1.5.3 The function of Example 1.5.1 has a derivative at every point. Since $f(z) = x^2 - y^2 + i(2xy)$, $\partial u/\partial x = 2x = \partial v/\partial y$ and $\partial u/\partial y = -2y = -(\partial v/\partial x)$ and these partial derivatives are continuous everywhere. We have $f'(z) = 2(x + iy) = 2z$. Notice the similarity with the differentiation formula

$$\frac{d}{dx}(x^2) = 2x$$

This is not just a coincidence.

EXAMPLE 1.5.4 Consider the function defined by $f(z) = |z|^2 = x^2 + y^2$. Here $u = x^2 + y^2$, $v = 0$. Then $\partial u/\partial x = 2x$, $\partial u/\partial y = 2y$, $\partial v/\partial x = 0$, $\partial v/\partial y = 0$. These partial derivatives are all continuous. However, $\partial u/\partial x = \partial v/\partial y$ and $\partial u/\partial y = -(\partial v/\partial x)$ at only one point: $x = y = 0$. This function is differentiable at the origin (where the derivative is zero) and at *no other point*.

EXAMPLE 1.5.5 Consider the function defined by $f(z) = |z| = \sqrt{x^2 + y^2}$. Here $u = \sqrt{x^2 + y^2}$, $v = 0$. Therefore,

$$\frac{\partial u}{\partial x} = \frac{x}{\sqrt{x^2 + y^2}} \qquad \frac{\partial u}{\partial y} = \frac{y}{\sqrt{x^2 + y^2}} \qquad \frac{\partial v}{\partial x} = \frac{\partial v}{\partial y} = 0$$

† For the more conventional approach using limits of difference quotients see Dettman, op. cit.

The Cauchy-Riemann equations are never satisfied when x and y are different from zero, and when $x = y = 0$, the $\partial u/\partial x$ and $\partial u/\partial y$ do not exist. Therefore, this function is never differentiable.

Later on we shall want to discuss complex-valued solutions of differential equations. Suppose, for example, that we wish to show that $f(t) = \cos t + i \sin t$ is a solution of the equation $d^2f/dt^2 + f = 0$. We can interpret this to mean we have some function f defined on the x axis parameterized by $z(t) = t + i0$. We wish to differentiate f along the x axis, and using our above definition, we have

$$\frac{df}{dt} = -\sin t + i \cos t$$

Differentiating again, we have

$$\frac{d^2f}{dt^2} = -\cos t - i \sin t = -f$$

Therefore, $d^2f/dt^2 + f = 0$, where the equality is to be interpreted in the sense that *both* the real and imaginary parts of the left-hand side are zero.

On the other hand, we may wish to show that a function satisfies a differential equation where the derivatives are to be interpreted in terms of the complex variable z. For example, $f(z) = z^2$ is differentiable everywhere in the complex plane, and it satisfies the differential equation $zf' - 2f = 0$. In any given situation the context of the problem will indicate which interpretation should be put on the differential equation.

EXERCISES 1.5

1 Consider the function defined by $f(z) = z^3$. What is its domain? Find its real and imaginary parts. Where is it differentiable? What is its derivative?

2 Show that $f(z) = \text{Re}(z)$ and $g(z) = \text{Im}(z)$ are nowhere differentiable.

3 Assuming that $f(z)$ and $g(z)$ are both differentiable at z_0, prove:

(*a*) $(f + g)'(z_0) = f'(z_0) + g'(z_0)$.

(*b*) $(cf)'(z_0) = cf'(z_0)$, where c is a complex constant.

(*c*) $(fg)'(z_0) = f(z_0)g'(z_0) + g(z_0)f'(z_0)$.

(*d*) $(c_1 f + c_2 g)'(z_0) = c_1 f'(z_0) + c_2 g'(z_0)$ where c_1 and c_2 are complex constants.

4 Using part (*c*) of Exercise 3 and mathematical induction, prove that

$$\frac{d}{dz} z^n = nz^{n-1} \qquad n = 0, 1, 2, 3, \ldots$$

5 What is the derivative of the polynomial

$$p(z) = a_n z^n + a_{n-1} z^{n-1} + \cdots + a_1 z + a_0$$

where the a's are complex constants?

6 Consider the function defined by $f(z) = e^x \cos y + ie^x \sin y$. What is its domain? Where is it differentiable? What is its derivative?

7 Consider the function defined by $f(z) = \cos x \cosh y - i \sin x \sinh y$. What is its domain? Where is it differentiable? What is its derivative?

8 Consider the function defined by $f(z) = 1/z$. What is its domain? Where is it differentiable? What is its derivative?

9 Let $f(z) = u(x,y) + iv(x,y)$ be differentiable at z_0. Show that u and v are continuous at $z_0 = x_0 + iy_0$. *Hint:* Use the mean-value theorem for functions of two real variables.

10 Use the result of Exercise 9 to show that the function of Example 1.5.2 is not differentiable on the positive x axis. Where is this function differentiable? What is its derivative?

11 Consider the function defined by $f(z) = \ln |z| + i \arg z, 0 \leq \arg z < 2\pi$. What is the domain? Where is it differentiable? What is its derivative?

12 Show that $f(t) = e^{at} \cos bt + ie^{at} \sin bt$ satisfies the differential equation $f'' - 2af' + (a^2 + b^2)f = 0$. Here prime means derivative with respect to t, and a and b are real constants.

13 Show that the function $f(z) = e^{kx}(\cos ky + i \sin ky)$, where k is a real constant, satisfies the equation $df/dz = kf$.

1.6 EXPONENTIAL FUNCTION

In this section we discuss the exponential function of the complex variable z. As our point of departure we begin with the power-series definition of the real exponential function

$$e^x = 1 + \frac{x}{1!} + \frac{x^2}{2!} + \frac{x^3}{3!} + \cdots = \sum_{k=0}^{\infty} \frac{x^k}{k!}$$

A natural way to extend this to the complex plane is to define e^z as follows:†

$$e^z = 1 + \frac{z}{1!} + \frac{z^2}{2!} + \frac{z^3}{3!} + \cdots = \sum_{k=0}^{\infty} \frac{z^k}{k!}$$

Of course, we must define what we mean by the infinite series of complex numbers. Consider the partial sum

$$\sum_{k=0}^{n} \frac{z^k}{k!} = \sum_{k=0}^{n} \frac{\text{Re}\,(z^k)}{k!} + i \sum_{k=0}^{n} \frac{\text{Im}\,(z^k)}{k!}$$

† We shall also use the notation $\exp z$.

The two series

$$\sum_{k=0}^{\infty} \frac{\operatorname{Re}(z^k)}{k!} \quad \text{and} \quad \sum_{k=0}^{\infty} \frac{\operatorname{Im}(z^k)}{k!}$$

both converge, as can be seen by comparison with the series

$$\sum_{k=0}^{\infty} \frac{|z|^k}{k!}$$

which converges for all $|z|$. The necessary inequalities to see this are

$$|\operatorname{Re}(z^k)| \le |z^k| = |z|^k \qquad |\operatorname{Im}(z^k)| \le |z^k| = |z|^k$$

We shall say that the series of complex numbers $\sum_{k=0}^{\infty} \frac{z^k}{k!}$ converges if and only if the real series

$$\sum_{k=0}^{\infty} \frac{\operatorname{Re}(z^k)}{k!} \quad \text{and} \quad \sum_{k=0}^{\infty} \frac{\operatorname{Im}(z^k)}{k!}$$

both converge and has the sum $S = U + iV$, where U is the sum of the real parts and V is the sum of the imaginary parts. In this case, it is clear that the complex series converges for all z.

Now let $z = iy$. Then the $(2n + 1)$st partial sum is

$$\sum_{k=0}^{2n+1} \frac{(iy)^k}{k!} = 1 - \frac{y^2}{2!} + \frac{y^4}{4!} - \cdots + \frac{(-1)^n y^{2n}}{(2n)!}$$
$$+ i\left[y - \frac{y^3}{3!} + \frac{y^5}{5!} - \cdots + \frac{(-1)^n y^{2n+1}}{(2n+1)!}\right]$$

The series of real parts converges to $\cos y$, and the series of imaginary parts converges to $\sin y$. This proves the important *Euler formula*:

$$e^{iy} = \cos y + i \sin y$$

We shall prove in the next section that the complex exponential function has the usual property

$$e^{z_1 + z_2} = e^{z_1} e^{z_2}$$

Assuming this for the moment, we now have

$$e^z = e^{x+iy} = e^x e^{iy} = e^x \cos y + i e^x \sin y$$

It is now clear that the exponential function is analytic for all z, because $u(x,y) = e^x \cos y$ and $v(x,y) = e^x \sin y$ are continuously differentiable and satisfy the Cauchy-Riemann equations for all z.

Many of the common transcendental functions of a real variable can now be defined for the complex variable z using the exponential function. For example, from $e^{iy} = \cos y + i \sin y$ and $e^{-iy} = \cos y - i \sin y$ it is easy to show that

$$\cos y = \frac{e^{iy} + e^{-iy}}{2}$$

$$\sin y = \frac{e^{iy} - e^{-iy}}{2i}$$

We then generalize for the complex variable case to

$$\cos z = \frac{e^{iz} + e^{-iz}}{2}$$

$$\sin z = \frac{e^{iz} - e^{-iz}}{2i}$$

Recalling the definitions of the hyperbolic functions

$$\cosh x = \frac{e^{x} + e^{-x}}{2}$$

$$\sinh x = \frac{e^{x} - e^{-x}}{2}$$

we can express $\cos z$ and $\sin z$ as follows:

$$\begin{aligned}\cos z &= \tfrac{1}{2}(e^{-y}e^{ix} + e^{y}e^{-ix}) \\ &= \tfrac{1}{2}e^{-y}(\cos x + i \sin x) + \tfrac{1}{2}e^{y}(\cos x - i \sin x) \\ &= \cos x \cosh y - i \sin x \sinh y\end{aligned}$$

$$\begin{aligned}\sin z &= \frac{-i}{2}(e^{-y}e^{ix} - e^{y}e^{-ix}) \\ &= \frac{-i}{2}e^{-y}(\cos x + i \sin x) + \frac{i}{2}e^{y}(\cos x - i \sin x) \\ &= \sin x \cosh y + i \cos x \sinh y\end{aligned}$$

It is now clear that $\cos z$ and $\sin z$ are analytic everywhere.

In the case of the real variable x, $\tan x = (\sin x)/(\cos x)$. Hence, we generalize to the complex variable case as follows:

$$\begin{aligned}\tan z &= \frac{\sin z}{\cos z} = \frac{\sin x \cosh y + i \cos x \sinh y}{\cos x \cosh y - i \sin x \sinh y} \\ &= \frac{(\sin x \cosh y + i \cos x \sinh y)(\cos x \cosh y + i \sin x \sinh y)}{\cos^2 x \cosh^2 y + \sin^2 x \sinh^2 y} \\ &= \frac{\sin x \cos x + i \sinh y \cosh y}{\cos^2 x + \sinh^2 y}\end{aligned}$$

$\tan z$ is analytic everywhere except where $\cos^2 x + \sinh^2 y = 0$. But $\sinh y = 0$ only when $y = 0$, and $\cos x = 0$ only when x is an odd multiple of $\pi/2$. Therefore, the points where $\tan z$ is not analytic are

$$z = \frac{2n+1}{2}\pi \qquad n = 0, \pm 1, \pm 2, \ldots$$

The other trigonometric functions are defined as follows:

$$\cot z = \frac{\cos z}{\sin z} = \frac{\cos x \sin x - i \cosh y \sinh y}{\sin^2 x + \sinh^2 y}$$

$$\sec z = \frac{1}{\cos z} = \frac{\cos x \cosh y + i \sin x \sinh y}{\cos^2 x + \sinh^2 y}$$

$$\csc z = \frac{1}{\sin z} = \frac{\sin x \cosh y - i \cos x \sinh y}{\sin^2 x + \sinh^2 y}$$

$\sec z$ is analytic everywhere except where $\cos z = 0$ while $\cot z$ and $\csc z$ are analytic everywhere except where $\sin z = 0$.

The hyperbolic functions are similarly defined in terms of the complex exponential function:

$$\cosh z = \frac{e^z + e^{-z}}{2} \qquad \sinh z = \frac{e^z - e^{-z}}{2}$$

$$\tanh z = \frac{\sinh z}{\cosh z} \qquad \coth z = \frac{\cosh z}{\sinh z}$$

$$\operatorname{sech} z = \frac{1}{\cosh z} \qquad \operatorname{csch} z = \frac{1}{\sinh z}$$

We normally think of the logarithmic function as the inverse of the exponential function. However, in the case of the complex exponential function we have difficulty defining the inverse because of the property $e^{z+2\pi k i} = e^z$ for any integer k (see Exercise 1.6.6). Therefore, if we wish to define an inverse of the exponential function, we must restrict the imaginary part of the dependent variable. We begin with the equation

$$e^{u+iv} = e^u(\cos v + i \sin v) = z = x + iy$$

Therefore, $x = e^u \cos v$ and $y = e^u \sin v$, from which we derive

$$u = \tfrac{1}{2} \ln (x^2 + y^2) = \ln |z|$$

$$v = \tan^{-1} \frac{y}{x} = \arg z$$

However, in order to make this single-valued we must restrict arg z to some interval of length 2π, say $0 \leq \arg z < 2\pi$. With this restriction we can then write

$$u + iv = \log z = \ln |z| + i \arg z$$

This function is analytic everywhere except at the origin and on the positive x axis. Where it is differentiable,

$$\frac{d}{dz} \log z = \frac{1}{z}$$

EXERCISES 1.6

1 Starting with the definition $e^z = e^x(\cos y + i \sin y)$, prove that $e^{z_1+z_2} = e^{z_1}e^{z_2}$.

2 Show that the real and imaginary parts of e^z satisfy the Cauchy-Riemann equations everywhere.

3 Show that $\overline{e^z}$ is never zero.

4 Show that $\overline{e^z} = e^{\bar{z}}$.

5 Show that $e^{-z} = 1/e^z$.

6 Show that $e^{z+2k\pi i} = e^z$ for every integer k.

7 Show that e^{az} satisfies the differential equation $f' = af$ for any complex constant a.

8 Letting $z = r(\cos \theta + i \sin \theta)$, show that

$$e^{1/z} = \exp\left(\frac{1}{r} \cos \theta\right) \left[\cos\left(\frac{1}{r} \sin \theta\right) - i \sin\left(\frac{1}{r} \sin \theta\right)\right]$$

Prove that $e^{1/z}$ takes on every complex value except zero within every circle centered on zero.

9 Show that $e^{2\pi ki/n}$, $k = 0, 1, 2, \ldots, n - 1$, are the only distinct solutions of $z^n = 1$.

10 Prove that $\cos z$ and $\sin z$ are analytic everywhere and obtain the formulas

$$\frac{d}{dz} \cos z = -\sin z \qquad \frac{d}{dz} \sin z = \cos z$$

11 Show that $\cos z$ and $\sin z$ satisfy the differential equation $f'' + f = 0$.

12 Show that $|\cos z|$ and $|\sin z|$ are not bounded in the complex plane.

13 Prove that $\cosh z$ and $\sinh z$ are analytic everywhere and obtain the formulas

$$\frac{d}{dz} \cosh z = \sinh z \qquad \frac{d}{dz} \sinh z = \cosh z$$

14 Find all the points where $\cosh z = 0$.

15 Find all the points where $\sinh z = 0$.

16 Determine where $\tanh z$ is analytic.
17 Determine where $\coth z$ is analytic.
18 Show that $\cosh z$ and $\sinh z$ satisfy the differential equation $f'' - f = 0$.
19 Obtain the formulas $\cosh iz = \cos z$ and $\sinh iz = i \sin z$.
20 Determine where $\log z = \ln |z| + i \arg z$, $0 \le \arg z < 2\pi$, is analytic and obtain the formula $d/dz \log z = 1/z$.
21 Define $z^a = \exp (a \log z)$, $\log z$ defined as in Exercise 20. Where is this function analytic, and what is its derivative? Assume a is complex.
22 Define $a^z = \exp (z \log a)$, $a \neq 0$. Where is this function analytic, and what is its derivative?

*1.7 POWER SERIES

The purpose of this section is to develop further the theory of series of complex numbers and, in particular, the theory of power series of the form $\sum_{k=0}^{\infty} a_k(z - z_0)^k$, where z_0 is a fixed complex number and the a_k's are complex constants. The reader will recall that we used a power series to define the exponential function

$$e^z = \sum_{k=0}^{\infty} \frac{z^k}{k!}$$

One of our goals in this section will be to prove the validity of the formula

$$e^{z_1+z_2} = e^{z_1}e^{z_2}$$

We begin by defining the general concept of convergence of a series of complex numbers $\sum_{k=0}^{\infty} w_k$. We define the *partial sums* $S_n = \sum_{k=0}^{n} w_k$. We say that the series $\sum_{k=0}^{\infty} w_k$ *converges* to the sum S if $\lim_{n\to\infty} S_n = S$. If the limit does not exist, then we say that the series *diverges*. The limit of a sequence of complex numbers $\{S_n\}$ exists and is equal to S if and only if, given any $\varepsilon > 0$, there exists an N such that $|S_n - S| < \varepsilon$ whenever $n > N$.

EXAMPLE 1.7.1 Consider the series $\sum_{k=0}^{\infty} c^k$, where c is a complex number. The partial sums are

$$S_n = \sum_{k=0}^{n} c^k = 1 + c + c^2 + \cdots + c^n$$

Multiplying by c, we have

$$cS_n = c + c^2 + c^3 + \cdots + c^{n+1}$$

Subtracting, we obtain

$$(1 - c)S_n = 1 - c^{n+1}$$

If $c \neq 1$, then $S_n = (1 - c^{n+1})/(1 - c)$. If $|c| < 1$, then $|c^{n+1}|$ tends to zero as $n \to \infty$. Hence, we conjecture that for $|c| < 1$ $\lim_{n\to\infty} S_n = 1/(1 - c)$. Let us prove this from the definition.

$$\left| S_n - \frac{1}{1 - c} \right| = \frac{|c|^{n+1}}{|1 - c|}$$

Given $\varepsilon > 0$, we wish to show that $|c|^{n+1} < \varepsilon|1 - c|$ for n sufficiently large, or $(n + 1) \ln |c| < \ln \varepsilon + \ln |1 - c|$. Dividing by $\ln |c|$, which is negative, we wish to show that for sufficiently large n,

$$n + 1 > \frac{\ln \varepsilon + \ln |1 - c|}{\ln |c|}$$

It is now clear that if n is greater than the largest integer in

$$\frac{\ln \varepsilon + \ln |1 - c|}{\ln |c|}$$

then $n + 1$ will be greater than $(\ln \varepsilon + \ln |1 - c|)/\ln |c|$. Therefore, for each $\varepsilon > 0$ we can find

$$N = \left[\frac{\ln \varepsilon + \ln |1 - c|}{\ln |c|} \right]$$

where the bracket stands for the "largest integer in," and when $n > N$,

$$\left| S_n - \frac{1}{1 - c} \right| < \varepsilon$$

This proves our conjecture. It is clear that when $|c| > 1$, then $|c|^{n+1}$ tends to ∞ as $n \to \infty$. Hence $|S_n - 1/(1 - c)|$ cannot be made small for large n. This shows that the series diverges for $|c| > 1$. This still leaves the case $|c| = 1$ to be considered. If $c = 1$, then

$$S_n = \sum_{k=0}^{n} 1^k = n + 1$$

In this case, $S_n \to \infty$ as $n \to \infty$, and the series diverges. If $|c| = 1$ and $c \neq 1$, then $c = \cos\theta + i \sin\theta$, $\theta \neq 0$, and

$$S_n - \frac{1}{1 - c} = \frac{-c^{n+1}}{1 - c}$$

cannot remain arbitrarily close to some fixed point for all sufficiently large n because $c^{n+1} = \cos(n+1)\theta + i \sin(n+1)\theta$ moves around the unit circle in jumps of arc θ. Therefore, the series diverges for $|c| = 1$. In summary,

$$\sum_{k=0}^{\infty} c^k = \frac{1}{1-c} \qquad \text{if } |c| < 1$$

and the series diverges for $|c| \geq 1$.

The convergence of a series of complex numbers is equivalent to the convergence of both the series of real parts and the series of imaginary parts. Let $w_k = u_k + iv_k$, $u_k = \text{Re}(w_k)$, $v_k = \text{Im}(w_k)$, and

$$S_n = \sum_{k=0}^{n} u_k + i \sum_{k=0}^{n} v_k = U_n + iV_n$$

If $\lim_{n\to\infty} S_n = S = U + iV$, then $\lim_{n\to\infty} U_n = U$ and $\lim_{n\to\infty} V_n = V$. This is because

$$|U_n - U| \leq |S_n - S|$$
$$|V_n - V| \leq |S_n - S|$$

and given any $\varepsilon > 0$, there is an N such that $|S_n - S| < \varepsilon$ for $n > N$ and therefore $|U_n - U| < \varepsilon$ and $|V_n - V| < \varepsilon$ for $n > N$. Conversely, suppose $\lim_{n\to\infty} U_n = U$ and $\lim_{n\to\infty} V_n = V$. Then, by the triangle inequality with $S = U + iV$

$$|S_n - S| \leq |U_n - U| + |V_n - V|$$

Given $\varepsilon > 0$, there is an N such that $|U_n - U| < \varepsilon/2$ and $|V_n - V| < \varepsilon/2$, and hence $|S_n - S| < \varepsilon$ for $n > N$. We have then proved the following theorem.

Theorem 1.7.1 Let $u_k = \text{Re}(w_k)$ and $v_k = \text{Im}(w_k)$. Then the series $\sum_{k=0}^{\infty} w_k$ converges if and only if both $\sum_{k=0}^{\infty} u_k$ and $\sum_{k=0}^{\infty} v_k$ converge, and $\sum_{k=0}^{\infty} w_k = \sum_{k=0}^{\infty} u_k + i \sum_{k=0}^{\infty} v_k$.

Next we define *absolute convergence*. A series of complex numbers $\sum_{k=0}^{\infty} w_k$ is said to converge absolutely if the series of real numbers $\sum_{k=0}^{\infty} |w_k|$ converges. Absolute convergence implies convergence, as we see in the next theorem. However, the converse is not true, as is illustrated by the series $\sum_{k=1}^{\infty} \frac{(-1)^k}{k}$, which converges although $\sum_{k=1}^{\infty} \frac{1}{k}$ diverges.

Theorem 1.7.2 If the series $\sum_{k=0}^{\infty} |w_k|$ converges, then so does $\sum_{k=0}^{\infty} w_k$.

PROOF Let $u_k = \text{Re}(w_k)$ and $v_k = \text{Im}(w_k)$. Then $|u_k| \le |w_k|$ and $|v_k| \le |w_k|$, and hence the series $\sum_{k=0}^{\infty} |u_k|$ and $\sum_{k=0}^{\infty} |v_k|$ converge by comparison with the convergent series $\sum_{k=0}^{\infty} |w_k|$. Now

$$0 \le |u_k| - u_k \le 2|u_k|$$
$$0 \le |v_k| - v_k \le 2|v_k|$$

Then again by comparison, $\sum_{k=0}^{\infty} (|u_k| - u_k)$ and $\sum_{k=0}^{\infty} (|v_k| - v_k)$ converge, and therefore by subtraction

$$\sum_{k=0}^{\infty} |u_k| - \sum_{k=0}^{\infty} (|u_k| - u_k) = \sum_{k=0}^{\infty} u_k$$
$$\sum_{k=0}^{\infty} |v_k| - \sum_{k=0}^{\infty} (|v_k| - v_k) = \sum_{k=0}^{\infty} v_k$$

the series of real parts and imaginary parts both converge and, by Theorem 1.7.1, the series $\sum_{k=0}^{\infty} w_k$ converges.

Before we consider power series, we need one more theorem which gives us a necessary condition for convergence. The failure of this condition gives us a test for divergence.

Theorem 1.7.3 If $\sum_{k=0}^{\infty} w_k$ converges, then $\lim_{n\to\infty} |w_n| = 0$.

PROOF Since $\sum_{k=0}^{\infty} w_k$ converges, the limit of the partial sums exists and $\lim_{n\to\infty} S_n = \lim_{n\to\infty} S_{n-1} = S$. Therefore, $\lim_{n\to\infty} (S_n - S_{n-1}) = 0$, and $\lim_{n\to\infty} |w_n| = \lim_{n\to\infty} |S_n - S_{n-1}| = 0$.

EXAMPLE 1.7.2 The divergence of the series $\sum_{k=0}^{\infty} c^k$ for $|c| \ge 1$ follows directly from Theorem 1.7.3. For suppose the series converged; then $\lim_{n\to\infty} |c|^n$ would be zero. However, $|c|^n \ge 1$.

That the condition $\lim_{n\to\infty} |w_n| = 0$ is not a sufficient condition for convergence is shown by the divergent series $\sum_{k=1}^{\infty} \frac{1}{k}$, although $\lim_{n\to\infty} (1/n) = 0$.

We can now begin our study of power series of the form $\sum_{k=0}^{\infty} a_k(z - z_0)^k$, where z_0 is a fixed complex number and the a_k's are complex constants. By introducing the variable $\zeta = z - z_0$ we can always put the series in the form $\sum_{k=0}^{\infty} a_k\zeta^k$. Hence, without loss of generality we study series only of the form $\sum_{k=0}^{\infty} a_k z^k$.

Theorem 1.7.4 If the power series $\sum_{k=0}^{\infty} a_k z^k$ converges for $z = c$, then it converges absolutely for all z such that $|z| < |c|$. If the power series diverges for $z = c$, then it diverges for all z such that $|z| > |c|$.

PROOF Assume that $\sum_{k=0}^{\infty} a_k c^k$ converges, $c \neq 0$. Then $\lim_{k\to\infty} |a_k c^k| = 0$. Therefore, there is a constant M such that $|a_k c^k| \leq M$ for all k. Now let $|z| < |c|$, and consider the series $\sum_{k=0}^{\infty} |a_k z^k|$. We have $|a_k z^k| = |a_k c^k| \times |z/c|^k \leq M|z/c|^k$. But the series $\sum_{k=0}^{\infty} M\left|\frac{z}{c}\right|^k$ converges since $|z/c| < 1$. Therefore, by comparison $\sum_{k=0}^{\infty} |a_k z^k|$ converges, and $\sum_{k=0}^{\infty} a_k z^k$ converges absolutely for $|z| < |c|$. This completes the first part of the proof. For the second part assume that $\sum_{k=0}^{\infty} a_k c^k$ diverges but that $\sum_{k=0}^{\infty} a_k z^k$ converges for $|z| > |c|$. But then $\sum_{k=0}^{\infty} |a_k c^k|$ converges and $\sum_{k=0}^{\infty} a_k c^k$ converges, which is a contradiction.

There are power series which converge for all z. The series used to define the exponential function is a good example. There are power series which converge only for $z = 0$ (all power series converge for $z = 0$). An example of such a series is $\sum_{k=0} k!\, z^k$, for if $z \neq 0$, $\lim_{k\to\infty} |k!\, z^k| = \infty$.

If a power series does not converge for all z but does converge for some $z \neq 0$, then Theorem 1.7.4 implies that there is a positive number R such that the power series converges for $|z| < R$ and diverges for $|z| > R$. In this case,

R is called the *radius of convergence* of the power series, and the circle $|z| = R$ is called the *circle of convergence*. The next two theorems give methods for determining the radius of convergence of a power series.

Theorem 1.7.5 Let $\lim_{n\to\infty} |a_{n+1}/a_n| = L \neq 0$. Then the radius of convergence of $\sum_{k=0}^{\infty} a_k z^k$ is $1/L$.

PROOF Consider

$$\lim_{n\to\infty} \left| \frac{a_{n+1}z^{n+1}}{a_n z^n} \right| = L|z|$$

Suppose $L|z| < 1$. Then there is an r such that $0 < r < 1$ and an N such that $|a_{n+1}z^{n+1}/a_n z^n| \leq r$ for all $n \geq N$. Hence $|a_{N+1}z^{N+1}| \leq r|a_N z^N|$, $|a_{N+2}z^{N+2}| \leq r|a_{N+1}z^{N+1}| \leq r^2|a_N z^N|$, etc. In general,

$$|a_{N+j}z^{N+j}| \leq r^j|a_N z^N| \qquad j = 1, 2, 3, \ldots$$

The series $\sum_{j=1}^{\infty} r^j$ converges since $0 < r < 1$. Hence, by comparison the series $\sum_{k=0}^{\infty} |a_k z^k|$ converges, and so does $\sum_{k=0}^{\infty} a_k z^k$. We have then proved that the power series converges for $|z| < 1/L$.

Now suppose $L|z| > 1$. Then there is a constant $p > 1$ and an M such that

$$\left| \frac{a_{n+1}z^{n+1}}{a_n z^n} \right| \geq p \qquad \text{for all } n \geq M$$

By an argument like the one above we have for $j = 1, 2, 3, \ldots$

$$|a_{M+j}z^{M+j}| \geq p^j|a_M z^M|$$

But since $p > 1$, $p^j \to \infty$ as $j \to \infty$ and therefore

$$|a_{M+j}z^{M+j}| \to \infty$$

as $j \to \infty$, which shows that the series diverges for $|z| > 1/L$.

Theorem 1.7.6 Let $\lim_{n\to\infty} |a_n|^{1/n} = L \neq 0$. Then the radius of convergence of $\sum_{k=0}^{\infty} a_k z^k$ is $1/L$.

PROOF The proof, which is very similar to that of Theorem 1.7.5, will be left for the reader.

EXAMPLE 1.7.3 Find the radius of convergence of the power series $\sum_{k=1}^{\infty} \frac{(-1)^k z^k}{k}$. Here the $a_k = (-1)^k/k$, and

$$\lim_{n\to\infty} \left|\frac{a_{n+1}}{a_n}\right| = \lim_{n\to\infty} \frac{n}{n+1} = 1$$

Therefore, the radius of convergence is 1. This example shows that on the circle of convergence we may have either convergence or divergence. For example, at $z = 1$ the series converges, but at $z = -1$ the series diverges.

EXAMPLE 1.7.4 Determine where the power series $\sum_{k=2}^{\infty} \left(\frac{z}{\ln k}\right)^k$ converges. Consider

$$\sqrt[n]{\left(\frac{|z|}{\ln n}\right)^n} = \frac{|z|}{\ln n} \to 0 \qquad \text{as } n \to \infty$$

It is clear that there is an r, $0 < r < 1$, and an N such that

$$\sqrt[n]{\left(\frac{|z|}{\ln n}\right)^n} \le r \qquad \text{for all } n \ge N$$

Hence, $[|z|/\ln n]^n \le r^n$. Then by comparison with the convergent series $\sum_{k=0}^{\infty} r^k$ we have absolute convergence for all z.

If we formally differentiate a power series $\sum_{k=0}^{\infty} a_k z^k$, we obtain the power series $\sum_{k=1}^{\infty} k a_k z^{k-1}$. Suppose

$$\lim_{n\to\infty} \left|\frac{a_{n+1}}{a_n}\right| = L \ne 0$$

Then

$$\lim_{n\to\infty} \frac{n+1}{n} \left|\frac{a_{n+1}}{a_n}\right| = L$$

Therefore, the differentiated series has the same radius of convergence as the original series. In fact, it can be shown in general that the differentiated series has the same radius of convergence as the original. Moreover, if $f(z) = \sum_{k=0}^{\infty} a_k z^k$, $|z| < R$, then $f(z)$ is differentiable† inside the circle of convergence and

$$f'(z) = \sum_{k=1}^{\infty} k a_k z^{k-1}$$

† For proofs see Dettman, op. cit., pp. 146 and 147.

By induction we can then differentiate as many times as we please and obtain for $n = 1, 2, 3, \ldots$

$$f^{(n)}(z) = \sum_{k=n}^{\infty} k(k-1)\cdots(k-n+1)a_k z^{k-n}$$

valid for $|z| < R$. Evaluating $f^{(n)}(0)$ we have, since all the terms are zero except for $k = n$,

$$f^{(n)}(0) = n!\, a_n$$

and hence, $a_n = f^{(n)}(0)/n!$.

We conclude this section by showing that in a certain sense, which we shall define presently, we may multiply two power series together within their common circle of convergence. If we multiplied two polynomials

$$p(z) = a_0 + a_1 z + a_2 z^2 + \cdots + a_n z^n$$
$$q(z) = b_0 + b_1 z + b_2 z^2 + \cdots + b_n z^n$$

together, we would have

$$p(z)q(z) = a_0 b_0 + (a_0 b_1 + a_1 b_0)z + (a_0 b_2 + a_1 b_1 + a_2 b_0)z^2 + \cdots + (a_0 b_n + a_1 b_{n-1} + \cdots + a_{n-1} b_1 + a_n b_0)z^n + \cdots$$

We use this formula to motivate the definition of the *Cauchy product* of two series $\sum_{k=0}^{\infty} u_k$ and $\sum_{k=0}^{\infty} v_k$ as

$$\sum_{n=0}^{\infty} \left(\sum_{k=0}^{n} u_k v_{n-k} \right)$$

We prove the following theorem.

Theorem 1.7.7 The Cauchy product of two absolutely convergent series converges absolutely to the product of the two series.

PROOF We first show that the Cauchy product converges absolutely. Let $a_k = |u_k|$, $b_k = |v_k|$, and $c_k = |w_k| = \left| \sum_{j=0}^{k} u_j v_{k-j} \right|$. Then

$$\sum_{k=0}^{n} c_k = \sum_{k=0}^{n} \left| \sum_{j=0}^{k} u_j v_{k-j} \right|$$
$$\le \sum_{k=0}^{n} \sum_{j=0}^{k} a_j b_{k-j} \le \sum_{k=0}^{n} a_k \sum_{k=0}^{n} b_k \le AB$$

where $A = \sum_{k=0}^{\infty} |u_k|$ and $B = \sum_{k=0}^{\infty} |v_k|$. Therefore, since the $\sum_{k=0}^{n} |w_k|$ are bounded above, the series $\sum_{k=0}^{\infty} w_k$ converges absolutely.

If a series converges, then the average of its partial sums converges to the sum of the series. We shall prove this fact and then use it to find the sum of our absolutely convergent Cauchy product. Suppose $\lim_{n\to\infty} S_n = S$ where S_n is the partial sum of a convergent series. Let $\sigma_n = \frac{1}{n}\sum_{k=0}^{n} S_k$. Then $\sigma_n - S = \frac{1}{n}\sum_{k=0}^{n}(S_k - S)$ and $|\sigma_n - S| \le \frac{1}{n}\sum_{k=0}^{n}|S_k - S|$. Given $\varepsilon > 0$, there is an M such that $|S_k - S| < \varepsilon/2$ when $k > M$. Therefore, if $n > M$,

$$|\sigma_n - S| \le \frac{1}{n}\sum_{k=0}^{M}|S_k - S| + \frac{1}{n}\sum_{k=M+1}^{n}|S_k - S|$$

Having chosen ε, we fix M, and then $\sum_{k=0}^{M}|S_k - S| = L$ is fixed. We choose n so large that $L/n < \varepsilon/2$. Then $|\sigma_n - S| \le L/n + (n - M)\varepsilon/2n < \varepsilon$. This shows that $\lim_{n\to\infty}\sigma_n = S$.

Now let $U_n = \sum_{k=0}^{n} u_k$, $V_n = \sum_{k=0}^{n} v_k$, and $W_n = \sum_{k=0}^{n} w_k$. We have that $U_n \to U$, the sum of the first series, that $V_n \to V$, the sum of the second series, and we wish to show that $W_n \to UV$. To do this we consider the average

$$\sigma_m = \frac{1}{m}\sum_{n=0}^{m} W_n$$

We leave it to the reader to show by a simple induction argument on m that

$$\sum_{n=0}^{m} W_n = \sum_{k=0}^{m} U_k V_{m-k}$$

Now let $U_k = U + \alpha_k$ and $V_k = V + \beta_k$. Then

$$\sigma_m = \frac{1}{m}\sum_{k=0}^{m}(U + \alpha_k)(V + \beta_{m-k})$$

$$= \frac{m+1}{m}UV + \frac{V}{m}\sum_{k=0}^{m}\alpha_k + \frac{U}{m}\sum_{k=0}^{m}\beta_k + \frac{1}{m}\sum_{k=0}^{m}\alpha_k\beta_{m-k}$$

As $m \to \infty$, the first term approaches UV, the second term approaches zero since $\alpha_k \to 0$, the third term approaches zero since $\beta_k \to 0$, and it can be shown that the last term approaches zero. In fact, given $\varepsilon > 0$, there is an N such that $|\alpha_k| < \varepsilon$ and $|\beta_k| < \varepsilon$ for $k > N$ and $|\alpha_k| < M$ and $|\beta_k| < M$ for all k. Then for $m > 2N$

$$\begin{aligned}
\frac{1}{m}\left|\sum_{k=0}^{m} \alpha_k\beta_{m-k}\right| &\leq \frac{1}{m}\sum_{k=0}^{m} |\alpha_k|\,|\beta_{m-k}| \\
&= \frac{1}{m}\left(\sum_{k=0}^{N} + \sum_{k=N+1}^{m-N} + \sum_{k=m-N+1}^{m}\right)|\alpha_k|\,|\beta_{m-k}| \\
&\leq \frac{(2N+1)M\varepsilon}{m} + \frac{(m-2N)\varepsilon^2}{m}
\end{aligned}$$

which can be made less than ε^2 for m sufficiently large. This completes the proof of the theorem since we have shown that $\lim_{m\to\infty} \sigma_m = UV$.

EXAMPLE 1.7.5 Prove that $e^{z_1}e^{z_2} = e^{z_1+z_2}$. We have defined e^z by a power series, that is,

$$e^z = \sum_{k=0}^{\infty} \frac{z^k}{k!}$$

and we now know that the series converges absolutely for any z. Hence, we may multiply $e^{z_1}e^{z_2}$ as a Cauchy product. Therefore,

$$\begin{aligned}
e^{z_1}e^{z_2} &= \sum_{n=0}^{\infty}\sum_{k=0}^{n} \frac{z_1{}^k}{k!}\frac{z_2{}^{n-k}}{(n-k)!} \\
&= \sum_{n=0}^{\infty}\frac{1}{n!}\sum_{k=0}^{n}\frac{n!}{k!\,(n-k)!}\, z_1{}^k z_2{}^{n-k} \\
&= \sum_{n=0}^{\infty}\frac{(z_1+z_2)^n}{n!} = e^{z_1+z_2}
\end{aligned}$$

We conclude this section with an obvious theorem about multiplying two power series together.

Theorem 1.7.8 Let $f(z) = \sum_{k=0}^{\infty} a_k z^k$ and $g(z) = \sum_{k=0}^{\infty} b_k z^k$ for $|z| < R$. Then $f(z)g(z) = \sum_{n=0}^{\infty}\left(\sum_{k=0}^{n} a_k b_{n-k}\right) z^n$ for $|z| < R$.

EXERCISES 1.7

1 Prove that a convergent series can have but one sum; that is, if $\lim_{n\to\infty} S_n$ exists, the limit is unique.

2 Show that $\sum_{k=1}^{\infty} k^{-z}$ converges absolutely for Re $(z) > 1$.

3 Find the radius of convergence of the power series $\sum_{k=1}^{\infty} \frac{z^k}{k^2}$. Does the series converge on the circle of convergence?

4 Let a_k, $k = 0, 1, 2, \ldots$, be a sequence such that $\lim_{k\to\infty} a_k = a \neq 0$. What is the radius of convergence of $\sum_{k=0}^{\infty} a_k z^k$?

5 Find the radius of convergence of the power series $\sum_{k=0}^{\infty} \frac{z^{2k}}{2^k}$. *Hint:* Let $w = z^2/2$.

6 Prove that absolute convergence of a power series on the circle of convergence holds either everywhere or nowhere.

7 Find the radius of convergence of the power series

$$1 + \frac{\alpha z}{1!} + \frac{\alpha(\alpha - 1)}{2!} z^2 + \frac{\alpha(\alpha - 1)(\alpha - 2)}{3!} z^3 + \cdots$$

where α is complex.

8 Prove Theorem 1.7.6.

9 Let $\lim_{n\to\infty} \sqrt[n]{|a_n|} = L \neq 0$. Show that the radius of convergence of $\sum_{k=1}^{\infty} k a_k z^{k-1}$ is $1/L$.

10 Prove that if $f(z) = \sum_{k=0}^{\infty} a_k z^k = \sum_{k=0}^{\infty} b_k z^k$ for $|z| < R$, then $a_k = b_k$, $k = 0, 1, 2, \ldots$. This exercise shows that a power-series representation of a function is necessarily unique.

11 Verify the formula $\sum_{n=0}^{m} W_n = \sum_{k=0}^{m} U_k V_{m-k}$ in the proof of Theorem 1.7.7.

2

LINEAR ALGEBRAIC EQUATIONS

2.1 INTRODUCTION

There are at least two main approaches to the study of linear algebra. The more abstract is to introduce the general discussion of vector spaces first. In this approach matrices come up in a natural way in the discussion of linear transformation, and the usual theorems about the solvability of systems of linear equations come out in due course. The other approach, which is more concrete, is to begin the discussion with matrices and linear algebraic equations. The advantage of this direction is that it introduces the general concepts with examples in a more familiar setting. We can then build up to the abstract concept of a vector space with a set of concrete examples to guide and motivate the student. This is the approach we shall take in this book.

We first introduce the algebra of matrices and show how to write systems of linear algebraic equations in compact notation. We shall discuss the solution of such systems by elimination methods and obtain the usual existence and uniqueness theorems for solutions. We introduce the concept of determinant of a square matrix and obtain the usual properties. This leads into a discussion of the inverse of a square matrix, where we teach the reduction method of inverting a matrix. The last section (which is starred) takes up rank of a matrix and collects all the existence and uniqueness theorems we shall need for the discussion of linear algebraic equations.

2.2 MATRICES

One of the most important applications of matrices is the treatment of linear algebraic equations. Therefore, we shall use this application to introduce the concept of a matrix, and in the process the algebra of matrices will come out in a very natural way. We begin with an example.

Consider the linear algebraic equations

$$\begin{aligned} x_1 + 2x_2 - 3x_3 &= 0 \\ 2x_1 \qquad + x_3 &= 3 \\ -x_1 + x_2 \qquad &= 0 \\ - 2x_2 + 4x_3 &= 2 \end{aligned}$$

We shall leave aside for the moment the question of whether these equations have a solution. Our immediate objective is to obtain a compact notation for this set of equations and others like it. Let A be the rectangular array of numbers with four rows and three columns

$$A = \begin{pmatrix} 1 & 2 & -3 \\ 2 & 0 & 1 \\ -1 & 1 & 0 \\ 0 & -2 & 4 \end{pmatrix}$$

Let B be the array with four rows and one column

$$B = \begin{pmatrix} 0 \\ 3 \\ 0 \\ 2 \end{pmatrix}$$

and let X stand for the unknowns x_1, x_2, and x_3, as follows:

$$X = \begin{pmatrix} x_1 \\ x_2 \\ x_3 \end{pmatrix}$$

A, B, and X are all examples of matrices; A is a 4×3 array, B is a 4×1 array, and X is a 3×1 array. We shall now define a multiplication operation between pairs of matrices so that it will be possible to write the system of algebraic equations in the compact form $AX = B$.

First, we should define what we mean by equality of matrices. Two matrices are said to be equal if they have the same number of rows and columns and each entry (element) of the one is equal to the corresponding entry (element) of the other. For example, if

$$C = \begin{pmatrix} 1 & 0 & -3 & 7 \\ -1 & 2 & 0 & 5 \\ 0 & 3 & -1 & -6 \\ 4 & -2 & 0 & 6 \\ 8 & 1 & -3 & 9 \end{pmatrix}$$

and $D = C$, then we know that D has five rows and four columns and

$$D = \begin{pmatrix} 1 & 0 & -3 & 7 \\ -1 & 2 & 0 & 5 \\ 0 & 3 & -1 & -6 \\ 4 & -2 & 0 & 6 \\ 8 & 1 & -3 & 9 \end{pmatrix}$$

The multiplication will be defined only in the case where the first factor has the same number of columns as the second factor has rows. Suppose the first factor has l rows and m columns while the second factor has m rows and n columns. We define the product to have l rows and n columns obtained by the following rule. To obtain the element in the ith row and the jth column of the product we multiply the elements of the ith row of the first factor by the corresponding elements of the jth column of the second factor and sum. In other words, suppose the elements of the ith row of the first factor are p_{ik}, $k = 1, 2, \ldots, m$, while the elements of the jth column of the second factor are q_{kj}, $k = 1, 2, \ldots, m$; then the element in the ith row, jth column of the product is defined to be

$$p_{i1}q_{1j} + p_{i2}q_{2j} + \cdots + p_{im}q_{mj}$$

Since i can take on l different values and j can take on n different values, this definition defines ln elements for the product. If the first factor is $l \times m$ and the second factor is $m \times n$, then the product is $l \times n$.

Using this definition of the product of two matrices and the above definition of equality of matrices, we have for the example

$$AX = \begin{pmatrix} 1 \cdot x_1 + & 2 \cdot x_2 + (-3) \cdot x_3 \\ 2 \cdot x_1 + & 0 \cdot x_2 + \quad 1 \cdot x_3 \\ (-1) \cdot x_1 + & 1 \cdot x_2 + \quad 0 \cdot x_3 \\ 0 \cdot x_1 + (-2) \cdot x_2 + & 4 \cdot x_3 \end{pmatrix} = \begin{pmatrix} 0 \\ 3 \\ 0 \\ 2 \end{pmatrix} = B$$

so we have achieved our compact notation. In fact, for any system of m linear algebraic equations in n unknowns

$$\begin{array}{l} a_{11}x_1 + a_{12}x_2 + \cdots + a_{1n}x_n = b_1 \\ a_{21}x_1 + a_{22}x_2 + \cdots + a_{2n}x_n = b_2 \\ \cdots\cdots\cdots\cdots\cdots\cdots\cdots\cdots \\ a_{m1}x_1 + a_{m2}x_2 + \cdots + a_{mn}x_n = b_m \end{array}$$

we can use the compact matrix notation $AX = B$, where

$$A = \begin{pmatrix} a_{11} & a_{12} \cdots a_{1n} \\ a_{21} & a_{22} \cdots a_{2n} \\ \cdots & \cdots \\ a_{m1} & a_{m2} \cdots a_{mn} \end{pmatrix} \qquad X = \begin{pmatrix} x_1 \\ x_2 \\ \cdot \\ x_n \end{pmatrix} \qquad B = \begin{pmatrix} b_1 \\ b_2 \\ \cdot \\ b_m \end{pmatrix}$$

In this case, we usually call A the *coefficient matrix* of the system. There is another matrix which plays an important role in the study of systems of linear equations. This is the *augmented matrix*, which is formed by inserting $b_1, b_2, \ldots, b_m$ as a new last column into the coefficient matrix. In other words, the augmented matrix is

$$\begin{pmatrix} a_{11} & a_{12} \cdots a_{1n} & b_1 \\ a_{21} & a_{22} \cdots a_{2n} & b_2 \\ \cdots & \cdots & \cdots \\ a_{m1} & a_{m2} \cdots a_{mn} & b_m \end{pmatrix}$$

If the coefficient matrix is $m \times n$, then the augmented matrix is $m \times (n + 1)$.

Next we introduce the concepts of addition and subtraction of matrices. Suppose a certain set $x_1, x_2, \ldots, x_n$ satisfies two systems of equations

$$\begin{array}{l} a_{11}x_1 + a_{12}x_2 + \cdots + a_{1n}x_n = c_1 \\ a_{21}x_1 + a_{22}x_2 + \cdots + a_{2n}x_n = c_2 \\ \cdots\cdots\cdots\cdots\cdots\cdots\cdots\cdots \\ a_{m1}x_1 + a_{m2}x_2 + \cdots + a_{mn}x_n = c_m \end{array}$$

and

$$\begin{array}{l} b_{11}x_1 + b_{12}x_2 + \cdots + b_{1n}x_n = d_1 \\ b_{21}x_1 + b_{22}x_2 + \cdots + b_{2n}x_n = d_2 \\ \cdots\cdots\cdots\cdots\cdots\cdots\cdots\cdots \\ b_{m1}x_1 + b_{m2}x_2 + \cdots + b_{mn}x_n = d_m \end{array}$$

If we add or subtract the corresponding c's and d's, we have the following results

$$\begin{array}{l} (a_{11} \pm b_{11})x_1 + (a_{12} \pm b_{12})x_2 + \cdots + (a_{1n} \pm b_{1n})x_n = c_1 \pm d_1 \\ (a_{21} \pm b_{21})x_1 + (a_{22} \pm b_{22})x_2 + \cdots + (a_{2n} \pm b_{2n})x_n = c_2 \pm d_2 \\ \cdots\cdots\cdots\cdots\cdots\cdots\cdots\cdots\cdots\cdots\cdots\cdots \\ (a_{m1} \pm b_{m1})x_1 + (a_{m2} \pm b_{m2})x_2 + \cdots + (a_{mn} \pm b_{mn})x_n = c_m \pm d_m \end{array}$$

We can state this in the compact matrix notation: $AX = C$, $BX = D$ implies $(A \pm B)X = C \pm D$ provided we define addition and subtraction in a consistent manner. A definition which makes this work is the following: to add (subtract) an $m \times n$ matrix B to (from) an $m \times n$ matrix A, form the $m \times n$ matrix $A + B$ $(A - B)$ by adding (subtracting) elements of B to (from) corresponding elements of A.

Finally, we define the operation of multiplication of a matrix by a number (scalar). Suppose we multiply every equation of the system $AX = C$ by the same number a. Then the system becomes

$$\begin{aligned}
aa_{11}x_1 + aa_{12}x_2 + \cdots + aa_{1n}x_n &= ac_1 \\
aa_{21}x_1 + aa_{22}x_2 + \cdots + aa_{2n}x_n &= ac_2 \\
\cdots\cdots\cdots\cdots\cdots\cdots\cdots\cdots\cdots\cdots\cdots \\
aa_{m1}x_1 + aa_{m2}x_2 + \cdots + aa_{mn}x_n &= ac_m
\end{aligned}$$

This can be written as $(aA)X = aC$ provided we define the matrices aA and aC properly. This can be done as follows: the matrix formed by multiplying a matrix A by a scalar a is obtained by multiplying each element of A by the same scalar a.

Let us summarize all the definitions we have made so far in a more precise manner.

Definition 2.2.1 A matrix is a rectangular array of numbers (real or complex). If the matrix has m rows and n columns, the matrix is said to be $m \times n$. If $m = n$, the matrix is said to be *square* and of order m. The number in the ith row and the jth column is called the (i,j)th element; in the double-subscript notation for elements the first subscript is the row subscript, and the second subscript is the column subscript.

Definition 2.2.2 Two $m \times n$ matrices A and B are equal if the (i,j)th element of A is equal to the corresponding (i,j)th element of B for all possible i and j.

Definition 2.2.3 If A is an $m \times n$ matrix with elements a_{ij} and a is a scalar (real or complex), then aA is the $m \times n$ matrix with elements aa_{ij}.

Definition 2.2.4 If A and B are $m \times n$ matrices with elements a_{ij} and b_{ij}, then $A + B$ is the $m \times n$ matrix with elements $a_{ij} + b_{ij}$ and $A - B$ is the $m \times n$ matrix with elements $a_{ij} - b_{ij}$.

Definition 2.2.5 If A is the $m \times n$ matrix with elements a_{ij} and B is the $n \times p$ matrix with elements b_{jk}, then AB is the $m \times p$ matrix with elements $\sum_{j=1}^{n} a_{ij}b_{jk}$.

Definition 2.2.6 The $m \times n$ zero matrix, denoted by 0, is the $m \times n$ matrix all of whose elements are zero.

Definition 2.2.7 The mth order identity matrix, denoted by I, is the mth order square matrix whose elements are 1 on the principal diagonal (upper left to lower right) and 0 elsewhere.

We conclude this section with some theorems whose proofs are straightforward applications of these definitions. The only part which is particularly hard is associativity of matrix multiplication (Theorem 2.2.4). In this case the reader should at least see what the calculations involve by multiplying out some small (but general) matrices.

Theorem 2.2.1 If A and B are $m \times n$ matrices and a and b are scalars, then:

(i) $(a \pm b)A = aA \pm bA$.
(ii) $a(A \pm B) = aA \pm aB$.
(iii) $(ab)A = a(bA) = b(aA)$.

Theorem 2.2.2 If A is an $m \times n$ matrix, B is an $n \times p$ matrix, and a is a scalar, then $a(AB) = (aA)B = A(aB)$.

Theorem 2.2.3 If A, B, and C are $m \times n$ matrices, then $A + B = B + A$ and $A + (B + C) = (A + B) + C$.

Theorem 2.2.4 If A is an $m \times n$ matrix, B is an $n \times p$ matrix, and C is a $p \times q$ matrix, then $A(BC) = (AB)C$.

Theorem 2.2.5 If A is an $m \times n$ matrix and B and C are $n \times p$ matrices, then $A(B \pm C) = AB \pm AC$.

The reader may be wondering at this point why no mention has been made of division. Actually, we shall be able later to define a kind of division by a certain type of square matrix. However, for the moment the reader should ponder the implication, as far as division is concerned, of the following product

$$\begin{pmatrix} 1 & 1 \\ 1 & 1 \end{pmatrix}\begin{pmatrix} 1 & 1 \\ -1 & -1 \end{pmatrix} = \begin{pmatrix} 0 & 0 \\ 0 & 0 \end{pmatrix}$$

Here we have a product $AB = 0$ where neither A nor B itself is a zero matrix.

EXERCISES 2.2

1 Consider the system of linear algebraic equations

$$\begin{aligned} x_1 - 2x_2 + 3x_5 &= 7 \\ 2x_2 + x_4 + 5x_5 &= -6 \\ x_1 - x_3 + x_4 &= 0 \end{aligned}$$

Identify the coefficient matrix and the augmented matrix of the system.

2 Let

$$A = \begin{pmatrix} 1 & 0 & -3 & 2 \\ 0 & -1 & 7 & 5 \\ 2 & 3 & -4 & 0 \end{pmatrix} \qquad B = \begin{pmatrix} -2 & 6 & 1 & 5 \\ 2 & 0 & -3 & 4 \\ 1 & -5 & 0 & 1 \end{pmatrix}$$

Compute $A + B$, $A - B$, $3A$, $-2B$, $5A - 7B$.

3 Let

$$C = \begin{pmatrix} 1 & 2 \\ -1 & 0 \\ 3 & -2 \\ 0 & 5 \end{pmatrix}$$

Compute AC and BC, where A and B are defined in Exercise 2.

4 Let

$$A = \begin{pmatrix} 1 & 0 & 1 \\ -1 & 2 & 0 \\ 0 & 3 & 5 \end{pmatrix} \qquad B = \begin{pmatrix} 0 & 2 & 3 \\ 1 & -2 & 4 \\ 5 & 0 & -7 \end{pmatrix}$$

Compute AB and BA. Is $AB = BA$?

5 Let A be an $m \times n$ matrix, and let 0 be an $m \times n$ zero matrix. Show that $A + 0 = 0 + A = A$ and that $A + (-1)A = 0$.

6 Let A be an $m \times n$ matrix, and let 0 be an $n \times p$ zero matrix. Find $A0$.

7 Let A be an mth order square matrix and I the mth order identity. Show that $AI = IA = A$.

8 Let

$$A = \begin{pmatrix} 1 & 1 \\ -1 & 1 \end{pmatrix}$$

Find a 2×2 matrix B such that $AB = I$. Compute BA. Can the same thing be done if

$$A = \begin{pmatrix} 1 & 1 \\ 1 & 1 \end{pmatrix}$$

9 Let

$$E_1 = \begin{pmatrix} 1 & 0 & 0 \\ 0 & k & 0 \\ 0 & 0 & 1 \end{pmatrix} \qquad E_2 = \begin{pmatrix} 0 & 0 & 1 \\ 0 & 1 & 0 \\ 1 & 0 & 0 \end{pmatrix} \qquad E_3 = \begin{pmatrix} 1 & 1 & 0 \\ 0 & 1 & 0 \\ 0 & 0 & 1 \end{pmatrix}$$

Compute E_1A, E_2A, E_3A where

$$A = \begin{pmatrix} a_{11} & a_{12} & a_{13} & a_{14} \\ a_{21} & a_{22} & a_{23} & a_{24} \\ a_{31} & a_{32} & a_{33} & a_{34} \end{pmatrix}$$

Generalize. *Hint:* E_1 is obtained from I by multiplying the second row by k, E_2 is obtained from I by interchanging rows 1 and 3, and E_3 is obtained from I by adding the second row to the first row.

10 Powers of square matrices are defined as follows: $A^1 = A$, $A^2 = AA$, $A^3 = AA^2$, etc. Prove that $A^2 - I = (A - I)(A + I) = (A + I)(A - I)$ and $A^3 - I = (A - I)(A^2 + A + I) = (A^2 + A + I)(A - I)$.

11 Let

$$\begin{aligned} y_1 &= a_{11}x_1 + a_{12}x_2 + a_{13}x_3 \\ y_2 &= a_{21}x_1 + a_{22}x_2 + a_{23}x_3 \\ x_1 &= b_{11}z_1 + b_{12}z_2 \\ x_2 &= b_{21}z_1 + b_{22}z_2 \\ x_3 &= b_{31}z_1 + b_{32}z_2 \end{aligned}$$

Find c_{11}, c_{12}, c_{21}, and c_{22}, where

$$\begin{aligned} y_1 &= c_{11}z_1 + c_{12}z_2 \\ y_2 &= c_{21}z_1 + c_{22}z_2 \end{aligned}$$

and verify the following matrix notation: $Y = AX$, $X = BZ$, $Y = CZ$ where $C = AB$.

2.3 ELIMINATION METHOD

In this section, we shall take up an elimination method which is general enough to find all solutions of a system of linear algebraic equations (if it has any) and which will, in fact, tell us whether a given system has solutions. The idea is quite simple. We try to eliminate the first variable from all but the first equation, the second variable from all but the first and second equations, the third variable from all but the first, second, and third equations, etc. This will not always be possible, but in the attempt we shall find out what is possible, and it will turn out that this is good enough to achieve our purpose.

Let us return to the example of Sec. 2.2:

$$\begin{aligned} x_1 + 2x_2 - 3x_3 &= 0 \\ -2x_1 - x_3 &= -3 \\ -x_1 + x_2 &= 0 \\ -2x_2 + 4x_3 &= 2 \end{aligned}$$

If we add 2 times the first line to the second line and then add the first line to the third line, we have

$$\begin{aligned} x_1 + 2x_2 - 3x_3 &= 0 \\ 4x_2 - 7x_3 &= -3 \\ 3x_2 - 3x_3 &= 0 \\ -2x_2 + 4x_3 &= 2 \end{aligned}$$

and we have eliminated x_1 from all but the first equation. Next we multiply the fourth equation by $(-\frac{1}{2})$ and interchange it with the second equation. This leads to the system

$$\begin{aligned} x_1 + 2x_2 - 3x_3 &= 0 \\ x_2 - 2x_3 &= -1 \\ 3x_2 - 3x_3 &= 0 \\ 4x_2 - 7x_3 &= -3 \end{aligned}$$

Next we add (-3) times the second equation to the third equation and (-4) times the second equation to the fourth equation, and we have

$$\begin{aligned} x_1 + 2x_2 - 3x_3 &= 0 \\ x_2 - 2x_3 &= -1 \\ 3x_3 &= 3 \\ x_3 &= 1 \end{aligned}$$

Finally, we can multiply the third equation by $(-\frac{1}{3})$ and add it to the fourth equation. The result is

$$\begin{aligned} x_1 + 2x_2 - 3x_3 &= 0 \\ x_2 - 2x_3 &= -1 \\ x_3 &= 1 \\ 0 &= 0 \end{aligned}$$

We call this the *reduced system*. It is now apparent that $x_3 = 1$, and substituting this value into the second equation, we find that $x_2 = 1$. Finally, substituting these values into the first equation, we have $x_1 = 1$. This is obviously the only solution of the reduced system. We can substitute it into the original system and verify that it is a solution. However, are there other solutions of the original system which we have not uncovered? The answer is no, and we shall prove this later when we show that any solution of the original system is a solution of the reduced system and vice versa.

We notice in the above example that there is redundancy in the system. This became apparent at the third stage of reduction, when we had two equations $x_3 = 1$ and $3x_3 = 3$, and it showed itself further in the fourth stage, when the last equation dropped out altogether. This redundancy was not obvious in the original system, and it became so only as we reduced the system. In this case, it did not keep us from having a solution. However, suppose the fourth equation in the original system had read $-2x_2 + 4x_3 = 4$. Then the reduced system would have been

$$\begin{aligned} x_1 + 2x_2 - 3x_3 &= 0 \\ x_2 - 2x_3 &= -2 \\ x_3 &= 2 \\ 0 &= 3 \end{aligned}$$

Since this is impossible, we would have to conclude that there are no solutions of the reduced system and therefore that there are no solutions of the original system. Hence, we can expect that the reduction method will uncover redundancy in a system of equations and will also tell us when a system has no solutions.

Now let us go back over the original example again for further insight and possible simplification. First we note that there were only three basic operations involved in the reduction: (1) multiplication of an equation by a nonzero constant, (2) interchange of a pair of equations, and (3) addition of one equation to another. Second, we note that all the information contained in the various systems obtained in the process is contained in their respective augmented matrices. In other words, writing down a sequence of augmented matrices is just as good as writing down the various systems of equations. Finally, as operations on matrices our three basic operations are respectively (1) multiplication of a row by a nonzero constant, (2) interchange of a pair of rows, and (3) addition of one row to another.

To make this all clear let us review the example again, this time working with the augmented matrices. The original augmented matrix is

$$\begin{pmatrix} 1 & 2 & -3 & 0 \\ -2 & 0 & -1 & -3 \\ -1 & 1 & 0 & 0 \\ 0 & -2 & 4 & 2 \end{pmatrix}$$

Adding 2 times the first row to the second and adding the first row to the third, we obtain

$$\begin{pmatrix} 1 & 2 & -3 & 0 \\ 0 & 4 & -7 & -3 \\ 0 & 3 & -3 & 0 \\ 0 & -2 & 4 & 2 \end{pmatrix}$$

Multiplying the fourth row by $(-\frac{1}{2})$ and interchanging it with the second row, we have

$$\begin{pmatrix} 1 & 2 & -3 & 0 \\ 0 & 1 & -2 & -1 \\ 0 & 3 & -3 & 0 \\ 0 & 4 & -7 & -3 \end{pmatrix}$$

Adding (-3) times the second row to the third and (-4) times the second to the fourth, we obtain

$$\begin{pmatrix} 1 & 2 & -3 & 0 \\ 0 & 1 & -2 & -1 \\ 0 & 0 & 3 & 3 \\ 0 & 0 & 1 & 1 \end{pmatrix}$$

Finally, we multiply the third row by $(-\frac{1}{3})$ and add it to the fourth and obtain the *augmented matrix of the reduced system*

$$\begin{pmatrix} 1 & 2 & -3 & 0 \\ 0 & 1 & -2 & -1 \\ 0 & 0 & 1 & 1 \\ 0 & 0 & 0 & 0 \end{pmatrix}$$

At this point we can reconstruct the reduced system from its augmented matrix and solve as we did above.

In making such calculations we shall want to write down the original augmented matrix, the final augmented matrix, and the various intermediate matrices with some kind of connective symbol to indicate that we go from one to the other by some combination of the three basic row operations. Of course these matrices are not equal, so we should not use an equal sign. We shall instead use an arrow to mean that one is obtained from the other by row operations. Hence,†

$$\begin{pmatrix} 1 & 2 & -3 & 0 \\ -2 & 0 & -1 & -3 \\ -1 & 1 & 0 & 0 \\ 0 & -2 & 4 & 2 \end{pmatrix} \to \begin{pmatrix} 1 & 2 & -3 & 0 \\ 0 & 4 & -7 & -3 \\ 0 & 3 & -3 & 0 \\ 0 & -2 & 4 & 2 \end{pmatrix} \to \begin{pmatrix} 1 & 2 & -3 & 0 \\ 0 & 1 & -2 & -1 \\ 0 & 3 & -3 & 0 \\ 0 & 4 & -7 & -3 \end{pmatrix}$$

$$\to \begin{pmatrix} 1 & 2 & -3 & 0 \\ 0 & 1 & -2 & -1 \\ 0 & 0 & 3 & 3 \\ 0 & 0 & 1 & 1 \end{pmatrix} \to \begin{pmatrix} 1 & 2 & -3 & 0 \\ 0 & 1 & -2 & -1 \\ 0 & 0 & 1 & 1 \\ 0 & 0 & 0 & 0 \end{pmatrix}$$

† Actually, a double arrow ↔ would make perfectly good sense because one can go either way by basic row operations. The relation is, in fact, an equivalence relation (see Exercise 2.3.5).

Before we state the general method, let us consider one more example to illustrate what may happen in other cases. Consider the system of linear equations

$$\begin{aligned} x_1 + 2x_2 - 5x_3 - x_4 + 2x_5 &= -3 \\ x_2 - 2x_3 + x_4 - 4x_5 &= 1 \\ 2x_1 - 3x_2 + 4x_3 + 2x_4 - x_5 &= 9 \end{aligned}$$

The augmented matrix of the system is

$$\begin{pmatrix} 1 & 2 & -5 & -1 & 2 & -3 \\ 0 & 1 & -2 & 1 & -4 & 1 \\ 2 & -3 & 4 & 2 & -1 & 9 \end{pmatrix}$$

If we add (-2) times the first row to the third, we have

$$\begin{pmatrix} 1 & 2 & -5 & -1 & 2 & -3 \\ 0 & 1 & -2 & 1 & -4 & 1 \\ 0 & -7 & 14 & 4 & -5 & 15 \end{pmatrix}$$

Next we add 7 times the second row to the third, to obtain

$$\begin{pmatrix} 1 & 2 & -5 & -1 & 2 & -3 \\ 0 & 1 & -2 & 1 & -4 & 1 \\ 0 & 0 & 0 & 11 & -33 & 22 \end{pmatrix}$$

Notice that this last step eliminated both the second and third variables from the last equation (as indicated by the zeros in the second and third columns), showing that the process may proceed faster than one variable per row.

Finally, we may multiply the last row by $\frac{1}{11}$, and we have

$$\begin{pmatrix} 1 & 2 & -5 & -1 & 2 & -3 \\ 0 & 1 & -2 & 1 & -4 & 1 \\ 0 & 0 & 0 & 1 & -3 & 2 \end{pmatrix}$$

This is as far as the reduction can be carried, and it leads to all solutions of the original system. The reduced system is

$$\begin{aligned} x_1 + 2x_2 - 5x_3 - x_4 + 2x_5 &= -3 \\ x_2 - 2x_3 + x_4 - 4x_5 &= 1 \\ x_4 - 3x_5 &= 2 \end{aligned}$$

The last equation tells us that $x_4 = 2 + 3x_5$. Substituting this into the second equation yields $x_2 = -1 + 2x_3 + x_5$. Finally, substituting x_2 and x_4 into the first equation yields $x_1 = 1 + x_3 - x_5$. Hence, we see that x_1, x_2, and x_4 can be expressed in terms of x_3 and x_5, which are completely arbitrary. Any particular values of x_3 and x_5 will lead to a solution of the system, and any solution can be obtained by some special values of x_3 and x_5.

We are now ready to state the elimination method in general terms. Given any coefficient matrix, at least one of the elements of the first column is nonzero (otherwise x_1 would not have appeared in the equations). By an interchange of rows this nonzero element can be put in the first row, and it can be made 1 by multiplying the first row by its reciprocal. Next, by a series of multiplications of the first row by nonzero constants and additions to other rows the rest of the first column can be made zero. By row operations alone, we have eliminated x_1 from all but the first equation.

The next step is to look at the second column. If all of the elements below the first row are zero, we go on to the third column. If not, some element of the second column (below the first row) is nonzero. By an interchange we put this element in the second row and make it 1 by multiplying the second row by its reciprocal. Next we make all elements of the second column below the second row zero by a series of multiplications and additions. At this stage we have eliminated x_2 from all but possibly the first and second equations.

We then go on to the third column and proceed as above until either we run out of rows or find that all the rows below a certain point consist entirely of zeros. It is clear that any coefficient matrix can be reduced to the following form:

1 The element in the first row first column is 1.
2 The first nonzero element in each row is 1.
3 The first nonzero element of any row is to the right of the first nonzero element of the row above.
4 Any row consisting entirely of zeros is below all rows with nonzero elements.

We shall call a matrix in this form a *reduced matrix*. We can now state the elimination method in precise terms.

Elimination Method of Solving Linear Algebraic Equations

1 Write down the augmented matrix of the system.
2 Using only the three basic row operations on the augmented matrix, change the coefficient matrix to a reduced matrix.
3 The equations will have a solution if and only if for every row of the coefficient matrix, consisting entirely of zeros, the corresponding row of the augmented matrix of the reduced system consists entirely of zeros.
4 If the equations have solutions, solve the reduced system by starting

with the last nontrivial equation (not of the form $0 = 0$). Solve this equation for the variable with minimum subscript. Substitute this value into the next equation above and solve it for the variable with minimum subscript. Proceed in this way upward until x_1 is obtained from the first equation.

A proper understanding of the elimination method leads to several theorems about the solution of systems of linear algebraic equations, but first let us settle the question of the equivalence of the original and reduced systems of equations.

Theorem 2.3.1 A system of linear algebraic equations has a solution if and only if the corresponding reduced system has the same solution.

PROOF It is clear from the way we reduce the system that if a certain set of numbers $x_1, x_2, \ldots, x_n$ satisfies the original system, then they also satisfy the reduced system. Now turn the roles of original and reduced systems around. If we start with the reduced system, the original system can be obtained from it by some combination of the three basic operations. Now it is clear that any solution of the reduced system is also a solution of the original. This completes the proof.

A system of linear algebraic equations $AX = B$, where $B \neq 0$, is said to be *nonhomogeneous*. If $B = 0$, then the equations are said to be *homogeneous*. The homogeneous equations $AX = 0$ always have the *trivial solution* $X = 0$. If there is a solution $X \neq 0$ of the homogeneous equations, then it is called a *nontrivial solution*.

Theorem 2.3.2 A homogeneous system of m linear algebraic equations in n unknowns has nontrivial solutions if $n > m$.

PROOF When the coefficient matrix has been reduced, the augmented matrix of the reduced system has a last column consisting entirely of zeros. Therefore, the system has a solution, but there may be no nontrivial solutions. However, consider the elements a_{ii}, $i = 1, 2, \ldots, m$, of the reduced coefficient matrix. They are either 0 or 1. Suppose $a_{kk} = 0$ for some k and k is the smallest integer for which this occurs. Then the solution can be written in terms of x_k and possibly some other variables. But these variables are arbitrary, and so by picking $x_k \neq 0$, we have a nontrivial solution. If all the a_{ii}, $i = 1, 2, \ldots, m$, are 1, then the last row of the augmented matrix of the reduced system is

$$0, 0, \ldots, 0, 1, a_{m,m+1}, a_{m,m+2}, \ldots, a_{mn}, 0$$

and

$$x_m = -a_{m,m+1}x_{m+1} - a_{m,m+2}x_{m+2} - \cdots - a_{mn}x_n$$

while $x_{m+1}, x_{m+2}, \ldots, x_n$ are arbitrary. Picking $x_{m+1} \neq 0$ will give us a nontrivial solution.

Theorem 2.3.3 A homogeneous system of m linear algebraic equations in m unknowns has no nontrivial solutions if and only if the reduced coefficient matrix has no row consisting entirely of zeros.

PROOF Consider the elements a_{ii}, $i = 1, 2, \ldots, m$, of the reduced coefficient matrix. If $a_{ii} = 1$ for all i, then the augmented matrix of the reduced system looks like

$$\begin{pmatrix} 1 & a_{12} & \cdots & \cdots & \cdots & \cdots & 0 \\ 0 & 1 & a_{23} & \cdots & \cdots & \cdots & 0 \\ 0 & 0 & 1 & a_{34} & \cdots & \cdots & 0 \\ \cdots & \cdots & \cdots & \cdots & \cdots & \cdots & \cdots \\ 0 & 0 & 0 & \cdots & \cdots & 1 & 0 \end{pmatrix}$$

and the only solution is $x_1 = x_2 = x_3 = \cdots = x_m = 0$. Some of the a_{ii} are zero if and only if the last row of this matrix consists entirely of zeros. Thus the system has nontrivial solutions if and only if the last row of the reduced coefficient matrix consists entirely of zeros.

Theorem 2.3.4 A solution of a system of linear algebraic equations $AX = B$ is unique if and only if the homogeneous equations $AX = 0$ have no nontrivial solutions.

PROOF Suppose X and Y are both solutions. Then $AX = B$ and $AY = B$, and by subtraction $A(X - Y) = 0$. However, if the homogeneous equations have no nontrivial solutions, then $X - Y = 0$ and $X = Y$. This shows uniqueness. Conversely, suppose $Z \neq 0$ is a solution of the homogeneous equations, that is, $AZ = 0$, while X is a solution of $AX = B$. But then $X + Z$ is also a solution, since $A(X + Z) = AX + AZ = B + 0 = B$. This is a contradiction to uniqueness and completes the proof.

Theorem 2.3.5 A nonhomogeneous system of m linear algebraic equations in m unknowns has a unique solution if and only if the reduced coefficient matrix has no row consisting entirely of zeros.

PROOF The coefficient matrix of the reduced system has no row consisting entirely of zeros if and only if the corresponding augmented matrix looks like

$$\begin{pmatrix} 1 & a_{12} & \cdots & \cdots & \cdots & \cdots & b_1 \\ 0 & 1 & a_{23} & \cdots & \cdots & \cdots & b_2 \\ 0 & 0 & 1 & a_{34} & \cdots & \cdots & b_3 \\ \cdots & \cdots & \cdots & \cdots & \cdots & \cdots & \cdots \\ 0 & 0 & 0 & \cdots & \cdots & 1 & b_m \end{pmatrix}$$

Clearly the system has a solution, and it is unique. If the last row of the reduced coefficient matrix were entirely zeros, the system would have no solution unless $b_m = 0$. However, in this case the solution would not be unique since, by Theorem 2.3.3, the corresponding homogeneous system would have nontrivial solutions. This completes the proof.

As we have seen, when a system of equations has solutions, it may have many solutions. In fact, the general situation (when solutions are not unique) is that certain of the variables can be written in terms of the others, which are completely arbitrary. We may think of these arbitrary variables as parameters which can be varied to generate various solutions. We shall say that we have the *general solution* of a system if we have all the variables expressed in terms of certain parameters, such that every possible particular solution can be obtained by assigning appropriate values to these parameters. For example, in the second example of this section we were able to express the general solution as

$$\begin{pmatrix} x_1 \\ x_2 \\ x_3 \\ x_4 \\ x_5 \end{pmatrix} = \begin{pmatrix} 1 + a - b \\ -1 + 2a + b \\ a \\ 2 + 3b \\ b \end{pmatrix} = \begin{pmatrix} 1 \\ -1 \\ 0 \\ 2 \\ 0 \end{pmatrix} + a\begin{pmatrix} 1 \\ 2 \\ 1 \\ 0 \\ 0 \end{pmatrix} + b\begin{pmatrix} -1 \\ 1 \\ 0 \\ 3 \\ 1 \end{pmatrix}$$

where a and b are arbitrary parameters. The numbers $(1,-1,0,2,0)$ form a particular solution corresponding to the choice of values $a = b = 0$. Let us substitute the values $(1,2,1,0,0)$ into the left-hand sides of the equations. Then

$$\begin{aligned} 1 + 2(2) - 5(1) - 0 + 2(0) &= 0 \\ 2 - 2(1) + 0 - 4(0) &= 0 \\ 2(1) - 3(2) + 4(1) + 2(0) - 0 &= 0 \end{aligned}$$

This shows that the numbers $(1,2,1,0,0)$ form a solution of the corresponding homogeneous equations. The reader should also verify that the numbers

$(-1,1,0,3,1)$ satisfy the homogeneous equations. In fact, the part of the general solution of the nonhomogeneous equations

$$a\begin{pmatrix}1\\2\\1\\0\\0\end{pmatrix}+b\begin{pmatrix}-1\\1\\0\\3\\1\end{pmatrix}$$

is the general solution of the homogeneous equations. This is the situation, in general, as indicated by the next theorem. This theorem shows that finding the general solution of the homogeneous system goes a long way toward solving the nonhomogeneous system.

Theorem 2.3.6 The general solution of the nonhomogeneous system of equations, $AX = B$, can be obtained by adding the general solution of the homogeneous system $AX = 0$ to any particular solution of the nonhomogeneous system.

PROOF Suppose Z is a particular solution of the nonhomogeneous system; then $AZ = B$. Suppose X is *any* other particular solution. Then $AX = B$, and

$$A(X - Z) = AX - AZ = B - B = 0$$

Therefore, $Y = X - Z$ is a solution of the homogeneous equations and so can be obtained from the general solution of the homogeneous equations by the appropriate choice of certain parameters. Hence, $X = Z + Y$, and since X was *any* particular solution, we can obtain the general solution of the nonhomogeneous system by adding the general solution of the homogeneous system to a particular solution of the nonhomogeneous system.

EXERCISES 2.3

1 Which of the following matrices are in reduced form?

$$(a)\ \begin{pmatrix}1&0\\0&1\end{pmatrix}\qquad(b)\ \begin{pmatrix}0&1\\1&0\end{pmatrix}\qquad(c)\ \begin{pmatrix}1&2&3&4\\0&0&1&0\\0&0&0&0\end{pmatrix}$$

$$(d)\ \begin{pmatrix}1&0&1&0&1\\0&1&0&1&0\\0&0&1&0&1\\0&1&0&1&0\end{pmatrix}\qquad(e)\ \begin{pmatrix}0&1&2&3&4\\0&0&1&2&3\\0&0&0&0&1\\0&0&0&0&0\end{pmatrix}\qquad(f)\ \begin{pmatrix}1&2&3\\0&1&2\\0&0&0\\0&0&0\end{pmatrix}$$

2 Using the three basic row operations only, change the following matrices to reduced form:

$$(a)\begin{pmatrix} 1 & 2 & 3 & 4 & 5 \\ -2 & 0 & 1 & 2 & 3 \\ 0 & 1 & 5 & -2 & 4 \end{pmatrix}$$

$$(b)\begin{pmatrix} 1 & 2 & 3 \\ 3 & 0 & 4 \\ -1 & 1 & 5 \\ -2 & 0 & 7 \end{pmatrix} \qquad (c)\begin{pmatrix} 1 & 2 & 3 & 4 \\ 5 & 6 & 7 & 8 \\ 9 & 10 & 11 & 12 \\ 13 & 14 & 15 & 16 \end{pmatrix}$$

3 The following matrix is in reduced form:

$$\begin{pmatrix} 1 & 2 & 3 & 4 & 5 \\ 0 & 1 & -1 & 0 & 2 \\ 0 & 0 & 0 & 1 & 1 \\ 0 & 0 & 0 & 0 & 1 \end{pmatrix}$$

Show that by using the basic row operations only, the matrix can be changed to the form

$$\begin{pmatrix} 1 & 0 & a & 0 & 0 \\ 0 & 1 & b & 0 & 0 \\ 0 & 0 & 0 & 1 & 0 \\ 0 & 0 & 0 & 0 & 1 \end{pmatrix}$$

Find a and b.

4 (a) If B can be obtained from A by multiplying a row of A by $k \neq 0$, can A be obtained from B by a basic row operation?
(b) If B can be obtained from A by interchanging two rows, can A be obtained from B by a basic row operation?
(c) If B can be obtained from A by adding one row to another, can A be obtained from B by basic row operations?

5 Let $A \to B$ stand for the property "B can be obtained from A by basic row operations." Prove that this is an equivalence relation. In other words, prove that:
(a) $A \to A$.
(b) If $A \to B$, then $B \to A$.
(c) If $A \to B$ and $B \to C$, then $A \to C$.

6 Find all possible solutions of the following systems of equations:

$$(a)\quad \begin{aligned} x_1 - 2x_2 + 3x_3 - x_4 &= 0 \\ -x_1 + 2x_3 + x_4 &= 0 \\ 2x_1 + x_2 - 2x_4 &= 0 \end{aligned} \qquad (b)\quad \begin{aligned} x_1 + x_2 - x_3 &= 1 \\ 2x_1 + x_2 + 3x_3 &= 2 \\ x_2 - 5x_3 &= 1 \end{aligned}$$

$$(c)\quad \begin{aligned} x_1 + 2x_3 - x_4 &= 0 \\ 2x_1 + x_2 - x_3 &= 5 \\ -x_1 + 2x_2 + x_3 + 2x_4 &= 3 \\ 3x_2 - 2x_3 + 5x_4 &= 1 \end{aligned}$$

(*d*)
$$\begin{aligned} x_1 - 2x_2 + x_3 - x_4 + x_5 &= 1 \\ 2x_2 - x_3 + 3x_4 \qquad &= 2 \\ 3x_1 + x_2 + 2x_3 \qquad - 2x_5 &= -1 \\ 4x_1 + x_2 + 2x_3 + 2x_4 - x_5 &= 2 \end{aligned}$$

(*e*)
$$\begin{aligned} x_1 - 2x_2 + x_3 - x_4 + 2x_5 &= -7 \\ x_2 + x_3 + 2x_4 - x_5 &= 5 \\ x_1 - x_2 + 2x_3 + 2x_4 + 2x_5 &= -1 \end{aligned}$$

7 For parts (*d*) and (*e*) of Exercise 6 find the general solutions of the corresponding homogeneous equations, and then find the general solution of the nonhomogeneous equations by adding the general solution of the homogeneous system to a particular solution of the nonhomogeneous system.

8 Let

$$\begin{pmatrix} 1 & 2 & 3 & 4 & 5 \\ 0 & 1 & 2 & 3 & 4 \\ 0 & 0 & 0 & 1 & 2 \\ 0 & 0 & 0 & 0 & 0 \end{pmatrix}$$

be the reduced coefficient matrix of a system of homogeneous equations. Find the general solution of the system. How many arbitrary parameters are there in the solution?

9 Let $AX = 0$ stand for a homogeneous system of linear algebraic equations. Show that if X_1 and X_2 are solutions, then $aX_1 + bX_2$ is a solution for any scalars a and b.

10 Referring to Exercise 3, suppose a reduced system is

$$\begin{aligned} x_1 + 2x_2 + 3x_3 + 4x_4 + 5x_5 &= 1 \\ x_2 - x_3 \qquad + 2x_5 &= 0 \\ x_4 + x_5 &= 2 \\ x_5 &= -1 \end{aligned}$$

Solve the system by changing (by row operations only) the coefficient matrix to the form given in Exercise 3.

11 The following are reduced coefficient matrices of systems of linear algebraic equations. Which has a unique solution?

$$(a)\ \begin{pmatrix} 1 & 2 & 3 & 4 \\ 0 & 1 & 2 & 3 \\ 0 & 0 & 1 & 2 \\ 0 & 0 & 0 & 1 \end{pmatrix} \qquad (b)\ \begin{pmatrix} 1 & 2 & 3 & 4 \\ 0 & 1 & 2 & 3 \\ 0 & 0 & 0 & 1 \\ 0 & 0 & 0 & 0 \end{pmatrix}$$

2.4 DETERMINANTS

Consider a system of two equations in two unknowns:

$$\begin{aligned} a_{11}x_1 + a_{12}x_2 &= b_1 \\ a_{21}x_1 + a_{22}x_2 &= b_2 \end{aligned}$$

If $a_{11}a_{22} - a_{12}a_{21} \neq 0$, then we can solve them by elimination, as explained in Sec. 2.3, and we have

$$x_1 = \frac{b_1a_{22} - b_2a_{12}}{a_{11}a_{22} - a_{12}a_{21}}$$

$$x_2 = \frac{b_2a_{11} - b_1a_{21}}{a_{11}a_{22} - a_{12}a_{21}}$$

The quantity $a_{11}a_{22} - a_{12}a_{21}$, which is associated with the 2×2 coefficient matrix of the system, has a special significance in the solution. In fact, we shall define the *determinant* of the coefficient matrix as

$$\begin{vmatrix} a_{11} & a_{12} \\ a_{21} & a_{22} \end{vmatrix} = a_{11}a_{22} - a_{12}a_{21}$$

The numerators $b_1a_{22} - b_2a_{12}$ and $b_2a_{11} - b_1a_{21}$ are also determinants. We shall discuss this solution later as a special case of Cramer's rule, but first we shall define determinants of square matrices in general.

A *permutation* of the integers $1, 2, \ldots, n$ is an arrangement of the n integers. For example, there are six different permutations of the integers 1, 2, 3; that is, 1, 2, 3; 2, 3, 1; 3, 1, 2; 1, 3, 2; 2, 1, 3; and 3, 2, 1. If we interchange any pair of integers in a given permutation, we shall have changed the permutation. We call this process *inversion*. Given any permutation, by a finite number of inversions, we can put the n integers in *normal order* $1, 2, 3, \ldots, n$. We classify permutations as *even* or *odd* according to whether it takes respectively an even or odd number of inversions to put them in normal order. From this definition, the permutations 1, 2, 3; 2, 3, 1; and 3, 1, 2 are even while 1, 3, 2; 2, 1, 3; and 3, 2, 1 are odd. It can be shown that evenness or oddness of a permutation is independent of the particular set of inversions used to put the permutation in normal order.

Definition 2.4.1 The determinant of a 1×1 matrix with element a_{11} is a_{11}. To compute the determinant of an $n \times n$ matrix, form all possible products of elements one from each row and no two of which come from the same column. To each such product affix a plus or minus sign according to whether the column subscripts form respectively an even or odd permutation of $1, 2, 3, \ldots, n$ when the factors are arranged so that the row subscripts are in normal order. Finally, form the sum of the terms so determined.

According to this definition, for a 2×2 matrix, the various products which can be formed are $a_{11}a_{22}$ and $a_{12}a_{21}$. With the row subscripts in normal order in each case, the permutations of the column subscripts are 1, 2 and 2, 1. Therefore the first product is introduced with a plus sign and the second with a minus sign. Hence,

$$\begin{vmatrix} a_{11} & a_{12} \\ a_{21} & a_{22} \end{vmatrix} = a_{11}a_{22} - a_{12}a_{21}$$

For a 3×3 matrix the possible products are $a_{11}a_{22}a_{33}$, $a_{12}a_{23}a_{31}$, $a_{13}a_{21}a_{32}$, $a_{11}a_{23}a_{32}$, $a_{12}a_{21}a_{33}$, $a_{13}a_{22}a_{31}$, with the row subscripts in normal order. The permutations of the column subscripts are respectively even, even, even, odd, odd, odd. Therefore, the determinant of the 3×3 matrix is

$$\begin{vmatrix} a_{11} & a_{12} & a_{13} \\ a_{21} & a_{22} & a_{23} \\ a_{31} & a_{32} & a_{33} \end{vmatrix} = \begin{aligned} & a_{11}a_{22}a_{33} + a_{12}a_{23}a_{31} + a_{13}a_{21}a_{32} \\ & - a_{11}a_{23}a_{32} - a_{12}a_{21}a_{33} - a_{13}a_{22}a_{31} \end{aligned}$$

Of course, the expansion of the determinant of a 4×4 matrix will have 24 terms in it, while for a 5×5, we shall have 120 terms. However, as we shall see, one seldom uses Definition 2.4.1 to compute determinants of large order matrices. In fact, one of the main purposes of this section is to find alternative ways of computing determinants. We shall first need, for the purpose of proving theorems, a more compact way of writing our basic definition. We define a function which takes on the values 0, 1, -1 as follows, depending on the n integers $i, j, k, \ldots, p$, which can each take on the values $1, 2, \ldots, n$:

$$e_{ijk\cdots p} = \begin{cases} 0 & \text{if any pair of subscripts are equal} \\ 1 & \text{if the } n \text{ integers } i, j, k, \ldots, p \text{ form an even} \\ & \text{permutation of } 1, 2, 3, \ldots, n \\ -1 & \text{if the } n \text{ integers } i, j, k, \ldots, p \text{ form an odd} \\ & \text{permutation of } 1, 2, 3, \ldots, n \end{cases}$$

We can now write, in general,

$$\begin{vmatrix} a_{11} & a_{12} & \cdots & a_{1n} \\ a_{21} & a_{22} & \cdots & a_{2n} \\ \cdots & \cdots & \cdots & \cdots \\ a_{n1} & a_{n2} & \cdots & a_{nn} \end{vmatrix} = \sum_{i,j,\ldots,p}^{n} e_{ijk\cdots p} a_{1i}a_{2j}a_{3k}\cdots a_{np}$$

where the sum is to be taken over all possible values of $i, j, k, \ldots, p$.

We are now ready to prove several theorems about determinants. In some cases, the proofs are self-evident from the definition. In other cases, where the proofs are more complicated, we shall do the 3×3 case, which is sufficiently complex to show what is involved and yet keep the details down to a level where the reader can see the basic idea in the general case.

Theorem 2.4.1 If every element in a given row of a square matrix is zero, its determinant is zero.

PROOF Since every term in the expansion contains a factor from the given row of zeros, every term in the expansion is zero.

Theorem 2.4.2 If every element in a given row of a square matrix is multiplied by the same number α, the determinant is multiplied by α.

PROOF Since every term in the expansion contains a factor from the given row, if the row is multiplied by α, every term in the expansion contains α, which can be factored out as a common multiplier.

Theorem 2.4.3 If two rows of a square matrix are interchanged, the sign of the determinant is changed.

PROOF Consider the 3 × 3 case. Let

$$A = \begin{pmatrix} a_{11} & a_{12} & a_{13} \\ a_{21} & a_{22} & a_{23} \\ a_{31} & a_{32} & a_{33} \end{pmatrix} \qquad B = \begin{pmatrix} a_{31} & a_{32} & a_{33} \\ a_{21} & a_{22} & a_{23} \\ a_{11} & a_{12} & a_{13} \end{pmatrix}$$

The expansion† of $|B|$ is

$$\begin{aligned} |B| &= \sum_{i,j,k}^{3} e_{ijk} a_{3i} a_{2j} a_{1k} \\ &= \sum_{i,j,k}^{3} e_{ijk} a_{1k} a_{2j} a_{3i} \\ &= \sum_{i,j,k}^{3} e_{kji} a_{1i} a_{2j} a_{3k} \end{aligned}$$

The last step is possible because in summing over all possible values of i, j, and k the expression is not changed if we formally change the names of the summation indices; for example,

$$\sum_{i=1}^{m} \alpha_i = \sum_{j=1}^{m} \alpha_j = \sum_{k=1}^{m} \alpha_k$$

Now compare e_{ijk} with e_{kji} for the same values of i, j, and k. The values of both are zero if any pair of indices are the same. If not both zero, the values are of opposite sign. This is because if i, j, k is an even permutation, then k, j, i is an odd permutation or vice versa, since the one permutation

† We use the symbol $|B|$ to denote the determinant of B. Note that this is not absolute value, even when B is a 1 × 1 matrix.

is one inversion away from the other. This shows that

$$|B| = \sum_{i,j,k}^{3} e_{kji}a_{1i}a_{2j}a_{3k} = -\sum_{i,j,k}^{3} e_{ijk}a_{1i}a_{2j}a_{3k} = -|A|$$

The same general principle is involved in the interchange of any pair of rows of a matrix.

Theorem 2.4.4 If two rows of a square matrix are proportional, then the determinant of the matrix is zero.

PROOF Proportional in this case means that we can obtain one row from the other by multiplying the whole row by a constant. From Theorem 2.4.2, the value of the determinant is a constant times the value of the determinant of a matrix in which two rows are equal. If we interchange two equal rows, we do not change the value of the determinant. However, Theorem 2.4.3 tells us that the sign of the determinant is changed and the only real or complex number which remains unchanged when its sign is changed is zero.

Theorem 2.4.5 If each element of a given row of a square matrix can be written as the sum of two terms, then its determinant can be written as the sum of two determinants of matrices each of which contains one of the terms of the corresponding row but is otherwise like the original matrix.

PROOF We again do the 3 × 3 case. Let

$$A = \begin{pmatrix} a_{11} & a_{12} & a_{13} \\ a_{21} & a_{22} & a_{23} \\ a_{31} & a_{32} & a_{33} \end{pmatrix} \qquad B = \begin{pmatrix} b_{11} & b_{12} & b_{13} \\ a_{21} & a_{22} & a_{23} \\ a_{31} & a_{32} & a_{33} \end{pmatrix}$$

$$C = \begin{pmatrix} a_{11} + b_{11} & a_{12} + b_{12} & a_{13} + b_{13} \\ a_{21} & a_{22} & a_{23} \\ a_{31} & a_{32} & a_{33} \end{pmatrix}$$

We shall show that $|C| = |A| + |B|$. Now

$$\begin{aligned} |C| &= \sum_{i,j,k}^{3} e_{ijk}(a_{1i} + b_{1i})a_{2j}a_{3k} \\ &= \sum_{i,j,k}^{3} (e_{ijk}a_{1i}a_{2j}a_{3k} + e_{ijk}b_{1i}a_{2j}a_{3k}) \\ &= |A| + |B| \end{aligned}$$

The same general principles are involved with other rows.

Theorem 2.4.6 If to each element of a given row of a square matrix is added α times the corresponding element of another row, the value of its determinant is unchanged.

PROOF From Theorem 2.4.5 we see that after the addition has been performed, the determinant of the new matrix can be written as the sum of two determinants, one of which is the determinant of the original matrix and the other of a matrix with two rows proportional. However, by Theorem 2.4.4 the second of these is zero.

EXAMPLE 2.4.1 Evaluate the determinant

$$\begin{vmatrix} 2 & 3 & 4 & 5 \\ 3 & 1 & 3 & 1 \\ 2 & 1 & 4 & 3 \\ 1 & 2 & 3 & 4 \end{vmatrix}$$

To avoid writing out the 24-term expansion, we use the above theorems to change the matrix without changing the value of the determinant. It is generally to our advantage to introduce zeros. Therefore, we add (-2) times the fourth row to the first, we add (-3) times the fourth row to the second, and we add (-2) times the fourth row to the third. Hence,

$$\begin{vmatrix} 2 & 3 & 4 & 5 \\ 3 & 1 & 3 & 1 \\ 2 & 1 & 4 & 3 \\ 1 & 2 & 3 & 4 \end{vmatrix} = \begin{vmatrix} 0 & -1 & -2 & -3 \\ 0 & -5 & -6 & -11 \\ 0 & -3 & -2 & -5 \\ 1 & 2 & 3 & 4 \end{vmatrix} = (-1)\begin{vmatrix} 0 & 1 & 2 & 3 \\ 0 & 5 & 6 & 11 \\ 0 & 3 & 2 & 5 \\ 1 & 2 & 3 & 4 \end{vmatrix}$$

Next we add (-5) times the first row to the second and (-3) times the first row to the third. We have

$$\begin{vmatrix} 2 & 3 & 4 & 5 \\ 3 & 1 & 3 & 1 \\ 2 & 1 & 4 & 3 \\ 1 & 2 & 3 & 4 \end{vmatrix} = (-1)\begin{vmatrix} 0 & 1 & 2 & 3 \\ 0 & 0 & -4 & -4 \\ 0 & 0 & -4 & -4 \\ 1 & 2 & 3 & 4 \end{vmatrix} = 0$$

We have immediately concluded that the determinant is zero because we have obtained two rows proportional.

The next theorem shows that the value of the determinant of a square matrix is unchanged if the rows and columns are interchanged. This makes it possible to restate Theorems 2.4.1 to 2.4.6 with the word "row" replaced by the word "column" everywhere and means that we can use column properties as well as row properties to evaluate determinants.

Theorem 2.4.7 The value of the determinant of a square matrix is unchanged if the rows and columns of the matrix are interchanged.

PROOF We do the 3 × 3 case. Let

$$A = \begin{pmatrix} a_{11} & a_{12} & a_{13} \\ a_{21} & a_{22} & a_{23} \\ a_{31} & a_{32} & a_{33} \end{pmatrix} \qquad B = \begin{pmatrix} a_{11} & a_{21} & a_{31} \\ a_{12} & a_{22} & a_{32} \\ a_{13} & a_{23} & a_{33} \end{pmatrix}$$

Then

$$|B| = \sum_{i,j,k}^{3} e_{ijk} a_{i1} a_{j2} a_{k3}$$

For the terms in this expansion which are not zero, i, j, and k are different. Therefore, 1, 2, and 3 are all represented as row subscripts in these terms. By rearranging the order of the factors, the row subscripts can be put in normal order, that is,

$$a_{i1} a_{j2} a_{k3} = a_{1l} a_{2m} a_{3n}$$

where l, m, n is the permutation of 1, 2, 3 induced by the inversions needed to put i, j, k into normal order. Clearly i, j, k and l, m, n are even or odd together as permutations. There is a term $a_{1l}a_{2m}a_{3n}$ in the expansion of the determinant of A for each and every term $a_{i1}a_{j2}a_{k3}$ in the expansion of the determinant of B, and vice versa. These terms appear with the same sign. Therefore,

$$|B| = \sum_{i,j,k}^{3} e_{ijk} a_{i1} a_{j2} a_{k3} = \sum_{l,m,n}^{3} e_{lmn} a_{1l} a_{2m} a_{3n} = |A|$$

This completes the proof.

Consider the expansion of the determinant of a 3 × 3 matrix

$$\begin{vmatrix} a_{11} & a_{12} & a_{13} \\ a_{21} & a_{22} & a_{23} \\ a_{31} & a_{32} & a_{33} \end{vmatrix} = \sum_{i,j,k}^{3} e_{ijk} a_{1i} a_{2j} a_{3k}$$

$$= \sum_{i=1}^{3} a_{1i} \sum_{j,k}^{3} e_{ijk} a_{2j} a_{3k}$$

$$= \sum_{i=1}^{3} a_{1i} c_{1i}$$

where $c_{1i} = \sum_{j,k}^{3} e_{ijk} a_{2j} a_{3k}$. The quantity c_{1i} is called the *cofactor* of a_{1i}. Let us look at the cofactors c_{1i}:

$$c_{11} = \sum_{j,k}^{3} e_{1jk}a_{2j}a_{3k} = a_{22}a_{33} - a_{23}a_{32} = \begin{vmatrix} a_{22} & a_{23} \\ a_{32} & a_{33} \end{vmatrix}$$

$$c_{12} = \sum_{j,k}^{3} e_{2jk}a_{2j}a_{3k} = a_{23}a_{31} - a_{21}a_{33} = -\begin{vmatrix} a_{21} & a_{23} \\ a_{31} & a_{33} \end{vmatrix}$$

$$c_{13} = \sum_{j,k}^{3} e_{3jk}a_{2j}a_{3k} = a_{21}a_{32} - a_{22}a_{31} = \begin{vmatrix} a_{21} & a_{22} \\ a_{31} & a_{32} \end{vmatrix}$$

Therefore, we can expand the determinant as follows

$$\begin{vmatrix} a_{11} & a_{12} & a_{13} \\ a_{21} & a_{22} & a_{23} \\ a_{31} & a_{32} & a_{33} \end{vmatrix} = a_{11}\begin{vmatrix} a_{22} & a_{23} \\ a_{32} & a_{33} \end{vmatrix} - a_{12}\begin{vmatrix} a_{21} & a_{23} \\ a_{31} & a_{33} \end{vmatrix} + a_{13}\begin{vmatrix} a_{21} & a_{22} \\ a_{31} & a_{32} \end{vmatrix}$$

Notice that the 2×2 determinants appearing in this expansion are obtained by striking out the first row of the original and respectively the first, second, and third columns. In the general case, we define the cofactors of a_{1i} as $(-1)^{1+i}$ times the $(n - 1) \times (n - 1)$ determinant formed by striking out the first row and ith column of the original. It is easy to verify that

$$\sum_{i,j,k,\ldots,p}^{n} e_{ij\cdots p}a_{1i}a_{2j}\cdots a_{np} = \sum_{i=1}^{n} a_{1i} \sum_{j,k,\ldots,p}^{n} e_{ijk\cdots p}a_{2j}a_{3k}\cdots a_{np}$$
$$= \sum_{i=1}^{n} a_{1i}c_{1i}$$

where c_{1i} are the cofactors of a_{1i}. We call this the *cofactor expansion by the first row*.

Suppose we wish to expand a determinant using a row other than the first, say the jth row ($j \neq 1$). Then, using Theorem 2.4.3, we can interchange the first and jth row and then use the above expansion. We must remember that the interchange of two rows has changed the sign of the determinant. With this method the elements of the original first row will not appear in the first row of the determinants in the cofactors. However, by interchanging the elements from the original first row [now appearing in the $(j - 1)$st row of the cofactor determinants] with the $j - 2$ rows above it, we have the rows in the natural order except for the missing jth row. Keeping track of all the sign changes, we find we have the factor

$$(-1)^{1+i+1+j-2} = (-1)^{j+i}$$

in front of the $(n - 1) \times (n - 1)$ determinants appearing in the expansion. We summarize this in the following theorem. The expansion obtained is called the *cofactor expansion by rows*.

Theorem 2.4.8 If we define the cofactor c_{ij} of the (i,j)th element a_{ij} of a square $n \times n$ matrix A as $(-1)^{i+j}$ times the $(n-1) \times (n-1)$ determinant formed by striking out the ith row and jth column, then

$$|A| = \sum_{j=1}^{n} a_{ij}c_{ij}$$

The next theorem follows immediately from Theorem 2.4.7. The resulting expansion is called the cofactor expansion by columns.

Theorem 2.4.9 If c_{ij} is the cofactor of the (i,j)th element a_{ij} of a square $n \times n$ matrix A, then

$$|A| = \sum_{i=1}^{n} a_{ij}c_{ij}$$

EXAMPLE 2.4.2 Evaluate the determinant

$$|A| = \begin{vmatrix} 1 & 2 & 3 & 4 & 5 \\ 2 & -1 & 3 & -2 & 0 \\ 0 & 3 & -1 & 4 & 2 \\ 3 & 1 & 2 & 5 & -2 \\ 1 & 0 & 5 & 12 & 3 \end{vmatrix}$$

We try to put zeros into the determinant in a systematic manner, beginning with the first column, by adding multiples of the first row to the other rows. We obtain

$$|A| = \begin{vmatrix} 1 & 2 & 3 & 4 & 5 \\ 0 & -5 & -3 & -10 & -10 \\ 0 & 3 & -1 & 4 & 2 \\ 0 & -5 & -7 & -7 & -17 \\ 0 & -2 & 2 & 8 & -2 \end{vmatrix} = 2\begin{vmatrix} 1 & 2 & 3 & 4 & 5 \\ 0 & 1 & -1 & -4 & 1 \\ 0 & 3 & -1 & 4 & 2 \\ 0 & 5 & 7 & 7 & 17 \\ 0 & 5 & 3 & 10 & 10 \end{vmatrix}$$

In the last step we have interchanged the second and fifth rows and taken out common factors from the second, fourth, and fifth rows. Next we add (-1) times the fourth row to the fifth and multiples of the second row to the third and fourth rows. We obtain

$$|A| = 2\begin{vmatrix} 1 & 2 & 3 & 4 & 5 \\ 0 & 1 & -1 & -4 & 1 \\ 0 & 0 & 2 & 16 & -1 \\ 0 & 0 & 12 & 27 & 12 \\ 0 & 0 & -4 & 3 & -7 \end{vmatrix} = 6\begin{vmatrix} 1 & 2 & 3 & 4 & 5 \\ 0 & 1 & -1 & -4 & 1 \\ 0 & 0 & 2 & 16 & -1 \\ 0 & 0 & 0 & -23 & 6 \\ 0 & 0 & 0 & 12 & -3 \end{vmatrix}$$

where we have taken a factor of 3 out of the fourth row, added the fourth to the fifth, and added (-2) times the third row to the fourth. Finally, we take a factor of 3 out of the fifth row and add 4 times the fifth column to the fourth column. We obtain

$$|A| = 18 \begin{vmatrix} 1 & 2 & 3 & 24 & 5 \\ 0 & 1 & -1 & 0 & 1 \\ 0 & 0 & 2 & 12 & -1 \\ 0 & 0 & 0 & 1 & 6 \\ 0 & 0 & 0 & 0 & -1 \end{vmatrix}$$

The last determinant is easy to evaluate. In fact, if we expand by the first column, expand the first cofactor by the first column, etc., we soon see that the value of the determinant is just the product of the elements along the principal diagonal. Therefore, $|A| = -36$.

Let A be a 3×3 matrix with elements a_{ij}, $i, j = 1, 2, 3$. Consider the quantities

$$q_{lmn} = \sum_{i,j,k}^{3} e_{ijk} a_{li} a_{mj} a_{nk}$$

for $l, m, n = 1, 2, 3$. If $l = 1$, $m = 2$, $n = 3$, then

$$q_{123} = \sum_{i,j,k}^{3} e_{ijk} a_{1i} a_{2j} a_{3k} = |A|$$

If any pair of l, m, and n are equal, then q_{lmn} represents the expansion of a determinant in which two rows are equal. Hence, the value is zero. If l, m, n is an even permutation of 1, 2, 3, then we have the expansion of $|A|$ but with the rows not necessarily in normal order. However, the rows can be put in normal order by an even number of interchanges, and by Theorem 2.4.3 the value is $|A|$. If l, m, n is an odd permutation, then it takes an odd number of interchanges of rows to put them in normal order. Therefore, the value is $-|A|$. In summary,

$$q_{lmn} = \sum_{i,j,k}^{3} e_{ijk} a_{li} a_{mj} a_{nk} = |A| e_{lmn}$$

Similarly, using Theorem 2.4.7,

$$\sum_{i,j,k}^{3} e_{ijk} a_{il} a_{jm} a_{kn} = |A| e_{lmn}$$

Now consider two 3×3 matrices A, with elements a_{ij}, and B, with elements b_{ij}. Then if $D = AB$, the elements of D are $d_{ij} = \sum_{k=1}^{3} a_{ik} b_{kj}$. Now we consider the product of the determinants of A and B.

$$\begin{aligned}
|A|\,|B| &= |A| \sum_{l,m,n}^{3} e_{lmn} b_{l1} b_{m2} b_{n3} \\
&= \sum_{l,m,n}^{3} |A| e_{lmn} b_{l1} b_{m2} b_{n3} \\
&= \sum_{l,m,n}^{3} \sum_{i,j,k}^{3} e_{ijk} a_{il} a_{jm} a_{kn} b_{l1} b_{m2} b_{n3} \\
&= \sum_{i,j,k}^{3} e_{ijk} \sum_{l=1}^{3} a_{il} b_{l1} \sum_{m=1}^{3} a_{jm} b_{m2} \sum_{n=1}^{3} a_{kn} b_{n3} \\
&= \sum_{i,j,k}^{3} e_{ijk} d_{i1} d_{j2} d_{k3} \\
&= |AB|
\end{aligned}$$

It is easy to see that the same type of proof will go through in the general case. Therefore, we have the following theorem.

Theorem 2.4.10 If A and B are $n \times n$ matrices, then

$$|AB| = |A|\,|B| = |BA|$$

We conclude this section with one more definition which we shall need in the next section.

Definition 2.4.2 The *transpose* of an $m \times n$ matrix A is the $n \times m$ matrix obtained from A by interchanging the rows and columns. We denote transpose of A by $\tilde{A}$.

If A is square, then $\tilde{A}$ is also square and Theorem 2.4.7 gives us the following theorem immediately.

Theorem 2.4.11 If A is a square matrix, then $|\tilde{A}| = |A|$.

Another important theorem has to do with multiplication of the transposes of two matrices.

Theorem 2.4.12 If A is an $m \times n$ matrix and B is an $n \times p$ matrix, then $\widetilde{AB} = \tilde{B}\tilde{A}$.

PROOF Since A is $m \times n$, $\tilde{A}$ is $n \times m$. Also $\tilde{B}$ is $p \times n$. Hence, $\tilde{B}\tilde{A}$ can be computed. If a_{ij} is the (i,j)th element of A, then $\tilde{a}_{ji} = a_{ij}$ is the (j,i)th element of $\tilde{A}$. If b_{ij} is the (i,j)th element of B, then $\tilde{b}_{ji}$ is the

(j,i)th element of $\tilde{B}$. Also $\sum_{k=1}^{n} a_{ik}b_{kj}$ is the (i,j)th element of AB. The (i,j)th element of $\widetilde{AB}$ is

$$\sum_{k=1}^{n} a_{jk}b_{ki} = \sum_{k=1}^{n} \tilde{a}_{kj}\tilde{b}_{ik} = \sum_{k=1}^{n} \tilde{b}_{ik}\tilde{a}_{kj}$$

which is the (i,j)th element of $\tilde{B}\tilde{A}$. This completes the proof.

EXERCISES 2.4

1 Show that there are $n! = n(n-1)(n-2)\cdots 2\cdot 1$ different permutations of the integers $1, 2, 3, \ldots, n$.

2 Given a permutation P_1 of n integers. Obtain P_2 from P_1 by one inversion. Obtain P_3 from P_1 by one inversion. Show that P_3 can be obtained from P_2 by an even number of inversions. Use this to show that evenness or oddness of a permutation is independent of the particular set of inversions used to put it in normal order.

3 Show that a permutation is even or odd according to whether it takes respectively an even or odd number of inversions to obtain it from the normal order.

4 Write out all permutations of 1, 2, 3, 4, and classify them according to whether they are even or odd.

5 Write out the complete expansion of the determinant of a general 4×4 matrix. There should be 24 terms.

6 Evaluate the following determinants

$$(a)\ \begin{vmatrix} 1 & 2 \\ 3 & 4 \end{vmatrix} \qquad (b)\ \begin{vmatrix} 1 & 2 & 3 \\ 4 & 5 & 6 \\ 7 & 8 & 9 \end{vmatrix} \qquad (c)\ \begin{vmatrix} 1 & 3 & -1 & 2 \\ 2 & 1 & 3 & 1 \\ -1 & 2 & -1 & 3 \\ -2 & 1 & 2 & -3 \end{vmatrix}$$

7 Evaluate the following determinant by showing that it is equal to an upper-triangular determinant (one in which all elements below the principal diagonal are zero).

$$\begin{vmatrix} 1 & 2 & -1 & 3 & -2 \\ 2 & 0 & 4 & -5 & 1 \\ -3 & 1 & 6 & 0 & -7 \\ 0 & 3 & 1 & -5 & 2 \\ -2 & 6 & 3 & -1 & 2 \end{vmatrix}$$

8 Consider the three basic row operations of Sec. 2.3. Show that if a square matrix has a zero determinant, after any number of basic row operations the resulting matrix will have a zero determinant. Also show that if a square matrix has a nonzero determinant, after any number of basic row operations the resulting matrix will have a nonzero determinant.

9 Prove that a square matrix has a zero determinant if and only if it can be reduced to upper-triangular form with at least one zero element on the principal diagonal.

10 Restate Theorems 2.3.3 and 2.3.5 in terms of the vanishing or nonvanishing of the determinant of the coefficient matrix.

11 Determine whether or not the following system of equations has a unique solution by evaluating the determinant of the coefficient matrix:

$$\begin{aligned} 2x_1 + x_2 - x_3 + x_4 &= -2 \\ x_1 - x_2 - x_3 + x_4 &= 1 \\ x_1 - 4x_2 - 2x_3 + 2x_4 &= 6 \\ 4x_1 + x_2 - 3x_3 + 3x_4 &= -1 \end{aligned}$$

12 Determine the values of λ for which the following system of equations has a nontrivial solution:

$$\begin{aligned} 9x_1 - 3x_2 \qquad &= \lambda x_1 \\ -3x_1 + 12x_2 - 3x_3 &= \lambda x_2 \\ - 3x_2 + 9x_3 &= \lambda x_3 \end{aligned}$$

13 Determine the values of λ for which the following system of equations has a nontrivial solution:

$$\begin{aligned} x_1 + x_2 &= \lambda x_1 \\ -x_1 + x_2 &= \lambda x_2 \end{aligned}$$

For what values of λ is there a *real* nontrivial solution?

14 Let A and B be $m \times n$ matrices. Show that

(*a*) $\tilde{\tilde{A}} = A$ (*b*) $\widetilde{A + B} = \tilde{A} + \tilde{B}$

(c) $\widetilde{A - B} = \tilde{A} - \tilde{B}$ (*d*) $\widetilde{aA} = a\tilde{A}$

2.5 INVERSE OF A MATRIX

One of the most important concepts in the matrix theory is the notion of inverse of a square matrix.

Definition 2.5.1 If A is an $n \times n$ matrix, then A has an inverse if there exists an $n \times n$ matrix A^{-1} such that $A^{-1}A = AA^{-1} = I$, where I is the nth order identity.

Theorem 2.5.1 If a square matrix has an inverse, it is unique.

PROOF Suppose $B \neq A^{-1}$ was also an inverse for A. Then $BA = AB = I$. But then

$$A^{-1} = A^{-1}I = A^{-1}(AB) = (A^{-1}A)B = IB = B$$

which is a contradiction, proving that there can be only one inverse for a given matrix.

One way to construct the inverse of a square matrix A (if it has one) is to work with the cofactor expansion of the determinant of A. Let the elements of A be a_{ij} and the cofactors of a_{ij} be c_{ij}. Consider the quantities

$$q_{ij} = \sum_{k=1}^{n} a_{ik}c_{jk}$$

If $i = j$, then q_{ii} is the cofactor expansion of $|A|$ by the ith row and so $q_{ii} = |A|$. If $i \neq j$, then $q_{ij} = 0$. This is because the expansion then involves the elements of the ith row and the cofactors of a different row, and what we have, in effect, is the expansion of a determinant in which two rows are the same. Therefore,

$$\sum_{k=1}^{n} a_{ik}c_{jk} = |A|\delta_{ij}$$

where $\delta_{ij} = 1$ if $i = j$ and $\delta_{ij} = 0$ if $i \neq j$. Let C be the matrix of cofactors with elements c_{ij}. Then $\tilde{C}$, the transpose of C, has elements $\tilde{c}_{ij} = c_{ji}$. Therefore,

$$\sum_{k=1}^{n} a_{ik}c_{jk} = \sum_{k=1}^{n} a_{ik}\tilde{c}_{kj} = |A|\delta_{ij}$$

or, in other words, $A\tilde{C} = |A|I$. Similarly, using the cofactor expansion of $|A|$ by columns, we have

$$\sum_{k=1}^{n} a_{ki}c_{kj} = \sum_{k=1}^{n} \tilde{c}_{jk}a_{ki} = |A|\delta_{ji}$$

or $\tilde{C}A = |A|I$. Therefore, if $|A| \neq 0$, we have

$$\frac{1}{|A|}\tilde{C}A = A\frac{1}{|A|}\tilde{C} = I$$

This proves the following theorem.

Theorem 2.5.2 Let A be a square matrix with nonzero determinant. Let C be the matrix of cofactors; that is, c_{ij} is the cofactor of a_{ij}. Then

$$A^{-1} = \frac{1}{|A|}\tilde{C}$$

Theorem 2.5.3 A square matrix has an inverse if and only if it has a nonzero determinant.

PROOF The first part of the theorem is covered by Theorem 2.5.2. On the other hand, suppose A has an inverse A^{-1}. Then $AA^{-1} = I$, and

$$|AA^{-1}| = |A|\,|A^{-1}| = |I| = 1$$

Therefore, $|A| \neq 0$. Incidentally, we have also shown that

$$|A^{-1}| = \frac{1}{|A|}$$

Definition 2.5.2 A square matrix is said to be *nonsingular* if it has an inverse (nonzero determinant). If it does not have an inverse (has a zero determinant), then it is said to be *singular*.

EXAMPLE 2.5.1 Find the inverse of the nonsingular matrix

$$A = \begin{pmatrix} 1 & 2 & 1 \\ -1 & 0 & 1 \\ 1 & 2 & 3 \end{pmatrix}$$

First,

$$|A| = \begin{vmatrix} 1 & 2 & 1 \\ -1 & 0 & 1 \\ 1 & 2 & 3 \end{vmatrix} = \begin{vmatrix} 1 & 2 & 1 \\ 0 & 2 & 2 \\ 0 & 0 & 2 \end{vmatrix} = 4$$

The matrix of cofactors is

$$C = \begin{pmatrix} -2 & 4 & -2 \\ -4 & 2 & 0 \\ 2 & -2 & 2 \end{pmatrix}$$

Then

$$\tilde{C} = \begin{pmatrix} -2 & -4 & 2 \\ 4 & 2 & -2 \\ -2 & 0 & 2 \end{pmatrix}$$

and

$$A^{-1} = \begin{pmatrix} -\frac{1}{2} & -1 & \frac{1}{2} \\ 1 & \frac{1}{2} & -\frac{1}{2} \\ -\frac{1}{2} & 0 & \frac{1}{2} \end{pmatrix}$$

There are, of course, other methods of computing inverses. We shall now take up one which is very closely connected with the methods of Sec. 2.3. We have already seen (see Exercise 2.2.9) that we can perform any one of the three basic row operations on a matrix by multiplying it on the left by a matrix obtained from the identity by the same row operation. Suppose E is obtained from the identity by (1) multiplication of a row by a nonzero constant, or (2) interchange of a pair of rows, or (3) addition of one row to another. Clearly since $|I| = 1$, then $|E| \neq 0$, so E is nonsingular. Now EA is obtained from A by one of the basic row operations, and since $|EA| = |E|\,|A|$, we have that EA is nonsingular if and only if A is nonsingular. The same goes for any number of basic row operations, which shows that if A is reduced to B by basic row operations, then A is nonsingular if and only if B is nonsingular.

Let us reconsider Example 2.5.1 in the light of these comments. Let

$$E_1 = \begin{pmatrix} 1 & 0 & 0 \\ 1 & 1 & 0 \\ 0 & 0 & 1 \end{pmatrix} \qquad E_2 = \begin{pmatrix} 1 & 0 & 0 \\ 0 & 1 & 0 \\ -1 & 0 & 1 \end{pmatrix}$$

Then

$$E_2E_1A = \begin{pmatrix} 1 & 2 & 1 \\ 0 & 2 & 2 \\ 0 & 0 & 2 \end{pmatrix}$$

Let

$$E_3 = \begin{pmatrix} 1 & 0 & 0 \\ 0 & \frac{1}{2} & 0 \\ 0 & 0 & 1 \end{pmatrix} \qquad E_4 = \begin{pmatrix} 1 & 0 & 0 \\ 0 & 1 & 0 \\ 0 & 0 & \frac{1}{2} \end{pmatrix}$$

Then

$$E_4E_3E_2E_1A = \begin{pmatrix} 1 & 2 & 1 \\ 0 & 1 & 1 \\ 0 & 0 & 1 \end{pmatrix}$$

This is in reduced form (as defined in Sec. 2.3), but for the present purposes we go further. Let

$$E_5 = \begin{pmatrix} 1 & -2 & 0 \\ 0 & 1 & 0 \\ 0 & 0 & 1 \end{pmatrix} \qquad E_6 = \begin{pmatrix} 1 & 0 & 0 \\ 0 & 1 & -1 \\ 0 & 0 & 1 \end{pmatrix}$$

$$E_7 = \begin{pmatrix} 1 & 0 & 1 \\ 0 & 1 & 0 \\ 0 & 0 & 1 \end{pmatrix}$$

Then

$$E_7E_6E_5E_4E_3E_2E_1A = \begin{pmatrix} 1 & 0 & 0 \\ 0 & 1 & 0 \\ 0 & 0 & 1 \end{pmatrix} = I$$

or

$$E_7E_6E_5E_4E_3E_2E_1IA = A^{-1}A$$

and multiplying on the right by A^{-1}, we have

$$E_7E_6E_5E_4E_3E_2E_1I = A^{-1}$$

This tells us that if we perform the same basic row operations on the identity, we shall produce A^{-1}. It will not be necessary to write out the E's if we arrange the calculation as follows:

$$\left(\begin{array}{ccc|ccc} 1 & 2 & 1 & 1 & 0 & 0 \\ -1 & 0 & 1 & 0 & 1 & 0 \\ 1 & 2 & 3 & 0 & 0 & 1 \end{array}\right)$$

$$\rightarrow \left(\begin{array}{ccc|ccc} 1 & 2 & 1 & 1 & 0 & 0 \\ 0 & 2 & 2 & 1 & 1 & 0 \\ 0 & 0 & 2 & -1 & 0 & 1 \end{array}\right)$$

$$\rightarrow \left(\begin{array}{ccc|ccc} 1 & 2 & 1 & 1 & 0 & 0 \\ 0 & 1 & 1 & \frac{1}{2} & \frac{1}{2} & 0 \\ 0 & 0 & 1 & -\frac{1}{2} & 0 & \frac{1}{2} \end{array}\right)$$

$$\rightarrow \left(\begin{array}{ccc|ccc} 1 & 0 & -1 & 0 & -1 & 0 \\ 0 & 1 & 0 & 1 & \frac{1}{2} & -\frac{1}{2} \\ 0 & 0 & 1 & -\frac{1}{2} & 0 & \frac{1}{2} \end{array}\right)$$

$$\rightarrow \left(\begin{array}{ccc|ccc} 1 & 0 & 0 & -\frac{1}{2} & -1 & \frac{1}{2} \\ 0 & 1 & 0 & 1 & \frac{1}{2} & -\frac{1}{2} \\ 0 & 0 & 1 & -\frac{1}{2} & 0 & \frac{1}{2} \end{array}\right)$$

We display A on the left and the identity on the right. We then change A into I by a sequence of basic row operations. Performing these same operations on the identity, we obtain A^{-1}, which appears in the last step on the right with the identity on the left.

If the coefficient matrix of a system of n linear homogeneous equations in n unknowns is nonsingular, then the equations have only the trivial solution. Conversely, if the coefficient matrix is singular, the equations have nontrivial solutions. These results follow from Theorem 2.3.3 (see Exercise 2.4.10).

Theorem 2.5.4 A system of n linear homogeneous algebraic equations in n unknowns has nontrivial solutions if and only if the determinant of the coefficient matrix is zero.

Theorem 2.5.5 A system of n linear algebraic equations in n unknowns has a unique solution if and only if the coefficient matrix is nonsingular.

PROOF This follows from Theorem 2.3.5 and our above remarks about the singularity or nonsingularity of the reduced coefficient matrix.

In the case covered by Theorem 2.5.5, we can find a formula for the solution. We can write the equations as $AX = B$, with A nonsingular. Therefore, A^{-1} exists, and

$$A^{-1}(AX) = IX = X = A^{-1}B$$

If C is the matrix of cofactors of elements of A, then

$$A^{-1} = \frac{1}{|A|}\tilde{C} \quad \text{and} \quad X = \frac{1}{|A|}\tilde{C}B$$

The product $\tilde{C}B$ has elements $\sum_{j=1}^{n} \tilde{c}_{ij}b_j = \sum_{j=1}^{n} c_{ji}b_j$. Therefore,

$$x_i = \frac{1}{|A|}\sum_{j=1}^{n} c_{ji}b_j$$

But the sum is just the expansion by columns of a determinant formed from the coefficient matrix A by replacing the ith column by the b's. This gives us *Cramer's rule*: if the coefficient matrix A of the system of n equations in n unknowns, $AX = B$, is nonsingular, then the value of the ith variable is given by $1/|A|$ times the determinant of the matrix formed by replacing the ith column of A by $b_1, b_2, \ldots, b_n$.

In terms of total calculations, Cramer's rule is not very practical. It involves the calculation of $n + 1$ determinants of order n. Compared with the elimination method it is marginal for $n = 3$ and is at a disadvantage for $n \geq 4$. However, if the value of only one of the unknowns is needed, only two determinants need be calculated.

We conclude this section with a couple of definitions of special nonsingular matrices which involve the concept of inverse.

Definition 2.5.3 Let A be an $n \times n$ matrix with real elements. Then A is said to be *orthogonal* if $A^{-1} = \tilde{A}$.

Definition 2.5.4 Let A be a matrix with complex elements. Then the *conjugate* of A (written $\bar{A}$) is the matrix formed from A by taking the conjugate of each of its elements.

Definition 2.5.5 Let A be an $n \times n$ matrix with complex elements. Then A is said to be *unitary* if $A^{-1} = \tilde{\bar{A}}$.

EXERCISES 2.5

1 Which of the following matrices have inverses?

$$(a)\begin{pmatrix}1 & 2\\ 2 & 1\end{pmatrix} \quad (b)\begin{pmatrix}1 & 1\\ 1 & 1\end{pmatrix} \quad (c)\begin{pmatrix}1 & 0 & 1\\ 0 & -1 & 0\\ 1 & 0 & -1\end{pmatrix}$$

$$(d)\begin{pmatrix}1 & 2 & 3\\ 0 & -1 & 0\\ 1 & 0 & -1\end{pmatrix} \quad (e)\begin{pmatrix}1 & 3 & -1 & 2\\ 2 & 1 & 3 & 1\\ -1 & 2 & -4 & 1\\ -2 & 1 & 2 & -3\end{pmatrix}$$

$$(f)\begin{pmatrix}1 & 2 & 3 & 4\\ 0 & 1 & 2 & 3\\ 0 & 0 & 1 & 2\\ 0 & 0 & 0 & 1\end{pmatrix}$$

2 Find inverses of matrices in Exercise 1 which are nonsingular.

3 Find the inverse of the diagonal matrix†

$$\begin{pmatrix}1 & 0 & 0 & 0\\ 0 & 2 & 0 & 0\\ 0 & 0 & 3 & 0\\ 0 & 0 & 0 & 4\end{pmatrix}$$

When does a diagonal matrix have an inverse? State a general rule for finding the inverse of a diagonal matrix.

4 Show that an upper-triangular matrix with nonzero elements on the principal diagonal has an inverse which is upper-triangular.

5 Solve the following system of equations using Cramer's rule.

$$\begin{aligned} x_1 + x_2 - x_3 &= 7\\ -x_1 + 2x_2 + x_3 &= -3\\ 2x_1 - x_2 + 3x_3 &= 5 \end{aligned}$$

6 Solve the matrix equation $AB = C$ for B if

$$A = \begin{pmatrix}1 & 1 & -1\\ -1 & 2 & 1\\ 2 & -1 & 3\end{pmatrix} \quad \text{and} \quad C = \begin{pmatrix}2 & 0 & 1\\ 3 & -1 & 4\\ 1 & 5 & -1\end{pmatrix}$$

7 Let $AX = B$ be a system of n equations in n unknowns with $|A| = 0$. Show that there are no solutions unless all n determinants of matrices formed from A by inserting B in the n columns of A are zero. If solutions exist, are they unique? *Hint:* Multiply both sides of $AX = B$ on the left by $\tilde{C}$, the transpose of the matrix of cofactors of A.

† A diagonal matrix is a square matrix with zero elements off the principal diagonal.

8 Let

$$A = \begin{pmatrix} \cos\theta & -\sin\theta \\ \sin\theta & \cos\theta \end{pmatrix}$$

Find A^{-1}. Is A orthogonal?

9 If A and B are nonsingular, show that AB is nonsingular. Show that

$$(AB)^{-1} = B^{-1}A^{-1}$$

10 Let A be nonsingular. Show that $\widetilde{A^{-1}} = (\tilde{A})^{-1}$.

11 Let A be nonsingular. Show that $(A^{-1})^{-1} = A$.

12 If $AB = 0$, $B \neq 0$, is A nonsingular?

13 If $A^2 = 0$, is A nonsingular?

14 If A is orthogonal, what are the possible values of $|A|$?

15 If A and B are orthogonal, is AB orthogonal? Is A^{-1} orthogonal?

16 If A and B are unitary, is AB unitary? Is A^{-1} unitary?

17 Define $A^0 = I$ and $A^{-n} = (A^{-1})^n$, n a positive integer. Let A be nonsingular. Prove the general exponential formulas $(A^p)^q = A^{pq}$ and $A^pA^q = A^{p+q}$, where p and q are integers.

18 If C is the matrix of cofactors of the elements of A, what is the value of $|C|$?

*2.6 EXISTENCE AND UNIQUENESS THEOREMS

The two main questions concerning systems of linear algebraic equations (other than methods of finding explicit solutions) are (1) whether solutions exist (*existence*) and (2) if a solution exists, is it unique (*uniqueness*)? We have dealt with these questions to some extent in Sec. 2.3. In this section, we shall give a more systematic discussion of these two questions, but first we must introduce a new concept about matrices which can be defined in terms of determinants.

Every matrix, whether square or not, has square matrices in it which can be obtained by deleting whole rows and/or whole columns. There are of course only a finite number of such square matrices. Suppose we compute the determinants of all these matrices. We define the *rank* of the original matrix in terms of these determinants.

Definition 2.6.1 The rank of a matrix A is the order of the largest order nonsingular matrix which can be obtained from A by deleting whole rows and/or whole columns. We denote this number by rank (A); rank $(0) = 0$.

EXAMPLE 2.6.1 Find the rank of the matrix

$$A = \begin{pmatrix} 1 & 2 & 3 & 4 \\ 1 & -1 & 0 & 1 \\ 2 & 1 & 3 & 5 \end{pmatrix}$$

Since A has some nonzero elements, rank $(A) \geq 1$. Also, since there are no 4×4 matrices, rank $(A) \leq 3$. It is easy to find a 2×2 matrix with nonzero determinant, so rank $(A) \geq 2$. Now the 3×3 matrices are all singular since

$$\begin{vmatrix} 1 & 2 & 3 \\ 1 & -1 & 0 \\ 2 & 1 & 3 \end{vmatrix} = 0 \qquad \begin{vmatrix} 1 & 2 & 4 \\ 1 & -1 & 1 \\ 2 & 1 & 5 \end{vmatrix} = 0$$

$$\begin{vmatrix} 1 & 3 & 4 \\ 1 & 0 & 1 \\ 2 & 3 & 5 \end{vmatrix} = 0 \qquad \begin{vmatrix} 2 & 3 & 4 \\ -1 & 0 & 1 \\ 1 & 3 & 5 \end{vmatrix} = 0$$

Therefore, rank $(A) = 2$.

The proofs of the next two theorems are left to the reader.

Theorem 2.6.1 If A is an $m \times n$ matrix, then $0 \leq \text{rank}\,(A) \leq \min\,[m,n]$.

Theorem 2.6.2 If A is an $n \times n$ matrix, then rank $(A) = n$ if and only if A is nonsingular.

Let us consider the effect, if any, on the rank of a matrix A when one of the basic row operations is performed on A. We shall consider each operation separately. Suppose rank $(A) = r$, and suppose the jth row of A is multiplied by $k \neq 0$ to form a new matrix B. There is some $r \times r$ matrix C in A such that $|C| \neq 0$. Now, if C contains all or part of the jth row of A, then there is a $r \times r$ matrix in B whose determinant is $k|C| \neq 0$. If C does not involve the jth row of A, then C is also in B and B again contains an rth order nonsingular matrix. If A contains any square matrix D of order larger than r, $|D| = 0$. Any square matrix in B of order larger than r either does not involve the jth row of A, and is therefore already in A, or it does involve the jth row of A, in which case its determinant is zero because it is k times a zero determinant from A. Therefore, multiplication of a row of a matrix by a nonzero constant does not change the rank.

Next we consider the interchange of two rows. Suppose B is obtained from A by an interchange of the ith and jth rows of A. Suppose rank $(A) = r$.

Then there is an $r \times r$ matrix C in A such that $|C| \neq 0$. If C involves both the ith and jth rows of A, then there is an $r \times r$ matrix C^* in B such that $|C^*| = -|C| \neq 0$. If C involves neither the ith nor jth row, then C is also contained in B. If C involves the ith but not the jth row, then there is an $r \times r$ matrix in A with the ith row deleted and containing the jth row, which after the interchange becomes C except for the order of the rows. This matrix is obviously nonsingular. If A contains any square matrices of order larger than r, they are singular. This is also true of B because any square matrix of order larger than r in B is either in A or can be obtained from such a matrix in A by an interchange of rows. Hence, interchanging two rows of a matrix does not alter the rank.

Finally, we consider the operation of adding one row of A to another row. Suppose B is obtained from A by adding the ith row to the jth row. If rank $(A) = r$, then there is a nonsingular matrix C of order r in A. If C involves both the ith and jth rows of A, then there is a matrix C^* in B obtained from C by adding two rows. Hence, $|C^*| = |C| \neq 0$. If C involves neither the ith nor the jth row, then C is also in B. If C involves the ith row but not the jth row, then C is also in B. If C involves the jth row but not the ith row, then there is an $r \times r$ matrix in B whose determinant is $|C| + |C^*|$, where C^* is obtained from C by replacing those elements from the jth row of A by the corresponding elements from the ith row. If $|C| + |C^*| = 0$, then $|C^*| \neq 0$ and C^* is an rth order nonsingular matrix in B. Any matrix in B of order larger than r will either already be in A, and hence be singular, or will have a determinant which is the sum of determinants of singular matrices in A. We have shown that none of the three basic row operations will change the rank of a matrix. We summarize this in the following theorem.

Theorem 2.6.3 Let A and B be $m \times n$ matrices, where B can be obtained from A by a sequence of basic row operations. Then rank (A) = rank (B).

We are now ready to state the fundamental existence theorem for systems of linear algebraic equations. However, first for convenience we shall consider one additional operation on matrices, the interchange of two columns. It is clear that this operation does not change the rank of a matrix. Now let us consider the effect on a system of equations if two columns of the coefficient matrix are interchanged. Suppose the ith and jth columns are interchanged. Then the coefficients of the x_i unknown become the coefficients of the x_j unknown and vice versa. So this has the effect of just relabeling the x_i and x_j unknowns.

Theorem 2.6.4 A system of linear algebraic equations has a solution if and only if the rank of the coefficient matrix is equal to the rank of the augmented matrix.

PROOF By reducing the coefficient matrix using basic row operations and possibly the interchange of columns (relabeling of unknowns), we can develop the equivalent system

$$\begin{pmatrix} 1 & a_{12} & a_{13} & \cdots & a_{1r} & a_{1r+1} & \cdots & a_{1n} \\ 0 & 1 & a_{23} & \cdots & a_{2r} & a_{2r+1} & \cdots & a_{2n} \\ 0 & 0 & 1 & \cdots & a_{3r} & a_{3r+1} & \cdots & a_{3n} \\ \cdots & \cdots & \cdots & \cdots & \cdots & \cdots & \cdots & \cdots \\ 0 & 0 & 0 & \cdots & 1 & a_{rr+1} & \cdots & a_{rn} \\ 0 & 0 & 0 & \cdots & 0 & 0 & \cdots & 0 \\ \cdots & \cdots & \cdots & \cdots & \cdots & \cdots & \cdots & \cdots \\ 0 & 0 & 0 & \cdots & 0 & 0 & \cdots & 0 \end{pmatrix} \begin{pmatrix} x_1 \\ x_2 \\ x_3 \\ \vdots \\ x_n \end{pmatrix} = \begin{pmatrix} b_1 \\ b_2 \\ b_3 \\ \cdot \\ b_r \\ b_{r+1} \\ \cdot \\ b_m \end{pmatrix}$$

We should keep in mind that the a's and b's are not the numbers in the original system, which have been changed in the reduction process. However, the right-hand sides are affected only by row operations. The last column of the augmented matrix is never interchanged with columns of the coefficient matrix. It is now clear that the rank of the reduced coefficient matrix is r. It is also clear that the system will have a solution if and only if $b_{r+1} = b_{r+2} = \cdots = b_m = 0$. In this case, and only in this case, the rank of the augmented matrix of the system will be r. By Theorem 2.6.3 and the comment about interchanging columns, the statements about equality of rank will hold for the original coefficient and augmented matrices. This completes the proof.

Since any solution of the original system can be found as a solution of the reduced system, in the case where the equations have solutions we can find the general solution from the reduced system. Let us examine this general solution. The rth reduced equation is

$$x_r + a_{rr+1}x_{r+1} + \cdots + a_{rn}x_n = b_r$$

We can assign arbitrary values to $x_{r+1}, x_{r+2}, \ldots, x_n$ and obtain a value for x_r. Substituting this into the $(r - 1)$st equation, we can obtain a value for x_{r-1}. Working up to the first equation, we can finally determine x_1. Generally, the values of $x_1, x_2, \ldots, x_r$ depend on our choice of $x_{r+1}, x_{r+2}, \ldots, x_n$, but in any case, the choice of the last $n - r$ variables is completely arbitrary. Therefore, the general solution will contain $n - r$ arbitrary parameters. From this we can conclude the fundamental uniqueness theorem.

Theorem 2.6.5 A solution of a system of m linear algebraic equations in n unknowns is unique if and only if the rank of the coefficient matrix is n.

The final theorem follows from Theorems 2.6.5 and 2.6.1.

Theorem 2.6.6 A solution of a system of m linear algebraic equations in n unknowns is never unique if $n > m$.

PROOF If A is the coefficient matrix, rank $(A) \leq m < n$.

EXAMPLE 2.6.2 Determine whether the following system of equations has a solution; if so, is it unique?

$$\begin{aligned} x_1 - x_2 + 2x_3 - x_4 &= 1 \\ 2x_1 + x_2 - x_3 + x_4 &= -2 \\ 2x_2 + x_3 - 3x_4 &= 1 \\ 3x_1 + x_3 &= -1 \\ 3x_1 + 2x_2 + 2x_3 - 3x_4 &= 0 \end{aligned}$$

The augmented matrix of the system is

$$\begin{pmatrix} 1 & -1 & 2 & -1 & 1 \\ 2 & 1 & -1 & 1 & -2 \\ 0 & 2 & 1 & -3 & 1 \\ 3 & 0 & 1 & 0 & -1 \\ 3 & 2 & 2 & -3 & 0 \end{pmatrix}$$

Using row operations, we have

$$\begin{pmatrix} 1 & -1 & 2 & -1 & 1 \\ 2 & 1 & -1 & 1 & -2 \\ 0 & 2 & 1 & -3 & 1 \\ 3 & 0 & 1 & 0 & -1 \\ 3 & 2 & 2 & -3 & 0 \end{pmatrix}$$

$$\rightarrow \begin{pmatrix} 1 & -1 & 2 & -1 & 1 \\ 0 & 3 & -5 & 3 & -4 \\ 0 & 2 & 1 & -3 & 1 \\ 0 & 3 & -5 & 3 & -4 \\ 0 & 5 & -4 & 0 & -3 \end{pmatrix} \rightarrow \begin{pmatrix} 1 & -1 & 2 & -1 & 1 \\ 0 & 2 & 1 & -3 & 1 \\ 0 & 3 & -5 & 3 & -4 \\ 0 & 0 & 0 & 0 & 0 \\ 0 & 5 & -4 & 0 & -3 \end{pmatrix}$$

$$\rightarrow \begin{pmatrix} 1 & -1 & 2 & -1 & 1 \\ 0 & 2 & 1 & -3 & 1 \\ 0 & 3 & -5 & 3 & -4 \\ 0 & 3 & -5 & 3 & -4 \\ 0 & 0 & 0 & 0 & 0 \end{pmatrix} \rightarrow \begin{pmatrix} 1 & -1 & 2 & -1 & 1 \\ 0 & 2 & 1 & -3 & 1 \\ 0 & 3 & -5 & 3 & -4 \\ 0 & 0 & 0 & 0 & 0 \\ 0 & 0 & 0 & 0 & 0 \end{pmatrix}$$

It is apparent that the rank of the coefficient matrix equals the rank of the augmented matrix, which is 3. Therefore, the system has a solution, but it is not unique because $3 < 4$, the number of unknowns. The general solution of the system will contain one arbitrary parameter.

EXERCISES 2.6

1 Prove that the rank of the augmented matrix of a system of linear algebraic equations cannot be less than the rank of the coefficient matrix.

2 Consider the three basic column operations on matrices: (1) multiplication of a column by $k \neq 0$, (2) interchange of two columns, (3) addition of one column to another. Prove that these column operations cannot change the rank of a matrix.

3 State Theorem 2.3.3 in terms of the rank of the coefficient matrix. Prove your version.

4 State Theorem 2.3.5 in terms of the rank of the coefficient matrix. Prove your version.

5 Determine whether the following systems of equations have solutions; if so, are they unique?

(*a*)
$$\begin{aligned} x_1 + x_2 - x_3 &= 1 \\ 2x_1 + x_2 + 3x_3 &= 2 \\ x_2 - 5x_3 &= 1 \end{aligned}$$

(*b*)
$$\begin{aligned} x_1 + + 2x_3 - x_4 &= 0 \\ 2x_1 + x_2 - x_3 &= 5 \\ -x_1 + 2x_2 + x_3 + 2x_4 &= 3 \\ 3x_2 - 2x_3 + 5x_4 &= 1 \end{aligned}$$

(*c*)
$$\begin{aligned} 2x_1 - x_2 + x_3 &= 1 \\ x_1 + 2x_2 - 3x_3 &= 0 \\ -x_1 + 3x_2 - x_3 &= 2 \\ x_1 - x_2 - 2x_3 &= -3 \end{aligned}$$

(*d*)
$$\begin{aligned} x_1 - 2x_2 + 3x_3 - x_4 + x_5 &= 5 \\ -x_1 + 3x_2 + 4x_3 + x_4 &= 2 \\ 2x_1 + x_3 - 2x_4 + 2x_5 &= -1 \\ x_2 + 5x_3 - 2x_4 - x_5 &= 0 \end{aligned}$$

6 Consider the system of equations $AX = B$, where A is $m \times m$. Suppose that rank $(A) = m - 1$. Prove that the system has a solution if and only if the m matrices formed from A by replacing the columns by B are all singular.

3

VECTOR SPACES

3.1 INTRODUCTION

We have already seen two examples of vector spaces, the complex numbers and the two-dimensional vectors of Sec. 1.4. There are many more examples, and one of the objectives of this chapter is to give some idea of the scope of this subject. Before introducing the abstract concept of a vector space, we take up the three-dimensional euclidean vectors, partly because of their intrinsic importance in applications and partly to give the reader one more concrete example before we begin the general discussion of vector spaces. After introducing the axioms of a vector space and proving some basic theorems, we take up the very important concepts of dependence and independence of vectors. We then define basis and dimension of vector spaces. The scalar product is then introduced, and this leads to a discussion of orthonormal base. The last section (which is starred) takes up some of the fundamental properties of infinite-dimensional vector spaces.

3.2 THREE-DIMENSIONAL VECTORS

We shall define the three-dimensional euclidean vectors (from now on we shall say simply *vector*) by a simple extension of the two-dimensional vectors of Sec. 1.4. A three-dimensional vector is defined by a triple of real numbers (x,y,z), and we shall write $\mathbf{v} = (x,y,z)$. Two vectors $\mathbf{v}_1 = (x_1,y_1,z_1)$ and $\mathbf{v}_2 = (x_2,y_2,z_2)$ are *equal* if and only if $x_1 = x_2$, $y_1 = y_2$, and $z_1 = z_2$. We define *addition* of two vectors $\mathbf{v}_1 = (x_1,y_1,z_1)$ and $\mathbf{v}_2 = (x_2,y_2,z_2)$ by $\mathbf{v}_1 + \mathbf{v}_2 = (x_1 + x_2, y_1 + y_2, z_1 + z_2)$. We see that the result is a vector and the operation is associative and commutative. We define the *zero vector* as $\mathbf{0} = (0,0,0)$, and it follows immediately that $\mathbf{v} + \mathbf{0} = (x,y,z) + (0,0,0) = (x,y,z) = \mathbf{v}$ for all vectors $\mathbf{v}$. The *negative* of a vector $\mathbf{v} = (x,y,z)$ is defined by $-\mathbf{v} = (-x,-y,-z)$ and the following is obviously true: $\mathbf{v} + (-\mathbf{v}) = \mathbf{0}$ for all vectors $\mathbf{v}$.

We define the operation of multiplication of a vector $\mathbf{v} = (x,y,z)$ by a real scalar a as follows: $a\mathbf{v} = (ax,ay,az)$. The result is a vector, and it is easy to verify that the operation has the following properties:

1 $a(\mathbf{v}_1 + \mathbf{v}_2) = a\mathbf{v}_1 + a\mathbf{v}_2$.
2 $(a + b)\mathbf{v} = a\mathbf{v} + b\mathbf{v}$.
3 $a(b\mathbf{v}) = (ab)\mathbf{v}$.
4 $1\mathbf{v} = \mathbf{v}$.

The geometrical interpretation of three-dimensional vectors is similar to that for the two-dimensional vectors of Sec. 1.4. Consider a three-dimensional euclidean space with two points (a,b,c) and (d,e,f) (see Fig. 15). Let $x = d - a$, $y = e - b$, $z = f - c$. A geometrical interpretation of the vector $\mathbf{v} = (x,y,z)$ is the arrow drawn from the point (a,b,c) to the point (d,e,f). The length of the vector is defined as

$$|\mathbf{v}| = \sqrt{x^2 + y^2 + z^2} = \sqrt{(d - a)^2 + (e - b)^2 + (f - c)^2}$$

The direction of the vector (if $|\mathbf{v}| \neq 0$) is specified by the least nonnegative angles $(\theta_1,\theta_2,\theta_3)$ from the positive coordinate axes to the arrow of the vector. The cosines of these angles are given by

$$\cos\theta_1 = \frac{x}{|\mathbf{v}|} \qquad \cos\theta_2 = \frac{y}{|\mathbf{v}|} \qquad \cos\theta_3 = \frac{z}{|\mathbf{v}|}$$

These cosines are usually called *direction cosines*. Since

$$x^2 + y^2 + z^2 = |\mathbf{v}|^2(\cos^2\theta_1 + \cos^2\theta_2 + \cos^2\theta_3)$$

then

$$\cos^2\theta_1 + \cos^2\theta_2 + \cos^2\theta_3 = 1$$

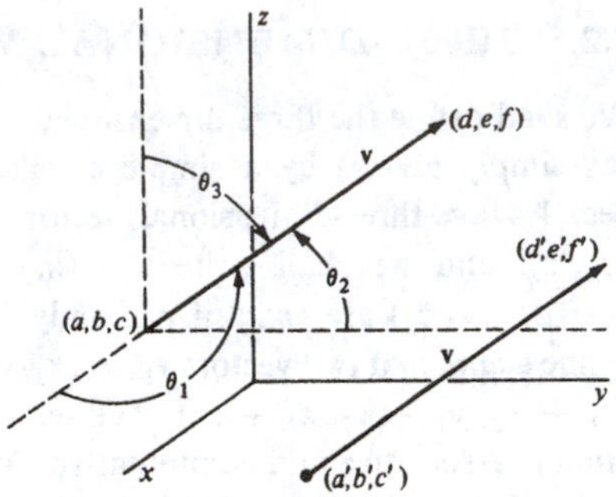

FIGURE 15

If there is another pair of points (a',b',c') and (d',e',f') such that $x = d' - a'$, $y = e' - b'$, $z = f' - c'$, then the arrow from (a',b',c') to (d',e',f') has the same length and direction as that from (a,b,c) to (d,e,f). Therefore, we say that the vector $\mathbf{v} = (x,y,z)$ is represented by *any* arrow with length $|\mathbf{v}|$ and direction specified by $(\theta_1,\theta_2,\theta_3)$. The zero vector has no direction and so has no arrows associated with it.

The geometrical interpretation of vector addition is as follows (see Fig. 16). Place an arrow representing $\mathbf{v}_1$ from point P to point Q. Place an arrow representing $\mathbf{v}_2$ from point Q to point R. Then the arrow from P to R represents the sum $\mathbf{v}_1 + \mathbf{v}_2$. If P and R coincide, then $\mathbf{v}_1 + \mathbf{v}_2 = \mathbf{0}$. Since the points P, Q, and R determine a plane, if they are distinct and not collinear, the triangle inequality simply states that

$$|\mathbf{v}_1 + \mathbf{v}_2| \le |\mathbf{v}_1| + |\mathbf{v}_2|$$

It is easy to see that the inequality continues to hold if P, Q, and R are not all distinct or are collinear.

Next let us give a geometrical interpretation of the operation of multiplication of a vector by a scalar. Let a be a scalar and $\mathbf{v} = (x,y,z)$ a vector. Then $a\mathbf{v} = (ax,ay,az)$ and

$$|a\mathbf{v}| = \sqrt{a^2x^2 + a^2y^2 + a^2z^2} = |a|\,|\mathbf{v}|$$

Therefore, multiplication by a modifies the length by a factor of $|a|$ if $|a| \neq 1$. Suppose $\cos\theta_1 = x/|\mathbf{v}|$, $\cos\theta_2 = y/|\mathbf{v}|$, $\cos\theta_3 = z/|\mathbf{v}|$. The new direction cosines are

$$\cos\theta_1' = \frac{a}{|a|}\cos\theta_1 \qquad \cos\theta_2' = \frac{a}{|a|}\cos\theta_2 \qquad \cos\theta_3' = \frac{a}{|a|}\cos\theta_3$$

If $a > 0$, the direction is unchanged. If $a < 0$, the direction is reversed. If $a = 0$, $a\mathbf{v} = \mathbf{0}$, which has no direction. The vector $-\mathbf{v} = (-1)\mathbf{v}$ is the negative

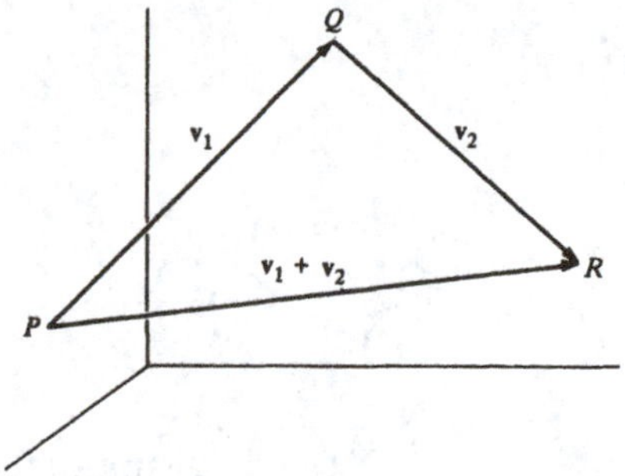

FIGURE 16

of $\mathbf{v}$, and we see that it has the same length as $\mathbf{v}$ but has the opposite direction (provided $\mathbf{v} \neq \mathbf{0}$). The geometrical interpretation of subtraction of two vectors can be obtained from that for addition by writing $\mathbf{v}_1 - \mathbf{v}_2 = \mathbf{v}_1 + (-\mathbf{v}_2)$.

We also have the notion of *scalar product* for three-dimensional vectors. Let $\mathbf{v}_1 = (x_1,y_1,z_1)$ and $\mathbf{v}_2 = (x_2,y_2,z_2)$. Then we define the scalar product (dot product) as

$$\mathbf{v}_1 \cdot \mathbf{v}_2 = x_1x_2 + y_1y_2 + z_1z_2$$

We see immediately that this definition leads to the following properties:

1 $\mathbf{v}_1 \cdot \mathbf{v}_2 = \mathbf{v}_2 \cdot \mathbf{v}_1$.
2 $\mathbf{v}_1 \cdot (\mathbf{v}_2 + \mathbf{v}_3) = (\mathbf{v}_1 \cdot \mathbf{v}_2) + (\mathbf{v}_1 \cdot \mathbf{v}_3)$.
3 $a\mathbf{v}_1 \cdot \mathbf{v}_2 = a(\mathbf{v}_1 \cdot \mathbf{v}_2)$.
4 $\mathbf{v} \cdot \mathbf{v} = |\mathbf{v}|^2 \geq 0$.
5 $\mathbf{v} \cdot \mathbf{v} = 0$ if and only if $\mathbf{v} = \mathbf{0}$.

Let us look for a geometrical interpretation of the scalar product. Suppose $\mathbf{v}_1$ and $\mathbf{v}_2$ are not zero and do not have the same or opposite directions. We place arrows representing the two vectors starting from the origin (see Fig. 17). The points O, P, Q determine a plane. Consider the triangle OPQ. The law of cosines gives the cosine of the angle θ between $\mathbf{v}_1$ and $\mathbf{v}_2$:

$$\cos\theta = \frac{|\mathbf{v}_1 - \mathbf{v}_2|^2 - |\mathbf{v}_1|^2 - |\mathbf{v}_2|^2}{-2|\mathbf{v}_1|\,|\mathbf{v}_2|}$$

Using the properties of the scalar product, we can evaluate the numerator

$$\begin{aligned} |\mathbf{v}_1 - \mathbf{v}_2|^2 - |\mathbf{v}_1|^2 - |\mathbf{v}_2|^2 &= (\mathbf{v}_1 - \mathbf{v}_2)\cdot(\mathbf{v}_1 - \mathbf{v}_2) - \mathbf{v}_1 \cdot \mathbf{v}_1 - \mathbf{v}_2 \cdot \mathbf{v}_2 \\ &= \mathbf{v}_1 \cdot \mathbf{v}_1 + \mathbf{v}_2 \cdot \mathbf{v}_2 - 2(\mathbf{v}_1 \cdot \mathbf{v}_2) - \mathbf{v}_1 \cdot \mathbf{v}_1 - \mathbf{v}_2 \cdot \mathbf{v}_2 \\ &= -2(\mathbf{v}_1 \cdot \mathbf{v}_2) \end{aligned}$$

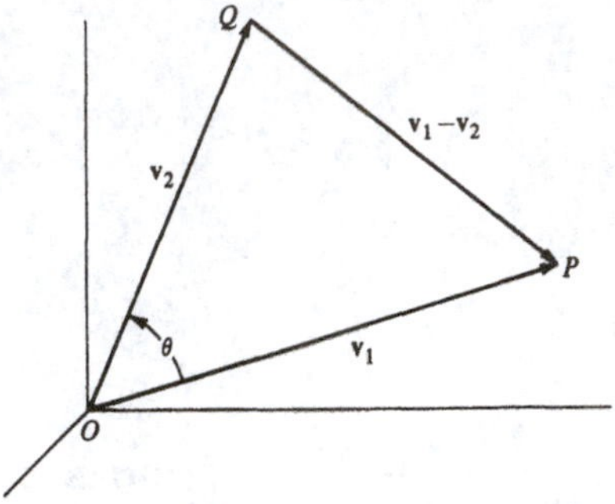

FIGURE 17

This gives us the formula we are seeking

$$\mathbf{v}_1 \cdot \mathbf{v}_2 = |\mathbf{v}_1|\,|\mathbf{v}_2| \cos\theta$$

If $\mathbf{v}_1$ and/or $\mathbf{v}_2$ are zero, it continues to hold because then both sides are zero. If $\mathbf{v}_1$ and $\mathbf{v}_2$ are parallel, then $\mathbf{v}_2 = a\mathbf{v}_1$ for some scalar a and

$$\mathbf{v}_1 \cdot \mathbf{v}_2 = a|\mathbf{v}_1|^2 = |a|\,|\mathbf{v}_1|\,|\mathbf{v}_1|\,\frac{a}{|a|}$$

$$= |\mathbf{v}_1|\,|\mathbf{v}_2| \cos\theta$$

where $\cos\theta = 1$ if $a > 0$ $(\theta = 0°)$ or $\cos\theta = -1$ if $a < 0$ $(\theta = 180°)$. Therefore, the formula holds in all cases. Two immediate consequences follow.

Theorem 3.2.1 $|\mathbf{v}_1 \cdot \mathbf{v}_2| \le |\mathbf{v}_1|\,|\mathbf{v}_2|$.

Theorem 3.2.2 If $|\mathbf{v}_1| \ne 0$ and $|\mathbf{v}_2| \ne 0$, then $\mathbf{v}_1 \cdot \mathbf{v}_2 = 0$ if and only if $\mathbf{v}_1$ and $\mathbf{v}_2$ are perpendicular.

One of the important applications of the three-dimensional vectors is the representation of lines and planes in space. Let us begin with lines.

EXAMPLE 3.2.1 Find an equation of the line through the point (x_0,y_0,z_0) having the direction of the vector (a,b,c). Let the vector $\mathbf{v} = (x,y,z)$ be the vector from the origin to the point (x,y,z) on the line (see Fig. 18). Clearly $\mathbf{v} = (x,y,z) = (x_0,y_0,z_0) + t(a,b,c)$, where t is a parameter. If t varies from $-\infty$ to ∞, we shall obtain the whole line.

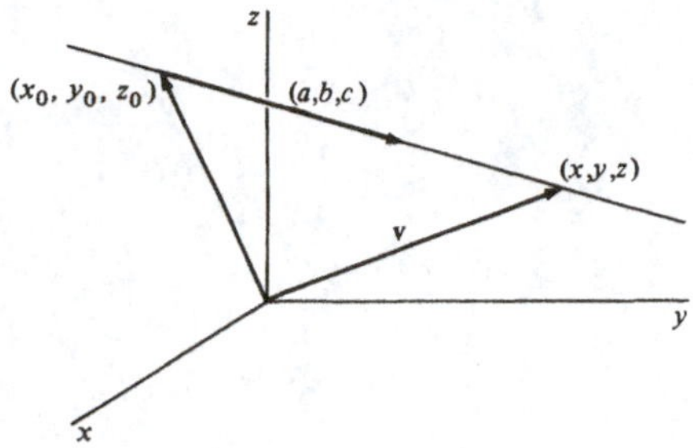

FIGURE 18

EXAMPLE 3.2.2 Find an equation of the line passing through the two points (x_1,y_1,z_1) and (x_2,y_2,z_2). A vector which specifies the direction of the line is $(a,b,c) = (x_2,y_2,z_2) - (x_1,y_1,z_1) = (x_2 - x_1, y_2 - y_1, z_2 - z_1)$. Therefore, following Example 3.2.1, an equation for the line is

$$(x,y,z) = (x_1,y_1,z_1) + t(x_2 - x_1, y_2 - y_1, z_2 - z_1) \qquad -\infty < t < \infty$$

To discuss planes effectively we need the notion of linear combinations of two vectors. Let $\mathbf{v}_1 = (a_1,b_1,c_1)$ and $\mathbf{v}_2 = (a_2,b_2,c_2)$ be nonzero vectors which are not parallel ($\mathbf{v}_2$ is not a scalar times $\mathbf{v}_1$ or vice versa). Let s and t be scalars. Consider arrows representing $\mathbf{v}_1$ and $\mathbf{v}_2$ starting from the origin (see Fig. 19). If all arrows originate from the origin, then $s\mathbf{v}_1$ is collinear with $\mathbf{v}_1$, $t\mathbf{v}_2$ is collinear with $\mathbf{v}_2$, and $s\mathbf{v}_1 + t\mathbf{v}_2$ is in the plane determined by $\mathbf{v}_1$ and $\mathbf{v}_2$. As s and t vary, we obtain different points in the plane determined by $\mathbf{v}_1$ and $\mathbf{v}_2$.

EXAMPLE 3.2.3 Find an equation of the plane containing the point (x_0,y_0,z_0) and parallel to the two vectors $\mathbf{v}_1 = (a_1,b_1,c_1)$ and $\mathbf{v}_2 = (a_2,b_2,c_2)$, which are not parallel to each other. Let (x,y,z) be the vector from the origin to a point on the plane. Then

$$(x,y,z) = (x_0,y_0,z_0) + s(a_1,b_1,c_1) + t(a_2,b_2,c_2)$$

As s and t range through values from $-\infty$ to ∞, we obtain all possible points in the given plane.

EXAMPLE 3.2.4 Find an equation of the plane containing three distinct noncollinear points (x_0,y_0,z_0), (x_1,y_1,z_1), and (x_2,y_2,z_2). Since the points are

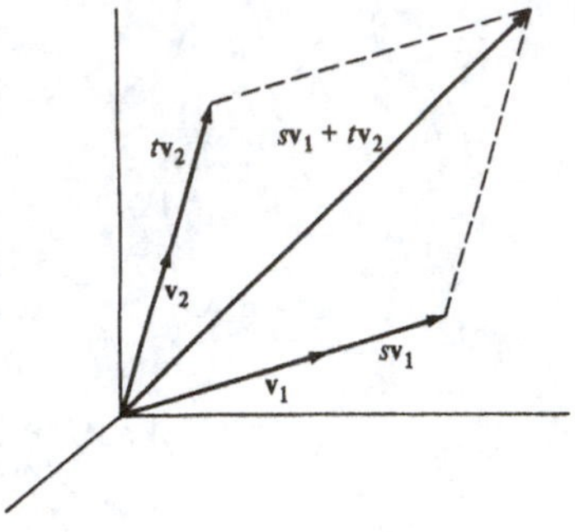

FIGURE 19

distinct and noncollinear, the vectors $(x_1 - x_0,\ y_1 - y_0,\ z_1 - z_0)$ and $(x_2 - x_0,\ y_2 - y_0,\ z_2 - z_0)$ are nonzero and not parallel. Therefore, if (x,y,z) is the vector from the origin to a point on the plane, an equation is

$$(x,y,z) = (x_0,y_0,z_0) + s(x_1 - x_0,\ y_1 - y_0,\ z_1 - z_0) \\ + t(x_2 - x_0,\ y_2 - y_0,\ z_2 - z_0)$$

where $-\infty < s < \infty$ and $-\infty < t < \infty$.

The equations given in Examples 3.2.1 to 3.2.4 are all *parametric equations*, where we have given the three coordinates of points on the geometrical figure in terms of one or two parameters. The fact that it took one parameter for the line and two parameters for the plane reflects the fact that the line is essentially a one-dimensional figure whereas the plane is essentially two-dimensional. There are also implicit representations for lines and planes. In the *implicit representation* we state one or more equations which the coordinates of a point on the figure must satisfy, while points not on the figure will not satisfy these equations.

EXAMPLE 3.2.5 Find an implicit representation of the plane containing the point (x_0,y_0,z_0) perpendicular to the vector (a,b,c). Let (x,y,z) be the displacement vector from the origin to a point in the plane (see Fig. 20). The vector $(x - x_0,\ y - y_0,\ z - z_0)$ is parallel to the plane and hence the scalar product

$$(a,b,c) \cdot (x - x_0,\ y - y_0,\ z - z_0) = a(x - x_0) + b(y - y_0) + c(z - z_0) = 0$$

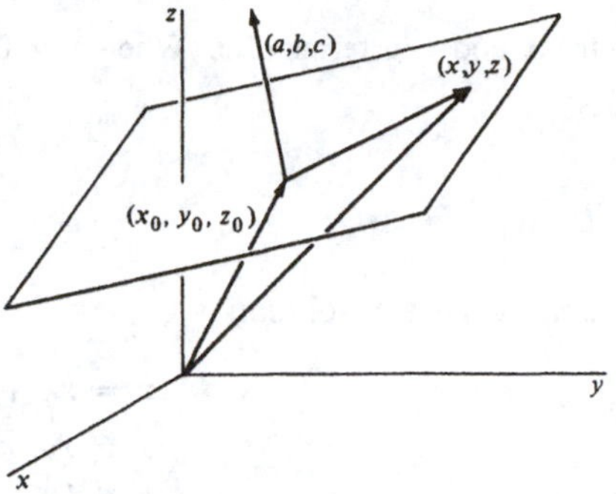

FIGURE 20

This is an implicit equation for the plane since the coordinates of all points on the plane will satisfy it, while the coordinates of points off the plane will not satisfy it. The equation can be put in the form

$$ax + by + cz = ax_0 + by_0 + cz_0 = d$$

Conversely, every equation of the form $\alpha x + \beta y + \gamma z = \delta$, where $(\alpha,\beta,\gamma) \neq \mathbf{0}$, is the equation of some plane. There is at least one point (x_0,y_0,z_0) such that $\alpha x_0 + \beta y_0 + \gamma z_0 = \delta$. But then $\alpha x + \beta y + \gamma z = \alpha x_0 + \beta y_0 + \gamma z_0$ and $\alpha(x - x_0) + \beta(y - y_0) + \gamma(z - z_0) = 0$ states the geometrical fact that the vector $(x - x_0, y - y_0, z - z_0)$ is perpendicular to (α,β,γ).

EXAMPLE 3.2.6 Suppose two planes given implicitly by $ax + by + cz = d$ and $\alpha x + \beta y + \gamma z = \delta$ intersect in a single line L. Find an equation for L. We must find the coordinates of all points which satisfy both equations simultaneously. The two equations do not represent the same plane or parallel planes; hence the vectors (a,b,c) and (α,β,γ) are not proportional. Therefore, at least one of the determinants

$$\begin{vmatrix} a & b \\ \alpha & \beta \end{vmatrix} \qquad \begin{vmatrix} b & c \\ \beta & \gamma \end{vmatrix} \qquad \text{or} \qquad \begin{vmatrix} a & c \\ \alpha & \gamma \end{vmatrix}$$

is different from zero. Suppose, for the sake of the argument, it is the first. Then we can solve the equations

$$ax + by = d - cz$$
$$\alpha x + \beta y = \delta - \gamma z$$

for x and y in terms of z. When $z = 0$

$$x_0 = \frac{\begin{vmatrix} d & b \\ \delta & \beta \end{vmatrix}}{\begin{vmatrix} a & b \\ \alpha & \beta \end{vmatrix}} \qquad y_0 = \frac{\begin{vmatrix} a & d \\ \alpha & \delta \end{vmatrix}}{\begin{vmatrix} a & b \\ \alpha & \beta \end{vmatrix}}$$

and the general solution is

$$\begin{aligned} x &= x_0 + tu \qquad &\text{where } u &= \frac{b\gamma - c\beta}{a\beta - b\alpha} \\ y &= y_0 + tv \\ z &= t & v &= \frac{c\alpha - a\gamma}{a\beta - b\alpha} \end{aligned}$$

In other words, $(x,y,z) = (x_0,y_0,0) + t(u,v,1)$, $-\infty < t < \infty$, and we have the equation of a line passing through $(x_0,y_0,0)$ and having the direction of $(u,v,1)$.

The concepts of a vector-valued function and derivative of a vector-valued function follow as extensions of the two-dimensional case of Sec. 1.4.

Definition 3.2.1 Suppose for each value of t in some set of real numbers D, called the *domain* of the function, a vector $\mathbf{v}(t)$ is unambiguously defined; then we say that $\mathbf{v}$ is a vector-valued function of t; t is called the *independent variable*, and $\mathbf{v}$ is called the *dependent variable*. The collection of all values of $\mathbf{v}(t)$, taken on for t in the domain, is called the *range* of the function.

EXAMPLE 3.2.7 Let $\mathbf{v}(t) = (x,y,z) = (a \cos t, a \sin t, bt)$, $0 \le t \le 2\pi$, where a and b are real constants. Then $\mathbf{v}$ is a vector-valued function of t. The domain is the interval $\{t \mid 0 \le t \le 2\pi\}$. If we think of $\mathbf{v}$ as the vector from the origin to a point in the three-dimensional space, then the range of the function is a spiral of radius $a = \sqrt{x^2 + y^2}$ joining the initial point $(a,0,0)$ and the final point $(a,0,2\pi b)$ (see Fig. 21).

Definition 3.2.2 Suppose for some t_0 and some $\delta > 0$, all t satisfying $t_0 - \delta < t < t_0 + \delta$ are in the domain of $\mathbf{v}(t)$ and there is a vector $\mathbf{v}'(t_0)$ such that

$$\lim_{t \to t_0} \left| \frac{\mathbf{v}(t) - \mathbf{v}(t_0)}{t - t_0} - \mathbf{v}'(t_0) \right| = 0$$

then $\mathbf{v}'(t_0)$ is the derivative of $\mathbf{v}(t)$ at t_0.

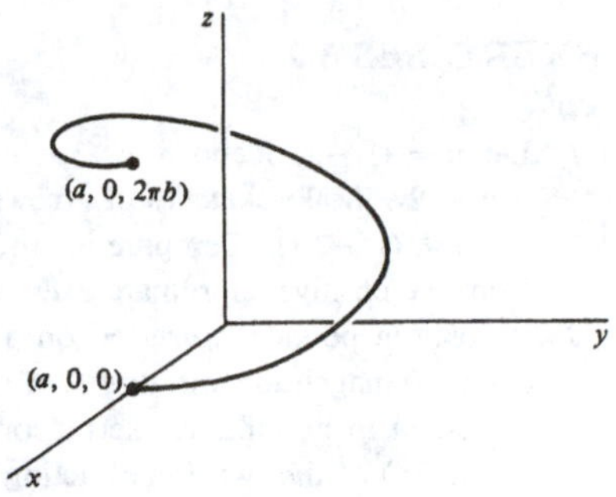

FIGURE 21

Just as in the two-dimensional case, we have the following theorem.

Theorem 3.2.3 The vector-valued function $\mathbf{v}$, with values $\mathbf{v}(t) = (x(t),y(t),z(t))$, has a derivative $\mathbf{v}'(t_0) = (a,b,c)$ if and only if $x(t)$, $y(t)$, and $z(t)$ have derivatives at t_0 such that $x'(t_0) = a$, $y'(t_0) = b$, and $z'(t_0) = c$.

EXAMPLE 3.2.8 The vector-valued function of Example 3.2.7 has a derivative at t_0 such that $0 < t_0 < 2\pi$. In fact,

$$\mathbf{v}'(t_0) = (-a \sin t_0, a \cos t_0, b)$$

Similarly, we have the notion of tangent to a curve in three-dimensional space.

Definition 3.2.3 Suppose a curve is given parametrically by the vector from the origin $\mathbf{v}(t) = (x(t),y(t),z(t))$, $\alpha \le t \le \beta$. If $\mathbf{v}(t)$ has a nonzero derivative $\mathbf{v}'(t_0)$ at t_0, then $\mathbf{v}'(t_0)$ is a tangent vector at the point $\mathbf{v}(t_0)$. The tangent line at $\mathbf{v}(t_0)$ is given parametrically by the equation

$$\mathbf{w}(t) = \mathbf{v}(t_0) + (t - t_0)\mathbf{v}'(t_0) \qquad -\infty < t < \infty$$

EXAMPLE 3.2.9 The tangent vector to the spiral of Example 3.2.7 at $t_0 = \pi/2$ is $\mathbf{v}'(\pi/2) = (-a,0,b)$. The tangent line at the point $\mathbf{v}(\pi/2) = (0,a,\pi b/2)$ is given by

$$\mathbf{w}(t) = \left(0,a,\frac{\pi b}{2}\right) + \left(t - \frac{\pi}{2}\right)(-a,0,b)$$

EXERCISES 3.2

1 Let $\mathbf{u} = (1,-2,1)$ and $\mathbf{v} = (3,1,-4)$. Compute $\mathbf{u} + \mathbf{v}$, $\mathbf{u} - \mathbf{v}$, $2\mathbf{u}$, $-\frac{1}{2}\mathbf{v}$, and $-\mathbf{u} + 2\mathbf{v}$. Make sketches of arrows representing each of these vectors.

2 Let $\mathbf{u} = (1,-2,1)$. Compute $|\mathbf{u}|$ and θ_1, θ_2, θ_3, the minimum nonnegative angles from the positive coordinate axes to the arrow of the vector.

3 A force in pounds is exerted on a body as designated by the vector $(3,1,-2)$. Find the magnitude and direction of this force. Another force of $(-4,5,3)$, also measured in pounds, is exerted on the same body. Find the combined effect (resultant) of the two forces acting together. Find the magnitude and direction of the resultant. *Hint*: The resultant is the vector sum of the two forces.

4 Let $\mathbf{u} = (2,0,-3)$ and $\mathbf{v} = (-1,4,5)$. Find the cosine of the angle between the two vectors. If $\cos\theta_1$, $\cos\theta_2$, $\cos\theta_3$ are the direction cosines of $\mathbf{u}$ and $\cos\phi_1$, $\cos\phi_2$, $\cos\phi_3$ are the direction cosines of $\mathbf{v}$, show that $\cos\theta = \cos\theta_1\cos\phi_1 + \cos\theta_2\cos\phi_2 + \cos\theta_3\cos\phi_3$, where θ is the angle between the two vectors.

5 Consider a nonzero vector $\mathbf{u}$ represented by the arrow from O to P. Consider the vector $\mathbf{v}$ represented by the arrow from O to Q. The projection of $\mathbf{v}$ on $\mathbf{u}$ is defined to be the vector O to N, where N is the foot of the perpendicular drawn from Q to the line of $\mathbf{u}$. Show that this projection is given by

$$\frac{\mathbf{u}\cdot\mathbf{v}}{|\mathbf{u}|^2}\mathbf{u}$$

Compute the projection of $(1,-2,1)$ on $(3,1,-4)$.

6 Consider a plane represented implicitly by $ax + by + cz = d$. Consider a vector $\mathbf{v}$ represented by an arrow from the point (x_0,y_0,z_0) in the plane to the point Q. The projection of $\mathbf{v}$ on the plane is defined to be the vector from (x_0,y_0,z_0) to N, the foot of the perpendicular drawn from Q to the plane. Let $\mathbf{u} = (a,b,c)$. Show that the projection of $\mathbf{v}$ on the plane is given by

$$\mathbf{v} - \frac{\mathbf{u}\cdot\mathbf{v}}{|\mathbf{u}|^2}\mathbf{u}$$

Compute the projection of $(2,3,-1)$ on the plane given by $x - 3y + 2z = 7$.

7 Show that the distance from the point (x_1,y_1,z_1) to the plane represented by $ax + by + cz = d$ is given by

$$\frac{|ax_1 + by_1 + cz_1 - d|}{\sqrt{a^2 + b^2 + c^2}}$$

Compute the distance from $(1,-2,3)$ to the plane represented by $2x - y + 3z = 7$.

8 Find an equation of the line through the point $(1,2,-3)$ in the direction of the vector $(-2,3,5)$.

9 Find an equation of the line through the two points $(3,-5,7)$ and $(-2,1,4)$.

10 Find an equation of the line through the point $(4,3,-5)$ and perpendicular to the plane given by $2x + 3y + 4z = 3$.

11 Find an equation of the line of intersection of the two planes given by $2x + 3y - 4z = 5$ and $-x + 7y + 5z = 2$.

12 Find an equation of the plane through the origin and parallel to the vectors $(1,-2,3)$ and $(5,0,7)$.

13 Find an implicit representation of the plane containing the three points $(1,-3,4)$, $(2,5,-1)$, $(0,7,-4)$.

14 Find an equation of the plane containing the two lines $(x,y,z) = (1,-3,4) + t(2,5,-1)$ and $(x,y,z) = (1,-3,4) + s(0,7,-4)$.

15 Find an equation of the plane containing the point $(1,2,3)$ and perpendicular to the line $(x,y,z) = (1,-3,4) + t(2,5,-1)$.

16 One way to compute a three-dimensional vector perpendicular to two given vectors is to use the *vector product.* Let $\mathbf{u} = (u_1,u_2,u_3)$ and $\mathbf{v} = (v_1,v_2,v_3)$ be two nonzero and nonparallel vectors. Show that

$$\mathbf{w} = \mathbf{u} \times \mathbf{v} = (u_2v_3 - u_3v_2, u_3v_1 - u_1v_3, u_1v_2 - u_2v_1)$$

is a vector perpendicular (orthogonal) to both $\mathbf{u}$ and $\mathbf{v}$. The vector product has no counterpart in other spaces.

17 Show that $|\mathbf{u} \times \mathbf{v}| = |\mathbf{u}|\,|\mathbf{v}|\,|\sin\theta|$, where θ is the angle between $\mathbf{u}$ and $\mathbf{v}$. *Hint:* Write $\mathbf{u} = |\mathbf{u}|(\cos\theta_1, \cos\theta_2, \cos\theta_3)$ and $\mathbf{v} = |\mathbf{v}|(\cos\phi_1, \cos\phi_2, \cos\phi_3)$.

18 Repeat Exercise 13, using the vector product to compute a vector perpendicular to the required plane.

19 Let $\mathbf{u}$, $\mathbf{v}$, $\mathbf{w}$ be three vectors whose arrow representations from the origin form the three edges of a parallelepiped. Show that $|\mathbf{u}\cdot(\mathbf{v} \times \mathbf{w})|$ is equal to the volume of the parallelepiped.

20 Show that:

(*a*) $$\mathbf{u}\cdot(\mathbf{v}\times\mathbf{w}) = \begin{vmatrix} u_1 & u_2 & u_3 \\ v_1 & v_2 & v_3 \\ w_1 & w_2 & w_3 \end{vmatrix}$$

(*b*) $\mathbf{u}\cdot(\mathbf{v}\times\mathbf{w}) = \mathbf{w}\cdot(\mathbf{u}\times\mathbf{v}) = \mathbf{v}\cdot(\mathbf{w}\times\mathbf{u})$.

(*c*) $\mathbf{u}\cdot(\mathbf{w}\times\mathbf{v}) = -\mathbf{u}\cdot(\mathbf{v}\times\mathbf{w})$.

21 Consider three planes given implicitly by

$$\begin{aligned} a_{11}x + a_{12}y + a_{13}z &= b_1 \\ a_{21}x + a_{22}y + a_{23}z &= b_2 \\ a_{31}x + a_{32}y + a_{33}z &= b_3 \end{aligned}$$

The intersection of these three planes could be (1) empty, (2) a line, (3) a plane, or (4) a point. In terms of the solutions of the three equations in three unknowns give an algebraic condition for each of the cases.

22 A point moves along a curve in three-dimensional space with its displacement vector from the origin given by the vector-valued function of time $\mathbf{r}(t) = (x(t),y(t),z(t))$, where x, y, and z have first and second derivatives. The first derivative $\mathbf{r}'(t) = \mathbf{v}(t) = (x'(t),y'(t),z'(t))$ is called the *velocity* of the point, and the second derivative $\mathbf{a}(t) = \mathbf{r}''(t) = \mathbf{v}'(t) = (x''(t),y''(t),z''(t))$ is called the *acceleration*.

If $\mathbf{v}(t) \neq \mathbf{0}$, show that the velocity is tangent to the curve. The magnitude of the velocity is called the *speed.* If the speed is constant and positive, show that the acceleration is *normal* to the curve (perpendicular to the tangent vector).

23 If the speed $s(t) = |\mathbf{v}(t)|$ of a point is not zero, show that $\mathbf{a}(t) = s'(t)\mathbf{T} + s(t)|\mathbf{T}'|\mathbf{n}$, where $\mathbf{T}$ is a unit tangent vector and $\mathbf{n}$ is a unit normal vector.

3.3 AXIOMS OF A VECTOR SPACE

We have already seen several examples of algebraic systems which, at least in certain respects, behave similarly. We have in mind those properties of complex numbers, two-dimensional euclidean vectors, three-dimensional euclidean vectors, and matrices with respect to addition and multiplication by a scalar. We now take the modern mathematical point of view and define abstract systems with those properties we wish to study. These systems we shall call *vector spaces.* This approach will have the distinct advantage that any properties we derive from this definition will be true of all vector spaces, and we shall not have to study each system separately as it comes up. We begin with the axioms for a vector space.

Definition 3.3.1 Consider a system V of objects, called *vectors*, for which we have defined two operations, addition and multiplication by a scalar, either real or complex. Then V is a vector space if these operations satisfy the following properties:

A1 If $\mathbf{u}$ and $\mathbf{v}$ are in V, then $\mathbf{u} + \mathbf{v}$ is in V.
A2 $\mathbf{u} + \mathbf{v} = \mathbf{v} + \mathbf{u}$.
A3 $\mathbf{u} + (\mathbf{v} + \mathbf{w}) = (\mathbf{u} + \mathbf{v}) + \mathbf{w}$.
A4 There is zero vector $\mathbf{0}$ in V such that $\mathbf{u} + \mathbf{0} = \mathbf{u}$ for all $\mathbf{u}$ in V.
A5 If $\mathbf{u}$ is in V, then there is a vector $-\mathbf{u}$ in V, called the negative of $\mathbf{u}$, such that $\mathbf{u} + (-\mathbf{u}) = \mathbf{0}$.
M1 If a is a scalar and $\mathbf{u}$ is in V, then $a\mathbf{u}$ is in V.
M2 $a(\mathbf{u} + \mathbf{v}) = a\mathbf{u} + a\mathbf{v}$.
M3 $(a + b)\mathbf{u} = a\mathbf{u} + b\mathbf{u}$.
M4 $(ab)\mathbf{u} = a(b\mathbf{u})$.
M5 $1\mathbf{u} = \mathbf{u}$.

If the set of scalars is the set of all real numbers, then we say that V is a *real vector space.* If the set of scalars is the set of all complex numbers, then we say that V is a *complex vector space.*

The reader should verify that the system of complex numbers is a real vector space. Here the vectors are complex numbers of the form $x + iy$ and the scalars are real numbers. Addition is the ordinary addition of complex numbers, and multiplication by a scalar is defined by $a(x + iy) = ax + iay$.

The reader should also verify that the systems of two-dimensional and three-dimensional euclidean vectors are real vector spaces.

EXAMPLE 3.3.1 Show that the system of $m \times n$ matrices with real elements is a real vector space, where addition is defined as ordinary matrix addition and multiplication by a real scalar is defined as in Sec. 2.2. Let A, B, and C be $m \times n$ matrices with real elements. Then clearly sums are defined, and the result is an $m \times n$ matrix with real elements. This verifies A1. A2 and A3 follow immediately from Sec. 2.2. A zero is the $m \times n$ zero matrix all of whose elements are zero. If A is an $m \times n$ matrix with real elements a_{ij}, then a negative $-A$ is an $m \times n$ matrix with real elements $-a_{ij}$. Then A4 and A5 follow at once. If A is an $m \times n$ matrix with real elements a_{ij}, then aA is an $m \times n$ matrix with elements aa_{ij}. Properties M2 to M5 follow easily.

EXAMPLE 3.3.2 Consider a system of n-tuples of real numbers $(u_1, u_2, u_3, \ldots, u_n)$. If $\mathbf{u} = (u_1, u_2, u_3, \ldots, u_n)$ and $\mathbf{v} = (v_1, v_2, v_3, \ldots, v_n)$ and a is a real scalar, then we define addition and multiplication by a real scalar as follows:†

$$\mathbf{u} + \mathbf{v} = (u_1 + v_1, u_2 + v_2, u_3 + v_3, \ldots, u_n + v_n)$$
$$a\mathbf{u} = (au_1, au_2, au_3, \ldots, au_n)$$

Show that this is a vector space. It is clear that A1 and M1 are already satisfied. For A2 we have

$$\begin{aligned}\mathbf{u} + \mathbf{v} &= (u_1 + v_1, u_2 + v_2, \ldots, u_n + v_n)\\ &= (v_1 + u_1, v_2 + u_2, \ldots, v_n + u_n)\\ &= \mathbf{v} + \mathbf{u}\end{aligned}$$

For A3 we have

$$\begin{aligned}&\mathbf{u} + (\mathbf{v} + \mathbf{w})\\ &\quad = (u_1 + (v_1 + w_1), u_2 + (v_2 + w_2), \ldots, u_n + (v_n + w_n))\\ &\quad = ((u_1 + v_1) + w_1, (u_2 + v_2) + w_2, \ldots, (u_n + v_n) + w_n)\\ &\quad = (\mathbf{u} + \mathbf{v}) + \mathbf{w}\end{aligned}$$

† It is understood that in each vector space the equality of two vectors is defined. In this case, $\mathbf{u} = \mathbf{v}$ if and only if $u_i = v_i$, $i = 1, 2, \ldots, n$.

A zero vector is defined to be $\mathbf{0} = (0, 0, \ldots, 0)$, and A4 follows immediately:

$$\begin{aligned}\mathbf{u} + \mathbf{0} &= (u_1 + 0, u_2 + 0, \ldots, u_n + 0)\\ &= (u_1, u_2, \ldots, u_n) = \mathbf{u}\end{aligned}$$

If $\mathbf{u} = (u_1, u_2, \ldots, u_n)$, then a negative $-\mathbf{u}$ is defined to be $(-u_1, -u_2, \ldots, -u_n)$ and A5 follows:

$$\begin{aligned}\mathbf{u} + (-\mathbf{u}) &= (u_1 - u_1, u_2 - u_2, \ldots, u_n - u_n)\\ &= (0, 0, \ldots, 0) = \mathbf{0}\end{aligned}$$

The rest of the properties, M2 to M5, follow easily:

M2:
$$\begin{aligned}a(\mathbf{u} + \mathbf{v}) &= (a(u_1 + v_1), a(u_2 + v_2), \ldots, a(u_n + v_n))\\ &= (au_1 + av_1, au_2 + av_2, \ldots, au_n + av_n)\\ &= a\mathbf{u} + a\mathbf{v}\end{aligned}$$

M3:
$$\begin{aligned}(a + b)\mathbf{u} &= ((a + b)u_1, (a + b)u_2, \ldots, (a + b)u_n)\\ &= (au_1 + bu_1, au_2 + bu_2, \ldots, au_n + bu_n)\\ &= a\mathbf{u} + b\mathbf{u}\end{aligned}$$

M4:
$$\begin{aligned}(ab)\mathbf{u} &= ((ab)u_1, (ab)u_2, \ldots, (ab)u_n)\\ &= (a(bu_1), a(bu_2), \ldots, a(bu_n)) = a(b\mathbf{u})\end{aligned}$$

M5:
$$1\mathbf{u} = (u_1, u_2, \ldots, u_n) = \mathbf{u}$$

The vector space of this example is called R^n.

The reader will be asked to show (see Exercise 3.3.3) that the system of n-tuples of complex numbers is a complex vector space, where addition and multiplication by a complex scalar is defined as in Example 3.3.2. This vector space we shall refer to as C^n.

EXAMPLE 3.3.3 Consider the collection P_n of all polynomials in the real variable x with real coefficients of degree n or less. If $P_n(x) = a_0 + a_1x + a_2x^2 + \cdots + a_nx^n$ and $q_n(x) = b_0 + b_1x + b_2x^2 + \cdots + b_nx^n$ are two such polynomials and a is a real scalar, we shall define addition and multiplication by a scalar as follows:

$$\begin{aligned}p_n(x) + q_n(x) &= (a_0 + b_0) + (a_1 + b_1)x + (a_2 + b_2)x^2 + \cdots + (a_n + b_n)x^n\\ ap_n(x) &= aa_0 + (aa_1)x + (aa_2)x^2 + \cdots + (aa_n)x^n\end{aligned}$$

Show that P_n is a real vector space. Before checking the axioms we must agree on the definition of equality. There are at least two definitions of equality which make good sense. One would be to say that two polynomials are equal if coefficients of like powers of x are equal. The other, since polynomials are functions, would be to say that polynomials are equal if their values are the same for every value of x. It turns out that these two definitions are equivalent. Clearly if coefficients of like powers of x are equal, then the two polynomials are equal for all values of x. Conversely, we can show that if the two polynomials are equal for all values of x, then coefficients of like powers of x are equal. Suppose $p_n(x) = q_n(x)$ for all x. Then $p_n(0) = a_0 = q_n(0) = b_0$. Polynomials are differentiable everywhere, so $p_n'(x) = q_n'(x)$. Hence, $p_n'(0) = a_1 = q_n'(0) = b_1$. Similarly, for

$$k = 2, 3, \ldots, n, \frac{p_n^{(k)}(0)}{k!} = a_k = \frac{q_n^{(k)}(0)}{k!} = b_k$$

We may now complete the example. Clearly A1 and M1 are satisfied. A zero polynomial is the constant function with the value zero everywhere. If $p_n(x) = a_0 + a_1x + \cdots + a_nx^n$, then a negative $-p_n(x) = (-a_0) + (-a_1)x + \cdots + (-a_n)x^n$. The rest of the axioms clearly follow from our definition.

EXAMPLE 3.3.4 Consider the collection of all real-valued continuous functions of the real variable x defined on the interval $\{x \mid 0 \le x \le 1\}$. By equality we shall mean that two such functions are equal if their values agree for all values of x in the interval; that is, $f = g$ if $f(x) = g(x)$ for all x satisfying $0 \le x \le 1$. By addition of two functions we shall mean the pointwise addition of their values; that is, $f + g = h$ if $h(x) = f(x) + g(x)$ for all x in the interval. By multiplication by a real scalar a we shall define the function af to have the values $af(x)$. We can show that this is a real vector space. It is clear that the operations of addition and multiplication by a real scalar will yield real-valued functions defined on the interval. That these functions are also continuous follows from two of the basic theorems of the calculus. This verifies A1 and M1. For A2 we simply have

$$f(x) + g(x) = g(x) + f(x)$$

and for A3 we have

$$f(x) + (g(x) + h(x)) = (f(x) + g(x)) + h(x)$$

A zero function is simply the constant function with value zero everywhere in the interval. Then for A4 we have $f(x) + 0 = f(x)$. If f has the value $f(x)$, then a negative $-f$ has the value $-f(x)$. A5 follows at once; $f(x) + (-f(x)) = 0$ for all x in the interval. For M2 to M5, we have

M2: $$a(f(x) + g(x)) = af(x) + ag(x)$$

M3: $$(a + b)f(x) = af(x) + bf(x)$$

M4: $$(ab)f(x) = a(bf(x))$$

M5: $$1f(x) = f(x)$$

Vector spaces of functions, as in this example, are sometimes called *function spaces*.

By now it should be clear that the concept of a vector space is very useful because of the many examples of vector spaces which occur in mathematics. Therefore, it is appropriate that we study abstract vector spaces in some detail. We begin by proving some basic theorems.

Theorem 3.3.1 There is only one zero vector in a given vector space V.

PROOF Suppose $\mathbf{0}$ and $\mathbf{0}^*$ are both zero vectors. Then, by A4, $\mathbf{0} = \mathbf{0} + \mathbf{0}^* = \mathbf{0}^*$.

Theorem 3.3.2 Corresponding to a given vector $\mathbf{u}$ in a vector space V, there is a unique negative.

PROOF Suppose $\mathbf{v}$ and $\mathbf{w}$ were both negatives for $\mathbf{u}$. Then $\mathbf{u} + \mathbf{v} = \mathbf{0}$ and $\mathbf{u} + \mathbf{w} = \mathbf{0}$. However, $(\mathbf{w} + \mathbf{u}) + \mathbf{v} = \mathbf{w} + \mathbf{0} = \mathbf{w}$. But $\mathbf{w} + \mathbf{u} = \mathbf{0}$ and $\mathbf{0} + \mathbf{v} = \mathbf{v}$. Hence, $\mathbf{v} = \mathbf{w}$.

Theorem 3.3.3 For all vectors $\mathbf{u}$ in a vector space V, $0\mathbf{u} = \mathbf{0}$.

PROOF By M3 and M5 we have $\mathbf{u} = 1\mathbf{u} = (1 + 0)\mathbf{u} = 1\mathbf{u} + 0\mathbf{u} = \mathbf{u} + 0\mathbf{u}$. Then adding $-\mathbf{u}$ to both sides, we have $\mathbf{0} = \mathbf{u} + (-\mathbf{u}) = \mathbf{u} + (-\mathbf{u}) + 0\mathbf{u} = 0\mathbf{u}$.

Theorem 3.3.4 For all vectors $\mathbf{u}$ in a vector space V, $(-1)\mathbf{u} = -\mathbf{u}$.

PROOF By M3, M5, and Theorem 3.3.3, we have $\mathbf{0} = 0\mathbf{u} = (1 - 1)\mathbf{u} = 1\mathbf{u} + (-1)\mathbf{u} = \mathbf{u} + (-1)\mathbf{u}$. Now adding $-\mathbf{u}$ to both sides, we have $-\mathbf{u} = -\mathbf{u} + \mathbf{u} + (-1)\mathbf{u} = \mathbf{0} + (-1)\mathbf{u} = (-1)\mathbf{u}$.

Theorem 3.3.5 For all vectors $\mathbf{u}$ in a vector space V and all scalars a, $-(a\mathbf{u}) = (-a)\mathbf{u}$.

PROOF By Theorem 3.3.4 and M4, we have $-(a\mathbf{u}) = (-1)(a\mathbf{u}) = (-a)\mathbf{u}$.

Theorem 3.3.6 For all scalars a, $a\mathbf{0} = \mathbf{0}$.

PROOF Let $\mathbf{u}$ be any vector in the vector space. Then $a\mathbf{u} + a\mathbf{0} = a(\mathbf{u} + \mathbf{0}) = a\mathbf{u}$. Then by Theorem 3.3.5, $a\mathbf{u} + (-a)\mathbf{u} = \mathbf{0} = a\mathbf{u} + (-a)\mathbf{u} + a\mathbf{0} = \mathbf{0} + a\mathbf{0} = a\mathbf{0}$.

We conclude this section with a discussion of *subspaces*.

Definition 3.3.2 Let U be a nonempty set of vectors from a vector space V. Then U is a subspace of V if whenever $\mathbf{u}$ and $\mathbf{v}$ belong to U, then $a\mathbf{u} + b\mathbf{v}$ belongs to U for all scalars a and b. If U is not all of V, then we say that U is a *proper subspace* of V.

There are two trivial subspaces in every vector space. One is the whole space, and the other is the *zero subspace* consisting of the zero vector alone.† In any case, a subspace is a vector space in its own right with the operations of addition and multiplication by a scalar inherited from the vector space V. Axioms A1 and M1 follow from the definition of subspace. A2 and A3 are inherited from V. The zero must be in U; and given $\mathbf{u}$ in U, the negative $-\mathbf{u} = (-1)\mathbf{u}$ must be in U. The rest of the axioms M2 to M5 are inherited from V.

EXAMPLE 3.3.5 Characterize all the subspaces of R^3. We shall visualize R^3 as points (x,y,z) in a three-dimensional euclidean space (or alternatively as vectors from the origin to points in the three-dimensional space). As we have already observed, the origin (0,0,0) and the whole space are subspaces. Also lines through the origin, that is, points of the form $(\alpha t,\beta t,\gamma t)$, form a subspace. Let us check this. Suppose $\mathbf{u} = (\alpha t_1,\beta t_1,\gamma t_1)$ and $\mathbf{v} = (\alpha t_2,\beta t_2,\gamma t_2)$; then

$$a\mathbf{u} + b\mathbf{v} = (\alpha(at_1 + bt_2),\beta(at_1 + bt_2),\gamma(at_1 + bt_2))$$

is also on the line. Another type of subspace is a plane through the origin. Suppose such a plane is given implicitly by the equation $\alpha x + \beta y + \gamma z = 0$. If $\mathbf{u} = (x_1,y_1,z_1)$ and $\mathbf{v} = (x_2,y_2,z_2)$ are points in the plane, then

$$\alpha x_1 + \beta y_1 + \gamma z_1 = 0$$
$$\alpha x_2 + \beta y_2 + \gamma z_2 = 0$$

† Every subspace must contain the zero vector since U must be nonempty and $0\mathbf{u} + 0\mathbf{v} = \mathbf{0}$ must be in U.

Multiplying the first equation by a and the second by b and adding, we have

$$\alpha(ax_1 + bx_2) + \beta(ay_1 + by_2) + \gamma(az_1 + bz_2) = 0$$

Therefore, $a\mathbf{u} + b\mathbf{v}$ is in the plane. We argue geometrically to show that these are all the subspaces. If there is a point $(x_0,y_0,z_0) \neq \mathbf{0}$ in the subspace, then the whole line (tx_0,ty_0,tz_0), $-\infty < t < \infty$, is in the subspace. If these are all the points in the subspace, then we have a line through the origin. If there are two points $\mathbf{u} = (x_1,y_1,z_1)$ and $\mathbf{v} = (x_2,y_2,z_2)$, such that $\mathbf{0}$, $\mathbf{u}$, and $\mathbf{v}$ are noncollinear, then the subspace contains the plane through the origin given by $a\mathbf{u} + b\mathbf{v}$, where a and b are any real numbers. If these are all the points, then we just have a plane through the origin. If there is a point $\mathbf{w}$ off the plane in the subspace, then we have the whole space because the subspace then contains all points of the form $a\mathbf{u} + b\mathbf{v} + c\mathbf{w}$, where a, b, and c are any real numbers.

EXAMPLE 3.3.6 Let $\mathbf{u}_1, \mathbf{u}_2, \mathbf{u}_3, \ldots, \mathbf{u}_n$ be a finite number of vectors from a vector space V. Consider the subset U of all vectors of the form

$$\mathbf{u} = c_1\mathbf{u}_1 + c_2\mathbf{u}_2 + c_3\mathbf{u}_3 + \cdots + c_n\mathbf{u}_n$$

where $c_1, c_2, c_3, \ldots, c_n$ is any set of scalars (real if V is a real vector space or complex if V is a complex vector space). Show that U is a subspace. Let $\mathbf{u}$ be as shown above and

$$\mathbf{v} = \gamma_1\mathbf{u}_1 + \gamma_2\mathbf{u}_2 + \gamma_3\mathbf{u}_3 + \cdots + \gamma_n\mathbf{u}_n$$

Then

$$a\mathbf{u} + b\mathbf{v} = (ac_1 + b\gamma_1)\mathbf{u}_1 + (ac_2 + b\gamma_2)\mathbf{u}_2 + \cdots + (ac_n + b\gamma_n)\mathbf{u}_n$$

so that $a\mathbf{u} + b\mathbf{v}$ is in V. In this case, we say that the set $\mathbf{u}_1, \mathbf{u}_2, \mathbf{u}_3, \ldots, \mathbf{u}_n$ *spans* the subspace U.

EXERCISES 3.3

1 (*a*) Consider a vector space consisting of one vector $\mathbf{0}$ with addition and multiplication defined by (i) $\mathbf{0} + \mathbf{0} = \mathbf{0}$ and (ii) $a\mathbf{0} = \mathbf{0}$. Prove that this space (the *zero space*) is a vector space.

(*b*) Show that all other vector spaces contain an infinite number of vectors.

2 Show that the system of $m \times n$ matrices with complex elements is a complex vector space where addition and multiplication by complex scalars are the usual matrix operations.

3 Show that the system of n-tuples of complex numbers is a complex vector space where addition and multiplication by a scalar are defined as in Example 3.3.2.

4 Show that the collection of all polynomials of degree n or less in the complex variable z with complex coefficients is a complex vector space, with addition and multiplication by a complex scalar as defined in Example 3.3.3.

5 Consider the collection of all real-valued Riemann-integrable functions of the real variable x defined on the interval $\{x \mid 0 \leq x \leq 1\}$. Show that this is a real vector space with addition and multiplication by a scalar as defined in Example 3.3.4. Is the space of Example 3.3.4 a subspace of this space? Is it a proper subspace?

6 Consider the collection of all real-valued differentiable functions of the real variable x defined on the interval $\{x \mid a \leq x \leq b\}$. Show that this is a real vector space with addition and multiplication by a scalar as defined in Example 3.3.4. Is this a subspace of real-valued continuous functions on $\{x \mid a \leq x \leq b\}$? Is it a proper subspace?

7 Given a vector space V. Prove that in V, $a\mathbf{u} = \mathbf{0}$ implies $a = 0$, $\mathbf{u} = \mathbf{0}$, or both.

8 Characterize all the subspaces of R^2.

9 Consider the system of homogeneous linear algebraic equations $AX = 0$ in the real variables $(x_1, x_2, \ldots, x_n)$ with real coefficients a_{ij}, $i = 1, 2, \ldots, m$; $j = 1, 2, \ldots, n$. Any solutions will be found in R^n. Prove that the set of all solutions is a subspace of R^n.

3.4 DEPENDENCE AND INDEPENDENCE OF VECTORS

We now come to the important concepts of dependence and independence of vectors. Suppose $\mathbf{u}_1, \mathbf{u}_2, \ldots, \mathbf{u}_k$ is some finite set of vectors from a vector space V. A *linear combination* of these vectors is a sum of the form

$$c_1\mathbf{u}_1 + c_2\mathbf{u}_2 + \cdots + c_k\mathbf{u}_k = \sum_{i=1}^{k} c_i\mathbf{u}_i$$

where $c_1, c_2, \ldots, c_k$ are scalars. Obviously such a linear combination is $\mathbf{0}$ if all the c's are zero. We say that $\mathbf{u}_1, \mathbf{u}_2, \ldots, \mathbf{u}_k$ are *dependent* if $\sum_{i=1}^{k} c_i\mathbf{u}_i = \mathbf{0}$ for some set of scalars, not all zero. If this is impossible, then we say that the set of vectors is *independent*.

Definition 3.4.1 A set of vectors $\mathbf{u}_1, \mathbf{u}_2, \ldots, \mathbf{u}_k$ in V is dependent if there is a linear combination $\sum_{i=1}^{k} c_i\mathbf{u}_i = \mathbf{0}$ with the scalars $c_1, c_2, \ldots, c_k$ not all zero. If $\sum_{i=1}^{k} c_i\mathbf{u}_i = \mathbf{0}$ only for $c_1 = c_2 = \cdots = c_k = 0$, then the set of vectors is independent.

EXAMPLE 3.4.1 Determine whether the set of vectors $\mathbf{u}_1 = (1,1,1)$, $\mathbf{u}_2 = (0,1,1)$, $\mathbf{u}_3 = (0,0,1)$ is dependent or independent in R^3. We write the linear combination

$$c_1\mathbf{u}_1 + c_2\mathbf{u}_2 + c_3\mathbf{u}_3 = \mathbf{0}$$

and see if we can determine possible values of c_1, c_2, and c_3. The three equations determined are

$$\begin{aligned} c_1 + c_2 + c_3 &= 0 \\ c_2 + c_3 &= 0 \\ c_3 &= 0 \end{aligned}$$

We immediately see that the only solution of these equations is $c_1 = c_2 = c_3 = 0$. Therefore, the set $\mathbf{u}_1$, $\mathbf{u}_2$, $\mathbf{u}_3$ is independent.

EXAMPLE 3.4.2 Determine whether the set of vectors $\mathbf{u}_1 = (1,-1,1,-1)$, $\mathbf{u}_2 = (2,3,-4,1)$, $\mathbf{u}_3 = (0,-5,6,-3)$ is dependent or independent in R^4. Let us represent the given vectors as column matrices:

$$\begin{pmatrix} 1 \\ -1 \\ 1 \\ -1 \end{pmatrix} \begin{pmatrix} 2 \\ 3 \\ -4 \\ 1 \end{pmatrix} \begin{pmatrix} 0 \\ -5 \\ 6 \\ -3 \end{pmatrix}$$

Then the equation $c_1\mathbf{u}_1 + c_2\mathbf{u}_2 + c_3\mathbf{u}_3 = \mathbf{0}$ becomes

$$c_1 \begin{pmatrix} 1 \\ -1 \\ 1 \\ -1 \end{pmatrix} + c_2 \begin{pmatrix} 2 \\ 3 \\ -4 \\ 1 \end{pmatrix} + c_3 \begin{pmatrix} 0 \\ -5 \\ 6 \\ -3 \end{pmatrix} = \begin{pmatrix} 0 \\ 0 \\ 0 \\ 0 \end{pmatrix}$$

or

$$\begin{pmatrix} 1 & 2 & 0 \\ -1 & 3 & -5 \\ 1 & -4 & 6 \\ -1 & 1 & -3 \end{pmatrix} \begin{pmatrix} c_1 \\ c_2 \\ c_3 \end{pmatrix} = \begin{pmatrix} 0 \\ 0 \\ 0 \end{pmatrix}$$

or $AC = 0$, where A is the 4×3 matrix formed by placing the given vectors in columns and C is the column matrix of the unknowns c_1, c_2, and c_3. The question of dependence and independence of the vectors then becomes a question of whether these homogeneous equations have nontrivial solutions or not. We reduce the coefficient matrix A using row operations:

$$\begin{pmatrix} 1 & 2 & 0 \\ -1 & 3 & -5 \\ 1 & -4 & 6 \\ -1 & 1 & -3 \end{pmatrix} \rightarrow \begin{pmatrix} 1 & 2 & 0 \\ 0 & 5 & -5 \\ 0 & -6 & 6 \\ 0 & 3 & -3 \end{pmatrix} \rightarrow \begin{pmatrix} 1 & 2 & 0 \\ 0 & 1 & -1 \\ 0 & 0 & 0 \\ 0 & 0 & 0 \end{pmatrix}$$

The reduced system of equations is then $c_1 + 2c_2 = 0$, $c_2 - c_3 = 0$. Let $c_3 = 1$, then $c_2 = 1$ and $c_1 = -2$. We have found a nontrivial solution so the given set of vectors is dependent.

Theorem 3.4.1 Any set of m vectors in R^n is dependent if $m > n$.

PROOF The m vectors are in the form of n-tuples of real numbers. Placing these m n-tuples in the columns of a matrix A, as in Example 3.4.2, we have a system $AC = 0$, where A is $n \times m$ and C is $m \times 1$. We have a system of n homogeneous equations in m unknowns with $m > n$. Therefore, by Theorem 2.3.2, the system has nontrivial solutions. Therefore, the set of vectors is dependent.

Theorem 3.4.2 Let $\mathbf{u}_1, \mathbf{u}_2, \ldots, \mathbf{u}_n$ be an independent set of vectors in R^n. Then any vector $\mathbf{u}$ in R^n can be written as a linear combination of $\mathbf{u}_1, \mathbf{u}_2, \ldots, \mathbf{u}_n$.

PROOF Consider the set $\mathbf{u}_1, \mathbf{u}_2, \ldots, \mathbf{u}_n, \mathbf{u}$. This is a set of $n + 1$ vectors in R^n. By Theorem 3.4.1, this set is dependent, and hence

$$c_1\mathbf{u}_1 + c_2\mathbf{u}_2 + \cdots + c_n\mathbf{u}_n + c_{n+1}\mathbf{u} = \mathbf{0}$$

where the c's are not all zero. If $c_{n+1} = 0$, then the set $\mathbf{u}_1, \mathbf{u}_2, \ldots, \mathbf{u}_n$ is dependent, contrary to assumption. Therefore, $c_{n+1} \neq 0$, and

$$\mathbf{u} = \frac{-c_1}{c_{n+1}}\mathbf{u}_1 + \frac{-c_2}{c_{n+1}}\mathbf{u}_2 + \cdots + \frac{-c_n}{c_{n+1}}\mathbf{u}_n$$

as we wished to prove.

Theorem 3.4.3 In a vector space V, a set $\mathbf{u}_1, \mathbf{u}_2, \ldots, \mathbf{u}_k$, $k \geq 2$, is dependent if and only if at least one of the vectors in the set can be written as a linear combination of the others.

PROOF Suppose $\mathbf{u}_1$ can be written as a linear combination of $\mathbf{u}_2, \mathbf{u}_3, \ldots, \mathbf{u}_k$. If not $\mathbf{u}_1$, then we can make it $\mathbf{u}_1$ by relabeling the vectors. Then

$$\mathbf{u}_1 = c_2\mathbf{u}_2 + c_3\mathbf{u}_3 + \cdots + c_k\mathbf{u}_k$$

and we have for $c_1 = -1$

$$c_1\mathbf{u}_1 + c_2\mathbf{u}_2 + \cdots + c_k\mathbf{u}_k = \mathbf{0}$$

showing that the set is dependent.

Conversely, suppose the set is dependent. Then

$$c_1\mathbf{u}_1 + c_2\mathbf{u}_2 + \cdots + c_k\mathbf{u}_k = \mathbf{0}$$

where the c's are not all zero. Suppose $c_1 \neq 0$. If not c_1, then by relabeling we can make it c_1. Hence,

$$\mathbf{u}_1 = \frac{-c_2}{c_1}\mathbf{u}_2 + \frac{-c_3}{c_1}\mathbf{u}_3 + \cdots + \frac{-c_k}{c_1}\mathbf{u}_k$$

which completes the proof.

EXAMPLE 3.4.3 Consider the vector space of continuous real-valued functions defined on the interval $\{x \mid -1 \leq x \leq 1\}$. Determine whether the functions $1, x, x^2$ are dependent or independent. We shall do the example two ways. We write $c_1 + c_2x + c_3x^2 = 0$, where the equality is to hold everywhere in the interval. Putting $x = 0$, we have $c_1 = 0$. Putting $x = 1$, we have $c_2 + c_3 = 0$. Putting $x = -1$, we have $-c_2 + c_3 = 0$. But these equations have the unique solution $c_1 = c_2 = c_3 = 0$. Therefore, the functions are independent. Alternatively, since the functions $1, x, x^2$ are differentiable on the given interval, we can proceed as follows. Let $p(x) = c_1 + c_2x + c_3x^2 = 0$. Then $p(0) = c_1 = 0$, $p'(0) = c_2 = 0$, and $p''(0) = 2c_3 = 0$.

In dealing with function spaces, the question of dependence and independence of sets of functions is related to the interval over which the space is defined. Consider the same interval as in Example 3.4.3, and consider the functions f, g, and h defined as follows:

$$f(x) = 1 \qquad -1 \leq x \leq 1$$

$$g(x) = \begin{cases} 0 & -1 \leq x \leq 0 \\ x & 0 \leq x \leq 1 \end{cases} \qquad h(x) = \begin{cases} 0 & -1 \leq x \leq 0 \\ x^2 & 0 \leq x \leq 1 \end{cases}$$

These functions are independent. Let $p(x) = c_1 f(x) + c_2 g(x) + c_3 h(x) = 0$. Then $p(0) = c_1 = 0$, $p(1) = c_2 + c_3 = 0$, and $p(\frac{1}{2}) = \frac{1}{2}c_2 + \frac{1}{4}c_3 = 0$. These equations have only the trivial solution $c_1 = c_2 = c_3 = 0$. Therefore, the functions are independent. However, if we restrict these same functions to the interval $\{x \mid -1 \leq x \leq 0\}$, then they are dependent because

$$0 \cdot f(x) + 1 \cdot g(x) + 1 \cdot h(x) = 0$$

for $-1 \leq x \leq 0$.

In dealing with sets of functions which have a certain number of derivatives on the interval of definition, the concept of the *Wronskian* is very useful. Suppose the set of real-valued functions $f_1(x), f_2(x), \ldots, f_k(x)$ are all defined and are differentiable† $k - 1$ times on the interval $\{x \mid a \leq x \leq b\}$. The

† We require only one-sided differentiability at the end points of the interval.

Wronskian of the set of functions is defined on the interval to be the determinant

$$W(x) = \begin{vmatrix} f_1(x) & f_2(x) & \cdots & f_k(x) \\ f_1'(x) & f_2'(x) & \cdots & f_k'(x) \\ f_1''(x) & f_2''(x) & \cdots & f_k''(x) \\ \cdots & \cdots & \cdots & \cdots \\ f_1^{(k-1)}(x) & f_2^{(k-1)}(x) & \cdots & f_k^{(k-1)}(x) \end{vmatrix}$$

Theorem 3.4.4 A set of real-valued functions $f_1(x), f_2(x), \ldots, f_k(x)$, differentiable $k - 1$ times on the interval $\{x \mid a \le x \le b\}$, are independent if $W(x_0) \neq 0$ at some point x_0 in the interval.

PROOF We write $c_1 f_1(x) + c_2 f_2(x) + \cdots + c_k f_k(x) = 0$, where this is to hold everywhere in the interval. Differentiating $k - 1$ times, we have

$$\begin{aligned} c_1 f_1(x) + c_2 f_2(x) + \cdots + c_k f_k(x) &= 0 \\ c_1 f_1'(x) + c_2 f_2'(x) + \cdots + c_k f_k'(x) &= 0 \\ c_1 f_1''(x) + c_2 f_2''(x) + \cdots + c_k f_k''(x) &= 0 \\ \cdots\cdots\cdots\cdots\cdots\cdots\cdots\cdots\cdots \\ c_1 f_1^{(k-1)}(x) + c_2 f_2^{(k-1)}(x) + \cdots + c_k f_k^{(k-1)}(x) &= 0 \end{aligned}$$

If we put $x = x_0$, then we have a system of homogeneous linear equations such that the coefficient matrix $W(x_0)$ is nonsingular. This implies that $c_1 = c_2 = \cdots = c_k = 0$ is the only solution, which implies that the set of functions is independent. Another way to state this theorem is to say that if the given set of functions is dependent, then $W(x) \equiv 0$.

The converse of Theorem 3.4.4 is not true. In other words, the Wronskian of an independent set of functions may be identically zero. Consider the interval $\{x \mid -1 \le x \le 1\}$, and consider the two functions defined by

$$f(x) = \begin{cases} 0 & -1 \le x \le 0 \\ x^2 & 0 \le x \le 1 \end{cases} \qquad g(x) = \begin{cases} x^2 & -1 \le x \le 0 \\ 0 & 0 \le x \le 1 \end{cases}$$

These functions are both differentiable, but the Wronskian is given by

$$W(x) = \begin{cases} \begin{vmatrix} 0 & x^2 \\ 0 & 2x \end{vmatrix} & -1 \le x \le 0 \\[2ex] \begin{vmatrix} x^2 & 0 \\ 2x & 0 \end{vmatrix} & 0 \le x \le 1 \end{cases}$$

So $W(x) \equiv 0$, and yet the set of functions is independent.

EXAMPLE 3.4.4 Show that the functions $1, x, x^2, \ldots, x^k$ are independent in the vector space of continuous real-valued functions on the interval $\{x \mid 0 \le x \le 1\}$. We use Theorem 3.4.4. The Wronskian of the set of functions is

$$W(x) = \begin{vmatrix} 1 & x & x^2 & \cdots & x^k \\ 0 & 1 & 2x & \cdots & kx^{k-1} \\ 0 & 0 & 2! & \cdots & k(k-1)x^{k-2} \\ \cdots & \cdots & \cdots & \cdots & \cdots \\ 0 & 0 & 0 & \cdots & k! \end{vmatrix}$$

$$= (2!)(3!)\cdots(k!)$$

Since $W(x) \neq 0$, the functions are independent. We note that the result is the same no matter how large the value of k. We shall see in the next section that this implies that this space of functions is infinite-dimensional.

EXAMPLE 3.4.5 Show that the functions $\sin x, \sin 2x, \sin 3x, \ldots, \sin kx$ are independent in the vector space of continuous real-valued functions on the interval $\{x \mid 0 \le x \le 2\pi\}$. We could use the Wronskian again, but this time, because of a special property of the trigonometric functions given, there is another method available. We write

$$c_1 \sin x + c_2 \sin 2x + c_3 \sin 3x + \cdots + c_k \sin kx \equiv 0$$

Multiplying by $\sin x$ and integrating, we have

$$c_1 \int_0^{2\pi} \sin^2 x \, dx + c_2 \int_0^{2\pi} \sin x \sin 2x \, dx + \cdots + c_k \int_0^{2\pi} \sin x \sin kx \, dx = 0$$

Now

$$\int_0^{2\pi} \sin^2 x \, dx = \frac{1}{2} \int_0^{2\pi} (1 - \cos 2x) \, dx = \pi$$

On the other hand, if $n \neq m$,

$$\int_0^{2\pi} \sin nx \sin mx \, dx = \frac{1}{2} \int_0^{2\pi} [\cos (n - m)x - \cos (n + m)x] \, dx = 0$$

Therefore, all the integrals are zero except the first, and this implies that $c_1 = 0$. Similarly, multiplying by $\sin nx$, $n = 2, 3, \ldots, k$ and integrating shows that $c_2 = c_3 = \cdots = c_k = 0$. Again k is an arbitrary positive integer.

EXERCISES 3.4

1 Show that any set of vectors from a vector space V is dependent if the set contains the zero vector.

2 Show that in R^3 a set of two nonzero vectors is dependent if and only if they are parallel.

3 Show that in R^3 a set of three nonzero vectors is dependent if and only if they are all parallel to a given plane.

4 Show that in R^3 any set of three mutually perpendicular vectors is independent. This shows by Theorem 3.4.2 that any vector in R^3 can be written as a linear combination of a given set of three mutually perpendicular vectors.

5 Determine whether the vectors (1,0,1), (0,1,1), (1,−1,1) are dependent or independent in R^3. Can the vector (1,2,3) be expressed as a linear combination of these vectors?

6 Determine whether the vectors (1,−1,1,−1), (2,0,−3,1), (0,1,2,−1), (4,−3,−3,0) are dependent or independent in R^4. Can the vector (1,2,3,4) be expressed as a linear combination of these vectors?

7 Determine whether the vectors (1,0,1,0), (0,2,−1,3), (1,4,2,−1) are dependent or independent in R^4. Can the vector (4,6,7,−5) be expressed as a linear combination of these vectors?

8 Determine whether the vectors $(1,i,-1)$, $(1 + i, 0, 1 - i)$, $(i,-1,-i)$ are dependent or independent in C^3.

9 In the space of continuous real-valued functions defined on the interval $\{x \mid 0 \leq x \leq 1\}$, are the functions x, $x^2 - 1$, and $x^2 + 2x + 1$ dependent or independent?

10 In the space of real-valued polynomials of degree 3 or less, show that the polynomials

$$\begin{aligned} p_0(x) &= -\tfrac{1}{6}(x - 1)(x - 2)(x - 3) \\ p_1(x) &= \tfrac{1}{2}x(x - 2)(x - 3) \\ p_2(x) &= -\tfrac{1}{2}x(x - 1)(x - 3) \\ p_3(x) &= \tfrac{1}{6}x(x - 1)(x - 2) \end{aligned}$$

are independent. Show that any real-valued polynomial $p(x)$ of degree 3 or less can be expressed uniquely by

$$p(x) = p(0)p_0(x) + p(1)p_1(x) + p(2)p_2(x) + p(3)p_3(x)$$

11 Let A be an $n \times n$ matrix with real elements. Show that the following statements are all equivalent:

(*a*) A is nonsingular.

(*b*) $AX = 0$ has only the trivial solution.

(*c*) The columns of A are independent in R^n.

(*d*) The rows of A are independent in R^n.

12 Show that the functions $1, x + 1, x^2 + x + 1, \ldots, x^k + x^{k-1} + \cdots + x + 1$ are independent on the interval $\{x \mid 0 \leq x \leq 1\}$. Does the result depend on k?

13 Show that the functions e^x, e^{2x}, e^{3x} are independent on the interval $\{x \mid 0 \leq x \leq 1\}$.

14 Are the functions e^x, xe^x, x^2e^x dependent or independent on the interval $\{x \mid 0 \le x \le 1\}$?

15 Are the functions $\sin x$, $\cos x$, $x \sin x$, $x \cos x$ dependent or independent on the interval $\{x \mid 0 \le x \le 2\pi\}$?

16 Suppose $f(x)$ and $g(x)$ satisfy the differential equation $y'' + p(x)y = 0$ on the interval $\{x \mid 0 \le x \le 1\}$, where $p(x)$ is continuous. All functions are real-valued. Show that the Wronskian of f and g is constant. If $f(0) = 1, f'(0) = 0, g(0) = 0, g'(0) = 1$, are f and g independent? *Hint*: Compute $(f'g - g'f)'$.

3.5 BASIS AND DIMENSION

We have already seen several examples of vector spaces in which every vector can be expressed as a linear combination of some finite set of vectors. The collection of polynomials of Example 3.3.3 can all be expressed as linear combinations of the polynomials $1, x, x^2, \ldots, x^n$. Theorem 3.4.2 shows that any vector in R^n can be expressed as a linear combination of some independent set of n vectors. In Example 3.3.6, we showed that the collection of all linear combinations of a given set of vectors in V forms a subspace of V. But a subspace is a vector space, so this is another example of a vector space with such a representation. We formalize this situation by giving the following definition.

Definition 3.5.1 A given set $\mathbf{u}_1, \mathbf{u}_2, \ldots, \mathbf{u}_k$ from a vector space V is said to *span* V if every vector in V can be written as a linear combination of $\mathbf{u}_1, \mathbf{u}_2, \ldots, \mathbf{u}_k$.

Theorem 3.5.1 If V is not the zero space and is spanned by a set $\mathbf{u}_1, \mathbf{u}_2, \ldots, \mathbf{u}_k$, then there is an independent subset which also spans V.

PROOF If V is not the zero space (consisting of the zero vector only), there is at least one nonzero vector in V. Therefore, there is at least one nonzero vector in the given spanning set. Hence, there are subsets of the spanning set which are independent. Now suppose the given set $\mathbf{u}_1, \mathbf{u}_2, \ldots, \mathbf{u}_k$ is dependent. Then $c_1\mathbf{u}_1 + c_2\mathbf{u}_2 + \cdots + c_k\mathbf{u}_k = \mathbf{0}$ with the c's not all zero. Suppose $c_k \neq 0$ (if $c_k = 0$, we can relabel the vectors so that the kth scalar is different from zero). Then

$$\mathbf{u}_k = \frac{-c_1}{c_k}\mathbf{u}_1 + \frac{-c_2}{c_k}\mathbf{u}_2 + \cdots + \frac{-c_{k-1}}{c_k}\mathbf{u}_{k-1}$$

Now since any vector in V can be written as a linear combination of $\mathbf{u}_1, \mathbf{u}_2, \ldots, \mathbf{u}_k$, and since $\mathbf{u}_k$ can be written in terms of $\mathbf{u}_1, \mathbf{u}_2, \ldots, \mathbf{u}_{k-1}$, the latter set actually spans V. In this way, if the original set is dependent, we can reduce by 1 the number of spanning vectors. Next, if the set $\mathbf{u}_1, \mathbf{u}_2, \ldots, \mathbf{u}_{k-1}$ is dependent, we can proceed in the same way to reduce the number of spanning vectors by 1. This process will continue until we obtain an independent subset of spanning vectors from the original set. This subset must contain at least one vector since V is not the zero space. This completes the proof.

We have a special name for an independent spanning set.

Definition 3.5.2 If a vector space V has an independent spanning set, we call such a set a *basis* for V.

EXAMPLE 3.5.1 Show that the set of vectors

$$\begin{aligned} \mathbf{e}_1 &= (1, 0, 0, \ldots, 0) \\ \mathbf{e}_2 &= (0, 1, 0, \ldots, 0) \\ \mathbf{e}_3 &= (0, 0, 1, \ldots, 0) \\ &\cdots\cdots\cdots\cdots \\ \mathbf{e}_n &= (0, 0, 0, \ldots, 1) \end{aligned}$$

is a basis for R^n. Clearly these vectors span the space since

$$(x_1, x_2, \ldots, x_n) = x_1\mathbf{e}_1 + x_2\mathbf{e}_2 + \cdots + x_n\mathbf{e}_n$$

Also, they are independent since

$$c_1\mathbf{e}_1 + c_2\mathbf{e}_2 + \cdots + c_n\mathbf{e}_n = (c_1, c_2, \ldots, c_n) = \mathbf{0}$$

implies that $c_1 = c_2 = \cdots = c_n = 0$. This basis is called the *standard basis* for R^n.

EXAMPLE 3.5.2 Show that the set of vectors

$$\mathbf{u}_1 = (1,1,1,1) \qquad \mathbf{u}_2 = (1,-1,1,-1)$$
$$\mathbf{u}_3 = (1,2,3,4) \qquad \mathbf{u}_4 = (1,0,2,0)$$

spans R^4. Consider an arbitrary vector in R^4, $\mathbf{v} = (b_1,b_2,b_3,b_4)$. We attempt to find a linear combination of the $\mathbf{u}$'s equal to $\mathbf{v}$. Hence, we look for c_1, c_2, c_3, c_4 such that

$$c_1\mathbf{u}_1 + c_2\mathbf{u}_2 + c_3\mathbf{u}_3 + c_4\mathbf{u}_4 = \mathbf{v}$$

Representing $\mathbf{u}_1$, $\mathbf{u}_2$, $\mathbf{u}_3$, $\mathbf{u}_4$, and $\mathbf{v}$ as column matrices, we have

$$c_1\begin{pmatrix}1\\1\\1\\1\end{pmatrix}+c_2\begin{pmatrix}1\\-1\\1\\-1\end{pmatrix}+c_3\begin{pmatrix}1\\2\\3\\4\end{pmatrix}+c_4\begin{pmatrix}1\\0\\2\\0\end{pmatrix}=\begin{pmatrix}b_1\\b_2\\b_3\\b_4\end{pmatrix}$$

$$\begin{pmatrix}1 & 1 & 1 & 1\\1 & -1 & 2 & 0\\1 & 1 & 3 & 2\\1 & -1 & 4 & 0\end{pmatrix}\begin{pmatrix}c_1\\c_2\\c_3\\c_4\end{pmatrix}=\begin{pmatrix}b_1\\b_2\\b_3\\b_4\end{pmatrix}$$

If the determinant of the coefficient matrix is not zero, there will be a unique solution for the c's for a given $\mathbf{v}$.

$$\begin{vmatrix}1 & 1 & 1 & 1\\1 & -1 & 2 & 0\\1 & 1 & 3 & 2\\1 & -1 & 4 & 0\end{vmatrix}=\begin{vmatrix}1 & 1 & 1 & 1\\1 & -1 & 2 & 0\\-1 & -1 & 1 & 0\\1 & -1 & 4 & 0\end{vmatrix}=(-1)\begin{vmatrix}1 & -1 & 2\\-1 & -1 & 1\\1 & -1 & 4\end{vmatrix}$$

$$=(-1)\begin{vmatrix}1 & 0 & 2\\-1 & -2 & 1\\1 & 0 & 4\end{vmatrix}=2\begin{vmatrix}1 & 2\\1 & 4\end{vmatrix}=4$$

This shows that the set $\mathbf{u}_1$, $\mathbf{u}_2$, $\mathbf{u}_3$, $\mathbf{u}_4$ spans R^4. The fact that the coefficient matrix is nonsingular shows that the set is independent, since if $\mathbf{v} = \mathbf{0}$, there is only the trivial solution $c_1 = c_2 = c_3 = c_4 = 0$.

EXAMPLE 3.5.3 Consider the space of $m \times n$ matrices with complex elements. Show that the set of matrices E_{ij}, with 1 as the (i,j)th element and 0 everywhere else, is a basis for the space. Clearly these matrices span the space, since

$$\begin{pmatrix}a_{11} & a_{12} & \cdots & a_{1n}\\a_{21} & a_{22} & \cdots & a_{2n}\\\cdots & \cdots & \cdots & \cdots\\a_{m1} & a_{m2} & \cdots & a_{mn}\end{pmatrix}=a_{11}E_{11}+a_{12}E_{12}+\cdots+a_{mn}E_{mn}$$

Also if $c_{11}E_{11} + c_{12}E_{12} + \cdots + c_{mn}E_{mn} = 0$, then

$$\begin{pmatrix}c_{11} & c_{12} & \cdots & c_{1n}\\c_{21} & c_{22} & \cdots & c_{2n}\\\cdots & \cdots & \cdots & \cdots\\c_{m1} & c_{m2} & \cdots & c_{mn}\end{pmatrix}=\begin{pmatrix}0 & 0 & \cdots & 0\\0 & 0 & \cdots & 0\\\cdots & \cdots & \cdots & \cdots\\0 & 0 & \cdots & 0\end{pmatrix}$$

and so all the c's are zero, showing that the given set of matrices is independent.

EXAMPLE 3.5.4 Show that the polynomials $p_1(x) = \frac{1}{2}x(x - 1)$, $p_2(x) = -(x - 1)(x + 1)$, $p_3(x) = \frac{1}{2}x(x + 1)$ form a basis for the space of real-valued polynomials of degree 2 or less. We first note that $p_1(-1) = 1$, $p_1(0) = 0$, $p_1(1) = 0$, $p_2(-1) = 0$, $p_2(0) = 1$, $p_2(1) = 0$, $p_3(-1) = 0$, $p_3(0) = 0$, and $p_3(1) = 1$. Let $p(x) = a + bx + cx^2$. We wish to find scalars c_1, c_2, c_3 such that

$$c_1p_1(x) + c_2p_2(x) + c_3p_3(x) = a + bx + cx^2$$

Substituting $x = -1$, then $x = 0$, then $x = 1$, we have $c_1 = a - b + c$, $c_2 = a$, and $c_3 = a + b + c$. This shows the given set spans the space because if two quadratic polynomials agree at three distinct points, they agree everywhere. If $a = b = c = 0$, then $c_1 = 0$, $c_2 = 0$, $c_3 = 0$, showing that the set is independent.

Theorem 3.5.2 The representation of a given vector $\mathbf{v}$ in the vector space V in terms of a given basis is unique.

PROOF Let $\mathbf{u}_1, \mathbf{u}_2, \ldots, \mathbf{u}_n$ be the given basis. Let $\mathbf{v}$ be a given vector $\mathbf{v}$. Then $\mathbf{v}$ can be expressed as a linear combination

$$\mathbf{v} = c_1\mathbf{u}_1 + c_2\mathbf{u}_2 + \cdots + c_n\mathbf{u}_n$$

Suppose $\mathbf{v}$ has another representation

$$\mathbf{v} = \gamma_1\mathbf{u}_1 + \gamma_2\mathbf{u}_2 + \cdots + \gamma_n\mathbf{u}_n$$

Subtracting, we have

$$\mathbf{0} = (c_1 - \gamma_1)\mathbf{u}_1 + (c_2 - \gamma_2)\mathbf{u}_2 + \cdots + (c_n - \gamma_n)\mathbf{u}_n$$

But this implies that $c_1 - \gamma_1 = c_2 - \gamma_2 = \cdots = c_n - \gamma_n = 0$ since the set $\mathbf{u}_1, \mathbf{u}_2, \ldots, \mathbf{u}_n$ is independent.

Definition 3.5.3 If V is a vector space with a basis $\mathbf{u}_1, \mathbf{u}_2, \ldots, \mathbf{u}_n$ and $\mathbf{v}$ is a vector in V such that $\mathbf{v} = c_1\mathbf{u}_1 + c_2\mathbf{u}_2 + \cdots + c_n\mathbf{u}_n$, then c_j, $j = 1, 2, \ldots, n$, is the jth *coordinate* of $\mathbf{v}$ with respect to the given basis.

EXAMPLE 3.5.5 Find the coordinates of the vector $(-2,0,3,1)$ in R^4 with respect to the basis of Example 3.5.2. Representing the vectors as column matrices, we have

$$c_1\begin{pmatrix}1\\1\\1\\1\end{pmatrix} + c_2\begin{pmatrix}1\\-1\\1\\-1\end{pmatrix} + c_3\begin{pmatrix}1\\2\\3\\4\end{pmatrix} + c_4\begin{pmatrix}1\\0\\2\\0\end{pmatrix} = \begin{pmatrix}-2\\0\\3\\1\end{pmatrix}$$

or

$$\begin{pmatrix} 1 & 1 & 1 & 1 \\ 1 & -1 & 2 & 0 \\ 1 & 1 & 3 & 2 \\ 1 & -1 & 4 & 0 \end{pmatrix} \begin{pmatrix} c_1 \\ c_2 \\ c_3 \\ c_4 \end{pmatrix} = \begin{pmatrix} -2 \\ 0 \\ 3 \\ 1 \end{pmatrix}$$

We know that these equations have a unique solution since the coefficient matrix is nonsingular. Solving by the method of Sec. 2.3, we change the augmented matrix by basic row operations:

$$\begin{pmatrix} 1 & 1 & 1 & 1 & -2 \\ 1 & -1 & 2 & 0 & 0 \\ 1 & 1 & 3 & 2 & 3 \\ 1 & -1 & 4 & 0 & 1 \end{pmatrix}$$

$$\rightarrow \begin{pmatrix} 1 & 1 & 1 & 1 & -2 \\ 0 & -2 & 1 & -1 & 2 \\ 0 & 0 & 2 & 1 & 5 \\ 0 & -2 & 3 & -1 & 3 \end{pmatrix} \rightarrow \begin{pmatrix} 1 & 1 & 1 & 1 & -2 \\ 0 & -2 & 1 & -1 & 2 \\ 0 & 0 & 2 & 1 & 5 \\ 0 & 0 & 2 & 0 & 1 \end{pmatrix}$$

$$\rightarrow \begin{pmatrix} 1 & 1 & 1 & 1 & -2 \\ 0 & -2 & 1 & -1 & 2 \\ 0 & 0 & 2 & 1 & 5 \\ 0 & 0 & 0 & -1 & -4 \end{pmatrix} \rightarrow \begin{pmatrix} 1 & 1 & 1 & 1 & -2 \\ 0 & 1 & -\frac{1}{2} & \frac{1}{2} & -1 \\ 0 & 0 & 1 & \frac{1}{2} & \frac{5}{2} \\ 0 & 0 & 0 & 1 & 4 \end{pmatrix}$$

Hence, the coordinates are $c_1 = -15/4$, $c_2 = -11/4$, $c_3 = \frac{1}{2}$, $c_4 = 4$.

Theorem 3.5.3 If a vector space V has a basis consisting of n vectors, then any other basis will also contain n vectors.

PROOF Suppose there are two bases in V, $\mathbf{u}_1, \mathbf{u}_2, \ldots, \mathbf{u}_n$ and $\mathbf{v}_1, \mathbf{v}_2, \ldots, \mathbf{v}_m$. Now $\mathbf{v}_1$ can be expressed as a linear combination

$$\mathbf{v}_1 = c_1\mathbf{u}_1 + c_2\mathbf{u}_2 + \cdots + c_n\mathbf{u}_n$$

with at least one of the c's not zero. Suppose $c_1 \neq 0$ (if not we can relabel the $\mathbf{u}$'s). Then

$$\mathbf{u}_1 = \frac{1}{c_1}\mathbf{v}_1 + \frac{-c_2}{c_1}\mathbf{u}_2 + \cdots + \frac{-c_n}{c_1}\mathbf{u}_n$$

Therefore, in any linear combination of the $\mathbf{u}$'s used to represent a vector we can substitute for $\mathbf{u}_1$ in terms of $\mathbf{v}_1, \mathbf{u}_2, \mathbf{u}_3, \ldots, \mathbf{u}_n$. This shows that the vectors $\mathbf{v}_1, \mathbf{u}_2, \mathbf{u}_3, \ldots, \mathbf{u}_n$ span the space. Hence, $\mathbf{v}_2$ can be written as a linear combination

$$\mathbf{v}_2 = \gamma_1\mathbf{v}_1 + \gamma_2\mathbf{u}_2 + \cdots + \gamma_n\mathbf{u}_n$$

with at least one of the γ's not zero. If $\gamma_1 \neq 0$ while $\gamma_2 = \gamma_3 = \cdots = \gamma_n = 0$, then $\mathbf{v}_2 = \gamma_1\mathbf{v}_1$ and the set of v's would be dependent, contradicting the fact that the v's form a basis. Therefore, at least one of $\gamma_2, \gamma_3, \ldots, \gamma_n$ is not zero. Suppose $\gamma_2 \neq 0$, then

$$\mathbf{u}_2 = \frac{1}{\gamma_2}\mathbf{v}_2 + \frac{-\gamma_1}{\gamma_2}\mathbf{v}_1 + \frac{-\gamma_3}{\gamma_2}\mathbf{u}_3 + \cdots + \frac{-\gamma_n}{\gamma_2}\mathbf{u}_n$$

and hence, the set $\mathbf{v}_1, \mathbf{v}_2, \mathbf{u}_3, \ldots, \mathbf{u}_n$ spans the space. We continue this process. If $m > n$, we shall eventually cast out all of the u's and end up with some v expressed as a linear combination of a subset of the v's. But this is impossible since the v's are independent. Therefore, $m \leq n$. Reversing the roles of the u's and the v's in the above discussion, we prove that $n \leq m$. Hence, $m = n$, as we wished to prove.

Since the number of basis vectors (when a vector space has a basis) is a characteristic of the space which does not depend on the particular basis chosen, we can use this number to define the *dimension* of the space.

Definition 3.5.4 The zero space or any vector space with a basis is said to be finite-dimensional. The zero space has dimension zero. The dimension of any vector space with a basis is the number of basis vectors.

From this definition it is clear that the dimension of R^n is n since the standard basis consists of n vectors. The space of real-valued polynomials of degree n or less has the dimension $n + 1$. The space of $m \times n$ matrices with complex elements is mn (see Example 3.5.3). We have not mentioned the dimension of the function space of Example 3.3.4 for a very good reason; it is not finite-dimensional. This will be implied by the next theorem.

Theorem 3.5.4 In a finite-dimensional vector space of dimension n, any set of m vectors with $m > n$ is dependent.

PROOF The proof is similar to that of Theorem 3.5.3. The given vector space V has a basis $\mathbf{u}_1, \mathbf{u}_2, \ldots, \mathbf{u}_n$. Suppose the set $\mathbf{v}_1, \mathbf{v}_2, \ldots, \mathbf{v}_m$ in V is independent, where $m > n$. We can represent $\mathbf{v}_1$ as a linear combination of the u's,

$$\mathbf{v}_1 = c_1\mathbf{u}_1 + c_2\mathbf{u}_2 + \cdots + c_n\mathbf{u}_n$$

with the c's not all zero. Suppose $c_1 \neq 0$ (we can relabel if necessary). Then

$$\mathbf{u}_1 = \frac{1}{c_1}\mathbf{v}_1 + \frac{-c_2}{c_1}\mathbf{u}_2 + \cdots + \frac{-c_n}{c_1}\mathbf{u}_n$$

which shows that $\mathbf{v}_1, \mathbf{u}_2, \mathbf{u}_3, \ldots, \mathbf{u}_n$ span V. Proceeding as in the proof of Theorem 3.5.3, casting out **u**'s and replacing them with **v**'s, we eventually end up with n of the **v**'s as a spanning set. But then, since $m > n$, there are **v**'s which can be expressed as linear combinations of n of the **v**'s, contradicting the independence of the **v**'s. This completes the proof.

There are vector spaces with arbitrarily large† independent sets of vectors (see Examples 3.4.4 and 3.4.5). These vector spaces cannot be finite-dimensional. We simply say that such spaces are infinite-dimensional.

Definition 3.5.5 A vector space with independent sets of arbitrarily many vectors is said to be infinite-dimensional.

We conclude this section with a theorem which will simplify the search for bases of finite-dimensional vector spaces.

Theorem 3.5.5 In an n-dimensional vector space $(n \geq 1)$ a set of n vectors is a basis if (i) it spans the space or (ii) it is independent.

PROOF (i) If a set of vectors spans the space but is dependent, then there is a subset of m vectors which spans the space and is independent with $m < n$. But this implies that there is a basis with fewer than n vectors, contradicting Theorem 3.5.3.

(ii) If a set of n vectors $\mathbf{u}_1, \mathbf{u}_2, \ldots, \mathbf{u}_n$ is independent but does not span the space, there is at least one vector $\mathbf{v}$ which cannot be written as a linear combination of the **u**'s. Consider a linear combination

$$c\mathbf{v} + c_1\mathbf{u}_1 + c_2\mathbf{u}_2 + \cdots + c_n\mathbf{u}_n = \mathbf{0}$$

If $c \neq 0$, then $\mathbf{v}$ is a linear combination of the **u**'s. Therefore, $c = 0$. If any of $c_1, c_2, \ldots, c_n$ is not zero, then the **u**'s are dependent. Therefore, the set $\mathbf{v}, \mathbf{u}_1, \mathbf{u}_2, \ldots, \mathbf{u}_n$ is independent. But this contradicts Theorem 3.5.4. Hence, $\mathbf{u}_1, \mathbf{u}_2, \ldots, \mathbf{u}_n$ span the space.

EXERCISES 3.5

1 Determine which of the following sets of vectors, if any, is a basis for R^3:

(*a*) (1,1,1), (1,−1,1), (0,1,0)
(*b*) (1,2,3), (1,0,1), (0,−1,2)
(*c*) (0,0,1), (0,1,−1), (0,−1,1)

† Here "large" refers to the number of vectors in the set.

2 Each of the following sets of vectors spans some subspace of R^4. Find the dimension of the subspace in each case.

(*a*) (1,1,1,1), (1,0,1,0), (0,1,0,1), (1, −1,1, −1)
(*b*) (1,2,3,4), (−1,0,1,3), (0,1, −1,2), (1,2, −1,4)
(*c*) (1,2,3,0), (1,0,1,0), (0, −1,2,0), (−1,1,3,0)
(*d*) (−1,3,4,2), (1, −3, −4, −2), (−2,6,8,4), (2, −6, −8, −4)

3 Show that the vectors (1,1,1), (1, −1,1), (2,0,3) form a basis for R^3. Find the coordinates of (4,5,6) with respect to this basis.

4 The vectors (1,1,1,1), (1,0,1,0), (0,1,0,1), (1, −1,1, −1) span a subspace of R^4. Is the vector (4, −2,4, −2) in that subspace? If so, express the vector as a linear combination of the given vectors.

5 Which of the following sets of vectors is a basis for C^4?

(*a*) (i,0,0,0), (0,i,0,0), (0,0,i,0), (0,0,0,i)
(*b*) (1,0,0,0), (1,1,0,0), (1,1,1,0), (1,1,1,1)
(*c*) (1,1,1,1), (i,i,i,i), (0,1,0,1), (i,0,i,0)

6 Show that the space of differentiable real-valued functions defined on the interval $\{x \mid 0 \le x \le 1\}$ is infinite-dimensional.

7 Show that the space of Riemann-integrable real-valued functions defined on the interval $\{x \mid 0 \le x \le 1\}$ is infinite-dimensional.

8 Prove that a vector space with an infinite-dimensional subspace is infinite-dimensional. Is the converse true?

9 Show that the set of vectors (1,1,1,1), (0,1,0,1), (1,0,2,0) is independent in R^4. Construct a basis in R^4 containing the three vectors.

10 Let V be an n-dimensional vector space. Given a set of vectors $\mathbf{u}_1, \mathbf{u}_2, \ldots, \mathbf{u}_k$, $k \le n$, which are independent, prove that there is a basis for V containing the given set.

3.6 SCALAR PRODUCT

We have already seen a couple of vector spaces in which it was useful to introduce a kind of scalar-valued product between pairs of vectors. We did this in the systems of two- and three-dimensional euclidean vectors when we defined a scalar product† (dot product). The concept is, in fact, so useful that we shall now postulate a set of properties for a scalar product in general and study the properties of such a product. Then any particular vector space which has a suitable scalar product will have these additional properties. It is not necessary to have a scalar product defined in the space in order to have a vector space, but in most cases of interest to us we shall have a scalar product.

† This is not to be confused with multiplication by a scalar, where the product is between a scalar and a vector with the result a vector.

We shall give the properties of a scalar product for a complex vector space. Here the product is between a pair of vectors, and its value is a complex number. The corresponding thing for a real vector space is for the scalar product to have real values. The corresponding properties for real vector spaces can be obtained by simply removing the conjugation symbol, since for real numbers conjugation has no effect.

Definition 3.6.1 Let V be a complex vector space. Let $\mathbf{u}$ and $\mathbf{v}$ be vectors in V. We define a complex-valued product $(\mathbf{u} \cdot \mathbf{v})$ to be a scalar product if it has the following properties:

(i) $(\mathbf{u} \cdot \mathbf{v}) = \overline{(\mathbf{v} \cdot \mathbf{u})}$.
(ii) $(\mathbf{u} \cdot \mathbf{v} + \mathbf{w}) = (\mathbf{u} \cdot \mathbf{v}) + (\mathbf{u} \cdot \mathbf{w})$.
(iii) $(a\mathbf{u} \cdot \mathbf{v}) = a(\mathbf{u} \cdot \mathbf{v})$.
(iv) $(\mathbf{u} \cdot \mathbf{u}) \geq 0$.
(v) $(\mathbf{u} \cdot \mathbf{u}) = 0$ if and only if $\mathbf{u} = \mathbf{0}$.

EXAMPLE 3.6.1 Consider the complex vector space C^n of n-tuples of complex numbers. Let $\mathbf{u} = (u_1, u_2, \ldots, u_n)$ and $\mathbf{v} = (v_1, v_2, \ldots, v_n)$. We define $(\mathbf{u} \cdot \mathbf{v}) = u_1\bar{v}_1 + u_2\bar{v}_2 + \cdots + u_n\bar{v}_n$. Show that this is a scalar product. We can easily verify the five required properties:

(i):
$$\begin{aligned}\overline{(\mathbf{v} \cdot \mathbf{u})} &= \overline{v_1\bar{u}_1 + v_2\bar{u}_2 + \cdots + v_n\bar{u}_n} \\ &= \bar{v}_1 u_1 + \bar{v}_2 u_2 + \cdots + \bar{v}_n u_n \\ &= u_1\bar{v}_1 + u_2\bar{v}_2 + \cdots + u_n\bar{v}_n = (\mathbf{u} \cdot \mathbf{v})\end{aligned}$$

(ii):
$$\begin{aligned}(\mathbf{u} \cdot \mathbf{v} + \mathbf{w}) &= u_1\overline{(v_1 + w_1)} + u_2\overline{(v_2 + w_2)} + \cdots + u_n\overline{(v_n + w_n)} \\ &= u_1(\bar{v}_1 + \bar{w}_1) + u_2(\bar{v}_2 + \bar{w}_2) + \cdots + u_n(\bar{v}_n + \bar{w}_n) \\ &= u_1\bar{v}_1 + u_2\bar{v}_2 + \cdots + u_n\bar{v}_n \\ &\quad + u_1\bar{w}_1 + u_2\bar{w}_2 + \cdots + u_n\bar{w}_n \\ &= (\mathbf{u} \cdot \mathbf{v}) + (\mathbf{u} \cdot \mathbf{w})\end{aligned}$$

(iii):
$$\begin{aligned}(a\mathbf{u} \cdot \mathbf{v}) &= au_1\bar{v}_1 + au_2\bar{v}_2 + \cdots + au_n\bar{v}_n \\ &= a(u_1\bar{v}_1 + u_2\bar{v}_2 + \cdots + u_n\bar{v}_n) \\ &= a(\mathbf{u} \cdot \mathbf{v})\end{aligned}$$

(iv):
$$\begin{aligned}(\mathbf{u} \cdot \mathbf{u}) &= u_1\bar{u}_1 + u_2\bar{u}_2 + \cdots + u_n\bar{u}_n \\ &= |u_1|^2 + |u_2|^2 + \cdots + |u_n|^2 \geq 0\end{aligned}$$

(v) If $\mathbf{u} = \mathbf{0}$, $u_1 = u_2 = \cdots = u_n = 0$ and $(\mathbf{u} \cdot \mathbf{u}) = |u_1|^2 + |u_2|^2 + \cdots + |u_n|^2 = 0$. Conversely, since $|u_j|^2 \geq 0$ for $j = 1, 2, \ldots, n$ and equal to zero only if $u_j = 0$, $(\mathbf{u} \cdot \mathbf{u}) = 0$ implies $u_1 = u_2 = \cdots = u_n = 0$.

If we consider, on the other hand, R^n the space of n-tuples of real numbers, we have a real-valued scalar product $(\mathbf{u} \cdot \mathbf{v}) = u_1 v_1 + u_2 v_2 + \cdots + u_n v_n$ and the verification of the properties is exactly like Example 3.6.1, where all conjugation symbols are removed.

EXAMPLE 3.6.2 Consider the vector space of real-valued continuous functions defined on the interval $\{x \mid 0 \leq x \leq 1\}$. Define a real-valued product by

$$(f \cdot g) = \int_0^1 f(x)g(x)\,dx$$

Show that this is a scalar product. The five properties are easily verified as follows:

(i):
$$(f \cdot g) = \int_0^1 f(x)g(x)\,dx = \int_0^1 g(x)f(x)\,dx = (g \cdot f)$$

(ii):
$$\begin{aligned}(f \cdot g + h) &= \int_0^1 f(x)[g(x) + h(x)]\,dx \\ &= \int_0^1 f(x)g(x)\,dx + \int_0^1 f(x)h(x)\,dx \\ &= (f \cdot g) + (f \cdot h)\end{aligned}$$

(iii):
$$\begin{aligned}(af \cdot g) &= \int_0^1 af(x)g(x)\,dx = a\int_0^1 f(x)g(x)\,dx \\ &= a(f \cdot g)\end{aligned}$$

(iv):
$$(f \cdot f) = \int_0^1 [f(x)]^2\,dx \geq 0$$

(v) If $f(x) \equiv 0$, then $\int_0^1 [f(x)]^2\,dx = 0$. Conversely, if $\int_0^1 [f(x)]^2\,dx = 0$, then $f(x) \equiv 0$. This is because if $[f(x_0)]^2 > 0$, then there would be (by the continuity) an interval containing x_0 where $[f(x)]^2 > 0$ and hence $\int_0^1 [f(x)]^2\,dx > 0$.

EXAMPLE 3.6.3 Consider the space P_n of real-valued polynomials of degree n or less in the real variable x. If $p_n(x) = a_0 + a_1 x + a_2 x^2 + \cdots + a_n x^n$ and $q_n(x) = b_0 + b_1 x + b_2 x^2 + \cdots + b_n x^n$, then we can define the product

$$(p_n \cdot q_n) = a_0 b_0 + a_1 b_1 + \cdots + a_n b_n$$

Show that this is a scalar product. We verify the five properties as follows:

(i):
$$\begin{aligned}(p_n \cdot q_n) &= a_0b_0 + a_1b_1 + \cdots + a_nb_n \\ &= b_0a_0 + b_1a_1 + \cdots + b_na_n = (q_n \cdot p_n)\end{aligned}$$

(ii):
$$\begin{aligned}(p_n \cdot q_n + r_n) &= a_0(b_0 + c_0) \\ &\quad + a_1(b_1 + c_1) + \cdots + a_n(b_n + c_n) \\ &= a_0b_0 + a_1b_1 + \cdots + a_nb_n \\ &\quad + a_0c_0 + a_1c_1 + \cdots + a_nc_n \\ &= (p_n \cdot q_n) + (p_n \cdot r_n)\end{aligned}$$

(iii):
$$\begin{aligned}(ap_n \cdot q_n) &= (aa_0)b_0 + (aa_1)b_1 + \cdots + (aa_n)b_n \\ &= a(a_0b_0 + a_1b_1 + \cdots + a_nb_n) \\ &= a(p_n \cdot q_n)\end{aligned}$$

(iv):
$$(p_n \cdot p_n) = a_0^2 + a_1^2 + \cdots + a_n^2 \geq 0$$

(v) If $p_n(x) \equiv 0$, then $a_0 = a_1 = \cdots = a_n = 0$ and $(p_n \cdot p_n) = a_0^2 + a_1^2 + \cdots + a_n^2 = 0$. Conversely, if $a_0^2 + a_1^2 + \cdots + a_n^2 = 0$, then $a_0 = a_1 = \cdots = a_n = 0$ and $p_n(x) \equiv 0$.

We can add to the list of properties of the scalar product by proving some theorems, assuming of course that we are dealing with a complex vector space with a scalar product.

Theorem 3.6.1 $(\mathbf{u} + \mathbf{v} \cdot \mathbf{w}) = (\mathbf{u} \cdot \mathbf{w}) + (\mathbf{v} \cdot \mathbf{w})$.

PROOF Using property (i) of the definition, we have

$$(\mathbf{u} + \mathbf{v} \cdot \mathbf{w}) = \overline{(\mathbf{w} \cdot \mathbf{u} + \mathbf{v})} = \overline{(\mathbf{w} \cdot \mathbf{u})} + \overline{(\mathbf{w} \cdot \mathbf{v})} = (\mathbf{u} \cdot \mathbf{w}) + (\mathbf{v} \cdot \mathbf{w})$$

Theorem 3.6.2 $(\mathbf{u} \cdot a\mathbf{v}) = \bar{a}(\mathbf{u} \cdot \mathbf{v})$.

PROOF Using property (i) of the definition, we have

$$(\mathbf{u} \cdot a\mathbf{v}) = \overline{(a\mathbf{v} \cdot \mathbf{u})} = \overline{a(\mathbf{v} \cdot \mathbf{u})} = \bar{a}\overline{(\mathbf{v} \cdot \mathbf{u})} = \bar{a}(\mathbf{u} \cdot \mathbf{v})$$

The quantity $(\mathbf{u} \cdot \mathbf{u})$ is nonnegative and is zero if and only if $\mathbf{u} = \mathbf{0}$. Therefore, we associate with it the square of the length of the vector. In fact, we define the length (or norm) of $\mathbf{u}$ to be $(\mathbf{u} \cdot \mathbf{u})^{1/2}$ and designate it by the symbol $\|\mathbf{u}\|$. Some of the properties of the norm are given by the next theorem.

Theorem 3.6.3 If V is a vector space with a scalar product $(\mathbf{u} \cdot \mathbf{v})$ then the norm $\|\mathbf{u}\| = (\mathbf{u} \cdot \mathbf{u})^{1/2}$ has the following properties:

(i) $\|\mathbf{u}\| \geq 0$.
(ii) $\|\mathbf{u}\| = 0$ if and only if $\mathbf{u} = \mathbf{0}$.
(iii) $\|a\mathbf{u}\| = |a|\,\|\mathbf{u}\|$.
(iv) $|(\mathbf{u} \cdot \mathbf{v})| \leq \|\mathbf{u}\|\,\|\mathbf{v}\|$ (Cauchy inequality).
(v) $\|\mathbf{u} + \mathbf{v}\| \leq \|\mathbf{u}\| + \|\mathbf{v}\|$ (triangle inequality).

PROOF (i) This follows from property (iv) of the scalar product.

(ii) This follows from property (v) of the scalar product.

(iii) $\|a\mathbf{u}\|^2 = (a\mathbf{u} \cdot a\mathbf{u}) = a\bar{a}(\mathbf{u} \cdot \mathbf{u}) = |a|^2\,\|\mathbf{u}\|^2$. The result follows by taking positive square roots.

(iv) For any complex scalar a

$$\begin{aligned} 0 \leq \|\mathbf{u} + a\mathbf{v}\|^2 &= (\mathbf{u} + a\mathbf{v} \cdot \mathbf{u} + a\mathbf{v}) \\ &= (\mathbf{u} \cdot \mathbf{u}) + (a\mathbf{v} \cdot \mathbf{u}) + (\mathbf{u} \cdot a\mathbf{v}) + (a\mathbf{v} \cdot a\mathbf{v}) \\ &= \|\mathbf{u}\|^2 + a\overline{(\mathbf{u} \cdot \mathbf{v})} + \bar{a}(\mathbf{u} \cdot \mathbf{v}) + |a|^2\,\|\mathbf{v}\|^2 \end{aligned}$$

If $(\mathbf{u} \cdot \mathbf{v}) = 0$, then the inequality (iv) is satisfied. Therefore, assume $(\mathbf{u} \cdot \mathbf{v}) \neq 0$ and let

$$a = \frac{\lambda(\mathbf{u} \cdot \mathbf{v})}{|(\mathbf{u} \cdot \mathbf{v})|}$$

where λ is real. Then

$$\begin{aligned} 0 \leq \|\mathbf{u} + a\mathbf{v}\|^2 &= \|\mathbf{u}\|^2 + 2\lambda|(\mathbf{u} \cdot \mathbf{v})| + \lambda^2\|\mathbf{v}\|^2 \\ &= \alpha + 2\lambda\beta + \lambda^2\gamma \end{aligned}$$

This is a nonnegative quadratic expression in the real variable λ. Therefore, the discriminant $4\beta^2 - 4\alpha\gamma$ must be nonpositive. Therefore, since $\alpha = \|\mathbf{u}\|^2$, $\beta = |(\mathbf{u} \cdot \mathbf{v})|$, and $\gamma = \|\mathbf{v}\|^2$, we have

$$|(\mathbf{u} \cdot \mathbf{v})|^2 \leq \|\mathbf{u}\|^2\,\|\mathbf{v}\|^2$$

and the Cauchy inequality follows when we take positive square roots.

(v)
$$\begin{aligned} \|\mathbf{u} + \mathbf{v}\|^2 &= (\mathbf{u} + \mathbf{v} \cdot \mathbf{u} + \mathbf{v}) \\ &= \|\mathbf{u}\|^2 + (\mathbf{u} \cdot \mathbf{v}) + \overline{(\mathbf{u} \cdot \mathbf{v})} + \|\mathbf{v}\|^2 \\ &= \|\mathbf{u}\|^2 + 2\,\mathrm{Re}\,(\mathbf{u} \cdot \mathbf{v}) + \|\mathbf{v}\|^2 \\ &\leq \|\mathbf{u}\|^2 + 2|(\mathbf{u} \cdot \mathbf{v})| + \|\mathbf{v}\|^2 \\ &\leq \|\mathbf{u}\|^2 + 2\|\mathbf{u}\|\,\|\mathbf{v}\| + \|\mathbf{v}\|^2 \\ &\leq (\|\mathbf{u}\| + \|\mathbf{v}\|)^2 \end{aligned}$$

The triangle inequality follows when we take positive square roots.

EXAMPLE 3.6.4 Let f be a complex-valued continuous† function of the real variable x defined on the interval $\{x \mid a \le x \le b\}$. Prove that

$$\left| \int_a^b f(x)\,dx \right| \le (b - a)M, \text{ where } M = \max |f(x)|$$

We can consider f as a vector in the complex vector space of complex-valued continuous functions defined on the interval $\{x \mid a \le x \le b\}$. The reader should verify that this is a vector space. In this space we introduce the scalar product

$$(f \cdot g) = \int_a^b f(x)\overline{g(x)}\,dx$$

The reader should check the five properties. Using this scalar product and the Cauchy inequality for f and $g = 1$, we have

$$\begin{aligned} \left| \int_a^b f(x)\,dx \right| &\le \left(\int_a^b |f(x)|^2\,dx \right)^{1/2} \left(\int_a^b dx \right)^{1/2} \\ &\le [(b - a)M^2]^{1/2}(b - a)^{1/2} = (b - a)M \end{aligned}$$

It is very common to refer to vectors in a vector space as *points*. For example, in R^3 if we have a vector (x,y,z), we could consider the three numbers as the coordinates of a point in three-dimensional euclidean space. Thinking, in general, of vectors as points in a vector space V with a scalar product, we can introduce the concept of distance between two points. Let $\mathbf{u}$ and $\mathbf{v}$ be in V; then we define the distance between $\mathbf{u}$ and $\mathbf{v}$ as $\|\mathbf{u} - \mathbf{v}\|$. This distance function has the following four desirable properties:

(i) $\|\mathbf{u} - \mathbf{v}\| = \|\mathbf{v} - \mathbf{u}\|$.
(ii) $\|\mathbf{u} - \mathbf{v}\| \ge 0$.
(iii) $\|\mathbf{u} - \mathbf{v}\| = 0$ if and only if $\mathbf{u} = \mathbf{v}$.
(iv) $\|\mathbf{u} - \mathbf{v}\| \le \|\mathbf{u} - \mathbf{w}\| + \|\mathbf{w} - \mathbf{v}\|$ (triangle inequality).

These properties follow easily from Theorem 3.6.3. For example, for (i), $\|\mathbf{u} - \mathbf{v}\| = \|(-1)(\mathbf{v} - \mathbf{u})\| = |-1|\,\|\mathbf{v} - \mathbf{u}\| = \|\mathbf{v} - \mathbf{u}\|$. For (iv), we have

$$\|\mathbf{u} - \mathbf{v}\| = \|(\mathbf{u} - \mathbf{w}) + (\mathbf{w} - \mathbf{v})\| \le \|\mathbf{u} - \mathbf{w}\| + \|\mathbf{w} - \mathbf{v}\|$$

Whenever a vector space has a distance between pairs of points defined satisfying properties (i) to (iv), we say it is a *metric space*. We have therefore shown that

† Continuous here means that both real and imaginary parts are continuous functions of x. If $f(x) = u(x) + iv(x)$, where u and v are real, then

$$\int_a^b f(x)\,dx = \int_a^b u(x)\,dx + i\int_a^b v(x)\,dx$$

every vector space with a scalar product is a metric space. There are, however, distance functions which are not derivable from a scalar product (see Exercise 3.6.11). There are even vector spaces which do not have a distance function, but such discussions are beyond the scope of this book.

EXERCISES 3.6

1 Let $\mathbf{u} = (1,-2,3,0)$ and $\mathbf{v} = (-2,4,5,-1)$. Compute $(\mathbf{u} \cdot \mathbf{v})$, $(\mathbf{v} \cdot \mathbf{u})$, $(2\mathbf{u} \cdot \mathbf{v})$, and $(\mathbf{u} \cdot 4\mathbf{u} + 3\mathbf{v})$.

2 Consider the space of continuous real-valued functions defined on the interval $\{x \mid 0 \le x \le 2\pi\}$. Let $f(x) = \sin x$ and $g(x) = \cos x$. Compute $(f \cdot g)$, $\|f\|$, and $\|g\|$.

3 Show that the space of Example 3.6.4 is a complex vector space. Show that the product of this example is a scalar product. Let $f(x) = e^{ix}$ and $g(x) = e^{2ix}$. Compute $(f \cdot g)$, where $a = 0$ and $b = 2\pi$.

4 Let V be a complex vector space with scalar product $(\mathbf{u} \cdot \mathbf{v})$. Let $\|\mathbf{u}\| = (\mathbf{u} \cdot \mathbf{u})^{1/2}$ be the norm. Show that $\|\mathbf{u} - \mathbf{v}\| \ge \big| \|\mathbf{u}\| - \|\mathbf{v}\| \big|$. *Hint*: Apply the triangle inequality to $\mathbf{u} = (\mathbf{u} - \mathbf{v}) + \mathbf{v}$ and $\mathbf{v} = (\mathbf{v} - \mathbf{u}) + \mathbf{u}$.

5 Let V be a real vector space with scalar product $(\mathbf{u} \cdot \mathbf{v})$. Let $\|\mathbf{u}\| = (\mathbf{u} \cdot \mathbf{u})^{1/2}$ be the norm. Prove the pythagorean theorem: $\|\mathbf{u} + \mathbf{v}\|^2 = \|\mathbf{u}\|^2 + \|\mathbf{v}\|^2$ if and only if $(\mathbf{u} \cdot \mathbf{v}) = 0$. Why is this called the pythagorean theorem?

6 Let V be a complex vector space with scalar product $(\mathbf{u} \cdot \mathbf{v})$. Let $\|\mathbf{u}\| = (\mathbf{u} \cdot \mathbf{u})^{1/2}$ be the norm. Prove the parallelogram rule: $\|\mathbf{u} + \mathbf{v}\|^2 + \|\mathbf{u} - \mathbf{v}\|^2 = 2\|\mathbf{u}\|^2 + 2\|\mathbf{v}\|^2$. Why is this called the parallelogram rule?

7 Show that Cauchy's inequality is an equality if and only if the two vectors are proportional. *Hint*: Consider the proof for the case when the discriminant is zero.

8 Show that the triangle inequality (Theorem 3.6.3) is an equality if and only if the two vectors are proportional and the constant of proportionality is a nonnegative real number.

9 Let f be a continuous real-valued function defined on the interval $\{x \mid a \le x \le b\}$. Prove that

$$\int_a^b |f(x)|^2\,dx \le \left(\int_a^b |f(x)|\,dx\right)^{1/2} \left(\int_a^b |f(x)|^3\,dx\right)^{1/2}$$

10 Let V be the vector space of n-tuples of real numbers. If $\mathbf{u} = (u_1, u_2, \ldots, u_n)$, let $\|\mathbf{u}\|^* = |u_1| + |u_2| + \cdots + |u_n|$. Show that $\|\mathbf{u} - \mathbf{v}\|^*$ satisfies the four properties of a distance function.

11 Show that $\|\mathbf{u}\|^*$ of Exercise 10 cannot be derived from a scalar product. *Hint*: See Exercise 6.

12 Prove that a scalar product can be defined for any finite-dimensional vector space. *Hint*: If the dimension is $n \geq 1$, there is a basis $\mathbf{u}_1, \mathbf{u}_2, \ldots, \mathbf{u}_n$. Then $(\mathbf{v} \cdot \mathbf{w}) = \sum_{i=1}^{n} v_i \bar{w}_i$ is a scalar product, where v_i and w_i are coordinates with respect to the basis.

3.7 ORTHONORMAL BASES

In finite-dimensional vector spaces with a scalar product, we can select bases with special properties. These are called *orthonormal bases*, and they have many desirable properties, which we shall bring out in this section.

Definition 3.7.1 Let V be a vector space with a scalar product $(\mathbf{u} \cdot \mathbf{v})$. Two nonzero vectors are orthogonal if $(\mathbf{u} \cdot \mathbf{v}) = 0$. A vector $\mathbf{u}$ is normalized if $\|\mathbf{u}\| = 1$. A set of vectors $\mathbf{u}_1, \mathbf{u}_2, \ldots, \mathbf{u}_k$ is orthonormal if $(\mathbf{u}_i \cdot \mathbf{u}_j) = \delta_{ij}$, $i = 1, 2, \ldots, k; j = 1, 2, \ldots, k$.

Theorem 3.7.1 A set of orthonormal vectors is independent.

PROOF Let $\mathbf{u}_1, \mathbf{u}_2, \ldots, \mathbf{u}_k$ be an orthonormal set. Consider $c_1\mathbf{u}_1 + c_2\mathbf{u}_2 + \cdots + c_k\mathbf{u}_k = \mathbf{0}$. Let $1 \leq j \leq k$. Then

$$0 = (c_1\mathbf{u}_1 + c_2\mathbf{u}_2 + \cdots + c_k\mathbf{u}_k \cdot \mathbf{u}_j) = c_j$$

Theorem 3.7.2 Every finite-dimensional vector space which is not the zero space has an orthonormal basis.

PROOF If V has dimension $n > 0$, then it has a basis $\mathbf{v}_1, \mathbf{v}_2, \ldots, \mathbf{v}_n$, none of which is zero. We shall now discuss a process for constructing an orthonormal basis from a given basis. We start with $\mathbf{v}_1$. Let $\mathbf{u}_1 = \mathbf{v}_1/\|\mathbf{v}_1\|$. Then $\|\mathbf{u}_1\| = 1$. Next let

$$\mathbf{u}_2 = \frac{\mathbf{v}_2 - c_1\mathbf{u}_1}{\|\mathbf{v}_2 - c_1\mathbf{u}_1\|}$$

where $c_1 = (\mathbf{v}_2 \cdot \mathbf{u}_1)$. Then

$$(\mathbf{u}_2 \cdot \mathbf{u}_1) = \frac{(\mathbf{v}_2 \cdot \mathbf{u}_1) - c_1(\mathbf{u}_1 \cdot \mathbf{u}_1)}{\|\mathbf{v}_2 - c_1\mathbf{u}_1\|} = 0$$

and $\|\mathbf{u}_2\| = 1$. We must check that $\|\mathbf{v}_2 - c_1\mathbf{u}_1\| \neq 0$. If not, $\mathbf{v}_2$ would be a multiple of $\mathbf{v}_1$ and the $\mathbf{v}$'s would not be independent. We now have $\mathbf{u}_1$ and $\mathbf{u}_2$ orthonormal. Next we let

$$\mathbf{u}_3 = \frac{\mathbf{v}_3 - c_2\mathbf{u}_1 - c_3\mathbf{u}_2}{\|\mathbf{v}_3 - c_2\mathbf{u}_1 - c_3\mathbf{u}_2\|}$$

where $c_2 = (\mathbf{v}_3 \cdot \mathbf{u}_1)$ and $c_3 = (\mathbf{v}_3 \cdot \mathbf{u}_2)$. Then

$$(\mathbf{u}_3 \cdot \mathbf{u}_1) = \frac{(\mathbf{v}_3 \cdot \mathbf{u}_1) - c_2}{\|\mathbf{v}_3 - c_2\mathbf{u}_1 - c_3\mathbf{u}_2\|} = 0$$

$$(\mathbf{u}_3 \cdot \mathbf{u}_2) = \frac{(\mathbf{v}_3 \cdot \mathbf{u}_2) - c_3}{\|\mathbf{v}_3 - c_2\mathbf{u}_1 - c_3\mathbf{u}_2\|} = 0$$

and $\|\mathbf{u}_3\| = 1$. Again if $\|\mathbf{v}_3 - c_2\mathbf{u}_1 - c_3\mathbf{u}_2\| = 0$, then $\mathbf{v}_3$ is a linear combination of $\mathbf{v}_1$ and $\mathbf{v}_2$ and that contradicts the independence of the **v**'s. This process, which is known as the *Gram-Schmidt process*, is continued until all the **v**'s are used up and as many **u**'s are computed as there were **v**'s. The **u**'s have to form a basis because there are n of them and they are independent (see Theorem 3.5.5).

EXAMPLE 3.7.1 The standard basis is an orthonormal basis in either R^n or C^n. Recall that the standard basis is $\mathbf{e}_1 = (1, 0, 0, \ldots, 0)$, $\mathbf{e}_2 = (0, 1, 0, \ldots, 0)$, $\mathbf{e}_3 = (0, 0, 1, \ldots, 0)$, $\mathbf{e}_n = (0, 0, 0, \ldots, 1)$. Clearly $\|\mathbf{e}_j\| = 1, j = 1, 2, \ldots, n$, and $(\mathbf{e}_i \cdot \mathbf{e}_j) = \delta_{ij}, i = 1, 2, \ldots, n; j = 1, 2, \ldots, n$.

EXAMPLE 3.7.2 Consider the basis $\mathbf{v}_1 = (1, 0, 0, \ldots, 0)$, $\mathbf{v}_2 = (1, 1, 0, \ldots, 0)$, $\mathbf{v}_3 = (1, 1, 1, \ldots, 0), \ldots, \mathbf{v}_n = (1, 1, 1, \ldots, 1)$ in R^n. Construct an orthonormal basis from it by the Gram-Schmidt process. We let $\mathbf{u}_1 = \mathbf{v}_1$ and then $\|\mathbf{u}_1\| = 1$. Now

$$\mathbf{u}_2 = \frac{\mathbf{v}_2 - c_1\mathbf{u}_1}{\|\mathbf{v}_2 - c_1\mathbf{u}_1\|}$$

where $c_1 = (\mathbf{v}_2 \cdot \mathbf{u}_1) = 1$. Then

$$\begin{aligned}\mathbf{u}_2 &= (1, 1, 0, \ldots, 0) - (1, 0, 0, \ldots, 0)\\ &= (0, 1, 0, \ldots, 0)\end{aligned}$$

Next

$$\mathbf{u}_3 = \frac{\mathbf{v}_3 - c_2\mathbf{u}_1 - c_3\mathbf{u}_2}{\|\mathbf{v}_3 - c_2\mathbf{u}_1 - c_3\mathbf{u}_2\|}$$

where $c_2 = (\mathbf{v}_3 \cdot \mathbf{u}_1) = 1$ and $c_3 = (\mathbf{v}_3 \cdot \mathbf{u}_2) = 1$. Hence,

$$\begin{aligned}\mathbf{u}_3 = (1, 1, 1, 0, \ldots, 0) - (1, 0, 0, \ldots, 0) - (0, 1, 0, \ldots, 0)&\\ = (0, 0, 1, 0, \ldots, 0)&\end{aligned}$$

In this case, the process leads us back to the standard basis. However, the reader should not get the impression that in R^n (or for that matter in C^n) the Gram-Schmidt process will always yield the standard basis (see Exercise 3.7.1).

The coordinates of a vector $\mathbf{v}$ relative to an orthonormal basis are particularly easy to calculate. In fact, if $\mathbf{v} = c_1\mathbf{u}_1 + c_2\mathbf{u}_2 + \cdots + c_n\mathbf{u}_n$, where $\mathbf{u}_1, \mathbf{u}_2, \ldots, \mathbf{u}_n$ is an orthonormal basis, then

$$c_j = (\mathbf{v} \cdot \mathbf{u}_j)$$

Also, since $\mathbf{u}_j$ is a normalized vector,

$$c_j\mathbf{u}_j = (\mathbf{v} \cdot \mathbf{u}_j)\mathbf{u}_j$$

is just the projection of $\mathbf{v}$ onto $\mathbf{u}_j$, so that the vector $\mathbf{v}$ is just the sum of the projections of $\mathbf{v}$ onto the basis vectors.

Let V be any n-dimensional complex vector space with an orthonormal basis $\mathbf{u}_1, \mathbf{u}_2, \ldots, \mathbf{u}_n$. Let $\mathbf{v}$ and $\mathbf{w}$ be two vectors in V such that

$$\mathbf{v} = v_1\mathbf{u}_1 + v_2\mathbf{u}_2 + \cdots + v_n\mathbf{u}_n$$
$$\mathbf{w} = w_1\mathbf{u}_1 + w_2\mathbf{u}_2 + \cdots + w_n\mathbf{u}_n$$

so that v_j is the jth coordinate of $\mathbf{v}$ and w_j is the jth coordinate of $\mathbf{w}$ with respect to the given basis. Now let us compute the scalar product of $\mathbf{v}$ and $\mathbf{w}$.

$$\begin{aligned}(\mathbf{v} \cdot \mathbf{w}) &= (v_1\mathbf{u}_1 + v_2\mathbf{u}_2 + \cdots + v_n\mathbf{u}_n \cdot w_1\mathbf{u}_1 + w_2\mathbf{u}_2 + \cdots + w_n\mathbf{u}_n) \\ &= (v_1\mathbf{u}_1 \cdot w_1\mathbf{u}_1) + (v_2\mathbf{u}_2 \cdot w_2\mathbf{u}_2) + \cdots + (v_n\mathbf{u}_n \cdot w_n\mathbf{u}_n) \\ &= v_1\bar{w}_1 + v_2\bar{w}_2 + \cdots + v_n\bar{w}_n\end{aligned}$$

Also, since $(\mathbf{v} \cdot \mathbf{w})$ does not depend on the particular basis used, the result must be independent of the basis. Incidentally, we have also proved the following theorem.

Theorem 3.7.3 If V is an n-dimensional complex vector space ($n \geq 1$), then V has a scalar product $(\mathbf{v} \cdot \mathbf{w}) = v_1\bar{w}_1 + v_2\bar{w}_2 + \cdots + v_n\bar{w}_n$, where v_j and w_j are coordinates of $\mathbf{v}$ and $\mathbf{w}$ with respect to any orthonormal basis in V. The result also holds without the conjugation symbol for real vector spaces.

This discussion suggests that somehow C^n characterizes all n-dimensional complex vector spaces and R^n similarly characterizes all n-dimensional real vector spaces. This is indeed the case. The underlying concept is *isomorphism*, which we shall now define.

Definition 3.7.2 Let V and V^* be two complex (real) vector spaces. Then V and V^* are isomorphic ($V \leftrightarrow V^*$) if there is a one-to-one correspondence between the vectors $\mathbf{v}$ of V and $\mathbf{v}^*$ of $V^*(\mathbf{v} \leftrightarrow \mathbf{v}^*)$ such that (i) if $\mathbf{v} \leftrightarrow \mathbf{v}^*$ and $\mathbf{w} \leftrightarrow \mathbf{w}^*$, then $\mathbf{v} + \mathbf{w} \leftrightarrow \mathbf{v}^* + \mathbf{w}^*$ and (ii) if $\mathbf{v} \leftrightarrow \mathbf{v}^*$, then $a\mathbf{v} \leftrightarrow a\mathbf{v}^*$ for every complex (real) scalar.

EXAMPLE 3.7.3 Show that R^{n+1} and P_n (the space of real-valued polynomials of degree n or less in the real variable x) are isomorphic. Let $p_n(x) = a_0 + a_1x + a_2x^2 + \cdots + a_nx^n$ be in P_n. Then $(a_0, a_1, a_2 \ldots, a_n)$ is an $(n + 1)$-tuple of real numbers in R^{n+1}. Conversely, if $(b_0, b_1, b_2, \ldots, b_n)$ is an $(n + 1)$-tuple of real numbers, there is a polynomial $q_n(x) = b_0 + b_1x + b_2x^2 + \cdots + b_nx^n$ in P_n. Therefore, there is a one-to-one correspondence between the polynomials of P_n and the $(n + 1)$-tuples of R^{n+1}. Now if

$$(a_0, a_1, a_2, \ldots, a_n) \leftrightarrow p_n(x)$$
$$(b_0, b_1, b_2, \ldots, b_n) \leftrightarrow q_n(x)$$

then

$$\begin{aligned}(a_0, a_1, a_2, \ldots, a_n) &+ (b_0, b_1, b_2, \ldots, b_n)\\ &= (a_0 + b_0, a_1 + b_1, \ldots, a_n + b_n)\\ &\leftrightarrow (a_0 + b_0) + (a_1 + b_1)x + \cdots + (a_n + b_n)x^2\\ &= p_n(x) + q_n(x)\end{aligned}$$

Also

$$\begin{aligned}a(a_0, a_1, a_2, \ldots, a_n) &= (aa_0, aa_1, aa_2, \ldots, aa_n)\\ &\leftrightarrow aa_0 + aa_1x + aa_2x^2 + \cdots + aa_nx^n\\ &= ap_n(x)\end{aligned}$$

Theorem 3.7.4 Every n-dimensional complex (real) vector space V is isomorphic to C^n (R^n), $n \geq 1$.

PROOF By Theorem 3.7.2, V has an orthonormal basis $\mathbf{u}_1, \mathbf{u}_2, \ldots, \mathbf{u}_n$. We set up the correspondence $\mathbf{u}_1 \leftrightarrow \mathbf{e}_1$, $\mathbf{u}_2 \leftrightarrow \mathbf{e}_2, \ldots, \mathbf{u}_n \leftrightarrow \mathbf{e}_n$ between the $\mathbf{u}$'s and the standard basis. If $\mathbf{v}$ is in V, then it has unique coordinates $(v_1, v_2, \ldots, v_n)$ with respect to the basis $\mathbf{u}_1, \mathbf{u}_2, \ldots, \mathbf{u}_n$. The n-tuple $(v_1, v_2, \ldots, v_n)$ is in C^n (R^n). Hence, $\mathbf{v}^* = (v_1, v_2, \ldots, v_n)$ is in C^n (R^n), and we set up the correspondence $\mathbf{v} \leftrightarrow \mathbf{v}^*$. This is clearly one to one. Also, $a\mathbf{v} = av_1\mathbf{u}_1 + av_2\mathbf{u}_2 + \cdots + av_n\mathbf{u}_n \leftrightarrow (av_1, av_2, \ldots, av_n) = a\mathbf{v}^*$. If $\mathbf{w} = w_1\mathbf{u}_1 + w_2\mathbf{u}_2 + \cdots + w_n\mathbf{u}_n$ in V, then

$$\begin{aligned}\mathbf{v} + \mathbf{w} &= (v_1 + w_1)\mathbf{u}_1 + (v_2 + w_2)\mathbf{u}_2 + \cdots + (v_n + w_n)\mathbf{u}_n\\ &\leftrightarrow (v_1 + w_1, v_2 + w_2, \ldots, v_n + w_n)\\ &= \mathbf{v}^* + \mathbf{w}^*\end{aligned}$$

where $\mathbf{w}^* = (w_1, w_2, \ldots, w_n)$ is in C^n (R^n). This isomorphism also has the advantage that it preserves scalar products (see Theorem 3.7.3).

EXERCISES 3.7

1 Test the following set of vectors in R^3 for independence and construct from it an orthonormal basis: (1,0,1), (1, −1,1), (0,1,1).

2 Test the following set of vectors in R^4 for independence and construct from it an orthonormal basis: (1,0,1,0), (1, −1,0,1), (0,1, −1,1), (1, −1,1, −1).

3 Consider the space of real-valued polynomials of degree 2 or less defined on the interval $\{x \mid -1 \le x \le 1\}$. Using the scalar product $(p \cdot q) = \int_{-1}^{1} p(x)q(x)\,dx$, construct an orthonormal basis from the independent polynomials 1, x, x^2.

4 Consider the n-dimensional real vector space V. Let $\mathbf{u}_1, \mathbf{u}_2, \ldots, \mathbf{u}_n$ and $\mathbf{v}_1, \mathbf{v}_2, \ldots, \mathbf{v}_n$ be two orthonormal bases for V such that $\mathbf{v}_i = \sum_{k=1}^{n} a_{ki}\mathbf{u}_k$, $i = 1, 2, \ldots, n$. Prove that the matrix A with elements a_{ij} is orthogonal. Express the **u**'s in terms of the **v**'s.

5 Consider the n-dimensional complex vector space V. Let $\mathbf{u}_1, \mathbf{u}_2, \ldots, \mathbf{u}_n$ and $\mathbf{v}_1, \mathbf{v}_2, \ldots, \mathbf{v}_n$ be two orthonormal bases for V such that $\mathbf{v}_i = \sum_{k=1}^{n} a_{ki}\mathbf{u}_k$, $i = 1, 2, \ldots, n$. Prove that the matrix A with elements a_{ij} is unitary. Express the **u**'s in terms of the **v**'s.

6 Given two arbitrary bases $\mathbf{u}_1, \mathbf{u}_2, \ldots, \mathbf{u}_n$ and $\mathbf{v}_1, \mathbf{v}_2, \ldots, \mathbf{v}_n$ in a vector space V such that $\mathbf{v}_i = \sum_{k=1}^{n} a_{ki}\mathbf{u}_k$. Prove that the matrix A with elements a_{ij} is nonsingular. Express the **u**'s in terms of the **v**'s.

7 Consider the plane given implicitly by the equation $x + y + z = 0$ in euclidean three-dimensional space R^3. Construct an orthonormal basis as follows: select an orthonormal basis for the subspace consisting of those points in the given plane and then find a third unit vector orthogonal to the given plane.

8 Consider the subspace of R^4 spanned by the two vectors $\mathbf{u}_1 = (1,0,1,0)$ and $\mathbf{u}_2 = (1,-1,1,-1)$. Construct an orthonormal basis $\mathbf{v}_1$, $\mathbf{v}_2$ for this subspace. Now construct an orthonormal basis for R^4 containing $\mathbf{v}_1$ and $\mathbf{v}_2$.

9 Given any subspace U of dimension $m \ge 1$ in an n-dimensional vector space V ($m < n$), prove that V has an orthonormal basis consisting of m vectors in U and $n - m$ vectors orthogonal to all vectors in U.

10 Given a vector **v** in an n-dimensional vector space V and given a subspace U of dimension m ($1 \le m < n$). Prove that **v** can be expressed uniquely as $\mathbf{v} = \mathbf{u} + \mathbf{w}$, where **u** is in U and **w** is orthogonal to U (orthogonal to all vectors in U). **u** is called the *projection* of **v** on U.

11 Find the projection of (1,2,3) on the plane given implicitly by $x + y + z = 0$ (see Exercise 7).

12 Find the projection of (1,2,3,4) on the subspace of R^4 spanned by $\mathbf{u}_1 = (1,0,1,0)$ and $\mathbf{u}_2 = (1,-1,1,-1)$ (see Exercise 8).

13 Show that the space of $m \times n$ real matrices is isomorphic to R^{nm}. Exhibit a one-to-one correspondence.

14 Show that the space of $m \times n$ complex matrices is isomorphic to C^{nm}. Exhibit a one-to-one correspondence.

15 Show that the space of complex-valued polynomials in the complex variable z of degree n or less is isomorphic to C^{n+1}.

16 Prove that two finite-dimensional vector spaces which are isomorphic have the same dimension.

*3.8 INFINITE-DIMENSIONAL VECTOR SPACES

We have already established the existence of infinite-dimensional vector spaces; for example, the space of real-valued continuous functions defined on the interval $\{x \mid 0 \leq x \leq 1\}$. However, we have not had much to say about such spaces for a couple of good reasons. One is that our primary concern in this book is with finite-dimensional vector spaces. The other is that the theory of infinite-dimensional spaces is quite a bit more complicated than that for finite-dimensional spaces. This theory is properly a part of the branch of mathematics called *functional analysis*. However, it is possible to give a very brief introduction to the subject, which we propose to do in this section.

One of the easiest ways to obtain an infinite-dimensional vector space is to extend from R^n, the space of n-tuples of real numbers, to the space of infinite sequences of real numbers (infinite-tuples). Let $\mathbf{u} = (u_1, u_2, u_3, \ldots)$ and $\mathbf{v} = (v_1, v_2, v_3, \ldots)$ be infinite sequences of real numbers. We shall say that $\mathbf{u} = \mathbf{v}$ if $u_i = v_i$, for all positive integers i. We define the sum $\mathbf{u} + \mathbf{v} = (u_1 + v_1, u_2 + v_2, u_3 + v_3, \ldots)$ and multiplication by a real scalar a as $a\mathbf{u} = (au_1, au_2, au_3, \ldots)$. The zero vector we can define as $\mathbf{0} = (0, 0, 0, \ldots)$ and the negative by $-\mathbf{u} = (-u_1, -u_2, -u_3, \ldots)$. It is easy to verify that we have a real vector space. However, since we shall want to have a scalar product in this space, we shall restrict the sequences somewhat. We shall want to define the scalar product

$$(\mathbf{u} \cdot \mathbf{v}) = u_1v_1 + u_2v_2 + u_3v_3 + \cdots = \sum_{i=1}^{\infty} u_iv_i$$

and hence the norm as

$$\|\mathbf{u}\| = \left(\sum_{i=1}^{\infty} u_i^2\right)^{1/2}$$

Since we are now dealing with infinite sequences, in order to ensure convergence we shall restrict our sequences to those such that $\sum_{i=1}^{\infty} u_i^2 < \infty$. Since we have put a restriction on the sequences which we have in the space, we shall have to recheck the axioms. The only ones which can cause trouble are A1 and

M1. For A1, we have to show that if $\sum_{i=1}^{\infty} u_i^2 < \infty$ and $\sum_{i=1}^{\infty} v_i^2 < \infty$, then $\sum_{i=1}^{\infty} (u_i + v_i)^2 < \infty$. Since $0 \leq (|u_i| - |v_i|)^2 = u_i^2 + v_i^2 - 2|u_i v_i|$, we have that $2|u_i v_i| \leq u_i^2 + v_i^2$. Therefore,

$$(u_i + v_i)^2 = u_i^2 + v_i^2 + 2u_i v_i \leq 2(u_i^2 + v_i^2)$$

and

$$\sum_{i=1}^{\infty} (u_i + v_i)^2 \leq 2 \sum_{i=1}^{\infty} u_i^2 + 2 \sum_{i=1}^{\infty} v_i^2 < \infty$$

For M1, we have

$$\sum_{i=1}^{\infty} (au_i)^2 = a^2 \sum_{i=1}^{\infty} u_i^2 < \infty$$

Checking the other axioms is completely straightforward. We also have to show that the scalar product is defined for a pair of vectors in the space. We have that $|u_i v_i| \leq \frac{1}{2}(u_i^2 + v_i^2)$. Therefore,

$$\sum_{i=1}^{\infty} |u_i v_i| \leq \frac{1}{2} \sum_{i=1}^{\infty} u_i^2 + \frac{1}{2} \sum_{i=1}^{\infty} v_i^2$$

which shows that $\sum_{i=1}^{\infty} u_i v_i$ converges absolutely. The five properties of a scalar product are easily checked. Hence, we have shown that we have a real vector space with a scalar product.

Consider the infinite set of vectors $\mathbf{e}_1 = (1, 0, 0, \ldots)$, $\mathbf{e}_2 = (0, 1, 0, \ldots)$, $\mathbf{e}_3 = (0, 0, 1, 0, \ldots)$, etc. Then if $\mathbf{u} = (u_1, u_2, u_3, \ldots)$, we have that

$$\mathbf{u} = u_1\mathbf{e}_1 + u_2\mathbf{e}_2 + u_3\mathbf{e}_3 + \cdots = \sum_{i=1}^{\infty} u_i\mathbf{e}_i$$

This is an infinite series of vectors, so we must define what we mean by convergence of such a series. Let $\mathbf{u}_n$ be the vector of partial sums $\mathbf{u}_n = \sum_{i=1}^{n} u_i\mathbf{e}_i$. Then

$$\|\mathbf{u}_n - \mathbf{u}\| = \left(\sum_{i=n+1}^{\infty} u_i^2 \right)^{1/2} \to 0$$

as $n \to \infty$, because the series $\sum_{i=1}^{n} u_i^2$ converges.

Definition 3.8.1 Let V be an infinite-dimensional vector space with a norm.† Then a sequence of vectors $\{\mathbf{u}_n\}$, $n = 1, 2, 3, \ldots$, converges to $\mathbf{u}$ if $\|\mathbf{u}_n - \mathbf{u}\| \to 0$ as $n \to \infty$. An infinite series $\sum_{i=1}^{\infty} \mathbf{v}_i$ converges to $\mathbf{v}$ if the sequence of partial sums $\sum_{i=1}^{\infty} \mathbf{v}_i$ converges to $\mathbf{v}$.

† The norm is to have properties (i), (ii), (iii), and (v) of Theorem 3.6.3.

In the above example, the vectors $\mathbf{e}_1, \mathbf{e}_2, \mathbf{e}_3, \ldots$ are orthonormal because $(\mathbf{e}_i \cdot \mathbf{e}_j) = \delta_{ij}$. The coordinate of $\mathbf{u}$ with respect to $\mathbf{e}_i$ is $u_i = (\mathbf{u} \cdot \mathbf{e}_i)$, and the series $\sum_{i=1}^{\infty} u_i \mathbf{e}_i$ converges to $\mathbf{u}$ for all $\mathbf{u}$. When these conditions all hold, we say that we have an *orthonormal basis*.†

Definition 3.8.2 Let V be an infinite-dimensional vector space with a scalar product. Then $\mathbf{v}_1, \mathbf{v}_2, \mathbf{v}_3, \ldots$ is an orthonormal basis for V if (i) $(\mathbf{v}_i \cdot \mathbf{v}_j) = \delta_{ij}$, $i = 1, 2, 3, \ldots$; $j = 1, 2, 3, \ldots$, (ii) the series $\sum_{i=1}^{\infty} u_i \mathbf{v}_i$ converges to $\mathbf{u}$ for all $\mathbf{u}$ in V, where $u_i = (\mathbf{u} \cdot \mathbf{v}_i)$ is the coordinate of $\mathbf{u}$ with respect to $\mathbf{v}_i$.

Definition 3.8.3 Let V be an infinite-dimensional vector space with a norm. A sequence of vectors $\{\mathbf{u}_n\}$, $n = 1, 2, 3, \ldots$, is a Cauchy sequence if $\lim_{\substack{n\to\infty \\ m\to\infty}} \|\mathbf{u}_n - \mathbf{u}_m\| = 0$; alternatively, given any $\varepsilon > 0$, there is an N such that $\|\mathbf{u}_n - \mathbf{u}_m\| < \varepsilon$ when $n > N$ and $m > N$.

Theorem 3.8.1 Let V be an infinite-dimensional vector space with a norm. If a sequence of vectors $\{\mathbf{u}_n\}$, $n = 1, 2, 3, \ldots$, converges to $\mathbf{u}$ in V, then the sequence is a Cauchy sequence.

PROOF We have for arbitrary $\varepsilon > 0$ an N such that $\|\mathbf{u}_n - \mathbf{u}\| < \frac{1}{2}\varepsilon$ for $n > N$. Therefore,

$$\begin{aligned} \|\mathbf{u}_n - \mathbf{u}_m\| &= \|\mathbf{u}_n - \mathbf{u} + \mathbf{u} - \mathbf{u}_m\| \\ &\le \|\mathbf{u}_n - \mathbf{u}\| + \|\mathbf{u}_m - \mathbf{u}\| < \varepsilon \end{aligned}$$

for $n > N$ and $m > N$.

The converse of this theorem is not, in general, true. That is, we may have a Cauchy sequence which does not converge to a vector in the space. For example, if in the example of infinite sequences of real numbers we restrict our space to infinite sequences of rational numbers, then we shall still have a vector space with all the properties we have listed so far. However, it will now be possible to have Cauchy sequences which do not converge to sequences of rational numbers. Suppose $\{\mathbf{r}^{(n)}\}$, $n = 1, 2, 3, \ldots$, is the Cauchy sequence with $\mathbf{r}^{(n)} = (r_1^{(n)}, r_2^{(n)}, r_3^{(n)}, \ldots)$. Then

$$|r_i^{(n)} - r_i^{(m)}| \le \|\mathbf{r}^{(n)} - \mathbf{r}^{(m)}\| \to 0$$

† It is possible to define more general bases, but for the sake of brevity we shall restrict our attention to orthonormal bases.

as $n \to \infty$ and $m \to \infty$ for each i. Therefore, $\{r_i^{(n)}\}$ is a Cauchy sequence of rationals for each coordinate. But there are Cauchy sequences of rationals which converge to a real number, *not a rational.*

There are vector spaces in which every Cauchy sequence converges to a vector in the space. Such spaces are called *complete spaces.*

Definition 3.8.4 Let V be a vector space with a norm. Then V is a complete space if every Cauchy sequence in V converges to a vector in V. A complete normed vector space is called a *Banach space.* If the norm is derived from a scalar product, the space is called a *Hilbert space.*

Theorem 3.8.2 The space of infinite sequences of real numbers, that is, $\mathbf{u} = (u_1, u_2, u_3, \ldots)$, where $\sum_{i=1}^{\infty} u_i^2 < \infty$, with scalar product $(\mathbf{u} \cdot \mathbf{v}) = \sum_{i=1}^{\infty} u_i v_i$ is complete.

PROOF Let $\{\mathbf{u}^{(n)}\}$ be a Cauchy sequence such that

$$\mathbf{u}^{(n)} = (u_1^{(n)}, u_2^{(n)}, u_3^{(n)}, \ldots)$$

Then $|u_i^{(n)} - u_i^{(m)}| \leq \|\mathbf{u}^{(n)} - \mathbf{u}^{(m)}\| \to 0$ as $n \to \infty$ and $m \to \infty$. Therefore, each coordinate sequence $\{u_i^{(n)}\}$ is a Cauchy sequence of real numbers. It is a well-known property of real numbers that a Cauchy sequence converges to a unique real number. Therefore, we can assume that

$$\lim_{n \to \infty} u_i^{(n)} = u_i$$

for each i. We now define $\mathbf{u} = (u_1, u_2, u_3, \ldots)$ and prove that $\mathbf{u}$ is in the space and that $\{\mathbf{u}^{(n)}\}$ converges to $\mathbf{u}$. For some fixed M consider $\sum_{i=1}^{M} (u_i - u_i^{(n)})^2$. Then

$$\begin{aligned}\sum_{i=1}^{M} (u_i - u_i^{(n)})^2 &= \sum_{i=1}^{M} (u_i - u_i^{(m)} + u_i^{(m)} - u_i^{(n)})^2 \\ &\leq 2 \sum_{i=1}^{M} (u_i - u_i^{(m)})^2 + 2 \sum_{i=1}^{M} (u_i^{(m)} - u_i^{(n)})^2\end{aligned}$$

We can find an N such that $\sum_{i=1}^{M} (u_i^{(m)} - u_i^{(n)})^2 < \frac{1}{4}\varepsilon^2$ and $(u_i - u_i^{(m)})^2 < \varepsilon^2/4M$ for $n > N$ and some $m > N$. Then $\sum_{i=1}^{M} (u_i - u_i^{(n)})^2 < \varepsilon^2$. This is possible for arbitrary M. Letting $M \to \infty$, we have

$$\|\mathbf{u} - \mathbf{u}^{(n)}\|^2 = \sum_{i=1}^{\infty} (u_i - u_i^{(n)})^2 \leq \varepsilon^2$$

when $n > N$. This shows that $\{\mathbf{u}^{(n)}\}$ converges to $\mathbf{u}$. Finally,

$$\sum_{i=1}^{\infty} u_i^2 \leq 2 \sum_{i=1}^{\infty} (u_i - u_i^{(n)})^2 + 2 \sum_{i=1}^{\infty} (u_i^{(n)})^2 < \infty$$

which shows that $\mathbf{u}$ is in the space.

Now let us consider some infinite-dimensional function spaces. We can find infinite sets of orthonormal functions.

EXAMPLE 3.8.1 Find a set of polynomials which are orthonormal in the space of continuous real-valued functions on the interval $\{x \mid -1 \leq x \leq 1\}$. The scalar product is similar to the one introduced in Example 3.6.2, $(f \cdot g) = \int_{-1}^{1} f(x)g(x)\,dx$. Starting with a constant function $\phi_1(x) = c$, we determine c so that $\int_{-1}^{1} [\phi_1(x)]^2\,dx = 1$. The result is $c = 1/\sqrt{2}$. Next we take a linear function $\phi_2(x) = ax + b$ and determine a and b from the two conditions $\int_{-1}^{1} [\phi_2(x)]^2\,dx = 1$ and $\int_{-1}^{1} \phi_1(x)\phi_2(x)\,dx = 0$. We have

$$\int_{-1}^{1} \frac{ax + b}{\sqrt{2}}\,dx = \sqrt{2}\,b = 0 \qquad \text{and} \qquad \int_{-1}^{1} a^2x^2\,dx = \frac{2a^2}{3} = 1$$

Hence, $a = \sqrt{3}/\sqrt{2}$. Next we take a quadratic function $\phi_3(x) = \alpha x^2 + \beta x + \gamma$ and determine the constants α, β, and γ from the three conditions

$$\int_{-1}^{1} [\phi_3(x)]^2\,dx = 1 \qquad \int_{-1}^{1} \phi_1(x)\phi_3(x)\,dx = 0$$

and

$$\int_{-1}^{1} \phi_2(x)\phi_3(x)\,dx = 0$$

We have

$$\int_{-1}^{1} \frac{\alpha x^2 + \beta x + \gamma}{\sqrt{2}}\,dx = \frac{\sqrt{2}\,\alpha}{3} + \sqrt{2}\,\gamma = 0$$

and

$$\int_{-1}^{1} \frac{\sqrt{3}}{\sqrt{2}}\,x(\alpha x^2 + \beta x + \gamma)\,dx = \frac{\sqrt{2}}{\sqrt{3}}\,\beta = 0$$

So $\alpha = -3\gamma$, and $\gamma^2 \int_{-1}^{1} (3x^2 - 1)^2\,dx = \frac{8}{5}\gamma^2 = 1$. Therefore, $\gamma = \sqrt{5}/(2\sqrt{2})$. The first three polynomials are then

$$\phi_1(x) = \frac{1}{\sqrt{2}} \qquad \phi_2(x) = \frac{\sqrt{3}}{\sqrt{2}}\,x \qquad \phi_3(x) = \frac{\sqrt{5}}{\sqrt{2}}\,\frac{3x^2 - 1}{2}$$

This process can be continued indefinitely. At the nth step there are n constants to determine from $n - 1$ orthogonality conditions plus a normalization condition. The general polynomial is

$$\phi_n(x) = \sqrt{\frac{2n-1}{2}}\, P_{n-1}(x)$$

where $P_n(x)$ is the Legendre polynomial† given by

$$P_n(x) = \frac{1}{2^n n!} \frac{d^n}{dx^n} (x^2 - 1)^n$$

$n = 0, 1, 2, \ldots$.

EXAMPLE 3.8.2 Show that set of functions $1/\sqrt{2\pi}$, $(1/\sqrt{\pi}) \cos x$, $(1/\sqrt{\pi}) \sin x$, $(1/\sqrt{\pi}) \cos 2x$, $(1/\sqrt{\pi}) \sin 2x, \ldots$ is orthonormal on the interval $\{x \mid 0 \le x \le 2\pi\}$. We first check the normalization:

$$\int_0^{2\pi} \frac{1}{2\pi}\, dx = 1$$

$$\frac{1}{\pi} \int_0^{2\pi} \cos^2 nx\, dx = \frac{1}{2\pi} \int_0^{2\pi} (1 + \cos 2nx)\, dx = 1$$

$$\frac{1}{\pi} \int_0^{2\pi} \sin^2 nx\, dx = \frac{1}{2\pi} \int_0^{2\pi} (1 - \cos 2nx)\, dx = 1$$

Next we check the orthogonality. If $n \neq m$,

$$\int_0^{2\pi} \cos nx \cos mx\, dx = \frac{1}{2} \int_0^{2\pi} [\cos (n + m)x + \cos (n - m)x]\, dx = 0$$

$$\int_0^{2\pi} \sin nx \sin mx\, dx = \frac{1}{2} \int_0^{2\pi} [\cos (n - m)x - \cos (n + m)x]\, dx = 0$$

$$\int_0^{2\pi} \cos nx \sin mx\, dx = \frac{1}{2} \int_0^{2\pi} [\sin (n + m)x - \sin (n - m)x]\, dx = 0$$

$$\int_0^{2\pi} \sin nx \cos nx\, dx = \frac{1}{2} \int_0^{2\pi} \sin 2nx\, dx = 0$$

† See J. W. Dettman, "Mathematical Methods in Physics and Engineering," 2d ed., p. 202, McGraw-Hill, New York, 1969.

Consider the space of continuous real-valued functions on the interval $\{x \mid a \le x \le b\}$. Suppose the set of functions $\phi_1(x), \phi_2(x), \phi_3(x), \ldots$ is orthonormal and $f(x)$ is any function in the space. Consider the integral

$$\int_a^b \left[f(x) - \sum_{i=1}^n c_i\phi_i(x) \right]^2 dx$$

where c_i is the coordinate of f with respect to ϕ_i; that is,

$$c_i = \int_a^b f(x)\phi_i(x)\, dx$$

Then we have

$$0 \le \int_a^b \left[f(x) - \sum_{i=1}^n c_i\phi_i(x) \right] \left[f(x) - \sum_{j=1}^n c_j\phi_j(x) \right] dx$$

$$0 \le \int_a^b [f(x)]^2\, dx - 2\int_a^b f(x) \sum_{i=1}^n c_i\phi_i(x)\, dx + \sum_{i=1}^n \sum_{j=1}^n c_i c_j \int_a^b \phi_i(x)\phi_j(x)\, dx$$

$$0 \le \int_a^b [f(x)]^2\, dx - 2\sum_{i=1}^n c_i^2 + \sum_{i=1}^n \sum_{j=1}^n c_i c_j \delta_{ij}$$

$$0 \le \int_a^b [f(x)]^2\, dx - \sum_{i=1}^n c_i^2$$

Therefore, $\sum_{i=1}^n c_i^2 \le \int_a^b [f(x)]^2\, dx$ and this holds for arbitrary n. Letting $n \to \infty$, we obtain *Bessel's inequality*

$$\sum_{i=1}^\infty c_i^2 \le \int_a^b [f(x)]^2\, dx$$

This shows that the series $\sum_{i=1}^\infty c_i^2$ converges, which in turn implies that $\lim_{n\to\infty} c_n = 0$. This does not imply, however, that the series converges to $\int_a^b [f(x)]^2\, dx$. If it were true that for *all* $f(x)$ in the space $\sum_{i=1}^\infty c_i^2 = \int_a^b [f(x)]^2\, dx$, then we would have that

$$\lim_{n\to\infty} \left\| f - \sum_{i=1}^n c_i\phi_i \right\|^2 = \lim_{n\to\infty} \int_a^b \left[f(x) - \sum_{i=1}^n c_i\phi_i(x) \right]^2 dx = 0$$

and this would imply that the set of functions $\phi_1(x), \phi_2(x), \phi_3(x), \ldots$ is an orthonormal basis for the space.

Whether a given orthonormal set of functions is an orthonormal basis depends on the choice of the set and the function space being considered. The sets of functions of Examples 3.8.1 and 3.8.2 are orthonormal bases for the space of continuous real-valued functions defined on the appropriate interval. This, however, is not easy to prove. Usually it is better to consider the given space as a subspace of a larger complete space.†

Theorem 3.8.3 Suppose V is an infinite-dimensional space of real-valued functions defined on the interval $\{x \mid a \le x \le b\}$, complete with respect to the norm derived from the scalar product $(f \cdot g) = \int_a^b f(x)g(x)\,dx$. Let $\phi_1(x), \phi_2(x), \phi_3(x), \ldots$ be an orthonormal set in V. Let $\{c_i\}$, $i = 1, 2, 3, \ldots$, be a sequence of real numbers such that $\sum_{i=1}^{\infty} c_i^2 < \infty$. Then $\sum_{i=1}^{\infty} c_i\phi_i(x)$ converges‡ to a function $f(x)$ in V such that $c_i = \int_a^b f(x)\phi_i(x)\,dx$.

PROOF Let $f_n(x) = \sum_{i=1}^{n} c_i\phi_i(x)$. Then if $n > m$,

$$\|f_n(x) - f_m(x)\|^2 = \int_a^b \left[\sum_{i=m+1}^{n} c_i\phi_i(x)\right]^2 dx$$

$$= \sum_{i=m+1}^{n} c_i^2 \to 0$$

as $n,m \to \infty$, since the series $\sum_{i=1}^{\infty} c_i^2 < \infty$. Therefore, $\{f_n(x)\}$ is a Cauchy sequence, and since V is complete, $f_n(x)$ converges to a function $f(x)$ in V. Hence, $\|f(x) - f_n(x)\| \to 0$ as $n \to \infty$, and by the Cauchy inequality

$$|(f - f_n \cdot \phi_k)| \le \|f - f_n\| \to 0$$

so that

$$c_k = \lim_{n\to\infty} (f_n \cdot \phi_k) = (f \cdot \phi_k) = \int_a^b f(x)\phi_k(x)\,dx$$

This completes the proof.

Theorem 3.8.3 does not say that an arbitrary orthonormal set in a complete space is a basis for that space. Consider an orthonormal set in a complete space (Hilbert space), $\phi_1(x), \phi_2(x), \phi_3(x), \ldots$. If we delete $\phi_1(x)$, the set $\phi_2(x), \phi_3(x), \phi_4(x), \ldots$ is still orthonormal. However, for $k = 2, 3, 4, \ldots$

$$c_k = (\phi_1 \cdot \phi_k) = 0$$

† The space of continuous real-valued functions defined on the interval $\{x \mid a \le x \le b\}$ is not complete. It is possible to find Cauchy sequences of continuous functions which do not converge to continuous functions.

‡ Convergence here means in the sense of Definition 3.8.1.

Now the function $0 = \sum_{k=2}^{\infty} c_k\phi_k(x)$ is in the space, but $f_n(x) = \sum_{k=2}^{n} c_k\phi_k(x)$ does not converge to $\phi_1(x)$. Therefore, the orthonormal set $\phi_2(x), \phi_3(x), \phi_4(x), \ldots$ cannot be a basis. However, there is an alternative characterization of an orthonormal basis.

Theorem 3.8.4 Let V be a Hilbert space. Then the orthonormal set $\phi_1, \phi_2, \phi_3, \ldots$ in V is an orthonormal basis if and only if there is no nonzero vector in V orthogonal to every member of the set.

PROOF Suppose $\|f\| > 0$ and $(f \cdot \phi_i) = c_i = 0$ for $i = 1, 2, 3, \ldots$. Now

$$\lim_{n\to\infty} \left\| f - \sum_{i=1}^{n} c_i\phi_i \right\|^2 = \|f\| \neq 0$$

showing that the orthonormal set is not a basis. Conversely, suppose the orthonormal set is not a basis. Then there is a vector f such that

$$\lim_{n\to\infty} \left\| f - \sum_{i=1}^{n} c_i\phi_i \right\|^2 = \|f\|^2 - \sum_{i=1}^{\infty} c_i^2 > 0$$

where $c_i = (f \cdot \phi_i)$. However, the sequence $g_n = \sum_{i=1}^{n} c_i\phi_i$ is a Cauchy sequence and, since the space is complete, converges to a vector g in V. Now consider the vector $h = g - f$, which is orthogonal to the set $\phi_1, \phi_2, \phi_3, \ldots$ since

$$(h \cdot \phi_i) = (g \cdot \phi_i) - (f \cdot \phi_i) = c_i - c_i = 0$$

However, $\|h\| = \|g - f\| \geq \big| \|g - g_n\| - \|f - g_n\| \big| > 0$ since $\|g - g_n\| \to 0$ and $\lim_{n\to\infty} \|f - g_n\| > 0$. This completes the proof.

In Theorem 3.8.3, we did not state which definition of the integral we were using. As a matter of fact, the Riemann integral is not good enough since the space of Riemann-integrable functions is not complete. In order for the theorem to be meaningful, we would have to use the Lebesgue definition of the integral. Since the constant function $g = 1$ is in the space, we require that $(f \cdot 1) = \int_a^b f(x)\,dx$ exists as a Lebesgue integral for each f in the space. Also we require that $\int_a^b [f(x)]^2\,dx$ exists as a Lebesgue integral. Therefore, the proper setting for the theorem is the space of Lebesgue square integrable functions, $L_2(a,b)$. A famous theorem of analysis, the Riesz-Fischer theorem, asserts that $L_2(a,b)$ is a Hilbert space. If an orthonormal set is a basis for $L_2(a,b)$, then it is also a basis for any subspace of $L_2(a,b)$, say the space of continuous functions defined on the interval $\{x \mid a \leq x \leq b\}$.

EXERCISES 3.8

1 Consider the space of infinite-tuples of complex numbers as an extension of C^n. If $\mathbf{u} = (u_1, u_2, u_3, \ldots)$ and $\mathbf{v} = (v_1, v_2, v_3, \ldots)$, then define $\mathbf{u} + \mathbf{v} = (u_1 + v_1, u_2 + v_2, u_3 + v_3, \ldots)$. Also define $a\mathbf{u} = (au_1, au_2, au_3, \ldots)$ where a is a complex scalar. Prove that this is a complex vector space. Restrict the space to those sequences such that $\sum_{i=1}^{\infty} |u_i|^2 < \infty$. Prove that the restricted space is a complex vector space with a scalar product $(\mathbf{u} \cdot \mathbf{v}) = \sum_{i=1}^{\infty} u_i \bar{v}_i$.

2 Prove that the restricted space of Exercise 1 is complete.

3 Prove that the limit of a sequence (if it exists) in a normed vector space is unique.

4 Prove that R^n is complete.

5 Prove that C^n is complete.

6 Prove that any finite-dimensional vector space is complete. *Hint*: Make use of the isomorphism with R^n or C^n.

7 Consider the space of infinite-tuples of complex numbers, $\mathbf{u} = (u_1, u_2, u_3, \ldots)$ such that $\sum_{i=1}^{\infty} |u_i| < \infty$. Show that this is normed vector space with norm defined by $\|\mathbf{u}\| = \sum_{i=1}^{\infty} |u_i|$. Prove that the space is complete with respect to this norm.

8 Consider the space of continuous real-valued functions defined on the interval $\{x \mid a \le x \le b\}$, with the norm $\|f\| = \max_{a \le x \le b} |f(x)|$. Prove that the space is complete.

9 Construct the first three of a set of orthonormal polynomials in the space of continuous real-valued functions defined on the interval $\{x \mid 0 \le x \le 1\}$ with the scalar product

$$(f \cdot g) = \int_0^1 f(x)g(x)\,dx$$

10 Let V be a complex Hilbert space with orthonormal basis $\phi_1, \phi_2, \phi_3, \ldots$. If $u_i = (f \cdot \phi_i)$ and $v_i = (g \cdot \phi_i)$, then show that $(f \cdot g) = \sum_{i=1}^{\infty} u_i \bar{v}_i$. *Hint*:

$$(f \cdot g) - \sum_{i=1}^{n} u_i \bar{v}_i = (f \cdot g - \sum_{i=1}^{n} v_i \phi_i)$$

11 Starting from the formula

$$P_n(x) = \frac{1}{2^n n!} \frac{d^n}{dx^n} (x^2 - 1)^n$$

prove that

$$\int_{-1}^{1} P_n(x) P_m(x)\,dx = \delta_{nm} \frac{2}{2n + 1}$$

12 Prove that $(1/\sqrt{\pi}) \sin nx$, $n = 1, 2, 3, \ldots$, is not an orthonormal basis for the space of continuous real-valued functions defined on the interval $\{x \mid 0 \le x \le 2\pi\}$.

4

LINEAR TRANSFORMATIONS

4.1 INTRODUCTION

A large part of linear algebra is the study of linear transformations from one vector space to another. We begin the discussion with some concrete examples. The next section takes up the fundamental theorem about representing a linear transformation in terms of a matrix. Then we consider how the representation depends on the bases used in the domain and range space of the linear transformation. The notion of change of basis then leads to a discussion of similarity and diagonalization of matrices. This will get us into a discussion of characteristic values and characteristic vectors. Not all matrices are similar to diagonal matrices. We shall prove some theorems which will tell us when they are. This will include a discussion of symmetric and hermitian matrices. The last section will take up (without complete proof) the Jordan form, which is the "best you can do" in the general case when a matrix is not similar to a diagonal matrix. This will be extremely important later, when we are discussing systems of differential equations.

4.2 DEFINITIONS AND EXAMPLES

A linear transformation is a linear function from a vector space U to a vector space V. To be more specific we give the following definitions.

Definition 4.2.1 Suppose that for each vector $\mathbf{u}$ in some subset of a vector space U (domain space of f) there is unambiguously defined a vector $\mathbf{v}$ in a vector space V (range space of f). Then we have a function f defined from U to V such that $\mathbf{v} = f(\mathbf{u})$. The subset of vectors $\mathbf{u}$ in U, on which the function is defined, is called the *domain* of f. The subset of vectors $\mathbf{v}$ in V, which are values of the function, is called the *range* of f.

Definition 4.2.2 A function f from U to V is a linear transformation if:

(i) U and V are both real vector spaces or both complex vector spaces.
(ii) The domain of f is all of U.
(iii) For all scalars a and b and all vectors $\mathbf{u}_1$ and $\mathbf{u}_2$ in U,

$$f(a\mathbf{u}_1 + b\mathbf{u}_2) = af(\mathbf{u}_1) + bf(\mathbf{u}_2)$$

EXAMPLE 4.2.1 Let U be C^n and V be C^m, and let $f(\mathbf{u}) = \mathbf{0}$ for all $\mathbf{u}$ in U. Then f is a linear transformation. The domain space is C^n, and the range space is C^m, both complex vector spaces. The domain is all of U, and

$$\mathbf{0} = f(a\mathbf{u}_1 + b\mathbf{u}_2) = a\mathbf{0} + b\mathbf{0} = af(\mathbf{u}_1) + bf(\mathbf{u}_2)$$

The range of f is the zero subspace of V.

EXAMPLE 4.2.2 Let U be C^n and V be C^n, and let $f(\mathbf{u}) = \mathbf{u}$ for all $\mathbf{u}$ in U. Then f is a linear transformation. The domain and range spaces are both C^n. The domain is all of U, and $f(a\mathbf{u}_1 + b\mathbf{u}_2) = a\mathbf{u}_1 + b\mathbf{u}_2 = af(\mathbf{u}_1) + bf(\mathbf{u}_2)$. The range of f is V.

EXAMPLE 4.2.3 Let U be R^2 and V be R^2, and let the value of $f(\mathbf{u})$ be the vector which is obtained by rotating $\mathbf{u}$ through an angle θ in the counterclockwise direction. We can show that f is a linear transformation by arguing either geometrically or algebraically. Consider Fig. 22. The geometric argument is simply this: if $\mathbf{u}_1$ and $\mathbf{u}_2$ are rotated through an angle θ, then so are $a\mathbf{u}_1$ and $b\mathbf{u}_2$.

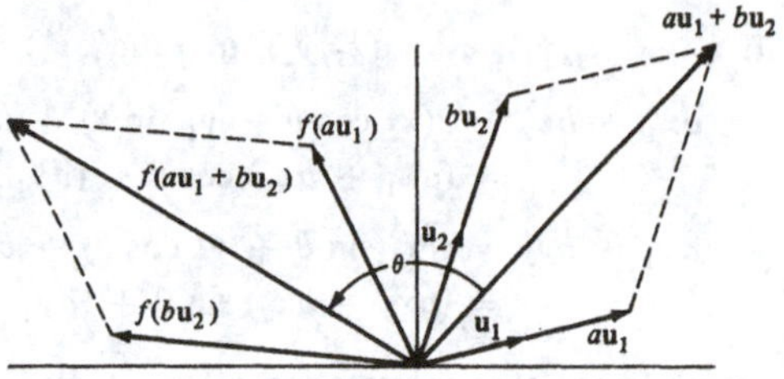

FIGURE 22

Therefore, the parallelogram formed by $a\mathbf{u}_1$ and $b\mathbf{u}_2$ is rotated through the angle θ, and so is the diagonal $a\mathbf{u}_1 + b\mathbf{u}_2$ of that parallelogram. Hence,

$$f(a\mathbf{u}_1 + b\mathbf{u}_2) = af(\mathbf{u}_1) + bf(\mathbf{u}_2)$$

since the left-hand side is the rotation of $a\mathbf{u}_1 + b\mathbf{u}_2$ and the right-hand side is the sum of the rotations of $a\mathbf{u}_1$ and $b\mathbf{u}_2$. The algebraic argument is the following (see Fig. 23). Let $\mathbf{u} = (x,y)$ and $\mathbf{u}' = (x',y')$, where $\mathbf{u}'$ is the rotation of $\mathbf{u}$ through the angle θ in the counterclockwise direction. Then

$$x = \|\mathbf{u}\| \cos \alpha$$

$$y = \|\mathbf{u}\| \sin \alpha$$

$$\begin{aligned} x' &= \|\mathbf{u}\| \cos (\theta + \alpha) = \|\mathbf{u}\| \cos \alpha \cos \theta - \|\mathbf{u}\| \sin \alpha \sin \theta \\ &= x \cos \theta - y \sin \theta \end{aligned}$$

$$\begin{aligned} y' &= \|\mathbf{u}\| \sin (\theta + \alpha) = \|\mathbf{u}\| \cos \alpha \sin \theta + \|\mathbf{u}\| \sin \alpha \cos \theta \\ &= x \sin \theta + y \cos \theta \end{aligned}$$

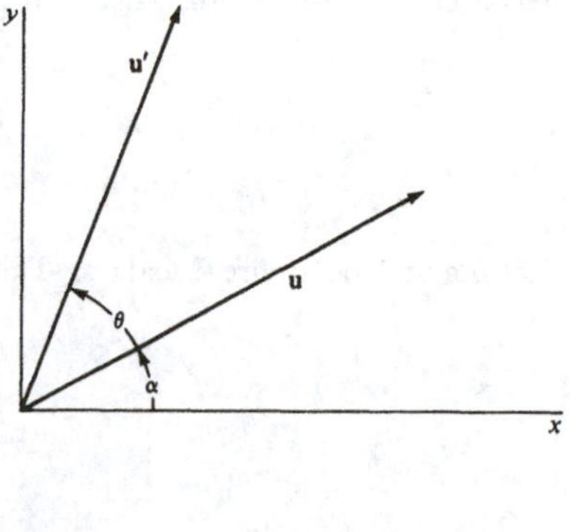

FIGURE 23

If $\mathbf{u}_1 = (x_1,y_1)$, $\mathbf{u}_2 = (x_2,y_2)$, $\mathbf{u}_1' = (x_1',y_1')$, and $\mathbf{u}_2' = (x_2',y_2')$, then

$$\begin{aligned} ax_1' + bx_2' &= a(x_1 \cos\theta - y_1 \sin\theta) + b(x_2 \cos\theta - y_2 \sin\theta) \\ &= (ax_1 + bx_2)\cos\theta - (ay_1 + by_2)\sin\theta \\ ay_1' + by_2' &= a(x_1 \sin\theta + y_1 \cos\theta) + b(x_2 \sin\theta + y_2 \cos\theta) \\ &= (ax_1 + bx_2)\sin\theta + (ay_1 + by_2)\cos\theta \end{aligned}$$

which shows that the rotation of $a\mathbf{u}_1 + b\mathbf{u}_2$ is equal to a times the rotation of $\mathbf{u}_1$ plus b times the rotation of $\mathbf{u}_2$. This calculation can be handled very nicely if we introduce the matrix

$$A = \begin{pmatrix} \cos\theta & -\sin\theta \\ \sin\theta & \cos\theta \end{pmatrix}$$

Then

$$\begin{pmatrix} x' \\ y' \end{pmatrix} = \begin{pmatrix} \cos\theta & -\sin\theta \\ \sin\theta & \cos\theta \end{pmatrix} \begin{pmatrix} x \\ y \end{pmatrix}$$

Now let $\mathbf{u}$ stand for the column matrix $\begin{pmatrix} x \\ y \end{pmatrix}$ and $\mathbf{v}$ stand for the column matrix $\begin{pmatrix} x' \\ y' \end{pmatrix}$. Then the rotation is given by $\mathbf{v} = A\mathbf{u}$, and clearly

$$A(a\mathbf{u}_1 + b\mathbf{u}_2) = aA\mathbf{u}_1 + bA\mathbf{u}_2 = a\mathbf{v}_1 + b\mathbf{v}_2$$

where $\mathbf{v}_1 = A\mathbf{u}_1$ and $\mathbf{v}_2 = A\mathbf{u}_2$.

EXAMPLE 4.2.4 Let $U = R^n$ and $V = R^m$. Let A be an $m \times n$ matrix with real elements. We shall represent a vector in U by a column matrix

$$\mathbf{u} = \begin{pmatrix} x_1 \\ x_2 \\ \vdots \\ x_n \end{pmatrix}$$

Now let $\mathbf{v} = f(\mathbf{u}) = A\mathbf{u}$, where $\mathbf{v}$ is in R^m, represented by a column matrix

$$\mathbf{v} = \begin{pmatrix} y_1 \\ y_2 \\ \vdots \\ y_m \end{pmatrix}$$

Let a and b be any real scalars. Then

$$\begin{aligned} f(a\mathbf{u}_1 + b\mathbf{u}_2) &= A(a\mathbf{u}_1 + b\mathbf{u}_2) \\ &= aA\mathbf{u}_1 + bA\mathbf{u}_2 \\ &= af(\mathbf{u}_1) + bf(\mathbf{u}_2) \end{aligned}$$

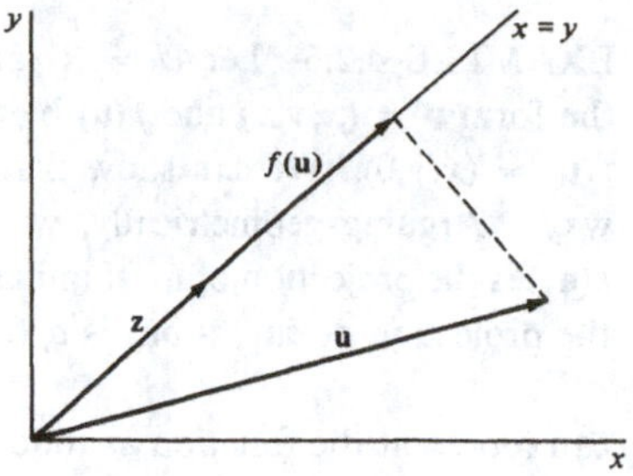

FIGURE 24

This shows that the transformation given by $f(\mathbf{u}) = A\mathbf{u}$ is linear. In fact, it shows that any time we can represent a function in this way we have a linear transformation.

EXAMPLE 4.2.5 Let $U = R^2$ and $V = R^2$. Let $\mathbf{u}$ be any vector in R^2 and $f(\mathbf{u})$ be the projection of $\mathbf{u}$ on the line $y = x$ in euclidean two-dimensional space (see Fig. 24). Let $\mathbf{z}$ be a unit vector along the line $x = y$; then $\mathbf{z} = (1/\sqrt{2}, 1/\sqrt{2})$, and if $\mathbf{u} = (x,y)$, $f(\mathbf{u}) = (\mathbf{u} \cdot \mathbf{z})\mathbf{z} = (x + y)(\frac{1}{2},\frac{1}{2})$. Let (x',y') be coordinates of $f(\mathbf{u})$ relative to the standard basis. Then

$$\begin{pmatrix} x' \\ y' \end{pmatrix} = \begin{pmatrix} \frac{1}{2} & \frac{1}{2} \\ \frac{1}{2} & \frac{1}{2} \end{pmatrix} \begin{pmatrix} x \\ y \end{pmatrix}$$

Therefore, by the result of Example 4.2.4, this is a linear transformation. The reader should try to verify the same result by a geometric argument.

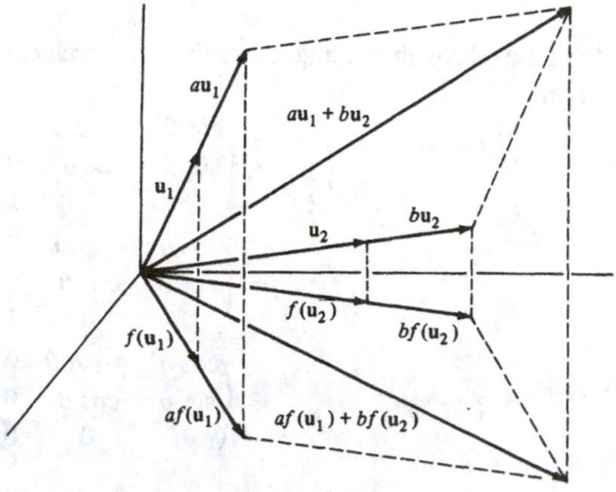

FIGURE 25

EXAMPLE 4.2.6 Let $U = R^3$ and $V = R^3$. Let $\mathbf{u}$ be any vector in R^3 of the form $\mathbf{u} = (x,y,z)$ and $f(\mathbf{u})$ be the projection of $\mathbf{u}$ on the xy plane; that is, $f(\mathbf{u}) = (x,y,0)$. We can show that f is a linear transformation in at least two ways. Arguing geometrically, we see that projection of $a\mathbf{u}_1$ is $af(\mathbf{u}_1)$, where $f(\mathbf{u}_1)$ is the projection of $\mathbf{u}_1$. Similarly, the projection of $b\mathbf{u}_2$ is $bf(\mathbf{u}_2)$, and finally the projection of $a\mathbf{u}_1 + b\mathbf{u}_2$ is $af(\mathbf{u}_1) + bf(\mathbf{u}_2)$ (see Fig. 25). Algebraically, we can represent the function as follows. $\mathbf{u} = \begin{pmatrix} x \\ y \\ z \end{pmatrix}$; then

$$\mathbf{v} = \begin{pmatrix} x' \\ y' \\ z' \end{pmatrix} = \begin{pmatrix} x \\ y \\ 0 \end{pmatrix} \qquad \text{or} \qquad \mathbf{v} = f(\mathbf{u}) = \begin{pmatrix} 1 & 0 & 0 \\ 0 & 1 & 0 \\ 0 & 0 & 0 \end{pmatrix} \mathbf{u}$$

and so the linearity follows from Example 4.2.4.

EXAMPLE 4.2.7 Let $U = R^3$ and $V = R^3$. Consider the linear transformation represented as follows:

$$\mathbf{v} = \begin{pmatrix} x' \\ y' \\ z' \end{pmatrix} = \begin{pmatrix} \cos\theta & -\sin\theta & 0 \\ \sin\theta & \cos\theta & 0 \\ 0 & 0 & 1 \end{pmatrix} \begin{pmatrix} x \\ y \\ z \end{pmatrix} = g(\mathbf{u})$$

where $\mathbf{u} = \begin{pmatrix} x \\ y \\ z \end{pmatrix}$. Since $x' = x\cos\theta - y\sin\theta$, $y' = x\sin\theta + y\cos\theta$, $z' = z$, this represents a rotation of the vector $\mathbf{u}$ about the z axis (see Example 4.2.3) through the angle θ in the counterclockwise direction. Now consider the function from R^3 to R^3 given by $\mathbf{w} = g[f(\mathbf{u})]$, where f is the linear transformation of Example 4.2.6. In other words, $\mathbf{w}$ is the vector obtained from $\mathbf{u}$ by first projecting $\mathbf{u}$ onto the xy plane and then rotating the resulting vector about the z axis through the angle θ in the counterclockwise direction. Let $\mathbf{w} = \begin{pmatrix} \tilde{x} \\ \tilde{y} \\ \tilde{z} \end{pmatrix}$. Then

$$\begin{aligned} \begin{pmatrix} \tilde{x} \\ \tilde{y} \\ \tilde{z} \end{pmatrix} &= \begin{pmatrix} \cos\theta & -\sin\theta & 0 \\ \sin\theta & \cos\theta & 0 \\ 0 & 0 & 1 \end{pmatrix} \begin{pmatrix} x' \\ y' \\ z' \end{pmatrix} \\ &= \begin{pmatrix} \cos\theta & -\sin\theta & 0 \\ \sin\theta & \cos\theta & 0 \\ 0 & 0 & 1 \end{pmatrix} \begin{pmatrix} 1 & 0 & 0 \\ 0 & 1 & 0 \\ 0 & 0 & 0 \end{pmatrix} \begin{pmatrix} x \\ y \\ z \end{pmatrix} \\ &= \begin{pmatrix} \cos\theta & -\sin\theta & 0 \\ \sin\theta & \cos\theta & 0 \\ 0 & 0 & 0 \end{pmatrix} \begin{pmatrix} x \\ y \\ z \end{pmatrix} \end{aligned}$$

This example illustrates how two linear transformations can be composed. In this case, it is obvious that the composition of two linear transformations is a linear transformation. In fact, this is a special case of Theorem 4.2.6. This example also illustrates the usefulness of the matrix algebra in composing a couple of linear transformations.

EXAMPLE 4.2.8 Let U be the space of all real-valued functions of the real variable x which are continuously differentiable on the interval $\{x \mid 0 \le x \le 1\}$. Let V be the space of real-valued continuous functions on the same interval. Consider the operation of differentiation. In other words, if $\mathbf{u} = f$ and $\mathbf{v} = g$, then $g(x) = f'(x)$ for all x satisfying $0 \le x \le 1$. Obviously, f is in U and g is in V, where U and V are real vector spaces. Also by a theorem from the calculus,

$$[af_1(x) + bf_2(x)]' = af_1'(x) + bf_2'(x)$$

which shows the linearity.

EXAMPLE 4.2.9 Let U be the space of real-valued Riemann-integrable functions defined on the interval $\{x \mid a \le x \le b\}$. Let V be R^1, and let the function be the operation of computing the Riemann integral; that is, if $\mathbf{u} = f$, then

$$\mathbf{v} = \int_a^b f(x)\,dx$$

Obviously $\mathbf{v}$ is in R^1, and linearity follows from a basic theorem of the calculus, that is,

$$\int_a^b [c_1 f_1(x) + c_2 f_2(x)]\,dx = c_1 \int_a^b f_1(x)\,dx + c_2 \int_a^b f_2(x)\,dx$$

Now that we have seen several examples of linear transformations, we can begin to study some of the important properties of these transformations.

Theorem 4.2.1 The range of a linear transformation is a subspace of the range space.

PROOF We have to show that if $\mathbf{v}_1$ and $\mathbf{v}_2$ are in the range, then so is $a\mathbf{v}_1 + b\mathbf{v}_2$ for any pair of scalars. Now if $\mathbf{v}_1$ and $\mathbf{v}_2$ are in the range, then there are vectors $\mathbf{u}_1$ and $\mathbf{u}_2$ in the domain such that $\mathbf{v}_1 = f(\mathbf{u}_1)$ and $\mathbf{v}_2 = f(\mathbf{u}_2)$. Therefore,

$$a\mathbf{v}_1 + b\mathbf{v}_2 = af(\mathbf{u}_1) + bf(\mathbf{u}_2) = f(a\mathbf{u}_1 + b\mathbf{u}_2)$$

and $a\mathbf{v}_1 + b\mathbf{v}_2$ is in the range because $a\mathbf{u}_1 + b\mathbf{u}_2$ is in the domain.

An important role in the theory is played by the set of vectors in the domain for which the value of the linear transformation is $\mathbf{0}$. We call this the *null space* of the linear transformation. The null space is always nonempty since $f(\mathbf{0}) = \mathbf{0}$ for any linear transformation (see Exercise 4.2.12).

Definition 4.2.3 Let f be a linear transformation from U to V. The null space of f is the set of vectors $\mathbf{u}$ in U such that $f(\mathbf{u}) = \mathbf{0}$.

Theorem 4.2.2 The null space of a linear transformation is a subspace of the domain.

PROOF We have to show that if $f(\mathbf{u}_1) = \mathbf{0}$ and $f(\mathbf{u}_2) = \mathbf{0}$, then $f(a\mathbf{u}_1 + b\mathbf{u}_2) = \mathbf{0}$. This is obvious, since

$$f(a\mathbf{u}_1 + b\mathbf{u}_2) = af(\mathbf{u}_1) + bf(\mathbf{u}_2) = a\mathbf{0} + b\mathbf{0} = \mathbf{0}$$

Theorem 4.2.3 If the domain of a linear transformation is finite-dimensional, then the dimension of the domain is equal to the dimension of the range plus the dimension of the null space.

PROOF By Theorem 4.2.2, the null space is a subspace of the domain, and therefore the null space is either the zero space or it has a finite basis $\mathbf{u}_1, \mathbf{u}_2, \ldots, \mathbf{u}_r$. By Exercise 3.7.9, the domain has an orthonormal basis $\mathbf{u}_1, \mathbf{u}_2, \ldots, \mathbf{u}_r, \mathbf{w}_1, \mathbf{w}_2, \ldots, \mathbf{w}_s$, where $r + s = n$, the dimension of the domain. If the null space is the zero space, then there are no $\mathbf{u}$'s and $r = 0$. Let $\mathbf{v}$ be any vector in the domain. Then

$$\mathbf{v} = c_1\mathbf{u}_1 + c_2\mathbf{u}_2 + \cdots + c_r\mathbf{u}_r + c_{r+1}\mathbf{w}_1 + c_{r+2}\mathbf{w}_2 + \cdots + c_{r+s}\mathbf{w}_s$$

Since the $\mathbf{u}$'s are in the null space, $f(\mathbf{u}_i) = \mathbf{0}$, $i = 1, 2, \ldots, r$. Therefore, $f(\mathbf{v}) = c_{r+1}f(\mathbf{w}_1) + c_{r+2}f(\mathbf{w}_2) + \cdots + c_{r+s}f(\mathbf{w}_s)$. Clearly any vector in the range can be expressed as a linear combination of $f(\mathbf{w}_1), f(\mathbf{w}_2), \ldots, f(\mathbf{w}_s)$. If these vectors are independent, then they form a basis for the range. Consider a linear combination

$$\gamma_1 f(\mathbf{w}_1) + \gamma_2 f(\mathbf{w}_2) + \cdots + \gamma_s f(\mathbf{w}_s) = f(\gamma_1\mathbf{w}_1 + \gamma_2\mathbf{w}_2 + \cdots + \gamma_s\mathbf{w}_s) = \mathbf{0}$$

But this implies that $\gamma_1\mathbf{w}_1 + \gamma_2\mathbf{w}_2 + \cdots + \gamma_s\mathbf{w}_s$ is in the null space, and therefore

$$\gamma_1\mathbf{w}_1 + \gamma_2\mathbf{w}_2 + \cdots + \gamma_s\mathbf{w}_s = \alpha_1\mathbf{u}_1 + \alpha_2\mathbf{u}_2 + \cdots + \alpha_r\mathbf{u}_r$$

Now we use the orthogonality of the $\mathbf{u}$'s and the $\mathbf{w}$'s to show that $\gamma_1 = \gamma_2 = \cdots = \gamma_s = 0$. Hence, $f(\mathbf{w}_1), f(\mathbf{w}_2), \ldots, f(\mathbf{w}_s)$ are independent, and the dimension of the range is s. This completes the proof.

EXAMPLE 4.2.10 Find the null space of the linear transformation of Example 4.2.5. We represent the transformation by

$$\begin{pmatrix} x' \\ y' \end{pmatrix} = \begin{pmatrix} \frac{1}{2} & \frac{1}{2} \\ \frac{1}{2} & \frac{1}{2} \end{pmatrix} \begin{pmatrix} x \\ y \end{pmatrix}$$

where (x,y) is the vector before projection and (x',y') is the vector after. To find the null space we put $(x',y') = (0,0)$ and solve the homogeneous equations

$$\begin{pmatrix} 0 \\ 0 \end{pmatrix} = \begin{pmatrix} \frac{1}{2} & \frac{1}{2} \\ \frac{1}{2} & \frac{1}{2} \end{pmatrix} \begin{pmatrix} x \\ y \end{pmatrix}$$

The null space consists of those vectors (x,y) such that $x + y = 0$. This is a line through the origin perpendicular to the line of projection, $x = y$. In this case, the range is the line $x = y$. The dimension of the domain is 2, the dimension of the null space is 1, and the dimension of the range is 1, in agreement with Theorem 4.2.3.

If there is a one-to-one correspondence between the vectors of the domain of a function and the vectors of the range, then the roles of the range and domain can be interchanged and we shall have a new function, which we call the *inverse*.

Definition 4.2.4 If f is a function which sets up a one-to-one correspondence between the vectors of its domain and range [for each $\mathbf{v}$ in the range there is precisely one $\mathbf{u}$ in the domain such that $\mathbf{v} = f(\mathbf{u})$], then f has an inverse f^{-1} defined by $\mathbf{u} = f^{-1}(\mathbf{v})$ when $\mathbf{v} = f(\mathbf{u})$. The domain of f^{-1} is the range of f and vice versa.

Theorem 4.2.4 A linear transformation has an inverse if and only if the null space is the zero space.

PROOF Suppose that the vector $\mathbf{u} = \mathbf{0}$ is the only vector such that $f(\mathbf{u}) = \mathbf{0}$. Suppose $\mathbf{v}$ is a vector in the range of f such that $f(\mathbf{u}_1) = f(\mathbf{u}_2) = \mathbf{v}$. Then $\mathbf{0} = f(\mathbf{u}_1) - f(\mathbf{u}_2) = f(\mathbf{u}_1 - \mathbf{u}_2)$. Therefore, $\mathbf{u}_1 = \mathbf{u}_2$, and f has an inverse. Conversely, suppose f has an inverse. If there was a nonzero $\mathbf{w}$ such that $f(\mathbf{w}) = \mathbf{0}$, then $f(\mathbf{u}) = f(\mathbf{u}) + \mathbf{0} = f(\mathbf{u}) + f(\mathbf{w}) = f(\mathbf{u} + \mathbf{w})$ with $\mathbf{u} \neq \mathbf{u} + \mathbf{w}$. This would contradict the assumption that f has an inverse.

EXAMPLE 4.2.11 Show that the linear transformation of Example 4.2.3 has an inverse. We represented the transformation by

$$\begin{pmatrix} x' \\ y' \end{pmatrix} = \begin{pmatrix} \cos\theta & -\sin\theta \\ \sin\theta & \cos\theta \end{pmatrix} \begin{pmatrix} x \\ y \end{pmatrix}$$

where (x,y) is the vector before rotation and (x',y') is the vector after rotation. We look for the null space when we set $(x',y') = (0,0)$. However, the homogeneous equations

$$\begin{pmatrix} 0 \\ 0 \end{pmatrix} = \begin{pmatrix} \cos\theta & -\sin\theta \\ \sin\theta & \cos\theta \end{pmatrix} \begin{pmatrix} x \\ y \end{pmatrix}$$

have only the trivial solution $x = y = 0$, because

$$\begin{vmatrix} \cos\theta & -\sin\theta \\ \sin\theta & \cos\theta \end{vmatrix} = \cos^2\theta + \sin^2\theta = 1$$

Therefore, the transformation has an inverse. If

$$A = \begin{pmatrix} \cos\theta & -\sin\theta \\ \sin\theta & \cos\theta \end{pmatrix}$$

then

$$A^{-1} = \begin{pmatrix} \cos\theta & \sin\theta \\ -\sin\theta & \cos\theta \end{pmatrix} \quad \text{and} \quad \begin{pmatrix} x \\ y \end{pmatrix} = \begin{pmatrix} \cos\theta & \sin\theta \\ -\sin\theta & \cos\theta \end{pmatrix} \begin{pmatrix} x' \\ y' \end{pmatrix}$$

and we have a representation of the inverse. This is a rotation through the angle θ in the clockwise direction.

Theorem 4.2.5 The inverse of a linear transformation is a linear transformation.

PROOF If f is a linear transformation with inverse f^{-1}, then the domain of f^{-1} is the range of f and vice versa. Therefore, conditions (i) and (ii) of Definition 4.2.2 are met. To check condition (iii), let $\mathbf{v}_1 = f(\mathbf{u}_1)$ and $\mathbf{v}_2 = f(\mathbf{u}_2)$. Then $f(a\mathbf{u}_1 + b\mathbf{u}_2) = af(\mathbf{u}_1) + bf(\mathbf{u}_2) = a\mathbf{v}_1 + b\mathbf{v}_2$, and $f^{-1}(a\mathbf{v}_1 + b\mathbf{v}_2) = a\mathbf{u}_1 + b\mathbf{u}_2 = af^{-1}(\mathbf{v}_1) + bf^{-1}(\mathbf{v}_2)$.

We conclude this section with a definition of composition of two linear transformations.

Definition 4.2.5 Suppose f is a linear transformation with domain U and range V. Suppose g is a linear transformation with domain V and range W. The composition $f \circ g$ is defined as follows: if $\mathbf{v} = f(\mathbf{u})$ and $\mathbf{w} = g(\mathbf{v})$, then $[f \circ g](\mathbf{u}) = \mathbf{w}$. The domain of $f \circ g$ is U, and the range of $f \circ g$ is W.

Theorem 4.2.6 The composition of two linear transformations is a linear transformation.

PROOF Clearly $f \circ g$ is defined on all of U. Also, if U and V are defined on the same scalars and so are V and W, then the same is true of U and W. Finally,

$$\begin{aligned}[f \circ g](a\mathbf{u}_1 + b\mathbf{u}_2) &= g[f(a\mathbf{u}_1 + b\mathbf{u}_2)] \\ &= g[af(\mathbf{u}_1) + bf(\mathbf{u}_2)] \\ &= ag[f(\mathbf{u}_1)] + bg[f(\mathbf{u}_2)] \\ &= a[f \circ g](\mathbf{u}_1) + b[f \circ g](\mathbf{u}_2)\end{aligned}$$

EXERCISES 4.2

1 Let U be R^2 and V be R^2. Let $f(\mathbf{u})$ be the reflection of $\mathbf{u}$ in the x axis; that is, if $\mathbf{u} = (x,y)$, then $f(\mathbf{u}) = (x,-y)$. Show that f is a linear transformation. Find the null space and range of f.

2 Let U be R^2 and V be R^2. Let $f(\mathbf{u})$ be the orthogonal complement of $\mathbf{u}$ with respect to the line $x = y$; that is, if $\mathbf{z}$ is a unit vector along the given line, then $(\mathbf{u} \cdot \mathbf{z})\mathbf{z}$ is the projection on the line and $\mathbf{u} - (\mathbf{u} \cdot \mathbf{z})\mathbf{z}$ is the orthogonal complement. Show that f is a linear transformation. Find the null space and range of f.

3 Let U be R^3 and V be R^3. Let $f(\mathbf{u})$ be the reflection of $\mathbf{u}$ in the xy plane; that is, if $\mathbf{u} = (x,y,z)$, then $f(\mathbf{u}) = (x,y,-z)$. Show that f is a linear transformation. Find the null space and range of f.

4 Let U be R^3 and V be R^3. Let $f(\mathbf{u})$ be the reflection of $\mathbf{u}$ in the z axis; that is, if $\mathbf{u} = (x,y,z)$, then $f(\mathbf{u}) = (-x,-y,z)$. Show that f is a linear transformation. Find the null space and range of f.

5 Let U be R^3 and V be R^3. Let $f(\mathbf{u})$ be the orthogonal complement of $\mathbf{u}$ with respect to the plane represented implicitly by $x + y + z = 0$. Show that f is a linear transformation. Find the null space and range of f.

6 Let U be C^n and V be C^n. Let $f(\mathbf{u}) = c\mathbf{u}$, where c is a complex number. Show that f is a linear transformation. Find the null space and range of f.

7 Let $U = R^4$ and $V = R^3$. Let (x_1,x_2,x_3,x_4) be coordinates of $\mathbf{u}$ relative to the standard basis in U. Let (y_1,y_2,y_3) be the coordinates of $f(\mathbf{u})$ relative to the standard basis in V, and

$$\begin{aligned}y_1 &= x_1 - x_2 + 2x_3 - x_4 \\ y_2 &= -x_1 + 2x_2 - 3x_3 + x_4 \\ y_3 &= x_1 - 3x_2 + 4x_3 - x_4\end{aligned}$$

Show that f is a linear transformation. Find the null space and range of f.

8 Let $U = C^n$ and $V = C^1$. Let $f(\mathbf{u}) = u_1$. Show that f is a linear transformation. Find the null space and range of f.

9 Let $U = C^n$ and $V = C^1$. Let $f(\mathbf{u}) = u_1 + u_2 + \cdots + u_n$. Show that f is a linear transformation. Find the null space and range of f.

10 Let U be the space of continuous real-valued functions defined on the interval $\{x \mid 0 \le x \le 1\}$. Let $T[f] = f(x_0)$, $0 \le x_0 \le 1$, where x_0 is fixed. Show that T is a linear transformation. Find the null space and range of T.

11 Find the null space and range of the linear transformations of Examples 4.2.8 and 4.2.9.

12 Prove that for any linear transformation f, $f(\mathbf{0}) = \mathbf{0}$.

13 Show that condition (iii) of Definition 4.2.2 can be replaced by the two conditions $f(a\mathbf{u}) = af(\mathbf{u})$ and $f(\mathbf{u}_1 + \mathbf{u}_2) = f(\mathbf{u}_1) + f(\mathbf{u}_2)$.

14 Find the most general linear transformation from R^1 to R^1.

15 Let f be a linear transformation from R^n to R^n represented by $f(\mathbf{u}) = A\mathbf{u}$, where A is $n \times n$, $\mathbf{u}$ is the column matrix of coordinates relative to the standard basis in U, and $f(\mathbf{u})$ is the column matrix of coordinates relative to the standard basis in V. Show that the following statements are all equivalent by citing the appropriate theorems:

(*a*) A is nonsingular.
(*b*) $|A| \neq 0$.
(c) The null space of f is the zero space.
(*d*) The dimension of the range of f is n.
(*e*) A is invertible.
(*f*) f has an inverse.
(*g*) The columns of A are independent.
(*h*) The rows of A are independent.
(*i*) The equations $AX = B$ have a unique solution.
(*j*) The equations $AX = 0$ have only the trivial solution.

16 Let f be a linear transformation from R^n to R^m represented by $f(\mathbf{u}) = A\mathbf{u}$, where A is $m \times n$. Prove that f is not invertible if $n > m$.

17 A linear transformation f is said to be *onto* if every vector in the range space is a value of $f(\mathbf{u})$ for at least one $\mathbf{u}$ in the domain. If the domain is finite-dimensional, show that f is onto if and only if the dimension of the domain is equal to the dimension of the null space plus the dimension of the range space.

18 Which of the linear transformations in Exercises 1 to 10 have inverses? Find the inverses where they exist.

19 If the domain and range of a linear transformation are the same and $f(\mathbf{u}) = \mathbf{u}$, then f is called the *identity transformation*. Show that the composition of an invertible linear transformation with its inverse (in either order) is the identity.

20 Find the compositions of the linear transformations of Example 4.2.5 and Exercise 1 in both orders. Is the operation of composition commutative? Is the operation of composition associative?

4.3 MATRIX REPRESENTATIONS

We saw in many examples of the previous section that it was possible to represent linear transformations in terms of matrices. This was convenient because we could then study the transformation using the methods developed in Chap. 2. In fact, whenever we are dealing with a linear transformation from a finite-dimensional vector space to another we can find a matrix representation. We shall prove this fundamental theorem and then show how the various aspects of the theory of linear transformations can be treated using matrix algebra.

Theorem 4.3.1 Let f be a linear transformation from C^n to C^m (or R^n to R^m). If X is the column matrix of coordinates of $\mathbf{u}$ relative to the standard basis in U and Y is the column matrix of coordinates of $f(\mathbf{u})$ relative to the standard basis in V, then the transformation can be represented by $Y = AX$, where A is the $m \times n$ complex (real) matrix such that the kth column of A is the set of coordinates of $f(\mathbf{e}_k)$ referred to the standard basis in V.

PROOF Let $\mathbf{u} = x_1\mathbf{e}_1 + x_2\mathbf{e}_2 + \cdots + x_n\mathbf{e}_n$. Then $f(\mathbf{u}) = x_1 f(\mathbf{e}_1) + x_2 f(\mathbf{e}_2) + \cdots + x_n f(\mathbf{e}_n)$. Now suppose

$$\begin{aligned} f(\mathbf{e}_1) &= a_{11}\mathbf{e}_1 + a_{21}\mathbf{e}_2 + \cdots + a_{m1}\mathbf{e}_m \\ f(\mathbf{e}_2) &= a_{12}\mathbf{e}_1 + a_{22}\mathbf{e}_2 + \cdots + a_{m2}\mathbf{e}_m \\ &\cdots\cdots\cdots\cdots\cdots\cdots \\ f(\mathbf{e}_n) &= a_{1n}\mathbf{e}_1 + a_{2n}\mathbf{e}_2 + \cdots + a_{mn}\mathbf{e}_m \end{aligned}$$

Then

$$\begin{aligned} f(\mathbf{u}) = (a_{11}x_1 &+ a_{12}x_2 + \cdots + a_{1n}x_n)\mathbf{e}_1 \\ &+ (a_{21}x_1 + a_{22}x_2 + \cdots + a_{2n})\mathbf{e}_2 + \cdots \\ &+ (a_{m1}x_1 + a_{m2}x_2 + \cdots + a_{mn}x_n)\mathbf{e}_n \end{aligned}$$

Therefore, if $f(\mathbf{u}) = y_1\mathbf{e}_1 + y_2\mathbf{e}_2 + \cdots + y_m\mathbf{e}_m$, then

$$\begin{aligned} y_1 &= a_{11}x_1 + a_{12}x_2 + \cdots + a_{1n}x_n \\ y_2 &= a_{21}x_1 + a_{22}x_2 + \cdots + a_{2n}x_n \\ &\cdots\cdots\cdots\cdots\cdots\cdots \\ y_m &= a_{m1}x_1 + a_{m2}x_2 + \cdots + a_{mn}x_n \end{aligned}$$

or $Y = AX$, as we wished to show. Clearly if the domain and range spaces are both complex vector spaces, then A will be complex. If the domain and range spaces are both real vector spaces, then A will be real.

The proof of Theorem 4.3.1 was carried out in such a way as to illustrate that the bases used in the domain and range spaces need not be the standard bases. In fact, the proof would be exactly the same if we substituted the basis $\mathbf{u}_1, \mathbf{u}_2, \ldots, \mathbf{u}_n$ for $\mathbf{e}_1, \mathbf{e}_2, \ldots, \mathbf{e}_n$ in U and the basis $\mathbf{v}_1, \mathbf{v}_2, \ldots, \mathbf{v}_m$ for $\mathbf{e}_1, \mathbf{e}_2, \ldots, \mathbf{e}_m$ in V. We therefore have the following theorem.

Theorem 4.3.2 Let f be a linear transformation from U to V. If X is the column matrix of coordinates of $\mathbf{u}$ relative to the basis $\mathbf{u}_1, \mathbf{u}_2, \ldots, \mathbf{u}_n$ in U and Y is the column matrix of coordinates of $f(\mathbf{u})$ relative to the basis $\mathbf{v}_1, \mathbf{v}_2, \ldots, \mathbf{v}_m$ in V, then the transformation can be represented by $Y = AX$, where A is the $m \times n$ matrix such that the kth column of A is the set of coordinates of $f(\mathbf{u}_k)$ referred to the basis $\mathbf{v}_1, \mathbf{v}_2, \ldots, \mathbf{v}_n$ in V.

EXAMPLE 4.3.1 Let $U = R^2$ and $V = R^2$. Let $\mathbf{u}$ be any vector in R^2 and $f(\mathbf{u})$ be the projection of $\mathbf{u}$ on the line $y = x$ in euclidean two-dimensional space. We showed in Example 4.2.5 that f is a linear transformation. If we refer $\mathbf{u}$ and $f(\mathbf{u})$ to the standard basis in R^2, then the transformation can be represented by

$$\begin{pmatrix} x' \\ y' \end{pmatrix} = A \begin{pmatrix} x \\ y \end{pmatrix}$$

where the first column of A is $f(\mathbf{e}_1)$ and the second column of A is $f(\mathbf{e}_2)$ (see Fig. 26). Clearly

$$f(\mathbf{e}_1) = \begin{pmatrix} \frac{1}{2} \\ \frac{1}{2} \end{pmatrix} \qquad f(\mathbf{e}_2) = \begin{pmatrix} \frac{1}{2} \\ \frac{1}{2} \end{pmatrix} \qquad \text{and} \qquad A = \begin{pmatrix} \frac{1}{2} & \frac{1}{2} \\ \frac{1}{2} & \frac{1}{2} \end{pmatrix}$$

Next let us refer the transformation to a different basis, namely $\mathbf{u}_1$ along the line $y = x$ and $\mathbf{u}_2$ perpendicular to the line. Referred to this basis, $f(\mathbf{u}_1) = \mathbf{u}_1$ and $f(\mathbf{u}_2) = \mathbf{0}$. Therefore, the transformation can be expressed as

$$\begin{pmatrix} y_1 \\ y_2 \end{pmatrix} = \begin{pmatrix} 1 & 0 \\ 0 & 0 \end{pmatrix} \begin{pmatrix} x_1 \\ x_2 \end{pmatrix}$$

where (x_1,x_2) and (y_1,y_2) are now coordinates relative to the new basis.

EXAMPLE 4.3.2 Let $U = R^3$ and $V = R^3$. Let $\mathbf{u}$ be any vector in R^3 and $f(\mathbf{u})$ be the projection of $\mathbf{u}$ on the plane given implicitly by $x - 2y + z = 0$. The unit vector $\mathbf{w} = (1,-2,1)/\sqrt{6}$ is perpendicular to the given plane, and if

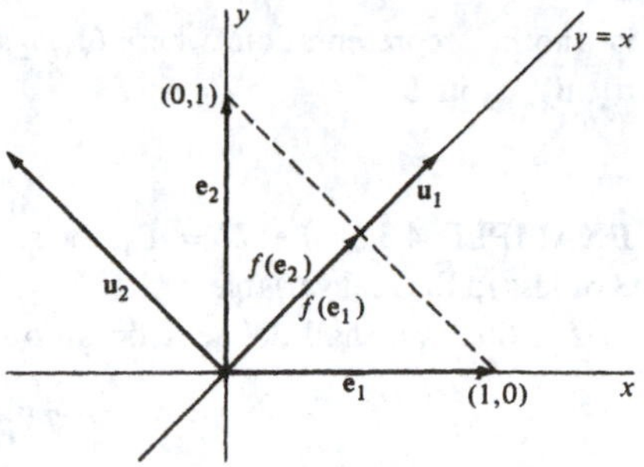

FIGURE 26

$\mathbf{u} = (x,y,z)$, then $(\mathbf{u}\cdot\mathbf{w})\mathbf{w}$ is the projection of $\mathbf{u}$ on $\mathbf{w}$. Therefore, $f(\mathbf{u}) = \mathbf{u} - (\mathbf{u}\cdot\mathbf{w})\mathbf{w}$, and

$$\begin{aligned} f(a\mathbf{u}_1 + b\mathbf{u}_2) &= a\mathbf{u}_1 + b\mathbf{u}_2 - (a\mathbf{u}_1 + b\mathbf{u}_2\cdot\mathbf{w})\mathbf{w} \\ &= a\mathbf{u}_1 - (a\mathbf{u}_1\cdot\mathbf{w})\mathbf{w} + b\mathbf{u}_2 - (b\mathbf{u}_2\cdot\mathbf{w})\mathbf{w} \\ &= af(\mathbf{u}_1) + bf(\mathbf{u}_2) \end{aligned}$$

This shows that the transformation is linear. To find a matrix representation, let us refer all vectors to the standard basis. Then

$$\begin{aligned} f(\mathbf{e}_1) &= (1,0,0) - (\tfrac{1}{6},-\tfrac{1}{3},\tfrac{1}{6}) = (\tfrac{5}{6},\tfrac{1}{3},-\tfrac{1}{6}) \\ f(\mathbf{e}_2) &= (0,1,0) + (\tfrac{1}{3},-\tfrac{2}{3},\tfrac{1}{3}) = (\tfrac{1}{3},\tfrac{1}{3},\tfrac{1}{3}) \\ f(\mathbf{e}_3) &= (0,0,1) - (\tfrac{1}{6},-\tfrac{1}{3},\tfrac{1}{6}) = (-\tfrac{1}{6},\tfrac{1}{3},\tfrac{5}{6}) \end{aligned}$$

and the transformation has the representation

$$\begin{pmatrix} x' \\ y' \\ z' \end{pmatrix} = \begin{pmatrix} \frac{5}{6} & \frac{1}{3} & -\frac{1}{6} \\ \frac{1}{3} & \frac{1}{3} & \frac{1}{3} \\ -\frac{1}{6} & \frac{1}{3} & \frac{5}{6} \end{pmatrix} \begin{pmatrix} x \\ y \\ z \end{pmatrix}$$

relative to the standard basis. Suppose we use the basis $\mathbf{u}_1 = (1,1,1)$, $\mathbf{u}_2 = (1,0,-1)$, $\mathbf{u}_3 = (1,-2,1)$ in the domain and the standard basis in the range space. Then

$$\begin{aligned} f(\mathbf{u}_1) &= (1,1,1) \\ f(\mathbf{u}_2) &= (1,0,-1) \\ f(\mathbf{u}_3) &= (0,0,0) \end{aligned}$$

and

$$\begin{pmatrix} x' \\ y' \\ z' \end{pmatrix} = \begin{pmatrix} 1 & 1 & 0 \\ 1 & 0 & 0 \\ 1 & -1 & 0 \end{pmatrix} \begin{pmatrix} u_1 \\ u_2 \\ u_3 \end{pmatrix}$$

is another representation, where (u_1,u_2,u_3) are coordinates relative to the basis $\mathbf{u}_1$, $\mathbf{u}_2$, $\mathbf{u}_3$ in U.

EXAMPLE 4.3.3 Let $U = P_4$, the space of real-valued polynomials of degree 3 or less in the real variable x. Let $V = P_3$. If $p(x) = a_0 + a_1x + a_2x^2 + a_3x^3$ in P_4, then we shall define a linear transformation by differentiation, namely

$$T[p(x)] = a_1 + 2a_2x + 3a_3x^2$$

We shall find a matrix representation of T relative to the basis $1, x, x^2, x^3$ in U and $1, x, x^2$ in V. Now

$$\begin{aligned} T[1] &= 0 &&= 0\cdot 1 + 0\cdot x + 0\cdot x^2 \\ T[x] &= 1 &&= 1\cdot 1 + 0\cdot x + 0\cdot x^2 \\ T[x^2] &= 2x &&= 0\cdot 1 + 2\cdot x + 0\cdot x^2 \\ T[x^3] &= 3x^2 &&= 0\cdot 1 + 0\cdot x + 3\cdot x^2 \end{aligned}$$

Therefore, the desired representation is

$$\begin{pmatrix} v_1 \\ v_2 \\ v_3 \end{pmatrix} = \begin{pmatrix} 0 & 1 & 0 & 0 \\ 0 & 0 & 2 & 0 \\ 0 & 0 & 0 & 3 \end{pmatrix} \begin{pmatrix} a_0 \\ a_1 \\ a_2 \\ a_3 \end{pmatrix}$$

Now that we see the intimate connection between linear transformations and matrices, we should try to exploit the matrix algebra in studying linear transformations. The pertinent operations in matrix algebra are addition, multiplication by a scalar, multiplication, and inversion, each of which has its counterpart in the theory of linear transformations.

Definition 4.3.1 Let f and g be linear transformations from U to V. The sum $f + g$ is defined by $[f + g](\mathbf{u}) = f(\mathbf{u}) + g(\mathbf{u})$.

Theorem 4.3.3 The sum of two linear transformations is a linear transformation.

PROOF Since f and g have the same domain U and range space V, $f + g$ has domain U and range space V. Hence, conditions (i) and (ii) of the definition are met. To check (iii) we have

$$\begin{aligned}[f+g](a\mathbf{u}_1+b\mathbf{u}_2) &= f(a\mathbf{u}_1+b\mathbf{u}_2)+g(a\mathbf{u}_1+b\mathbf{u}_2)\\ &= af(\mathbf{u}_1)+bf(\mathbf{u}_2)+ag(\mathbf{u}_1)+bg(\mathbf{u}_2)\\ &= a[f(\mathbf{u}_1)+g(\mathbf{u}_1)]+b[f(\mathbf{u}_2)+g(\mathbf{u}_2)]\\ &= a[f+g](\mathbf{u}_1)+b[f+g](\mathbf{u}_2)\end{aligned}$$

Theorem 4.3.4 Let f and g be linear transformations with finite-dimensional domain U and range space V. If A is the matrix representation of f with respect to given bases in U and V, and if B is the matrix representation of g relative to the same bases, then the matrix representation of $f + g$ is $A + B$.

PROOF If U has dimension n and V has dimension m, then both A and B are $m \times n$ and can be added. Let $\mathbf{u}$ have the coordinates X relative to the given basis in U. Then $f(\mathbf{u})$ and $g(\mathbf{u})$ have the coordinates $Y_1 = AX$ and $Y_2 = BX$, respectively, with respect to the given basis in V. Therefore, $[f + g](\mathbf{u})$ has the coordinates $Y_1 + Y_2 = AX + BX = (A + B)X$.

Definition 4.3.2 Let f be a linear transformation from the complex (real) vector space U to the complex (real) vector space V. The transformation f can be multiplied by a complex (real) scalar c to give the new function cf defined by $[cf](\mathbf{u}) = cf(\mathbf{u})$.

Theorem 4.3.5 The function cf is a linear transformation.

PROOF Clearly cf has the same domain and range space as f. Therefore, conditions (i) and (ii) of the definition are met. To check condition (iii) we have

$$\begin{aligned}[cf](a\mathbf{u}_1+b\mathbf{u}_2) &= cf(a\mathbf{u}_1+b\mathbf{u}_2)\\ &= c(af(\mathbf{u}_1)+bf(\mathbf{u}_2))\\ &= acf(\mathbf{u}_1)+bcf(\mathbf{u}_2)\\ &= a[cf](\mathbf{u}_1)+b[cf](\mathbf{u}_2)\end{aligned}$$

Theorem 4.3.6 Let f be a linear transformation with finite-dimensional domain U and range space V. If A is the matrix representation of f relative to given bases in U and V, then cA is the matrix representation of cf relative to the same bases in U and V.

PROOF Let X be the coordinates of $\mathbf{u}$ relative to the given basis in U. Then $Y = AX$ are the coordinates of $f(\mathbf{u})$ relative to the given basis in V. Therefore, the coordinates of $[cf](\mathbf{u})$ relative to the same bases are $cY = c(AX) = (cA)X$.

The difference between two linear transformations can be defined in the obvious manner, combining the operations of addition and multiplication by a scalar; that is, $f - g = f + (-1)g$. If f and g have matrix representation A and B, respectively, with respect to given bases in the domain and range spaces, then the difference $f - g$ will have the representation $A + (-1)B = A - B$ with respect to the same bases.

The next theorem, which deals with the collection of all linear transformations from a given domain U to a given range space V, is one level of abstraction beyond that of the basic definition of a vector space. No special use will be made of it in this book, and therefore it may be omitted on the first reading.

Theorem 4.3.7 Let U and V be complex (real) vector spaces. Then W, the collection of all linear transformations from U to V, is a complex (real) vector space. If U is of dimension n and V is of dimension m, then W is of dimension nm.

PROOF To define a vector space we need two operations, addition and multiplication by a scalar. These operations in W are defined in Definitions 4.3.1 and 4.3.2. The closure properties, A1 and M1, are verified by Theorems 4.3.3 and 4.3.5. We verify the other axioms as follows:†

A2: $[f + g](\mathbf{u}) = f(\mathbf{u}) + g(\mathbf{u}) = g(\mathbf{u}) + f(\mathbf{u}) = [g + f](\mathbf{u})$

A3: $[f + (g + h)](\mathbf{u})$

$$= f(\mathbf{u}) + [g + h](\mathbf{u}) = f(\mathbf{u}) + g(\mathbf{u}) + h(\mathbf{u})$$
$$= [f + g](\mathbf{u}) + h(\mathbf{u}) = [(f + g) + h](\mathbf{u})$$

A4: The zero transformation is the transformation which takes every vector of U to the zero vector of V; that is, $0(\mathbf{u}) = \mathbf{0}$ for all $\mathbf{u}$ in U. Clearly $[f + 0](\mathbf{u}) = f(\mathbf{u}) + 0(\mathbf{u}) = f(\mathbf{u}) + \mathbf{0} = f(\mathbf{u})$.

A5: The negative of f is the linear transformation $-f$ defined by the following: if $f(\mathbf{u}) = \mathbf{v}$, then $[-f](\mathbf{u}) = -\mathbf{v}$. Then $[f + (-f)](\mathbf{u}) = f(\mathbf{u}) + [-f](\mathbf{u}) = \mathbf{v} - \mathbf{v} = \mathbf{0}$ for all $\mathbf{u}$ in U.

M2:
$$a[f + g](\mathbf{u}) = a[f(\mathbf{u}) + g(\mathbf{u})] = af(\mathbf{u}) + ag(\mathbf{u}) = [af](\mathbf{u}) + [ag](\mathbf{u})$$

M3:
$$[(a + b)f](\mathbf{u}) = (a + b)f(\mathbf{u}) = af(\mathbf{u}) + bf(\mathbf{u}) = [af](\mathbf{u}) + [bf](\mathbf{u})$$

† By equality of two linear transformations we mean $f = g$ if $f(\mathbf{u}) = g(\mathbf{u})$ for all $\mathbf{u}$ in U.

M4: $[(ab)f](\mathbf{u}) = (ab)f(\mathbf{u}) = a(bf(\mathbf{u})) = a[bf](\mathbf{u})$

M5: $[1f](\mathbf{u}) = 1 \cdot f(\mathbf{u}) = f(\mathbf{u})$

To verify the statement about dimension, we note that if U is n-dimensional and V is m-dimensional, then for each linear transformation there is a unique $m \times n$ matrix A, assuming given bases in U and V (see Theorem 4.3.2 and Exercise 4.3.11). Conversely, for each $m \times n$ matrix A there is a linear transformation f (see Example 4.2.4). Therefore, there is a one-to-one correspondence between the collection of linear transformations and the collection of $m \times n$ matrices. Therefore, the vector space of linear transformations from U to V has dimension nm (see Exercise 3.7.16).

Next we come to the correspondence between composition of two linear transformations and the product of two matrices.

Theorem 4.3.8 Let f be a linear transformation from the n-dimensional space U to the m-dimensional space V, represented by the $m \times n$ matrix A with respect to the basis $\mathbf{u}_1, \mathbf{u}_2, \ldots, \mathbf{u}_n$ in U and the basis $\mathbf{v}_1, \mathbf{v}_2, \ldots, \mathbf{v}_m$ in V. Let g be a linear transformation from V to the p-dimensional space W, represented by the $p \times m$ matrix B with respect to the basis $\mathbf{v}_1, \mathbf{v}_2, \ldots, \mathbf{v}_m$ in V and the basis $\mathbf{w}_1, \mathbf{w}_2, \ldots, \mathbf{w}_p$ in W. Then the composition $f \circ g$ has the representation BA relative to the basis $\mathbf{u}_1, \mathbf{u}_2, \ldots, \mathbf{u}_n$ in U and $\mathbf{w}_1, \mathbf{w}_2, \ldots, \mathbf{w}_p$ in W.

PROOF Let $\mathbf{u}$ have coordinates X relative to the given basis in U and $f(\mathbf{u})$ have the coordinates $Y = AX$ relative to the given basis in V. If $\mathbf{v}$ has the coordinates Y with respect to the given basis in V, then $Z = BY$ are the coordinates of $g(\mathbf{v})$ relative to the given basis in W. Therefore, $Z = B(AX) = (BA)X$ are the coordinates of $g[f(\mathbf{u})]$ with respect to the given basis in W. But $[f \circ g](\mathbf{u}) = g[f(\mathbf{u})]$, which completes the proof.

Finally, we come to the correspondence between the inverse of a linear transformation and the inverse of a matrix. By Exercise 4.2.16, a linear transformation from an n-dimensional vector space to an m-dimensional vector space is never invertible if $n > m$. On the other hand, even if $m > n$, the range cannot have dimension greater than n by Theorem 4.2.3. Therefore, if we wish to study the invertibility of transformations with n-dimensional domains, we may as well consider only $n \times n$ matrix representations. This does not mean, of course, that every linear transformation from an n-dimensional space to an n-dimensional space is invertible.

Theorem 4.3.9 Let f be a linear transformation from the n-dimensional space U to the n-dimensional space V, represented by the $n \times n$ matrix A with respect to the bases $\mathbf{u}_1, \mathbf{u}_2, \ldots, \mathbf{u}_n$ in U and $\mathbf{v}_1, \mathbf{v}_2, \ldots, \mathbf{v}_n$ in V. Then f has an inverse f^{-1} if and only if A is invertible. The matrix representation of f^{-1} is A^{-1} with respect to the given bases in U and V.

PROOF Let X be the coordinates of $\mathbf{u}$ with respect to $\mathbf{u}_1, \mathbf{u}_2, \ldots, \mathbf{u}_n$. Then $Y = AX$ are the coordinates of $f(\mathbf{u})$ with respect to the basis $\mathbf{v}_1, \mathbf{v}_2, \ldots, \mathbf{v}_n$. Furthermore, f is invertible if and only if the dimension of the null space is zero. Therefore, f is invertible if and only if $AX = 0$ has only the trivial solution, and $AX = 0$ has only the trivial solution if and only if A is invertible. Now suppose A^{-1} exists. Then $A^{-1}Y = A^{-1}(AX) = (A^{-1}A)X = X$. Therefore, $X = A^{-1}Y$ expresses the coordinates of $\mathbf{u}$ in terms of the coordinates of $\mathbf{v} = f(\mathbf{u})$. Therefore, A^{-1} is the matrix representation of f^{-1} relative to the given bases in V and U.

EXAMPLE 4.3.4 Let $U = R^3$ and $V = R^3$. Let $f(\mathbf{u})$ be the vector obtained from $\mathbf{u}$ by first rotating $\mathbf{u}$ about the z axis through an angle of 90° in the counterclockwise direction and then through an angle of 90° in the counterclockwise direction about the x axis. We shall find a matrix representation of f with respect to the standard bases in U and V.

$$f(\mathbf{e}_1) = (0,0,1)$$
$$f(\mathbf{e}_2) = (-1,0,0)$$
$$f(\mathbf{e}_3) = (0,-1,0)$$

Therefore, f has the representation

$$\begin{pmatrix} x' \\ y' \\ z' \end{pmatrix} = \begin{pmatrix} 0 & -1 & 0 \\ 0 & 0 & -1 \\ 1 & 0 & 0 \end{pmatrix} \begin{pmatrix} x \\ y \\ z \end{pmatrix}$$

Now the matrix of the transformation is orthogonal and is therefore invertible. The inverse of the transformation has the representation

$$\begin{pmatrix} x \\ y \\ z \end{pmatrix} = \begin{pmatrix} 0 & 0 & 1 \\ -1 & 0 & 0 \\ 0 & -1 & 0 \end{pmatrix} \begin{pmatrix} x' \\ y' \\ z' \end{pmatrix}$$

In the next section, we study the question of how the representation of a linear transformation changes when we change the bases in the domain and range spaces.

EXERCISES 4.3

1 Find the matrix representation of the linear transformation of Example 4.2.1 with respect to the standard bases. Find the representation with respect to arbitrary bases.

2 Find the matrix representation of the linear transformation of Example 4.2.2 with respect to arbitrary bases.

3 Let $U = R^3$ and $V = R^2$. Let $\mathbf{u}$ be any vector in R^3, and let $f(\mathbf{u})$ be the projection of $\mathbf{u}$ on the xy plane. Find the matrix representation of f with respect to the standard bases.

4 Let U be the space of real-valued polynomials of degree n or less in the real variable x. Let f be the operation of integration over the interval $\{x \mid 0 \leq x \leq 1\}$. Find the matrix representation of f with respect to the basis $1, x, x^2, \ldots, x^n$ in U.

5 Find the matrix representation of the linear transformation of Exercise 4.2.1 with respect to the standard bases. Find the representation with respect to the basis (1,1), (1, −1) in both domain and range spaces.

6 Find the matrix representation of the linear transformation of Exercise 4.2.2 with respect to the standard bases. Find the representation with respect to the basis (1,1), (1, −1) in both domain and range spaces.

7 Find the matrix representation of the linear transformation of Exercise 4.2.4 with respect to the standard bases. Find the representation of the inverse with respect to the standard bases.

8 Find the matrix representation of the linear transformation of Exercise 4.2.7 with respect to the standard bases. Find a basis for the null space of the transformation and a basis for the domain consisting of this basis and other vectors orthogonal to the null space. Find a representation of the linear transformation with respect to this new basis and the standard basis in the range space.

9 Find matrix representations of the linear transformations in Exercises 4.2.8 and 4.2.9 with respect to the standard bases in domain and range spaces.

10 Let $U = R^3$ and $V = R^3$. Let A be the representation of a linear transformation f with respect to the standard bases. Which transformations are invertible? Find the inverse if its exists.

$$(a)\quad A = \begin{pmatrix} 1 & 1 & 1 \\ 1 & -1 & 1 \\ 2 & 0 & 3 \end{pmatrix} \qquad (b)\quad A = \begin{pmatrix} 1 & 1 & 1 \\ 1 & -1 & 1 \\ -1 & 5 & -1 \end{pmatrix}$$

11 Show that the representation of a linear transformation from a finite-dimensional domain U to a finite-dimensional range space V with respect to given bases in U and V is unique. *Hint*: If $Y = AX$ and $Y = BX$, then $0 = (A - B)X$ for *all* vectors X.

12 Let f be a linear transformation from U with basis $\mathbf{u}_1, \mathbf{u}_2, \ldots, \mathbf{u}_n$ to V with basis $\mathbf{v}_1, \mathbf{v}_2, \ldots, \mathbf{v}_n$. Let g be a linear transformation from V to W with basis $\mathbf{w}_1, \mathbf{w}_2, \ldots, \mathbf{w}_n$. If f has the representation A and g has the representation B with respect to the given bases, where A and B are nonsingular, show that $(f \circ g)^{-1}$ has the representation $A^{-1}B^{-1}$.

4.4 CHANGE OF BASES

In this section, we shall again consider linear transformation with a finite-dimensional domain and range space. If we pick a basis in the domain and a basis in the range space, we shall have a unique matrix representation of the transformation. If we change the bases in the domain and the range spaces, we shall, in general, change the representation. Our purpose is to find an easy way to find the new representation. Our approach will be the following. We shall first show that a change of basis in an n-dimensional vector space can be interpreted as an invertible linear transformation from C^n to C^n or R^n to R^n, depending on whether the space is complex or real. Then we shall show that the change of representation of a linear transformation can be obtained by composing three linear transformations.

Theorem 4.4.1 Let U be an n-dimensional vector space with a basis $\mathbf{u}_1, \mathbf{u}_2, \ldots, \mathbf{u}_n$. Let $\mathbf{u}_1', \mathbf{u}_2', \ldots, \mathbf{u}_n'$ be another basis for U, such that

$$\begin{aligned}
\mathbf{u}_1 &= p_{11}\mathbf{u}_1' + p_{21}\mathbf{u}_2' + \cdots + p_{n1}\mathbf{u}_n' \\
\mathbf{u}_2 &= p_{12}\mathbf{u}_1' + p_{22}\mathbf{u}_2' + \cdots + p_{n2}\mathbf{u}_n' \\
&\cdots\cdots\cdots\cdots\cdots\cdots\cdots\cdots \\
\mathbf{u}_n &= p_{1n}\mathbf{u}_1' + p_{2n}\mathbf{u}_2' + \cdots + p_{nn}\mathbf{u}_n'
\end{aligned}$$

If X is the column matrix of coordinates of $\mathbf{u}$ with respect to $\mathbf{u}_1, \mathbf{u}_2, \ldots, \mathbf{u}_n$ and X' is the column matrix of coordinates of $\mathbf{u}$ with respect to $\mathbf{u}_1', \mathbf{u}_2', \ldots, \mathbf{u}_n'$, then $X' = PX$, where P is the $n \times n$ matrix with elements p_{ij}. Also, P is invertible and $X = P^{-1}X'$.

PROOF Let $\mathbf{u} = x_1\mathbf{u}_1 + x_2\mathbf{u}_2 + \cdots + x_n\mathbf{u}_n$. Then

$$\begin{aligned}
\mathbf{u} &= x_1(p_{11}\mathbf{u}_1' + p_{21}\mathbf{u}_2' + \cdots + p_{n1}\mathbf{u}_n') \\
&\quad + x_2(p_{12}\mathbf{u}_1' + p_{22}\mathbf{u}_2' + \cdots + p_{n2}\mathbf{u}_n') \\
&\quad + \cdots + x_n(p_{1n}\mathbf{u}_1' + p_{2n}\mathbf{u}_2' + \cdots + p_{nn}\mathbf{u}_n') \\
&= (p_{11}x_1 + p_{12}x_2 + \cdots + p_{1n}x_n)\mathbf{u}_1' \\
&\quad + (p_{21}x_1 + p_{22}x_2 + \cdots + p_{2n}x_n)\mathbf{u}_2' \\
&\quad + \cdots + (p_{n1}x_1 + p_{n2}x_2 + \cdots + p_{nn}x_n)\mathbf{u}_n' \\
&= x_1'\mathbf{u}_1' + x_2'\mathbf{u}_2' + \cdots + x_n'\mathbf{u}_n'
\end{aligned}$$

Therefore,

$$\begin{aligned}
x_1' &= p_{11}x_1 + p_{12}x_2 + \cdots + p_{1n}x_n \\
x_2' &= p_{21}x_1 + p_{22}x_2 + \cdots + p_{2n}x_n \\
&\cdots\cdots\cdots\cdots\cdots\cdots\cdots\cdots \\
x_n' &= p_{n1}x_1 + p_{n2}x_2 + \cdots + p_{nn}x_n
\end{aligned}$$

or $X' = PX$. To show that P is invertible, we show that $PX = 0$ has only the trivial solution $X = 0$. Suppose there was a nontrivial solution $(\gamma_1, \gamma_2, \ldots, \gamma_n) \neq \mathbf{0}$. Then

$$\gamma_1\mathbf{u}_1 + \gamma_2\mathbf{u}_2 + \cdots + \gamma_n\mathbf{u}_n = 0\mathbf{u}_1' + 0\mathbf{u}_2' + \cdots + 0\mathbf{u}_n' = \mathbf{0}$$

But $\mathbf{u}_1, \mathbf{u}_2, \ldots, \mathbf{u}_n$ is a basis, and therefore, $\gamma_1 = \gamma_2 = \cdots = \gamma_n = 0$, contrary to our assumption. Finally, solving for X in terms of X', we have $X = P^{-1}X'$.

EXAMPLE 4.4.1 Let U be R^3, let $\mathbf{u}_1, \mathbf{u}_2, \mathbf{u}_3$ be the standard basis, and let $\mathbf{u}_1', \mathbf{u}_2', \mathbf{u}_3'$ be the independent vectors (1,0,1), (1,−1,1), (1,1,−1). Find the matrix P which represents the change of basis. Let us write

$$\begin{aligned}
\mathbf{u}_1' &= p_{11}^{-1}\mathbf{e}_1 + p_{21}^{-1}\mathbf{e}_2 + p_{31}^{-1}\mathbf{e}_3 \\
\mathbf{u}_2' &= p_{12}^{-1}\mathbf{e}_1 + p_{22}^{-1}\mathbf{e}_2 + p_{32}^{-1}\mathbf{e}_3 \\
\mathbf{u}_3' &= p_{13}^{-1}\mathbf{e}_1 + p_{23}^{-1}\mathbf{e}_2 + p_{33}^{-1}\mathbf{e}_3
\end{aligned}$$

Then it is clear that the first column of P^{-1} is the set of coordinates (with respect to the standard basis) of $\mathbf{u}_1'$, the second column the coordinates of $\mathbf{u}_2'$, and the third column the coordinates of $\mathbf{u}_3'$. Therefore,

$$P^{-1} = \begin{pmatrix} 1 & 1 & 1 \\ 0 & -1 & 1 \\ 1 & 1 & -1 \end{pmatrix}$$

Computing the inverse, we have

$$P = \begin{pmatrix} 0 & 1 & 1 \\ \frac{1}{2} & -1 & -\frac{1}{2} \\ \frac{1}{2} & 0 & -\frac{1}{2} \end{pmatrix}$$

Let $\mathbf{u} = (1,2,3)$ be a vector referred to the standard basis. Then

$$\begin{pmatrix} x_1' \\ x_2' \\ x_3' \end{pmatrix} = \begin{pmatrix} 0 & 1 & 1 \\ \frac{1}{2} & -1 & -\frac{1}{2} \\ \frac{1}{2} & 0 & -\frac{1}{2} \end{pmatrix} \begin{pmatrix} 1 \\ 2 \\ 3 \end{pmatrix} = \begin{pmatrix} 5 \\ -3 \\ -1 \end{pmatrix}$$

In other words,

$$(1,2,3) = 5(1,0,1) - 3(1,-1,1) - (1,1,-1)$$

Theorem 4.4.2 Let U be an n-dimensional real vector space with an orthonormal basis $\mathbf{u}_1, \mathbf{u}_2, \ldots, \mathbf{u}_n$. Let $\mathbf{u}_1', \mathbf{u}_2', \ldots, \mathbf{u}_n'$ be another orthonormal basis for U such that $\mathbf{u}_i = \sum_{j=1}^{n} p_{ji}\mathbf{u}_j'$, $i = 1, 2, \ldots, n$. Then the matrix P with elements p_{ij} is orthogonal.

PROOF Since both bases are orthonormal, we have the following computation:

$$\begin{aligned}\delta_{ij} = (\mathbf{u}_i \cdot \mathbf{u}_j) &= \left(\sum_{k=1}^{n} p_{ki}\mathbf{u}'_k \cdot \sum_{m=1}^{n} p_{mj}\mathbf{u}'_m\right) \\ &= \sum_{k=1}^{n} \sum_{m=1}^{n} p_{ki}p_{mj}(\mathbf{u}'_k \cdot \mathbf{u}'_m) \\ &= \sum_{k=1}^{n} \sum_{m=1}^{n} p_{ki}p_{mj}\delta_{km} \\ &= \sum_{k=1}^{n} p_{ki}p_{kj}\end{aligned}$$

which shows that the columns of the matrix P are orthonormal vectors in R^n. This is enough to show that P is orthogonal and hence that $P^{-1} = \tilde{P}$ (transpose). The corresponding theorem for complex vector spaces results in the conclusion that P is unitary. The proof of this will be left for the reader (see Exercise 4.4.3).

We now come to the most important theorem of this section, which shows how the matrix representation of a linear transformation changes when we change the basis in both the domain and the range space.

Theorem 4.4.3 Let f be a linear transformation from the n-dimensional vector space U to the m-dimensional vector space V. Let A be the matrix representation of f relative to the basis $\mathbf{u}_1, \mathbf{u}_2, \ldots, \mathbf{u}_n$ in U and the basis $\mathbf{v}_1, \mathbf{v}_2, \ldots, \mathbf{v}_m$ in V. Let P be the matrix which represents the change of basis in U from $\mathbf{u}_1, \mathbf{u}_2, \ldots, \mathbf{u}_n$ to $\mathbf{u}'_1, \mathbf{u}'_2, \ldots, \mathbf{u}'_n$. Let Q be the matrix which represents the change of basis in V from $\mathbf{v}_1, \mathbf{v}_2, \ldots, \mathbf{v}_m$ to $\mathbf{v}'_1, \mathbf{v}'_2, \ldots, \mathbf{v}'_m$. Then the matrix which represents f relative to the new bases is QAP^{-1}.

PROOF Let X be the column matrix of coordinates of $\mathbf{u}$ with respect to the basis $\mathbf{u}_1, \mathbf{u}_2, \ldots, \mathbf{u}_n$, and let Y be the column matrix of coordinates of $f(\mathbf{u})$ with respect to the basis $\mathbf{v}_1, \mathbf{v}_2, \ldots, \mathbf{v}_m$. Then $Y = AX$. Let X' be the column matrix of coordinates of $\mathbf{u}$ with respect to the basis $\mathbf{u}'_1, \mathbf{u}'_2, \ldots, \mathbf{u}'_n$. Let Y' be the column matrix of coordinates of $f(\mathbf{u})$ with respect to the basis $\mathbf{v}'_1, \mathbf{v}'_2, \ldots, \mathbf{v}'_m$. Then $X' = PX$, $X = P^{-1}X'$, $Y' = QY$, and $Y = Q^{-1}Y'$. Therefore, $Q^{-1}Y' = A(P^{-1}X') = (AP^{-1})X'$, and $Y' = (QAP^{-1})X'$. This completes the proof.

EXAMPLE 4.4.2 In Example 4.3.2, we considered a linear transformation f from R^3 to R^3 consisting of projection on the plane given implicitly by $x - 2y + z = 0$. Relative to the standard basis in both domain and range space, the representation of f was

$$\begin{pmatrix} x' \\ y' \\ z' \end{pmatrix} = \begin{pmatrix} \frac{5}{6} & \frac{1}{3} & -\frac{1}{6} \\ \frac{1}{3} & \frac{1}{3} & \frac{1}{3} \\ -\frac{1}{6} & \frac{1}{3} & \frac{5}{6} \end{pmatrix} \begin{pmatrix} x \\ y \\ z \end{pmatrix}$$

If we now change the basis in both domain and range space to (1,1,1), (1,0,−1), (1,−2,1), we change the representation. In this case, $Q = P$, and

$$P^{-1} = \begin{pmatrix} 1 & 1 & 1 \\ 1 & 0 & -2 \\ 1 & -1 & 1 \end{pmatrix}$$

Computing the inverse, we have

$$P = \begin{pmatrix} \frac{1}{3} & \frac{1}{3} & \frac{1}{3} \\ \frac{1}{2} & 0 & -\frac{1}{2} \\ \frac{1}{6} & -\frac{1}{3} & \frac{1}{6} \end{pmatrix}$$

The new representation of f is then

$$B = PAP^{-1} = \begin{pmatrix} \frac{1}{3} & \frac{1}{3} & \frac{1}{3} \\ \frac{1}{2} & 0 & -\frac{1}{2} \\ \frac{1}{6} & -\frac{1}{3} & \frac{1}{6} \end{pmatrix} \begin{pmatrix} \frac{5}{6} & \frac{1}{3} & -\frac{1}{6} \\ \frac{1}{3} & \frac{1}{3} & \frac{1}{3} \\ -\frac{1}{6} & \frac{1}{3} & \frac{5}{6} \end{pmatrix} \begin{pmatrix} 1 & 1 & 1 \\ 1 & 0 & -2 \\ 1 & -1 & 1 \end{pmatrix}$$

$$= \begin{pmatrix} \frac{1}{3} & \frac{1}{3} & \frac{1}{3} \\ \frac{1}{2} & 0 & -\frac{1}{2} \\ 0 & 0 & 0 \end{pmatrix} \begin{pmatrix} 1 & 1 & 1 \\ 1 & 0 & -2 \\ 1 & -1 & 1 \end{pmatrix}$$

$$= \begin{pmatrix} 1 & 0 & 0 \\ 0 & 1 & 0 \\ 0 & 0 & 0 \end{pmatrix}$$

The case illustrated in this example will turn out to be especially important, that is, the case where $U = V$, A is the representation of a linear transformation with respect to some basis $\mathbf{u}_1, \mathbf{u}_2, \ldots, \mathbf{u}_n$ in both the domain and range space, and P is the matrix which gives the change of coordinates when a new basis $\mathbf{u}'_1, \mathbf{u}'_2, \ldots, \mathbf{u}'_n$ is introduced in both domain and range spaces. The new representation of the linear transformation is $B = PAP^{-1}$. In this case, all the matrices are $n \times n$, where n is the common dimension of the domain and range spaces. The transformation of an $n \times n$ matrix according to the equation $B = PAP^{-1}$ is called a *similarity transformation*, and we say that B is *similar* to A.

Definition 4.4.1 Let A and B be $n \times n$ matrices. If there exists an invertible $n \times n$ matrix P such that $B = PAP^{-1}$, then we say that B is similar to A and B is obtained from A by a similarity transformation.

Theorem 4.4.4† Let A, B, and C be $n \times n$ matrices. Then (i) A is similar to A, for all A, (ii) if A is similar to B, then B is similar to A, and (iii) if A is similar to B, and B is similar to C, then A is similar to C.

PROOF (i) The $n \times n$ identity matrix is invertible, and $A = IAI^{-1}$.

(ii) If A is similar to B, there is an invertible matrix P such that $A = PBP^{-1}$. But then $B = P^{-1}AP = P^{-1}A(P^{-1})^{-1}$.

(iii) If A is similar to B, there is an invertible matrix P such that $A = PBP^{-1}$. If B is similar to C, there is an invertible matrix Q such that $B = QCQ^{-1}$. Then

$$A = PBP^{-1} = P(QCQ^{-1})P^{-1} = (PQ)C(Q^{-1}P^{-1}) = SCS^{-1}$$

where $S = PQ$.

Some of the other important properties of similarity transformations are given by the next theorem.

Theorem 4.4.5

(i) If A is similar to B, then $|A| = |B|$.

(ii) If A_1 is similar to B_1 and A_2 is similar to B_2 under the same similarity transformation, then $A_1 + A_2$ is similar to $B_1 + B_2$.

(iii) If A is similar to B, then A^k is similar to B^k under the same similarity transformation for any positive integer k.

(iv) If A is similar to B, then $p(A)$ is similar to $p(B)$ under the same similarity transformation, where p is a polynomial.‡

(v) If A is similar to B and A is nonsingular, then B is nonsingular and A^{-1} is similar to B^{-1}.

PROOF (i) There exists a nonsingular matrix P such that $A = PBP^{-1}$. Hence $|A| = |P|\,|B|\,|P^{-1}| = |B|\,|P|\,|P^{-1}| = |B|$, since $|P|\,|P^{-1}| = |PP^{-1}| = |I| = 1$.

(ii) There exists a nonsingular matrix P such that $A_1 = PB_1P^{-1}$ and $A_2 = PB_2P^{-1}$. Then

$$A_1 + A_2 = PB_1P^{-1} + PB_2P^{-1} = P(B_1 + B_2)P^{-1}$$

† This theorem shows that similarity is an equivalence relation.

‡ If A and B are real, p is to have real coefficients; while if A and B are complex, p is to have complex coefficients.

(iii) There exists a nonsingular matrix P such that $A = PBP^{-1}$. Then $A^2 = (PBP^{-1})(PBP^{-1}) = PB(P^{-1}P)BP^{-1} = P(BB)P^{-1} = PB^2P^{-1}$ The rest follows by induction.

(iv) There is a nonsingular matrix P such that $A = PBP^{-1}$. Let c be a scalar. Then $cA = c(PBP^{-1}) = P(cB)P^{-1}$. Therefore, similarity is preserved under multiplication by a scalar. Now let $p(A) = a_0I + a_1A + a_2A^2 + \cdots + a_kA^k$. Then by (ii) and (iii) of this theorem,

$$\begin{aligned} Pp(B)P^{-1} &= a_0PIP^{-1} + a_1PBP^{-1} + a_2PB^2P^{-1} + \cdots + a_kPB^kP^{-1} \\ &= a_0I + a_1A + a_2A^2 + \cdots + a_kA^k = p(A) \end{aligned}$$

(v) There is a nonsingular matrix P such that $A = PBP^{-1}$. By (i), $|B| = |A| \neq 0$. Therefore, B is nonsingular. Also $A^{-1} = (PBP^{-1})^{-1} = PB^{-1}P^{-1}$, showing that A^{-1} is similar to B^{-1}.

EXERCISES 4.4

1 Let $U = V = R^2$, and let $f(\mathbf{u})$ be the reflection of $\mathbf{u}$ in the line $x = y$. Find the matrix representation of f:

(*a*) Relative to the standard basis in both U and V.

(*b*) Relative to the standard basis in U and the basis (1,1), (1, −1) in V.

(*c*) Relative to the standard basis in V and the basis (1,1), (1, −1) in U.

(*d*) Relative to the basis (1,1), (1, −1) in both U and V.

2 Let $U = V = R^3$, and let $f(\mathbf{u})$ be the reflection of $\mathbf{u}$ in the plane given implicitly by $x + y + z = 0$. Find the matrix representation of f:

(*a*) Relative to the standard basis in both U and V.

(*b*) Relative to the standard basis in U and the basis (1,0, −1), (1, −2,1), (1,1,1) in V.

(*c*) Relative to the standard basis in V and the basis (1,0, −1), (1, −2,1), (1,1,1) in U.

(*d*) Relative to the basis (1,0, −1), (1, −2,1), (1,1,1) in both U and V.

3 Show that the matrix representing the change of basis from one orthonormal set to another in a complex vector space is a unitary matrix.

4 Consider the linear transformation of Example 4.3.4. Find a vector which is transformed into itself. Use this vector and two other vectors orthogonal to it and to each other as a basis. Find the representation with respect to the new basis in both domain and range space.

5 Show that if A is similar to B and A is nonsingular, then A^k is similar to B^k for all integers k.

6 Suppose A is similar to a diagonal matrix D with diagonal elements $\lambda_1, \lambda_2, \ldots, \lambda_n$ such that $|\lambda_i| < 1$ for $i = 1, 2, \ldots, n$. Let $p_k(A) = I + A + A^2 + \cdots + A^k = P(I + D + D^2 + \cdots + D^k)P^{-1}$, since $A = PDP^{-1}$. Consider $\lim_{k \to \infty} p_k(A)$. Show that this limit exists. If we denote the series $I + A + A^2 + \cdots$ by B, prove that $B = (I - A)^{-1}$. *Hint*: Consider $\lim_{k\to\infty} (I - A)p_k(A)$ and $\lim_{k\to\infty} p_k(A)(I - A)$.

7 Suppose A is similar to a diagonal matrix D with diagonal elements $\lambda_1, \lambda_2, \ldots, \lambda_n$. Let

$$\begin{aligned} p_k(A) &= I + \frac{A}{1!} + \frac{A^2}{2!} + \cdots + \frac{A^k}{k!} \\ &= P\left(I + \frac{D}{1!} + \frac{D^2}{2!} + \cdots + \frac{D^k}{k!}\right)P^{-1} \end{aligned}$$

since $A = PDP^{-1}$. Show that $\lim_{k\to\infty} p_k(A)$ exists and is equal to

$$P\begin{pmatrix} e^{\lambda_1} & 0 & \cdots & 0 \\ 0 & e^{\lambda_2} & \cdots & 0 \\ \cdots & \cdots & \cdots & \cdots \\ 0 & 0 & \cdots & e^{\lambda_n} \end{pmatrix}P^{-1}$$

4.5 CHARACTERISTIC VALUES AND CHARACTERISTIC VECTORS

In this section, we consider only linear transformations for which the domain is a subspace of the range space. Suppose the range space of f is a complex vector space, and suppose there is a complex number λ and a nonzero vector $\mathbf{u}$ such that $f(\mathbf{u}) = \lambda\mathbf{u}$. Then we say that λ is a *characteristic value* (eigenvalue) of f and $\mathbf{u}$ is a *characteristic vector* (eigenvector) of f corresponding to λ.

Definition 4.5.1 Let f be a linear transformation from the complex (real) vector space U to V, where U is contained in V. Let λ be a complex (real) number and $\mathbf{u}$ be a nonzero vector in U such that $f(\mathbf{u}) = \lambda\mathbf{u}$. Then λ is a characteristic value of f, and $\mathbf{u}$ is a characteristic vector of f corresponding to λ.

EXAMPLE 4.5.1 Let f be the identity transformation from the complex vector space U to U. Then $f(\mathbf{u}) = \mathbf{u}$, and clearly $\lambda = 1$ is a characteristic value of f with corresponding characteristic vector $\mathbf{u} \neq \mathbf{0}$. Therefore, every nonzero vector is a characteristic vector of f.

EXAMPLE 4.5.2 Let f be the zero transformation from the complex vector space U to U. Then $f(\mathbf{u}) = \mathbf{0} = 0\mathbf{u}$, and clearly 0 is a characteristic value of f

with corresponding characteristic vector $\mathbf{u} \neq \mathbf{0}$. Therefore, every nonzero vector is a characteristic vector of f.

EXAMPLE 4.5.3 Let f be the rotation from R^2 to R^2 of Example 4.2.3. Since R^2 is a real vector space, any characteristic values must be real. However, it is clear geometrically that there are no nonzero vectors which are rotated into multiples of themselves unless θ is a multiple of 180°. If θ is an even multiple of 180°, then every nonzero vector is rotated into itself and is therefore a characteristic vector corresponding to the characteristic value 1. If θ is an odd multiple of 180°, then every nonzero vector is reversed in direction and is a characteristic vector corresponding to the characteristic value -1. For all other values of θ there are no characteristic values or characteristic vectors. We can reach the same conclusions algebraically by the following method. Relative to the standard basis we have the representation

$$\begin{pmatrix} x' \\ y' \end{pmatrix} = \begin{pmatrix} \cos\theta & -\sin\theta \\ \sin\theta & \cos\theta \end{pmatrix} \begin{pmatrix} x \\ y \end{pmatrix}$$

Now if $f(\mathbf{u}) = \lambda\mathbf{u}$, $\mathbf{u} = (x,y)$, then

$$\begin{pmatrix} \lambda x \\ \lambda y \end{pmatrix} = \begin{pmatrix} \cos\theta & -\sin\theta \\ \sin\theta & \cos\theta \end{pmatrix} \begin{pmatrix} x \\ y \end{pmatrix}$$

or

$$\begin{pmatrix} \cos\theta - \lambda & -\sin\theta \\ \sin\theta & \cos\theta - \lambda \end{pmatrix} \begin{pmatrix} x \\ y \end{pmatrix} = \begin{pmatrix} 0 \\ 0 \end{pmatrix}$$

These equations have nontrivial solutions if and only if

$$\begin{vmatrix} \cos\theta - \lambda & -\sin\theta \\ \sin\theta & \cos\theta - \lambda \end{vmatrix} = \lambda^2 - 2\lambda\cos\theta + 1 = 0$$

or $\lambda = \cos\theta \pm \sqrt{\cos^2\theta - 1}$. But $\cos^2\theta \leq 1$, and so all solutions will be complex unless $\cos\theta = \pm 1$ or, in other words, θ is a multiple of 180°. If θ is an even multiple of 180°, then $\lambda = 1$ is the only characteristic value. If θ is an odd multiple of 180°, then $\lambda = -1$ is the only characteristic value. In either case,

$$\begin{pmatrix} \cos\theta - \lambda & -\sin\theta \\ \sin\theta & \cos\theta - \lambda \end{pmatrix} = \begin{pmatrix} 0 & 0 \\ 0 & 0 \end{pmatrix}$$

and so (x,y) is arbitrary, confirming that every nonzero vector is a characteristic vector.

EXAMPLE 4.5.4 Let f be the linear transformation of Example 4.2.5. Find all characteristic values and characteristic vectors of f. Here the representation relative to the standard basis is

$$\begin{pmatrix} x' \\ y' \end{pmatrix} = \begin{pmatrix} \frac{1}{2} & \frac{1}{2} \\ \frac{1}{2} & \frac{1}{2} \end{pmatrix} \begin{pmatrix} x \\ y \end{pmatrix}$$

If $f(\mathbf{u}) = \lambda\mathbf{u}$, $\mathbf{u} = (x,y)$, then

$$\begin{pmatrix} \lambda x \\ \lambda y \end{pmatrix} = \begin{pmatrix} \frac{1}{2} & \frac{1}{2} \\ \frac{1}{2} & \frac{1}{2} \end{pmatrix} \begin{pmatrix} x \\ y \end{pmatrix}$$

or

$$\begin{pmatrix} \frac{1}{2} - \lambda & \frac{1}{2} \\ \frac{1}{2} & \frac{1}{2} - \lambda \end{pmatrix} \begin{pmatrix} x \\ y \end{pmatrix} = \begin{pmatrix} 0 \\ 0 \end{pmatrix}$$

which has nontrivial solutions if and only if

$$\begin{vmatrix} \frac{1}{2} - \lambda & \frac{1}{2} \\ \frac{1}{2} & \frac{1}{2} - \lambda \end{vmatrix} = \lambda^2 - \lambda = \lambda(\lambda - 1) = 0$$

There are two characteristic values, 0 and 1. If $\lambda = 0$, the equations reduce to $x + y = 0$ or $y = -x$. Hence, any nonzero multiple of $(1,-1)$ is a characteristic vector corresponding to $\lambda = 0$. If $\lambda = 1$, the equations reduce to $x - y = 0$ or $x = y$. In this case, any nonzero multiple of $(1,1)$ is a characteristic vector corresponding to $\lambda = 1$.

EXAMPLE 4.5.5 Let U be the space of all real-valued continuous functions of the real variable x with a continuous derivative. Let V be the space of continuous functions of x. Let T be the operation of differentiation; that is, $T[f(x)] = f'(x)$. Find all the characteristic values and characteristic vectors of T. Let λ be real and consider the equation $T[f(x)] = f'(x) = \lambda f(x)$. Multiplying by $e^{-\lambda x}$, we have

$$e^{-\lambda x}f'(x) = \lambda e^{-\lambda x}f(x)$$

or

$$[e^{-\lambda x}f(x)]' = e^{-\lambda x}f'(x) - \lambda e^{-\lambda x}f(x) = 0$$

$$e^{-\lambda x}f(x) = K$$

where K is a constant. Hence, $f(x) = Ke^{\lambda x}$, for $K \neq 0$, is a characteristic vector corresponding to the characteristic value λ, where λ is any real number. Finally, we show that we have found all the characteristic vectors. Suppose that corresponding to λ there is a vector $g(x)$ such that $g'(x) = \lambda g(x)$. Let $g(0) = c$. There is a K such that $f(0) = K = g(0) = c$ (for this part of the discussion K

may be zero). Now consider $h = f - g$. Then $h' = f' - g' = \lambda(f - g) = \lambda h$, and $h(0) = f(0) - g(0) = 0$. Multiplying by $e^{-\lambda x}$, we have

$$e^{-\lambda x}h'(x) - \lambda e^{-\lambda x}h(x) = [e^{-\lambda x}h(x)]' = 0$$

and $e^{-\lambda x}h(x) = \gamma$ (a constant). But $\gamma = 0$ since $h(0) = 0$. Therefore, $h(x) = 0$ and $f(x) = g(x)$.

Now let us consider the case where the domain and range space of the linear transformation are the same and of dimension n. If we use the basis $\mathbf{u}_1, \mathbf{u}_2, \ldots, \mathbf{u}_n$ in both domain and range space, then there is a matrix representation $Y = AX$, where X is the column matrix of coordinates of $\mathbf{u}$ with respect to the given basis and Y is the column matrix of coordinates of $f(\mathbf{u})$. Now suppose we look for characteristic values of f. Suppose $f(\mathbf{u}) = \lambda\mathbf{u}$. Then $Y = \lambda X = AX$, or $(A - \lambda I)X = 0$. These equations have nontrivial solutions if and only if

$$|A - \lambda I| = c_0 + c_1\lambda + c_2\lambda^2 + \cdots + (-1)^n\lambda^n = 0$$

The polynomial $p(\lambda) = c_0 + c_1\lambda + c_2\lambda^2 + \cdots + (-1)^n\lambda^n$ is called the *characteristic polynomial* of A, and the equation $p(\lambda) = 0$ is called the *characteristic equation.* From the theory of such equations, we know that the characteristic equation must have at least one solution and can have at most n distinct solutions. We also know that $p(\lambda)$ can be factored as follows:

$$p(\lambda) = (\lambda_1 - \lambda)^{k_1}(\lambda_2 - \lambda)^{k_2} \cdots (\lambda_r - \lambda)^{k_r}$$

where $\lambda_1, \lambda_2, \ldots, \lambda_r$ are the r distinct roots of $p(\lambda)$ and the positive integers $k_1, k_2, \ldots, k_r$ are the multiplicities of $\lambda_1, \lambda_2, \ldots, \lambda_r$, respectively. Also $k_1 + k_2 + \cdots + k_r = n$. Therefore, if we can find the factorization of $p(\lambda)$, we know all the distinct characteristic values of the linear transformation f with representation A. If λ_j is a characteristic value of f, with multiplicity k_j in the characteristic polynomial, we say that λ_j is a characteristic value with multiplicity k_j.

Definition 4.5.2 Let A be an $n \times n$ complex matrix. Let $p(\lambda) = |A - \lambda I|$ be the characteristic polynomial of A. If λ_j is a root of $p(\lambda)$ with multiplicity k_j, then we say that λ_j is a characteristic value of A with multiplicity k_j. The nontrivial solutions of $(A - \lambda I)X = 0$ are called characteristic vectors of A.

It is important to note that, according to our definition, real matrices can have complex characteristic values (see Example 4.5.8). After all, real numbers are just complex numbers with imaginary part zero. Furthermore, a linear transformation defined on a complex vector space could have a real representation with respect to some basis, and we would definitely be interested in the complex characteristic values of the transformation. However, if the linear transformation is defined on a real vector space, then it can have only real characteristic values; and if its representative matrix A has a complex characteristic value, it cannot be a characteristic value of the transformation (see Example 4.5.3 for the case where θ is not a multiple of 180°).

Theorem 4.5.1 Let f be a linear transformation from the n-dimensional vector space U to U, with matrix representation A with respect to some basis. All the characteristic values and characteristic vectors of f can be found by finding the characteristic values and vectors of A. If U is a complex vector space, f will have a characteristic value λ if and only if λ is a characteristic value of A. If U is a real vector space, f will have a characteristic value λ if and only if λ is a real characteristic value of A.

PROOF The proof is included in the above discussion except for the problem of showing that the characteristic values are independent of the particular representation of the transformation. If we use a different basis, the representation changes to $B = PAP^{-1}$, where P is nonsingular. Consider the characteristic polynomial for B:

$$|B - \lambda I| = |PAP^{-1} - \lambda PP^{-1}| = |P|\,|P^{-1}|\,|A - \lambda I| = |A - \lambda I|$$

This shows that A and B have the same characteristic polynomial, and so the characteristic values are independent of the representation.

EXAMPLE 4.5.6 Find all the characteristic values and characteristic vectors of the matrix

$$A = \begin{pmatrix} 8 & 9 & 9 \\ 3 & 2 & 3 \\ -9 & -9 & -10 \end{pmatrix}$$

The characteristic equation is

$$\begin{vmatrix} 8-\lambda & 9 & 9 \\ 3 & 2-\lambda & 3 \\ -9 & -9 & -10-\lambda \end{vmatrix} = -\lambda^3 + 3\lambda + 2 = (\lambda + 1)^2(2 - \lambda) = 0$$

The characteristic values are $\lambda_1 = 2$, with multiplicity 1, and $\lambda_2 = -1$, with multiplicity 2. If $\lambda = \lambda_1 = 2$, we can find characteristic vectors by solving

$$\begin{pmatrix} 6 & 9 & 9 \\ 3 & 0 & 3 \\ -9 & -9 & -12 \end{pmatrix} \begin{pmatrix} x \\ y \\ z \end{pmatrix} = \begin{pmatrix} 0 \\ 0 \\ 0 \end{pmatrix}$$

or $2x + 3y + 3z = 0$, $x + z = 0$. This system has a one-parameter family of nontrivial solutions of the form $\mathbf{u} = a(3,1,-3)$. Therefore, we have a characteristic vector $\mathbf{u}_1 = (3,1,-3)$, and any other characteristic vector corresponding to λ_1 will be a multiple of $\mathbf{u}_1$. If $\lambda = \lambda_2 = -1$, we must solve the following system

$$\begin{pmatrix} 9 & 9 & 9 \\ 3 & 3 & 3 \\ -9 & -9 & -9 \end{pmatrix} \begin{pmatrix} x \\ y \\ z \end{pmatrix} = \begin{pmatrix} 0 \\ 0 \\ 0 \end{pmatrix}$$

or $x + y + z = 0$. This equation has a two-parameter family of solutions, which can be written as $\mathbf{u} = a(1,-1,0) + b(0,1,-1)$. Therefore, corresponding to λ_2 we have two independent characteristic vectors $\mathbf{u}_2 = (1,-1,0)$ and $\mathbf{u}_3 = (0,1,-1)$, and any other characteristic vector corresponding to λ_2 will be a linear combination of $\mathbf{u}_2$ and $\mathbf{u}_3$.

EXAMPLE 4.5.7 Find all the characteristic values and characteristic vectors of the matrix

$$A = \begin{pmatrix} 1 & 2 & 3 \\ 0 & 2 & 3 \\ 0 & 0 & 2 \end{pmatrix}$$

The characteristic equation is

$$\begin{vmatrix} 1-\lambda & 2 & 3 \\ 0 & 2-\lambda & 3 \\ 0 & 0 & 2-\lambda \end{vmatrix} = (1-\lambda)(2-\lambda)^2 = 0$$

The characteristic values are $\lambda_1 = 1$, with multiplicity 1, and $\lambda_2 = 2$, with multiplicity 2. If $\lambda = \lambda_1 = 1$, we can find characteristic vectors by solving

$$\begin{pmatrix} 0 & 2 & 3 \\ 0 & 1 & 3 \\ 0 & 0 & 1 \end{pmatrix} \begin{pmatrix} x \\ y \\ z \end{pmatrix} = \begin{pmatrix} 0 \\ 0 \\ 0 \end{pmatrix}$$

or $2y + 3z = 0$, $y + 3z = 0$, $z = 0$. This implies that $y = z = 0$. However,

x is arbitrary, so any vector of the form $\mathbf{u} = a(1,0,0)$ is a characteristic vector corresponding to λ_1 if $a \neq 0$. If $\lambda = \lambda_2 = 2$, we must solve

$$\begin{pmatrix} -1 & 2 & 3 \\ 0 & 0 & 3 \\ 0 & 0 & 0 \end{pmatrix} \begin{pmatrix} x \\ y \\ z \end{pmatrix} = \begin{pmatrix} 0 \\ 0 \\ 0 \end{pmatrix}$$

or $-x + 2y + 3z = 0$, $z = 0$. Therefore, $x = 2y$, and we have a one-parameter family of solutions of the form $\mathbf{u} = a(2,1,0)$. Hence, any characteristic vector corresponding to λ_2 will be a multiple of (2,1,0).

EXAMPLE 4.5.8 Find all the characteristic values and characteristic vectors of the matrix

$$A = \begin{pmatrix} 1 & -2 \\ 1 & -1 \end{pmatrix}$$

The characteristic equation is

$$\begin{vmatrix} 1 - \lambda & -2 \\ 1 & -1 - \lambda \end{vmatrix} = \lambda^2 + 1 = (\lambda - i)(\lambda + i) = 0$$

The characteristic values are $\lambda_1 = i$ and $\lambda_2 = -i$. If $\lambda = \lambda_1 = i$, we must solve

$$\begin{pmatrix} 1 - i & -2 \\ 1 & -1 - i \end{pmatrix} \begin{pmatrix} x \\ y \end{pmatrix} = \begin{pmatrix} 0 \\ 0 \end{pmatrix}$$

or $x = (1 + i)y$. Therefore, characteristic vectors corresponding to λ_1 are of the form $\mathbf{u} = a(1 + i, 1)$, $a \neq 0$. If $\lambda = \lambda_2 = -i$, we must solve

$$\begin{pmatrix} 1 + i & -2 \\ 1 & -1 + i \end{pmatrix} \begin{pmatrix} x \\ y \end{pmatrix} = \begin{pmatrix} 0 \\ 0 \end{pmatrix}$$

or $x = (1 - i)y$. Therefore, characteristic vectors corresponding to λ_2 are of the form $\mathbf{u} = a(1 - i, 1)$, $a \neq 0$.

Let us return to Example 4.5.6. We found a set of three independent characteristic vectors $\mathbf{u}_1 = (3,1,-3)$, $\mathbf{u}_2 = (1,-1,0)$, and $\mathbf{u}_3 = (0,1,-1)$. Let us assume for the moment that the matrix

$$A = \begin{pmatrix} 8 & 9 & 9 \\ 3 & 2 & 3 \\ -9 & -9 & -10 \end{pmatrix}$$

is the representation of a linear transformation from R^3 to R^3 relative to the standard basis. Suppose we introduce the basis $\mathbf{u}_1, \mathbf{u}_2, \mathbf{u}_3$. We wish to find the representation of f relative to this basis. Since $\mathbf{u}_1, \mathbf{u}_2, \mathbf{u}_3$ are characteristic vectors, we have

$$\begin{aligned} f(\mathbf{u}_1) &= \lambda_1\mathbf{u}_1 = 2\mathbf{u}_1 + 0\mathbf{u}_2 + 0\mathbf{u}_3 \\ f(\mathbf{u}_2) &= \lambda_2\mathbf{u}_2 = 0\mathbf{u}_1 - \mathbf{u}_2 + 0\mathbf{u}_3 \\ f(\mathbf{u}_3) &= \lambda_2\mathbf{u}_3 = 0\mathbf{u}_1 + 0\mathbf{u}_2 - \mathbf{u}_3 \end{aligned}$$

Therefore, the representation we seek is given by the matrix

$$B = \begin{pmatrix} 2 & 0 & 0 \\ 0 & -1 & 0 \\ 0 & 0 & -1 \end{pmatrix}$$

which is diagonal. This is not just a coincidence, as we see from the next theorem.

Theorem 4.5.2 Let f be a linear transformation from the n-dimensional space U to U. Then f has a diagonal representation if and only if f has n independent characteristic vectors.

PROOF Suppose relative to the basis $\mathbf{u}_1, \mathbf{u}_2, \ldots, \mathbf{u}_n$, f has the representation matrix

$$A = \begin{pmatrix} \lambda_1 & 0 & 0 & \cdots & 0 \\ 0 & \lambda_2 & 0 & \cdots & 0 \\ 0 & 0 & \lambda_3 & \cdots & 0 \\ \cdots & \cdots & \cdots & \cdots & \cdots \\ 0 & 0 & 0 & \cdots & \lambda_n \end{pmatrix}$$

where the λ's are not necessarily distinct. This means that $f(\mathbf{u}_1) = \lambda_1\mathbf{u}_1$, $f(\mathbf{u}_2) = \lambda_2\mathbf{u}_2, \ldots, f(\mathbf{u}_n) = \lambda_n\mathbf{u}_n$, which means that $\mathbf{u}_1, \mathbf{u}_2, \ldots, \mathbf{u}_n$ are characteristic vectors corresponding to the characteristic values $\lambda_1, \lambda_2, \ldots, \lambda_n$. Conversely, suppose $\mathbf{u}_1, \mathbf{u}_2, \ldots, \mathbf{u}_n$ are independent characteristic vectors corresponding to characteristic values $\lambda_1, \lambda_2, \ldots, \lambda_n$ (not necessarily distinct). We obtain the representation relative to $\mathbf{u}_1, \mathbf{u}_2, \ldots, \mathbf{u}_n$ as a basis.

$$\begin{aligned} f(\mathbf{u}_1) &= \lambda_1\mathbf{u}_1 = \lambda_1\mathbf{u}_1 + 0\mathbf{u}_2 + \cdots + 0\mathbf{u}_n \\ f(\mathbf{u}_2) &= \lambda_2\mathbf{u}_2 = 0\mathbf{u}_1 + \lambda_2\mathbf{u}_2 + \cdots + 0\mathbf{u}_n \\ &\cdots\cdots\cdots\cdots\cdots\cdots\cdots\cdots \\ f(\mathbf{u}_n) &= \lambda_n\mathbf{u}_n = 0\mathbf{u}_1 + 0\mathbf{u}_2 + \cdots + \lambda_n\mathbf{u}_n \end{aligned}$$

and so the representation is diagonal.

Theorem 4.5.3 Let f be a linear transformation from the n-dimensional space U to U. Then f will have a diagonal representation if f has n distinct characteristic values.

PROOF Let the distinct characteristic values be $\lambda_1, \lambda_2, \ldots, \lambda_n$ and the corresponding characteristic vectors $\mathbf{u}_1, \mathbf{u}_2, \ldots, \mathbf{u}_n$. We shall show that the $\mathbf{u}$'s are independent. Suppose that the $\mathbf{u}$'s are dependent. Then there is an independent set $\mathbf{u}_1, \mathbf{u}_2, \ldots, \mathbf{u}_k$ such that

$$\mathbf{u}_{k+1} = c_1\mathbf{u}_1 + c_2\mathbf{u}_2 + \cdots + c_k\mathbf{u}_k$$

and

$$f(\mathbf{u}_{k+1}) = \lambda_{k+1}\mathbf{u}_{k+1} = \sum_{j=1}^{k} c_j f(\mathbf{u}_j) = \sum_{j=1}^{k} \lambda_j c_j \mathbf{u}_j$$

Also $\lambda_{k+1}\mathbf{u}_{k+1} = \sum_{j=1}^{k} \lambda_{k+1} c_j \mathbf{u}_j$; subtracting, we have

$$\mathbf{0} = \sum_{j=1}^{k} c_j(\lambda_j - \lambda_{k+1})\mathbf{u}_j$$

But $\lambda_j \neq \lambda_{k+1}$, $j = 1, 2, \ldots, k$, and so $c_1 = c_2 = \cdots = c_k = 0$ because $\mathbf{u}_1, \mathbf{u}_2, \ldots, \mathbf{u}_k$ are independent. Therefore, $\mathbf{u}_{k+1} = \mathbf{0}$, which is clearly impossible since $\mathbf{u}_{k+1}$ is a characteristic vector. Hence, $\mathbf{u}_1, \mathbf{u}_2, \ldots, \mathbf{u}_n$ are independent and the theorem follows from Theorem 4.5.2.

EXAMPLE 4.5.9 Let f be a linear transformation from a three-dimensional real vector space U to U. Suppose relative to some basis $\mathbf{u}_1, \mathbf{u}_2, \mathbf{u}_3$, f has the representation

$$A = \begin{pmatrix} 9 & -3 & 0 \\ -3 & 12 & -3 \\ 0 & -3 & 9 \end{pmatrix}$$

Find, if possible, a diagonal representation for f. We look for the characteristic values of A. The characteristic equation is

$$\begin{vmatrix} 9-\lambda & -3 & 0 \\ -3 & 12-\lambda & -3 \\ 0 & -3 & 9-\lambda \end{vmatrix} = (6-\lambda)(9-\lambda)(15-\lambda) = 0$$

The characteristic values are $\lambda_1 = 6$, $\lambda_2 = 9$, $\lambda_3 = 15$. They are real and distinct. Therefore, by Theorem 4.5.3, there is a diagonal representation

$$B = \begin{pmatrix} 6 & 0 & 0 \\ 0 & 9 & 0 \\ 0 & 0 & 15 \end{pmatrix}$$

Let us find the basis (characteristic vectors) relative to which this is the representation. If $\lambda = \lambda_1 = 6$, we must solve

$$\begin{pmatrix} 3 & -3 & 0 \\ -3 & 6 & -3 \\ 0 & -3 & 3 \end{pmatrix} \begin{pmatrix} x \\ y \\ z \end{pmatrix} = \begin{pmatrix} 0 \\ 0 \\ 0 \end{pmatrix}$$

or $x - y = 0$, $y - z = 0$. In other words, $x = y = z$ and a characteristic vector is $\mathbf{v}_1 = \mathbf{u}_1 + \mathbf{u}_2 + \mathbf{u}_3$. If $\lambda = \lambda_2 = 9$, we must solve

$$\begin{pmatrix} 0 & -3 & 0 \\ -3 & 3 & -3 \\ 0 & -3 & 0 \end{pmatrix} \begin{pmatrix} x \\ y \\ z \end{pmatrix} = \begin{pmatrix} 0 \\ 0 \\ 0 \end{pmatrix}$$

or $y = 0$, $x + z = 0$, and a corresponding characteristic vector is $\mathbf{v}_2 = \mathbf{u}_1 - \mathbf{u}_3$. If $\lambda = \lambda_3 = 15$, we must solve

$$\begin{pmatrix} -6 & -3 & 0 \\ -3 & -3 & -3 \\ 0 & -3 & -6 \end{pmatrix} \begin{pmatrix} x \\ y \\ z \end{pmatrix} = \begin{pmatrix} 0 \\ 0 \\ 0 \end{pmatrix}$$

or $2x + y = 0$, $x + y + z = 0$. A corresponding characteristic vector is $\mathbf{v}_3 = \mathbf{u}_1 - 2\mathbf{u}_2 + \mathbf{u}_3$.

In Example 4.5.7, we had two characteristic values and only two independent characteristic vectors in R^3. Therefore, it will not always be possible to find a diagonal representation. On the other hand, Example 4.5.6 illustrates that there may be n independent characteristic vectors even when there are not n distinct characteristic values. Hence, Theorem 4.5.3 gives a sufficient but not necessary condition for a diagonal representation. In the next section, we take up a couple of special cases where it will always be possible to obtain a diagonal representation.

EXERCISES 4.5

1 Find the characteristic values and characteristic vectors of the following matrices:

(*a*) $\begin{pmatrix} 2 & 1 \\ 1 & 2 \end{pmatrix}$ (*b*) $\begin{pmatrix} 1 & 1 \\ 0 & 1 \end{pmatrix}$ (*c*) $\begin{pmatrix} 1 & 1 \\ 0 & 2 \end{pmatrix}$ (*d*) $\begin{pmatrix} 1 & 3 \\ -2 & 1 \end{pmatrix}$

2 Find the characteristic values and characteristic vectors of the following matrices:

$$(a)\ \begin{pmatrix} 3 & 1 & 0 \\ 1 & 3 & 0 \\ 0 & 0 & 2 \end{pmatrix} \qquad (b)\ \begin{pmatrix} 2 & 1 & 2 \\ 0 & -1 & 3 \\ 0 & 0 & 3 \end{pmatrix} \qquad (c)\ \begin{pmatrix} 5 & 1 & 1 \\ -3 & 1 & -3 \\ -2 & -2 & -2 \end{pmatrix}$$

$$(d)\ \begin{pmatrix} 1 & 1 & 1 \\ 0 & 1 & 1 \\ 0 & 0 & 2 \end{pmatrix} \qquad (e)\ \begin{pmatrix} -8 & 5 & 4 \\ 5 & 3 & 1 \\ 4 & 1 & 0 \end{pmatrix}$$

3 Let $U = V = R^2$, and let $f(\mathbf{u})$ be the reflection of $\mathbf{u}$ in the line $y = -x$. Find all the characteristic values and characteristic vectors of f. Find a representation of f which is diagonal.

4 Let $U = V = R^3$, and let $f(\mathbf{u})$ be the projection of $\mathbf{u}$ on the plane given implicitly by $x + 2y - z = 0$. Find all the characteristic values and characteristic vectors of f. Find a representation of f which is diagonal.

5 Let $U = V = R^3$, and let f be a linear transformation with the representation matrix relative to the standard basis

$$A = \begin{pmatrix} 4 & -20 & -10 \\ -2 & 10 & 4 \\ 6 & -30 & -13 \end{pmatrix}$$

Find a basis with respect to which the representation is diagonal.

6 If λ is a characteristic value of a square matrix A, show that λ^n is a characteristic value of A^n, where n is a positive integer.

7 Show that a square matrix A is invertible if and only if $\lambda = 0$ is not a characteristic value of A.

8 Show that if λ is a characteristic value of an invertible matrix A, then λ^{-1} is a characteristic value of A^{-1}.

9 If λ is a characteristic value of a square matrix A, show that $\lambda^3 - 3\lambda^2 + \lambda - 2$ is a characteristic value of $A^3 - 3A^2 + A - 2I$.

10 If $p(\lambda) = 0$ is the characteristic equation of the $n \times n$ matrix A and A has n independent characteristic vectors $X_1, X_2, \ldots, X_n$, prove that $p(A) = 0$. *Hint*: Show that $p(A)X_i = 0$ for $i = 1, 2, \ldots, n$.

11 Show that if $\mathbf{u}_1, \mathbf{u}_2, \ldots, \mathbf{u}_k$ are a set of characteristic vectors of a linear transformation f corresponding to the same characteristic value λ, then they span a subspace S such that for any $\mathbf{u}$ in S, $f(\mathbf{u}) = \lambda\mathbf{u}$. *Note*: Such subspaces are called *invariant subspaces*.

12 Let f be a linear transformation from U to U, where U is n-dimensional. Show that f has a diagonal representation if and only if the sum of the dimensions of its invariant subspaces is n.

13 Suppose we want to find a matrix C such that $C^2 = A$. (C might be called a square root of A.) Suppose A is similar to a diagonal matrix B with diagonal elements $\lambda_1, \lambda_2, \ldots, \lambda_n$, with $\lambda_i \geq 0$. Then $B = PAP^{-1}$. Let D be a diagonal

matrix with diagonal elements $\pm\sqrt{\lambda_1}, \pm\sqrt{\lambda_2}, \ldots, \pm\sqrt{\lambda_n}$. Then $B = D^2$. Let $C = P^{-1}DP$. Show that $C^2 = A$. Use this method to find square roots of

$$A = \begin{pmatrix} 1 & 2 & 1 \\ 0 & 2 & 1 \\ 0 & 0 & 3 \end{pmatrix}$$

14 Solve the following system of equations:

$$\frac{dx}{dt} = 2x + y$$

$$\frac{dy}{dt} = x + 2y$$

Hint: Let $\mathbf{u} = \begin{pmatrix} x \\ y \end{pmatrix}$, and write the equations as $d\mathbf{u}/dt = A\mathbf{u}$. Find P such that $\mathbf{v} = P\mathbf{u}$. Hence, $P^{-1}\, d\mathbf{v}/dt = (AP^{-1})\mathbf{v}$ and $d\mathbf{v}/dt = (PAP^{-1})\mathbf{v}$. If

$$PAP^{-1} = \begin{pmatrix} \lambda_1 & 0 \\ 0 & \lambda_2 \end{pmatrix}$$

then the equations are separated.

15 Solve the following system of equations:

$$\frac{dx}{dt} = 3x + z$$

$$\frac{dy}{dt} = 3y + z$$

$$\frac{dz}{dt} = x + y + 2z$$

16 Consider the differential equation

$$\frac{d^2x}{dt^2} - 3\frac{dx}{dt} - 4x = 0$$

Look for solutions of the form $x = e^{\lambda t}$. Show that λ must be a root of the equation $\lambda^2 - 3\lambda - 4 = 0$. Show that the given equation is equivalent to the system $dx/dt = y$, $dy/dt = 4x + 3y$. Compare with Exercise 14.

4.6 SYMMETRIC AND HERMITIAN MATRICES

We saw, in the last section, that an $n \times n$ matrix is similar to a diagonal matrix if and only if it has n independent characteristic vectors. We also saw square matrices which are not similar to diagonal matrices. In this section, we shall study two types of matrices, real symmetric and complex hermitian, which are always similar to diagonal matrices. We shall begin with real symmetric matrices.

Definition 4.6.1 An $n \times n$ matrix A is symmetric if $A = \tilde{A}$.

Theorem 4.6.1 All the characteristic values of a real symmetric $n \times n$ matrix A are real.

PROOF Let X be a characteristic vector of A corresponding to the characteristic value λ. Then, if the bar stands for a complex conjugate,

$$AX = \lambda X$$
$$A\bar{X} = \bar{\lambda}\bar{X}$$
$$\tilde{X}A\bar{X} = \bar{\lambda}\tilde{X}\bar{X} = \bar{\lambda} \sum_{k=1}^{n} |x_k|^2$$
$$\tilde{\bar{X}}AX = \lambda\tilde{\bar{X}}X = \lambda \sum_{k=1}^{n} |x_k|^2$$

Subtracting, we have

$$(\lambda - \bar{\lambda}) \sum_{k=1}^{n} |x_k|^2 = \tilde{\bar{X}}AX - \tilde{X}A\bar{X} = 0$$

since $\widetilde{\tilde{X}A\bar{X}} = \tilde{\bar{X}}AX$. Therefore, since $\sum_{k=1}^{n} |x_k|^2 > 0$, $\lambda = \bar{\lambda}$, which proves the theorem.

Theorem 4.6.2 Characteristic vectors corresponding to different characteristic values of a real symmetric matrix A are orthogonal.

PROOF Let X_i and X_j be characteristic vectors corresponding to characteristic values λ_i and λ_j, where $\lambda_i \neq \lambda_j$. We can assume that X_i and X_j are real since A and λ_i and λ_j are all real. Then

$$AX_i = \lambda_i X_i$$
$$AX_j = \lambda_j X_j$$
$$\tilde{X}_j A X_i = \lambda_i \tilde{X}_j X_i = \lambda_i (X_i \cdot X_j)$$
$$\tilde{X}_i A X_j = \lambda_j \tilde{X}_i X_j = \lambda_j (X_i \cdot X_j)$$

where $(X_i \cdot X_j)$ is the scalar product of X_i and X_j. Now since $\widetilde{\tilde{X}_j A X_i} = \tilde{X}_i A X_j$, we have

$$(\lambda_i - \lambda_j)(X_i \cdot X_j) = 0$$

But $\lambda_i - \lambda_j \neq 0$ and hence, $(X_i \cdot X_j) = 0$.

Theorem 4.6.3 If a linear transformation f from R^n to R^n has a representation A with respect to the standard basis which is symmetric, then there is an orthogonal matrix P such that the representation $PA\tilde{P} = D$ is diagonal, with diagonal elements the characteristic values of A.

PROOF We prove the theorem by induction on n. If $n = 1$, A is already diagonal and $P = I$. Let us assume that the theorem is true for all spaces of dimension up to and including $n - 1$. Let X_1 be a unit characteristic vector corresponding to the characteristic value λ_1 (A has at least one characteristic value, and it is real). Then $AX_1 = \lambda_1 X_1$. Let S be the subspace of R^n orthogonal to X_1. S has a basis $Z_2, Z_3, \ldots, Z_n$, which we can assume is orthonormal. The change from the original basis (with respect to which the representation is A) to the basis $X_1, Z_2, Z_3, \ldots, Z_n$ is via the orthogonal matrix R. With respect to the new basis the representation is

$$B = RA\tilde{R} = \begin{pmatrix} \lambda_1 & 0 & 0 & \cdots & 0 \\ 0 & b_{22} & b_{23} & \cdots & b_{2n} \\ 0 & b_{32} & b_{33} & \cdots & b_{3n} \\ \cdots & \cdots & \cdots & \cdots & \cdots \\ 0 & b_{n2} & b_{n3} & \cdots & b_{nn} \end{pmatrix}$$

$$= \begin{pmatrix} \lambda_1 & 0 \; 0 \; \cdots \; 0 \\ 0 & \\ 0 & \begin{pmatrix} B^* \end{pmatrix} \\ \cdot & \\ 0 & \end{pmatrix}$$

B is symmetric since $\tilde{B} = \widetilde{RA\tilde{R}} = RA\tilde{R}$. The matrix B^* is $(n - 1) \times (n - 1)$ and is real and symmetric. By the induction hypothesis there is an orthogonal change of basis in S, represented by Q^*, which diagonalizes B^*. Therefore,

$$Q^*B^*\tilde{Q}^* = \begin{pmatrix} \lambda_2 & 0 & \cdots & 0 \\ 0 & \lambda_3 & \cdots & 0 \\ \cdots & \cdots & \cdots & \cdots \\ 0 & 0 & \cdots & \lambda_n \end{pmatrix}$$

is an $(n - 1) \times (n - 1)$ diagonal matrix. Now consider the $n \times n$ matrix

$$Q = \begin{pmatrix} 1 & 0 \; 0 \; \cdots \; 0 \\ 0 & \\ 0 & \begin{pmatrix} Q^* \end{pmatrix} \\ \cdot & \\ 0 & \end{pmatrix}$$

Q is orthogonal since Q^* is orthogonal, and

$$QB\tilde{Q} = Q(RA\tilde{R})\tilde{Q} = (QR)A(\widetilde{QR}) = PA\tilde{P}$$

where $P = QR$ is orthogonal, since $P^{-1} = (QR)^{-1} = R^{-1}Q^{-1} = \tilde{R}\tilde{Q} = \widetilde{QR} = P$. Now we multiply out $QB\tilde{Q}$, and we find

$$QB\tilde{Q} = \begin{pmatrix} \lambda_1 & 0 & 0 & \cdots & 0 \\ 0 & \lambda_2 & 0 & \cdots & 0 \\ 0 & 0 & \lambda_3 & \cdots & 0 \\ \cdots & \cdots & \cdots & \cdots & \cdots \\ 0 & 0 & 0 & \cdots & \lambda_n \end{pmatrix}$$

which proves the theorem.

Theorem 4.6.4 If a linear transformation f from U to U, where U is an n-dimensional real vector space, has a representation matrix A which is symmetric, then f has a diagonal representation with respect to n independent characteristic vectors.

PROOF Let $\mathbf{u}_1, \mathbf{u}_2, \ldots, \mathbf{u}_n$ be the basis with respect to which the representation matrix is A. Let $\mathbf{u} = x_1\mathbf{u}_1 + x_2\mathbf{u}_2 + \cdots + x_n\mathbf{u}_n$. Then X, the column matrix of coordinates of $\mathbf{u}$, is in R^n. Therefore, we can view the transformation as a transformation from R^n to R^n. Hence, Theorem 4.6.3 applies. If $PA\tilde{P} = D$, D diagonal, then $A\tilde{P} = \tilde{P}D$ and the columns of P are characteristic vectors of A. They are independent since $\tilde{P}$ is orthogonal. Hence, A has n orthonormal characteristic vectors. Suppose $(x_1', x_2', \ldots, x_n')$ is a characteristic vector of A. Then $\mathbf{v} = x_1'\mathbf{u}_1 + x_2'\mathbf{u}_2 + \cdots + x_n'\mathbf{u}_n$ is a characteristic vector of f. Therefore, f has independent characteristic vectors, and, using these vectors as a basis, we have a diagonal representation for f.

EXAMPLE 4.6.1 Let f be a linear transformation from the three-dimensional real vector space U to U with the representation matrix

$$A = \begin{pmatrix} 7 & -16 & -8 \\ -16 & 7 & 8 \\ -8 & 8 & -5 \end{pmatrix}$$

with respect to the basis $\mathbf{u}_1, \mathbf{u}_2, \mathbf{u}_3$. Find a basis with respect to which the representation of f is diagonal. The characteristic equation of A is

$$|A - \lambda I| = \begin{vmatrix} 7-\lambda & -16 & -8 \\ -16 & 7-\lambda & 8 \\ -8 & 8 & -5-\lambda \end{vmatrix} = -\lambda^3 + 9\lambda^2 + 405\lambda + 2{,}187 = 0$$

The characteristic values are $\lambda_1 = 27$, $\lambda_2 = \lambda_3 = -9$. To obtain a characteristic vector corresponding to λ_1 we must solve $(A - \lambda_1 I)X = 0$ or $5x + 4y +$

$2z = 0, 4x + 5y - 2z = 0$. A solution is $(2, -2, -1)$. To obtain characteristic vectors corresponding to $\lambda_2 = \lambda_3 = -9$, we must solve $(A - \lambda_2 I)X = 0$, or $2x - 2y - z = 0$. One solution is $(1,0,2)$. Another solution orthogonal to the first is $(4,5,-2)$. Notice that both these vectors are orthogonal to $(2,-2,-1)$. A set of independent characteristic vectors of f is

$$\begin{aligned} \mathbf{v}_1 &= 2\mathbf{u}_1 - 2\mathbf{u}_2 - \mathbf{u}_3 \\ \mathbf{v}_2 &= \mathbf{u}_1 + 2\mathbf{u}_3 \\ \mathbf{v}_3 &= 4\mathbf{u}_1 + 5\mathbf{u}_2 - 2\mathbf{u}_3 \end{aligned}$$

With respect to these vectors as a basis the representation of f is

$$B = \begin{pmatrix} 27 & 0 & 0 \\ 0 & -9 & 0 \\ 0 & 0 & -9 \end{pmatrix}$$

Next we consider the case of hermitian matrices, where the situation is quite similar.

Definition 4.6.2 An $n \times n$ matrix A is hermitian if $A = \tilde{\bar{A}}$.

Theorem 4.6.5 All the characteristic values of a hermitian $n \times n$ matrix A are real.

PROOF Let Z be a characteristic vector of A corresponding to the characteristic value λ. Then if the bar stands for a complex conjugate,

$$\begin{aligned} AZ &= \lambda Z \\ \bar{A}\bar{Z} &= \bar{\lambda}\bar{Z} \\ \tilde{Z}\bar{A}\bar{Z} &= \bar{\lambda}\tilde{Z}\bar{Z} = \bar{\lambda} \sum_{k=1}^{n} |z_k|^2 \\ \tilde{\bar{Z}}AZ &= \lambda\tilde{\bar{Z}}Z = \lambda \sum_{k=1}^{n} |z_k|^2 \end{aligned}$$

Subtracting, we have

$$(\lambda - \bar{\lambda}) \sum_{k=1}^{n} |z_k|^2 = \tilde{\bar{Z}}AZ - \tilde{Z}\bar{A}\bar{Z} = 0$$

since $\widetilde{\tilde{Z}\bar{A}\bar{Z}} = \tilde{\bar{Z}}\tilde{\bar{A}}Z = \tilde{\bar{Z}}AZ$. Therefore, since $\sum_{k=1}^{n} |z_k|^2 > 0$, $\lambda = \bar{\lambda}$, which proves the theorem.

Theorem 4.6.6 Characteristic vectors corresponding to different characteristic values of a hermitian matrix A are orthogonal.

PROOF Let Z_i and Z_j be characteristic vectors corresponding to characteristic values λ_i and λ_j, where $\lambda_i \neq \lambda_j$. Then

$$AZ_i = \lambda_i Z_i$$
$$AZ_j = \lambda_j Z_j$$
$$\tilde{\bar{Z}}_j A Z_i = \lambda_i (Z_i \cdot Z_j)$$
$$\tilde{Z}_i \bar{A} \bar{Z}_j = \lambda_j (Z_i \cdot Z_j)$$

where $(Z_i \cdot Z_j)$ is the scalar product of Z_i and Z_j. Subtracting, we have

$$(\lambda_i - \lambda_j)(Z_i \cdot Z_j) = \tilde{\bar{Z}}_j A Z_i - \tilde{Z}_i \bar{A} \bar{Z}_j = 0$$

since $\widetilde{\tilde{Z}_i \bar{A} \bar{Z}_j} = \tilde{\bar{Z}}_j \tilde{\bar{A}} Z_i = \tilde{\bar{Z}}_j A Z_i$. But $\lambda_i - \lambda_j \neq 0$ and hence, $(Z_i \cdot Z_j) = 0$.

Theorem 4.6.7 If a linear transformation f from C^n to C^n has a representation A with respect to the standard basis which is hermitian, then there is a unitary matrix P such that the representation $PA\tilde{\bar{P}} = D$ is diagonal, with diagonal elements the characteristic values of A.

PROOF The proof will be left to the reader. It can be done as an induction on n very much like that for Theorem 4.6.3. The reader should make the necessary changes in that proof.

Theorem 4.6.8 If a linear transformation f from U to U, where U is an n-dimensional complex vector space, has a representation matrix A which is hermitian, then f has a diagonal representation with respect to n independent characteristic vectors.

PROOF The proof will be left to the reader.

EXAMPLE 4.6.2 Show that the matrix

$$A = \begin{pmatrix} 2 & i & i \\ -i & 1 & 0 \\ -i & 0 & 1 \end{pmatrix}$$

is similar to a diagonal matrix with real elements. A is hermitian since $\tilde{\bar{A}} = A$. Therefore, Theorem 4.6.7 applies. The characteristic equation of A is

$$|A - \lambda I| = \begin{vmatrix} 2-\lambda & i & i \\ -i & 1-\lambda & 0 \\ -i & 0 & 1-\lambda \end{vmatrix} = -\lambda(\lambda - 1)(\lambda - 3) = 0$$

The characteristic values are $\lambda_1 = 0$, $\lambda_2 = 1$, $\lambda_3 = 3$. To find a characteristic vector corresponding to λ_1 we must solve $AZ = 0$,

$$\begin{pmatrix} 2 & i & i \\ -i & 1 & 0 \\ -i & 0 & 1 \end{pmatrix} \begin{pmatrix} z_1 \\ z_2 \\ z_3 \end{pmatrix} = \begin{pmatrix} 0 \\ 0 \\ 0 \end{pmatrix}$$

or $2z_1 + iz_2 + iz_3 = 0$, $-iz_1 + z_2 = 0$. A solution is $\mathbf{u}_1 = (1,i,i)$. We must solve

$$\begin{pmatrix} 1 & i & i \\ -i & 0 & 0 \\ -i & 0 & 0 \end{pmatrix} \begin{pmatrix} z_1 \\ z_2 \\ z_3 \end{pmatrix} = \begin{pmatrix} 0 \\ 0 \\ 0 \end{pmatrix}$$

corresponding to λ_2, or $z_1 = 0$, $z_2 + z_3 = 0$. A solution is $\mathbf{u}_2 = (0,1,-1)$. We must solve

$$\begin{pmatrix} -1 & i & i \\ -i & -2 & 0 \\ -i & 0 & -2 \end{pmatrix} \begin{pmatrix} z_1 \\ z_2 \\ z_3 \end{pmatrix} = \begin{pmatrix} 0 \\ 0 \\ 0 \end{pmatrix}$$

corresponding to λ_3, or $z_1 - iz_2 - iz_3 = 0$, $iz_1 + 2z_2 = 0$. A solution is $\mathbf{u}_3 = (2i,1,1)$. If we normalize $\mathbf{u}_1$, $\mathbf{u}_2$, and $\mathbf{u}_3$ and place them in columns, we have a unitary matrix

$$P^{-1} = \begin{pmatrix} \frac{1}{\sqrt{3}} & 0 & \frac{2i}{\sqrt{6}} \\ \frac{i}{\sqrt{3}} & \frac{1}{\sqrt{2}} & \frac{1}{\sqrt{6}} \\ \frac{i}{\sqrt{3}} & -\frac{1}{\sqrt{2}} & \frac{1}{\sqrt{6}} \end{pmatrix}$$

and its inverse

$$P = \begin{pmatrix} \frac{1}{\sqrt{3}} & -\frac{i}{\sqrt{3}} & -\frac{i}{\sqrt{3}} \\ 0 & \frac{1}{\sqrt{2}} & -\frac{1}{\sqrt{2}} \\ -\frac{2i}{\sqrt{6}} & \frac{1}{\sqrt{6}} & \frac{1}{\sqrt{6}} \end{pmatrix}$$

The diagonal matrix similar to A is

$$PAP^{-1} = \begin{pmatrix} 0 & 0 & 0 \\ 0 & 1 & 0 \\ 0 & 0 & 3 \end{pmatrix}$$

We conclude this section with a couple of applications of the similarity of symmetric matrices to diagonal matrices. The first of these is in the study of quadratic forms. The most general quadratic form in two real variables is $q(x,y) = ax^2 + 2bxy + cy^2$. If we let $X = \begin{pmatrix} x \\ y \end{pmatrix}$ and A be the real matrix

$$A = \begin{pmatrix} a & b \\ b & c \end{pmatrix}$$

then we can write $q(x,y) = \tilde{X}AX$. By analogy we shall define the general quadratic form in n real variables as $q(x_1, x_2, \ldots, x_n) = \tilde{X}AX$, where A is a real symmetric matrix. Since A is real and symmetric, Theorem 4.6.3 applies and there is an orthogonal coordinate transformation $Y = PX$ such that $PA\tilde{P} = D$ is diagonal. We have $X = \tilde{P}Y$ and

$$\begin{aligned} q(x_1, x_2, \ldots, x_n) &= \tilde{X}AX = \tilde{Y}(PA\tilde{P})Y = \tilde{Y}DY \\ &= \lambda_1 {y_1}^2 + \lambda_2 {y_2}^2 + \cdots + \lambda_n {y_n}^2 \end{aligned}$$

where $\lambda_1, \lambda_2, \ldots, \lambda_n$ are the characteristic values of A. We have reduced the quadratic form to diagonal form.

EXAMPLE 4.6.3 Identify the figure in the xy plane defined by the equation $x^2 + 4xy - 2y^2 = 6$. We have a quadratic form $q(x,y) = x^2 + 4xy - 2y^2 = \tilde{X}AX = 6$, where

$$A = \begin{pmatrix} 1 & 2 \\ 2 & -2 \end{pmatrix}$$

The characteristic equation of A is

$$\begin{vmatrix} 1 - \lambda & 2 \\ 2 & -2 - \lambda \end{vmatrix} = (\lambda - 2)(\lambda + 3) = 0$$

The characteristic values are $\lambda_1 = 2$, $\lambda_2 = -3$. The diagonal form after the change of coordinates to (x', y') is

$$q(x,y) = 2(x')^2 - 3(y')^2 = 6$$

or

$$\frac{(x')^2}{3} - \frac{(y')^2}{2} = 1$$

This is the equation of a hyperbola. To locate the axes of symmetry of the hyperbola we must examine the coordinate transformation $X' = PX$, where the rows of P (columns of $\tilde{P}$) are the normalized characteristic vectors $\mathbf{u}_1$ and $\mathbf{u}_2$ of A corresponding respectively to λ_1 and λ_2:

$$\mathbf{u}_1 = \left(\frac{2}{\sqrt{5}}, \frac{1}{\sqrt{5}}\right) \qquad \text{and} \qquad \mathbf{u}_2 = \left(-\frac{1}{\sqrt{5}}, \frac{2}{\sqrt{5}}\right)$$

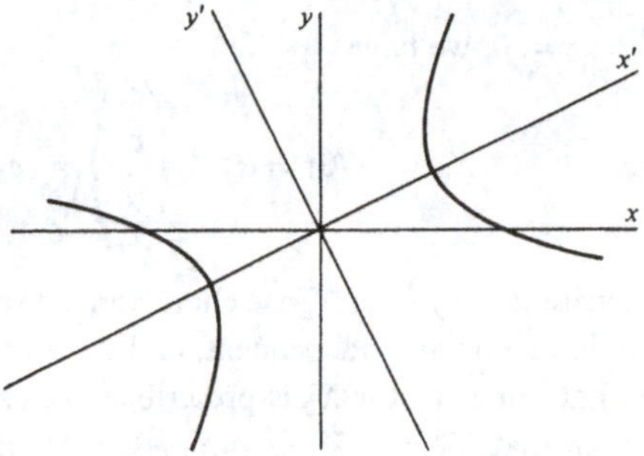

FIGURE 27

The coordinate transformation is

$$\begin{pmatrix} x' \\ y' \end{pmatrix} = \begin{pmatrix} \dfrac{2}{\sqrt{5}} & \dfrac{1}{\sqrt{5}} \\ -\dfrac{1}{\sqrt{5}} & \dfrac{2}{\sqrt{5}} \end{pmatrix} \begin{pmatrix} x \\ y \end{pmatrix} = \begin{pmatrix} \cos\theta & \sin\theta \\ -\sin\theta & \cos\theta \end{pmatrix} \begin{pmatrix} x \\ y \end{pmatrix}$$

where $\cos\theta = 2/\sqrt{5}$ and $\sin\theta = 1/\sqrt{5}$, or $\theta = \tan^{-1}\frac{1}{2}$ is in the first quadrant. The positive x' axis makes an angle of θ with the positive x axis and points in the direction of $\mathbf{u}_1$. The positive y' axis points in the direction of $\mathbf{u}_2$. These directions serve to locate the given hyperbola (see Fig. 27).

The second application is to the solution of certain systems of differential equations. Let $X(t)$ be a column matrix with elements $x_1(t), x_2(t), \ldots, x_n(t)$. Consider the system of differential equations $X' = AX$, where A is real and symmetric and the prime refers to differentiation with respect to t. We wish to find a solution such that $X(0) = X_0$, a given column matrix. We make a coordinate transformation $Y = PX$, where P is independent of t. Then $Y' = PX'$, $X' = P^{-1}Y' = AP^{-1}Y$, and $Y' = (PAP^{-1})Y$. By Theorem 4.6.3, there is a P such that $PAP^{-1} = D$, a diagonal matrix. Hence, the new system is

$$\frac{dy_1}{dt} = \lambda_1 y_1, \frac{dy_2}{dt} = \lambda_2 y_2, \ldots, \frac{dy_n}{dt} = \lambda_n y_n$$

where $\lambda_1, \lambda_2, \ldots, \lambda_n$ are the characteristic values of A. A solution is $y_1 = c_1 e^{\lambda_1 t}$, $y_2 = c_2 e^{\lambda_2 t}, \ldots, y_n = c_n e^{\lambda_n t}$, or

$$X(t) = P^{-1} \begin{pmatrix} c_1 e^{\lambda_1 t} \\ c_2 e^{\lambda_2 t} \\ \vdots \\ c_n e^{\lambda_n t} \end{pmatrix} = c_1 e^{\lambda_1 t} X_1 + c_2 e^{\lambda_2 t} X_2 + \cdots + c_n e^{\lambda_n t} X_n$$

At $t = 0$, we have

$$X_0 = X(0) = P^{-1}\begin{pmatrix} c_1 \\ c_2 \\ \vdots \\ c_n \end{pmatrix} = c_1X_1 + c_2X_2 + \cdots + c_nX_n$$

where $X_1, X_2, \ldots, X_n$ are characteristic vectors of A. In this case, the characteristic vectors are independent, and therefore they form a basis for R^n. No matter what initial vector X_0 is prescribed, we can always find constants $c_1, c_2, \ldots, c_n$ such that $X_0 = c_1X_1 + c_2X_2 + \cdots + c_nX_n$.

EXAMPLE 4.6.4 Find a solution of the following system of differential equations

$$\frac{dx_1}{dt} = 2x_1 + 3x_2 + 3x_3$$

$$\frac{dx_2}{dt} = 3x_1 - x_2$$

$$\frac{dx_3}{dt} = 3x_1 - x_3$$

satisfying the initial conditions $x_1(0) = 1$, $x_2(0) = -2$, $x_3(0) = 0$. The system can be written as $X' = AX$, where

$$A = \begin{pmatrix} 2 & 3 & 3 \\ 3 & -1 & 0 \\ 3 & 0 & -1 \end{pmatrix}$$

is real and symmetric. The characteristic equation is

$$\begin{vmatrix} 2-\lambda & 3 & 3 \\ 3 & -1-\lambda & 0 \\ 3 & 0 & -1-\lambda \end{vmatrix} = (\lambda + 1)(\lambda + 4)(-\lambda + 5) = 0$$

The characteristic values are $\lambda_1 = -1$, $\lambda_2 = -4$, $\lambda_3 = 5$. The corresponding characteristic vectors are

$$X_1 = \begin{pmatrix} 0 \\ 1 \\ -1 \end{pmatrix} \qquad X_2 = \begin{pmatrix} -1 \\ 1 \\ 1 \end{pmatrix} \qquad X_3 = \begin{pmatrix} 2 \\ 1 \\ 1 \end{pmatrix}$$

Therefore, a solution is

$$x_1(t) = -c_2e^{-4t} + 2c_3e^{5t}$$

$$x_2(t) = c_1e^{-t} + c_2e^{-4t} + c_3e^{5t}$$

$$x_3(t) = -c_1e^{-t} + c_2e^{-4t} + c_3e^{5t}$$

In order for $X(0) = (1,-2,0)$, we must have

$$\begin{pmatrix} 1 \\ -2 \\ 0 \end{pmatrix} = \begin{pmatrix} 0 & -1 & 2 \\ 1 & 1 & 1 \\ -1 & 1 & 1 \end{pmatrix} \begin{pmatrix} c_1 \\ c_2 \\ c_3 \end{pmatrix}$$

Solving, we have $c_1 = -1$, $c_2 = -1$, $c_3 = 0$. Our solution satisfying the initial conditions is

$$x_1(t) = e^{-4t}$$
$$x_2(t) = -e^{-t} - e^{-4t}$$
$$x_3(t) = e^{-t} - e^{-4t}$$

EXERCISES 4.6

1 Find similarity transformations which reduce each of the following matrices to diagonal form:

(*a*) $\begin{pmatrix} 3 & 0 & 1 \\ 0 & 2 & 0 \\ 1 & 0 & 3 \end{pmatrix}$ (*b*) $\begin{pmatrix} 4 & 1 & 0 \\ 1 & 4 & 0 \\ 0 & 0 & 3 \end{pmatrix}$ (*c*) $\begin{pmatrix} 4 & -2 & 2 \\ -2 & 1 & -1 \\ 2 & -1 & 1 \end{pmatrix}$

2 Find solutions of the system of differential equations $X' = AX$, where A is each of the matrices of Exercise 1, subject to the initial conditions $X(0) = (1,2,3)$.

3 Find similarity transformations which reduce each of the following matrices to diagonal form:

(*a*) $\begin{pmatrix} 1 & i \\ -i & 1 \end{pmatrix}$ (*b*) $\begin{pmatrix} 2 & 2-2i \\ 2+2i & 0 \end{pmatrix}$ (*c*) $\begin{pmatrix} 2 & 0 & 2i \\ 0 & 2 & -2i \\ -2i & 2i & 4 \end{pmatrix}$

4 Find a solution of the system of differential equations $Z' = AZ$, where A is the matrix of Exercise 3(*c*), subject to the initial conditions $Z(0) = (i,0,i)$.

5 Identify the figure in the xy plane given by the equation $3x^2 + 2xy + 3y^2 = 1$. Find the axes of symmetry.

6 Identify the surface in R^3 given by the equation

$$9x^2 + 12y^2 + 9z^2 - 6xy - 6yz = 1$$

Hint: An equation of the form $\lambda_1 x^2 + \lambda_2 y^2 + \lambda_3 z^2 = 1$ represents an ellipsoid if $\lambda_1, \lambda_2, \lambda_3$ are all positive.

7 A quadratic form $q(x_1, x_2, \ldots, x_n) = \tilde{X}AX$, where A is real and symmetric, is called positive-definite if $q > 0$ for all $X \neq 0$. Prove that $q = \tilde{X}AX$ is positive-definite if and only if all the characteristic values of A are positive.

8 Let $q = \tilde{X}AX$ be a positive-definite quadratic form. Show that $(X \cdot Y) = \tilde{X}AY$ is a scalar product for R^n.

9 A hermitian form $h(z_1, z_2, \ldots, z_n) = \tilde{\bar{Z}}AZ$, where A is hermitian, is called positive-definite if $h > 0$ for all $Z \neq 0$. Prove that $h = \tilde{\bar{Z}}AZ$ is positive-definite if and only if all the characteristic values of A are positive.

10 Let $\tilde{\bar{Z}}AZ$ be a positive-definite hermitian form. Show that $(Z_1 \cdot Z_2) = \tilde{\bar{Z}}_2AZ_1$ is a scalar product for C^n.

*4.7 JORDAN FORMS

Theorem 4.5.2 gives necessary and sufficient conditions for an $n \times n$ matrix to be similar to a diagonal matrix, namely that it should have n independent characteristic vectors. We have also seen square matrices which are similar to no diagonal matrix. In this section, we shall discuss the so-called *Jordan canonical form*, a form of matrix to which every square matrix is similar. The Jordan form is not quite as simple as a diagonal matrix but is nevertheless simple enough to make it very applicable, particularly in the solution of systems of differential equations.

Before we embark upon the discussion of Jordan forms, it will be convenient to introduce the concept of partitioning of matrices. Suppose we write

$$M = \begin{pmatrix} A & B \\ C & D \end{pmatrix}$$

where A is an $m \times n$ matrix, B is an $m \times p$ matrix, C is a $q \times n$ matrix, and D is a $q \times p$ matrix. In other words, M is an $(m + q) \times (n + p)$ matrix partitioned into blocks of matrices so that each block in a given row has the same number of rows and each block in a given column has the same number of columns. The reader should convince himself that the following product rule is valid for partitioned matrices. Let

$$P = \begin{pmatrix} A_{11} & A_{12} & \cdots & A_{1n} \\ A_{21} & A_{22} & \cdots & A_{2n} \\ \cdots & \cdots & \cdots & \cdots \\ A_{m1} & A_{m2} & \cdots & A_{mn} \end{pmatrix} \qquad Q = \begin{pmatrix} B_{11} & B_{12} & \cdots & B_{1p} \\ B_{21} & B_{22} & \cdots & B_{2p} \\ \cdots & \cdots & \cdots & \cdots \\ B_{n1} & B_{n2} & \cdots & B_{np} \end{pmatrix}$$

Then

$$PQ = \begin{pmatrix} C_{11} & C_{12} & \cdots & C_{1p} \\ C_{21} & C_{22} & \cdots & C_{2p} \\ \cdots & \cdots & \cdots & \cdots \\ C_{m1} & C_{m2} & \cdots & C_{mp} \end{pmatrix}$$

where $C_{ij} = \sum_{k=1}^{n} A_{ik}B_{kj}$ provided P and Q are partitioned in such a way that A_{ik} has the same number of columns as B_{kj} has rows for all i, j, and k.

Theorem 4.7.1 Every square matrix A is similar to an upper-triangular matrix with the characteristic values of A along the principal diagonal.

PROOF Let A be an $n \times n$ matrix. We shall show that there exists an invertible matrix P such that $PAP^{-1} = T$, where T is upper-triangular; that is, $t_{ij} = 0$ for $i > j$. We shall prove it by an induction on n. For $n = 1$, A is already upper-triangular; and so $P = I$. Now let us assume that the theorem is true for all $(n-1) \times (n-1)$ matrices. We know that A has at least one characteristic value λ_1 and characteristic vector X_1 such that $AX_1 = \lambda_1 X_1$. We pick any basis for C^n of the form $X_1, Z_2, Z_3, \ldots, Z_n$. If Q is the matrix representing the coordinate transformation from the standard basis to the new basis, we have

$$QAQ^{-1} = \begin{pmatrix} \lambda_1 & U \\ 0 & B \end{pmatrix}$$

where U is $1 \times (n-1)$, 0 is $(n-1) \times 1$, and B is $(n-1) \times (n-1)$. Now B is $(n-1) \times (n-1)$, and so there is an $(n-1) \times (n-1)$ nonsingular matrix R such that

$$RBR^{-1} = \begin{pmatrix} \lambda_2 & v_{23} & \cdots & v_{2n} \\ 0 & \lambda_3 & \cdots & v_{3n} \\ \cdots & \cdots & \cdots & \cdots \\ 0 & 0 & \cdots & \lambda_n \end{pmatrix}$$

is upper-triangular. Now let

$$S = \begin{pmatrix} 1 & 0 \\ 0 & R \end{pmatrix}$$

and $P = SQ$. We have

$$\begin{aligned} PAP^{-1} &= S(QAQ^{-1})S^{-1} \\ &= \begin{pmatrix} 1 & 0 \\ 0 & R \end{pmatrix}\begin{pmatrix} \lambda_1 & U \\ 0 & B \end{pmatrix}\begin{pmatrix} 1 & 0 \\ 0 & R^{-1} \end{pmatrix} \\ &= \begin{pmatrix} 1 & 0 \\ 0 & R \end{pmatrix}\begin{pmatrix} \lambda_1 & UR^{-1} \\ 0 & BR^{-1} \end{pmatrix} \\ &= \begin{pmatrix} \lambda_1 & UR^{-1} \\ 0 & RBR^{-1} \end{pmatrix} \\ &= \begin{pmatrix} \lambda_1 & w_{12} & w_{13} & \cdots & w_{1n} \\ 0 & \lambda_2 & v_{23} & \cdots & v_{2n} \\ 0 & 0 & \lambda_3 & \cdots & v_{3n} \\ \cdots & \cdots & \cdots & \cdots & \cdots \\ 0 & 0 & 0 & \cdots & \lambda_n \end{pmatrix} \end{aligned}$$

Since T is upper-triangular, its characteristic values are $\lambda_1, \lambda_2, \ldots, \lambda_n$. But A has the same characteristic values. This completes the proof.

Theorem 4.7.2 Every square matrix A is similar to an upper-triangular matrix of the form

$$\begin{pmatrix} T_1 & 0 & 0 & \cdots & 0 \\ 0 & T_2 & 0 & \cdots & 0 \\ 0 & 0 & T_3 & \cdots & 0 \\ \cdots & \cdots & \cdots & \cdots & \cdots \\ 0 & 0 & 0 & \cdots & T_k \end{pmatrix}$$

where each T_i is upper-triangular with diagonal elements λ_i, the order of T_i is the multiplicity of λ_i as a characteristic value of A, and k is the number of distinct characteristic values.

PROOF Theorem 4.7.1 tells us that A is similar to an upper-triangular matrix. In fact, we can assume that there is a P such that

$$PAP^{-1} = \begin{pmatrix} T_{11} & T_{12} & \cdots & T_{1k} \\ 0 & T_{22} & \cdots & T_{2k} \\ \cdots & \cdots & \cdots & \cdots \\ 0 & 0 & \cdots & T_{kk} \end{pmatrix} = T$$

where T_{ii} is upper-triangular with all of its diagonal elements λ_i. The order of T_{ii} is the multiplicity of λ_i as a characteristic value of A, and $\lambda_1, \lambda_2, \ldots, \lambda_k$ are all distinct. This assumption is justified by the proof of Theorem 4.7.1. Clearly we can place any characteristic value λ_1 in the upper left-hand corner of T. If λ_1 has multiplicity greater than 1, we can place λ_1 in the upper left-hand corner of RBR^{-1}. If λ_1 has multiplicity greater than 2, we place it again in the third position along the diagonal T. This process can be repeated as many times as the multiplicity of λ_1, then with λ_2, λ_3, etc. Consider a given nonzero element t_{pq} for $p < q$. Let $Q = I + C$, where C is a matrix all of whose elements are zero except the (p,q)th element, which is c. The inverse of Q is $Q^{-1} = I - C$. If we multiply out QTQ^{-1}, we find that the element t_{pq} has been changed to $t_{pq} - c(t_{pp} - t_{qq})$. If $t_{pp} - t_{qq} \neq 0$, we can choose c so that the (p,q)th element is now zero. Otherwise the transformation QTQ^{-1} affects only the elements in the pth row to the right of t_{pq} and in the qth column above t_{pq}. By using a finite sequence of such similarity transformations we can reduce T to the form required by the theorem.

We now define a special upper-triangular matrix known as a *Jordan block*. A Jordan block is a matrix whose elements on the principal diagonal are all equal, whose elements on the first superdiagonal above the principal diagonal are all 1s, and whose elements otherwise are 0s.

EXAMPLE 4.7.1 The most general Jordan block of order four is

$$J = \begin{pmatrix} \lambda & 1 & 0 & 0 \\ 0 & \lambda & 1 & 0 \\ 0 & 0 & \lambda & 1 \\ 0 & 0 & 0 & \lambda \end{pmatrix}$$

where λ is a complex number.

Theorem 4.7.3 Every upper-triangular matrix T all of whose diagonal elements are equal to λ is similar to an upper triangular matrix

$$\begin{pmatrix} J_1 & 0 & \cdots & 0 \\ 0 & J_2 & \cdots & 0 \\ \cdots & \cdots & \cdots & \cdots \\ 0 & 0 & \cdots & J_l \end{pmatrix}$$

where J_i is a Jordan block with diagonal elements λ, and l is the number of independent characteristic vectors of T.

PROOF We shall not prove this theorem.†

Theorem 4.7.4 Jordan canonical form Every square matrix A is similar to an upper-triangular matrix of the form

$$\begin{pmatrix} J_1 & 0 & \cdots & 0 \\ 0 & J_2 & \cdots & 0 \\ \cdots & \cdots & \cdots & \cdots \\ 0 & 0 & \cdots & J_m \end{pmatrix}$$

where J_i is a Jordan block with diagonal element λ_i, a characteristic value of A. A characteristic value λ_i may occur in more than one block, but the number of blocks which contain λ_i on the diagonal is equal to the number of independent characteristic vectors corresponding to λ_i.

†The interested reader should consult a book like B. Noble and J. W. Daniel, "Applied Linear Algebra," 2d. ed., Prentice-Hall, Englewood Cliffs, N.J., 1977.

PROOF By Theorem 4.7.2, there is a nonsingular matrix Q such that

$$T = QAQ^{-1} = \begin{pmatrix} T_1 & 0 & \cdots & 0 \\ 0 & T_2 & \cdots & 0 \\ \cdots & \cdots & \cdots & \cdots \\ 0 & 0 & \cdots & T_k \end{pmatrix}$$

where T_i is upper-triangular with the diagonal elements λ_i, the order of T_i is the multiplicity of λ_i as a characteristic value of A, and k is the number of distinct characteristic values of A. By Theorem 4.7.3, there are k nonsingular matrices R_i such that

$$R_i T_i R_i^{-1} = \begin{pmatrix} J_{i1} & 0 & \cdots & 0 \\ 0 & J_{i2} & \cdots & 0 \\ \cdots & \cdots & \cdots & \cdots \\ 0 & 0 & \cdots & J_{il_i} \end{pmatrix}$$

where $J_{i1}, J_{i2}, \ldots, J_{il_i}$ are Jordan blocks with diagonal elements λ_i, and l_i is the number of independent characteristic vectors corresponding to λ_i. Let

$$R = \begin{pmatrix} R_1 & 0 & \cdots & 0 \\ 0 & R_2 & \cdots & 0 \\ \cdots & \cdots & \cdots & \cdots \\ 0 & 0 & \cdots & R_k \end{pmatrix}$$

Then

$$R^{-1} = \begin{pmatrix} R_1^{-1} & 0 & \cdots & 0 \\ 0 & R_2^{-1} & \cdots & 0 \\ \cdots & \cdots & \cdots & \cdots \\ 0 & 0 & \cdots & R_k^{-1} \end{pmatrix}$$

We have

$$R(QAQ^{-1})R^{-1} = RTR^{-1} = \begin{pmatrix} R_1 T_1 R_1^{-1} & 0 & \cdots & 0 \\ 0 & R_2 T_2 R_2^{-1} & \cdots & 0 \\ \cdots & \cdots & \cdots & \cdots \\ 0 & 0 & \cdots & R_k T_k R_k^{-1} \end{pmatrix}$$

Clearly RTR^{-1} is in the form required by the theorem where $m = \sum_{i=1}^{k} l_i$.

EXAMPLE 4.7.2 Find a matrix in Jordan canonical form which is similar to

$$A = \begin{pmatrix} 2 & 1 & 0 & 1 \\ 1 & 3 & -1 & 3 \\ 0 & 1 & 2 & 1 \\ 1 & -1 & -1 & -1 \end{pmatrix}$$

The characteristic equation is $|A - \lambda I| = \lambda(\lambda - 2)^3 = 0$. The characteristic values are $\lambda_1 = 2$ of multiplicity 3 and $\lambda_2 = 0$ of multiplicity 1. To find characteristic vectors corresponding to $\lambda_1 = 2$ we must solve

$$\begin{pmatrix} 0 & 1 & 0 & 1 \\ 1 & 1 & -1 & 3 \\ 0 & 1 & 0 & 1 \\ 1 & -1 & -1 & -3 \end{pmatrix} \begin{pmatrix} x_1 \\ x_2 \\ x_3 \\ x_4 \end{pmatrix} = \begin{pmatrix} 0 \\ 0 \\ 0 \\ 0 \end{pmatrix}$$

The null space of the coefficient matrix has dimension 1. We can find a characteristic vector $X_1 = (1,0,1,0)$, but all other solutions will be a multiple of X_1 and hence not independent. We must solve

$$\begin{pmatrix} 2 & 1 & 0 & 1 \\ 1 & 3 & -1 & 3 \\ 0 & 1 & 2 & 1 \\ 1 & -1 & -1 & -1 \end{pmatrix} \begin{pmatrix} x_1 \\ x_2 \\ x_3 \\ x_4 \end{pmatrix} = \begin{pmatrix} 0 \\ 0 \\ 0 \\ 0 \end{pmatrix}$$

corresponding to $\lambda_2 = 0$. Again the dimension of the null space is 1, and we have a characteristic vector $X_4 = (0,1,0,-1)$. There are two independent characteristic vectors, and therefore there are two Jordan blocks in the canonical form which (except possibly for the order of the blocks) must look like

$$J = \begin{pmatrix} 2 & 1 & 0 & 0 \\ 0 & 2 & 1 & 0 \\ 0 & 0 & 2 & 0 \\ 0 & 0 & 0 & 0 \end{pmatrix}$$

This answers the question posed in the example. However, let us continue to demonstrate explicitly a similarity transformation which will produce J. Let $P^{-1} = (X_1,X_2,X_3,X_4)$, where $PAP^{-1} = J$. Then $AP^{-1} = P^{-1}J$, or

$$\begin{aligned} A(X_1,X_2,X_3,X_4) &= (AX_1,AX_2,AX_3,AX_4) \\ &= (X_1,X_2,X_3,X_4)J \\ &= (\lambda_1 X_1,\ X_1 + \lambda_1 X_2,\ X_2 + \lambda_1 X_3,\ \lambda_2 X_4) \end{aligned}$$

$(A - \lambda_1 I)X_1 = 0$, $(A - \lambda_1 I)X_2 = X_1$, $(A - \lambda_1 I)X_3 = X_2$, and $(A - \lambda_2 I)X_4 = 0$.

The first and last of these equations we have already solved. The second equation takes the form

$$\begin{pmatrix} 0 & 1 & 0 & 1 \\ 1 & 1 & -1 & 3 \\ 0 & 1 & 0 & 1 \\ 1 & -1 & -1 & -3 \end{pmatrix} \begin{pmatrix} x_1 \\ x_2 \\ x_3 \\ x_4 \end{pmatrix} = \begin{pmatrix} 1 \\ 0 \\ 1 \\ 0 \end{pmatrix}$$

A solution is $\tilde{X}_2 = (0,\frac{3}{2},0,-\frac{1}{2})$. The third equation takes the form

$$\begin{pmatrix} 0 & 1 & 0 & 1 \\ 1 & 1 & -1 & 3 \\ 0 & 1 & 0 & 1 \\ 1 & -1 & -1 & -3 \end{pmatrix} \begin{pmatrix} x_1 \\ x_2 \\ x_3 \\ x_4 \end{pmatrix} = \begin{pmatrix} 0 \\ \frac{3}{2} \\ 0 \\ -\frac{1}{2} \end{pmatrix}$$

and it has a solution $\tilde{X}_3 = (0,-\frac{1}{2},-\frac{1}{2},\frac{1}{2})$. Therefore,

$$P^{-1} = \begin{pmatrix} 1 & 0 & 0 & 0 \\ 0 & \frac{3}{2} & -\frac{1}{2} & 1 \\ 1 & 0 & -\frac{1}{2} & 0 \\ 0 & -\frac{1}{2} & \frac{1}{2} & -1 \end{pmatrix} \qquad P = \begin{pmatrix} 1 & 0 & 0 & 0 \\ 0 & 1 & 0 & 1 \\ 2 & 0 & -2 & 0 \\ 1 & -\frac{1}{2} & -1 & -\frac{3}{2} \end{pmatrix}$$

and multiplying out PAP^{-1}, we find that $PAP^{-1} = J$.

There is a certain amount of ambiguity in Theorem 4.7.4. To illustrate the problem, suppose A is a 4×4 matrix with a single characteristic value λ of multiplicity 4. Also suppose that there are only two independent characteristic vectors corresponding to λ. Then we know from Theorem 4.7.4 that there are two Jordan blocks on the diagonal of J, the Jordan form similar to A. However the Jordan form could look like (aside from the order of the blocks)

$$\begin{pmatrix} \lambda & 0 & 0 & 0 \\ 0 & \lambda & 1 & 0 \\ 0 & 0 & \lambda & 1 \\ 0 & 0 & 0 & \lambda \end{pmatrix} \quad \text{or} \quad \begin{pmatrix} \lambda & 1 & 0 & 0 \\ 0 & \lambda & 0 & 0 \\ 0 & 0 & \lambda & 1 \\ 0 & 0 & 0 & \lambda \end{pmatrix}$$

On the other hand, if we write $P^{-1} = (X_1,X_2,X_3,X_4)$, then we have from $AP^{-1} = P^{-1}J$, $A(X_1,X_2,X_3,X_4) = (X_1,X_2,X_3,X_4)J$. In the first case, we have the equations $(A - \lambda I)X_1 = 0$, $(A - \lambda I)X_2 = 0$, $(A - \lambda I)X_3 = X_2$, and $(A - \lambda I)X_4 = X_3$. In the second case, we have the equations $(A - \lambda I)X_1 = 0$, $(A - \lambda I)X_2 = X_1$, $(A - \lambda I)X_3 = 0$, and $(A - \lambda I)X_4 = X_3$. Given A, only one of these sets of equations will have four independent solutions X_1, X_2, X_3, and X_4. This will then determine the appropriate Jordan form.

We conclude this section with an application to the solution of systems of differential equations. Suppose we have a system of differential equations which can be written in the form

$$\frac{dX}{dt} = AX$$

where $\tilde{X} = (x_1(t), x_2(t), \ldots, x_n(t))$ and A is an $n \times n$ matrix of constants. We wish to find a solution of the system satisfying the initial conditions $X(0) = X_0$, a given vector. Suppose we make a change of coordinates $Y = PX$, where P does not depend on t. Then $X = P^{-1}Y$ and $X' = P^{-1}Y' = AP^{-1}Y$, or $Y' = (PAP^{-1})Y$. Now suppose $PAP^{-1} = J$, a Jordan canonical form; then

$$y_i' = \lambda_i y_i + \mu_i y_{i+1} \qquad i = 1, 2, \ldots, n$$

where λ_i is a characteristic value† of A and μ_i is either 1 or 0 depending on J. In every case, $\mu_n = 0$, and the last equation is simply $y_n' = \lambda_n y_n$. Therefore, $y_n = c_n e^{\lambda_n t}$, where c_n is a constant; putting this into the $(n - 1)$st equation, we can solve for y_{n-1}. Working upward in the system, we can find all the y_i. Finally, we find X from $Y = PX$ and evaluate the constants of integration using the condition $X(0) = X_0$. We shall show in Chap. 9 that such a system always has a solution satisfying a given set of initial conditions and that the solution is unique.

EXAMPLE 4.7.3 Find a solution of the system

$$\frac{dx_1}{dt} = 2x_1 + x_2 + x_4$$

$$\frac{dx_2}{dt} = x_1 + 3x_2 - x_3 + 3x_4$$

$$\frac{dx_3}{dt} = x_2 + 2x_3 + x_4$$

$$\frac{dx_4}{dt} = x_1 - x_2 - x_3 - x_4$$

satisfying $\tilde{X}(0) = \tilde{X}_0 = (1,0,0,0)$. We can write the system as $X' = AX$, where A is the matrix of Example 4.7.2. We have already found a P such that

$$PAP^{-1} = \begin{pmatrix} 2 & 1 & 0 & 0 \\ 0 & 2 & 1 & 0 \\ 0 & 0 & 2 & 0 \\ 0 & 0 & 0 & 0 \end{pmatrix}$$

† In general, the characteristic values are not all distinct. Therefore, it will be expected that λ_i is the same for different values of i.

Under the change of variables $Y = PX$, the equations become $y_1' = 2y_1 + y_2$, $y_2' = 2y_2 + y_3$, $y_3' = 2y_3$, $y_4' = 0$. Therefore, we have $y_4 = c_4$, $y_3 = c_3e^{2t}$, and

$$\frac{d}{dt}(y_2e^{-2t}) = y_3e^{-2t} = c_3$$

so that $y_2 = c_3te^{2t} + c_2e^{2t}$. Finally,

$$\frac{d}{dt}(y_1e^{-2t}) = y_2e^{-2t} = c_3t + c_2$$

and $y_1 = (c_3t^2/2 + c_2t + c_1)e^{2t}$. Let $\tilde{Y}_0 = \tilde{Y}(0) = (c_1,c_2,c_3,c_4)$. Then

$$Y(t) = \begin{pmatrix} e^{2t} & te^{2t} & \frac{1}{2}t^2e^{2t} & 0 \\ 0 & e^{2t} & te^{2t} & 0 \\ 0 & 0 & e^{2t} & 0 \\ 0 & 0 & 0 & 1 \end{pmatrix} \begin{pmatrix} c_1 \\ c_2 \\ c_3 \\ c_4 \end{pmatrix}$$

If $Y_0 = PX_0$, then

$$\begin{pmatrix} c_1 \\ c_2 \\ c_3 \\ c_4 \end{pmatrix} = \begin{pmatrix} 1 & 0 & 0 & 0 \\ 0 & 1 & 0 & 1 \\ 2 & 0 & -2 & 0 \\ 1 & -\frac{1}{2} & -1 & -\frac{3}{2} \end{pmatrix} \begin{pmatrix} 1 \\ 0 \\ 0 \\ 0 \end{pmatrix} = \begin{pmatrix} 1 \\ 0 \\ 2 \\ 1 \end{pmatrix}$$

$$X(t) = P^{-1}Y(t) = \begin{pmatrix} (1 + t^2)e^{2t} \\ 3te^{2t} - e^{2t} + 1 \\ (1 + t^2)e^{2t} - e^{2t} \\ -te^{2t} + e^{2t} - 1 \end{pmatrix}$$

EXERCISES 4.7

1 Let J be a Jordan block with λ on the diagonal. Show that the null space of $J - \lambda I$ is of dimension 1.

2 Show that a Jordan canonical matrix with k Jordan blocks on the diagonal has exactly k independent characteristic vectors.

3 Show that if $PAP^{-1} = J$, a Jordan canonical matrix, then A has the same number of independent characteristic vectors as J.

4 Find Jordan canonical matrices similar to each of the following:

(a) $\begin{pmatrix} 5 & 4 & 3 \\ -1 & 0 & -3 \\ 1 & -2 & 1 \end{pmatrix}$ (b) $\begin{pmatrix} 2 & 1 & 0 \\ 1 & -1 & -1 \\ 0 & 1 & 2 \end{pmatrix}$ (c) $\begin{pmatrix} 2 & 2 & -1 \\ -1 & -1 & 1 \\ -1 & -2 & 2 \end{pmatrix}$

5 Solve each of the systems of differential equations $X' = AX$, where A is one of the matrices of Exercise 4, subject to the initial conditions $\tilde{X}(0) = \tilde{X}_0 = (1,2,3)$.

6 Show that a solution of $X' = JX$, where J is an $n \times n$ Jordan block with λ on the diagonal, is

$$X(t) = \begin{pmatrix} 1 & t & \dfrac{t^2}{2!} & \cdots & \dfrac{t^{n-1}}{(n-1)!} \\ 0 & 1 & t & \cdots & \dfrac{t^{n-2}}{(n-2)!} \\ \cdots & \cdots & \cdots & \cdots & \cdots \\ 0 & 0 & 0 & \cdots & 1 \end{pmatrix} \begin{pmatrix} c_1 \\ c_2 \\ \cdot \\ c_n \end{pmatrix} e^{\lambda t}$$

Show that the system has a solution for arbitrary $\tilde{X}(0) = (c_1, c_2, \ldots, c_n)$.

7 Let

$$T_i = \begin{pmatrix} 1 & t & \dfrac{t^2}{2!} & \cdots & \dfrac{t^{k-1}}{(k-1)!} \\ 0 & 1 & t & \cdots & \dfrac{t^{k-2}}{(k-2)!} \\ \cdots & \cdots & \cdots & \cdots & \cdots \\ 0 & 0 & 0 & \cdots & 1 \end{pmatrix} e^{\lambda_i t}$$

Let

$$J = \begin{pmatrix} J_1 & 0 & \cdots & 0 \\ 0 & J_2 & \cdots & 0 \\ \cdots & \cdots & \cdots & \cdots \\ 0 & 0 & \cdots & J_m \end{pmatrix}$$

be a Jordan canonical matrix with J_i a $k_i \times k_i$ Jordan block with diagonal elements λ_i. Show that the system of differential equations $X' = JX$ has a solution

$$X(t) = \begin{pmatrix} T_1 & 0 & \cdots & 0 \\ 0 & T_2 & \cdots & 0 \\ \cdots & \cdots & \cdots & \cdots \\ 0 & 0 & \cdots & T_m \end{pmatrix} \begin{pmatrix} c_1 \\ c_2 \\ \cdot \\ c_n \end{pmatrix}$$

Show that the system always has a solution for arbitrary $\tilde{X}(0) = (c_1, c_2, \ldots, c_n)$.

8 Show that the system $X' = AX$ always has a solution for arbitrary $\tilde{X}(0) = (c_1, c_2, \ldots, c_n)$.

9 Prove that every $n \times n$ matrix satisfies its own characteristic equation. *Hint*: Show that $(A - \lambda_1 I)(A - \lambda_2 I) \cdots (A - \lambda_n I)X_i = 0$ for a set of n independent vectors X_i, $i = 1, 2, \ldots, n$.

5

FIRST ORDER DIFFERENTIAL EQUATIONS

5.1 INTRODUCTION

This chapter begins our study of ordinary differential equations with first order equations. In a sense, the last section should come first because there we take up the fundamental existence and uniqueness theorem for first order equations. The proof is the traditional Picard iteration argument. However, because the argument is more sophisticated than the general level of this book, this material is more appropriately placed in a starred section for the more ambitious students. After a section giving an elementary example of how differential equations arise in, and are related to, applied mathematics, there is a section on some of the basic definitions in the subject. Next we take up the solution of first order linear equations. The following section deals with a few of the specific types of nonlinear equations which can be solved in closed form. The last two unstarred sections, Secs. 5.6 and 5.7, are on applications of first order equations and numerical methods, in that order.

5.2 AN EXAMPLE

The broad area of applied mathematics usually breaks down into four parts:

1 Formulation of a mathematical model to describe some physical situation.
2 Precise statement and analysis of an appropriate mathematical problem.
3 Approximate numerical calculation of important physical quantities.
4 Comparison of physical quantities with experimental data to check the validity of the model.

The lines of demarcation are never clear, but generally parts 1 and 4 are the province of the physicist or engineer while parts 2 and 3 are the province of the mathematician. Part 1 is very difficult; at the very least it requires an intimate knowledge of existing physical principles, and at the most it may require the formulation of some new theory to cover the given situation. It requires great insight into the question of which effects are the principal effects in the problem and which are secondary, hence can be neglected. This is because nature is usually too complicated to be described precisely, and even if we understood completely all physical principles, we would probably not be able to solve the resulting mathematical problems with enough precision to make this knowledge pay off. Therefore, when a mathematical model is formulated, one must take into account both the inability to describe the physical situation precisely and also the inability to analyze the mathematical model which may be forthcoming.

In a sense, once the mathematical model has been formulated, the analysis of part 2 has nothing to do with physics. The question of whether the problem is well formulated, whether it has a solution, and how to find the solution are purely mathematical in nature. The mathematician cannot, for example, argue that the solution exists and is unique because the physical situation indicates this, because by the time the problem reaches him, it is no longer a precise description of nature but only an approximate model which at best retains only the principal effects to be studied. Therefore, the mathematician must decide questions of existence and uniqueness within the framework of the mathematical model, which will do what it is supposed to only if it has been well formulated in part 1. This is not to say that physical intuition is never valuable to the mathematician. It may suggest methods of analysis which would not otherwise be apparent, but the mathematician must not rely on some sort of vague physical intuition to replace sound mathematical analysis.

The mathematician's work is not done when the mathematical model has been analyzed to the extent of deciding that a solution exists and is unique. For knowing that a unique solution exists is of little help to the physicist or engineer

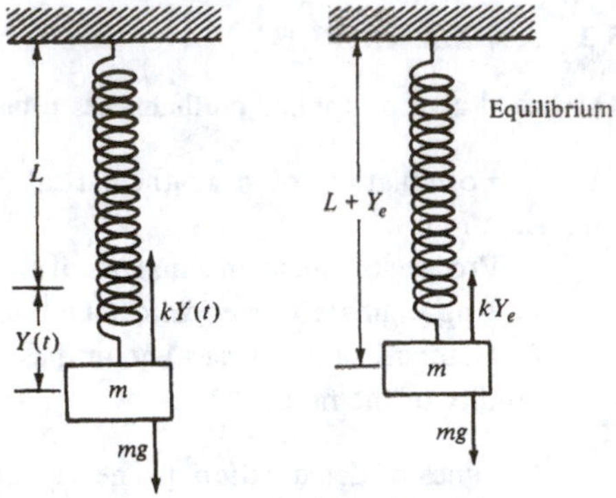

FIGURE 28

if he cannot find it. In many cases, a numerical approximation to some physical quantity is the best that can be hoped for. In other cases, a more complete description of a solution may be available, but not in a form from which data can be quickly and easily extracted. Hence, in either case part 3 may involve considerable numerical analysis before meaningful answers can be derived. This will involve finding algorithms from which data can be computed and analyzing the degree of accuracy of these results compared with some analytical solution which is known to exist.

Part 4 is the province of the experimentalist. It is his job to devise experiments to check the data which the mathematical model has provided against the physical situation. This book is a mathematics book and therefore will not deal with parts 1 and 4 of this outline. We shall treat parts 2 and 3 especially as they relate to those mathematical models which involve the solution of ordinary differential equations. Before getting on with a systematic study of ordinary differential equations, we shall illustrate some of the foregoing remarks in relation to a simple mass-spring system.

Consider a physical system consisting of a mass of m slugs† hanging on a helical spring (see Fig. 28). We assume that the spring has a certain natural length L. If the spring is stretched by a small amount ε to the length $L + \varepsilon$, there is a restoring force $k\varepsilon$ in the spring which opposes the extension. If the spring is compressed by a small amount ε to the length $L - \varepsilon$, then there is a force $k\varepsilon$ which opposes the compression. If we measure force in pounds and length in feet, the spring constant k is measured in pounds per foot.

† One slug is a unit of mass such that a force of one pound exerted on it will produce an acceleration of one foot per second per second.

Suppose at a given time t the spring has length $L + Y(t)$; then $Y(t)$ is the amount the spring is stretched beyond the natural length. If $Y(t)$ is negative, then the spring is under compression. There is a downward force of mg due to gravity acting on the mass, where g is the acceleration of gravity measured in feet per second per second. The spring exerts a force $kY(t)$ on the mass, which is upward when $Y(t)$ is positive and downward when $Y(t)$ is negative. According to Newton's law of motion,

$$m\frac{d^2Y}{dt^2} = mg - kY$$

where d^2Y/dt^2 is the acceleration of the mass m and time t is measured in seconds. We are neglecting any force due to air resistance, and we are assuming that the mass has no horizontal motion. We see that the displacement $Y(t)$ satisfies a *differential equation.*†

As is the case in many problems, the differential equation can be simplified by introducing a new variable. Consider the equilibrium position of the mass on the spring, that is, where the mass will hang without motion so that the downward force of gravity is just balanced by the upward reaction of the spring. If Y_e is the amount the spring is stretched in the equilibrium position, then $kY_e = mg$. Now let $y(t)$ be the displacement of the mass measured positively downward from equilibrium. Then $Y(t) = y(t) + Y_e$ and

$$\frac{d^2Y}{dt^2} = \frac{d^2y}{dt^2}$$

and the differential equation becomes

$$m\frac{d^2y}{dt^2} = mg - k[y(t) + Y_e] = -ky(t)$$

Therefore, $y(t)$ satisfies $\ddot{y} + \omega^2 y = 0$, where $\omega^2 = k/m$ and the two dots over y stand for the second derivative of y with respect to t.

So far we have said nothing about how the mass will be set into motion. Suppose at time $t = 0$ the mass is given an initial displacement $y_0 = y(0)$ from equilibrium and an initial velocity $\dot{y}_0 = \dot{y}(0)$, measured in feet per second. Then the problem‡ is to find a function $y(t)$ satisfying $\ddot{y} + \omega^2 y = 0$ for $t \geq 0$ such that $y(0) = y_0$ and $\dot{y}(0) = \dot{y}_0$. Since the displacement, the velocity, and

† We shall define more precisely in the next section what we mean by a differential equation.

‡ This type of problem is called an *initial-value problem.* All the data are given at a single time $t = 0$. Later we shall consider boundary-value problems, where data are given at more than one value of the independent variable.

the forces acting on the mass cannot become infinite or change discontinuously, we further specify that $y(t)$, $\dot{y}(t)$, and $\ddot{y}(t)$ should be continuous for $t \geq 0$. In our terminology this means that these functions are all continuous for $t > 0$ and that $\lim_{t\to 0^+} y(t)$, $\lim_{t\to 0^+} \dot{y}(t)$, and $\lim_{t\to 0^+} \ddot{y}(t)$ all exist.

We now have a mathematical model to go with the mass-spring system described above. To reemphasize that this only approximately describes the physical situation let us list some of the assumptions we have tacitly made.

1 The elastic behavior of the spring is such that the restoring force is proportional to the displacement.
2 The displacement is small enough to ensure that the elastic limit of the spring is not exceeded.
3 There is no air resistance.
4 The acceleration of gravity does not vary with height.
5 The motion is in a straight vertical line.

In addition to these, we could mention that Newton's law is only an approximate theory which assumes that relativistic effects are negligible. The point is that the mathematical model is only an idealization of the actual physical system. If it turns out to be a good approximation to the physical system, it is because we have been clever enough to include all the major effects and have neglected only secondary effects.

One of the first things we should do with the mathematical model is prove that there exists a solution to the problem. In this case, we shall do so by actually finding a solution.† Let $z = dy/dt$. Then

$$\frac{d^2y}{dt^2} = \frac{dz}{dt} = \frac{dz}{dy}\frac{dy}{dt} = z\frac{dz}{dy}$$

Hence,

$$z\frac{dz}{dy} + \omega^2 y = 0$$

$$z\,dz + \omega^2 y\,dy = 0$$

$$\int z\,dz + \omega^2 \int y\,dy = \frac{z^2}{2} + \frac{\omega^2 y^2}{2} = \frac{c^2}{2}$$

where c^2 is a constant. Then

$$z = \frac{dy}{dt} = \pm\sqrt{c^2 - \omega^2 y^2}$$

† This type of existence theorem is called *constructive* since it actually constructs a solution. There are *nonconstructive* existence theorems where solutions are proved to exist without showing how to find them.

Ignoring, for the moment, the ambiguity in sign, we have

$$\frac{dy}{\sqrt{c^2 - \omega^2 y^2}} = dt$$

$$\frac{dy}{\sqrt{a^2 - y^2}} = \omega\, dt$$

$$\sin^{-1}\frac{y}{a} = \omega t + \phi$$

where $a = c/\omega$ and ϕ is another constant of integration. Finally, we have

$$y = a \sin(\omega t + \phi)$$

where a and ϕ are constants of integration. We should go back and consider the other sign. But we see that this is not necessary since

$$\dot{y} = \omega a \cos(\omega t + \phi)$$

$$\ddot{y} = -\omega^2 a \sin(\omega t + \phi) = -\omega^2 y$$

and we see that this function satisfies the differential equation for arbitrary a and ϕ. Now

$$y(0) = a \sin\phi = y_0$$

$$\dot{y}(0) = \omega a \cos\phi = \dot{y}_0$$

can be satisfied by

$$a = \sqrt{y_0^{\,2} + \frac{\dot{y}_0^{\,2}}{\omega^2}}$$

$$\phi = \tan^{-1}\frac{\omega y_0}{\dot{y}_0}$$

where $0 \leq \phi < \pi$. Therefore, we have found an explicit solution to our problem.

Next we prove that the solution is unique. Suppose that there are two solutions $y_1(t)$ and $y_2(t)$. Then $\ddot{y}_1 + \omega^2 y_1 = 0$, $\ddot{y}_2 + \omega^2 y_2 = 0$, $y_1(0) = y_2(0) = y_0$, $\dot{y}_1(0) = \dot{y}_2(0) = \dot{y}_0$. We form the difference $w(t) = y_1(t) - y_2(t)$. Then $\ddot{w} + \omega^2 w = 0$, $w(0) = \dot{w}(0) = 0$. We compute

$$E(t) = \tfrac{1}{2} m \dot{w}^2 + \tfrac{1}{2} k w^2$$

Then

$$\dot{E}(t) = m\dot{w}\ddot{w} + kw\dot{w}$$

$$= m\dot{w}\frac{-k}{m} w + kw\dot{w} = 0$$

This means that $E(t)$ is constant. But $E(0) = 0$ and, therefore, $E(t) \equiv 0$. However, we notice that $E(t)$ is the sum of nonnegative quantities. Therefore, $w \equiv 0$ and $y_1(t) \equiv y_2(t)$. This completes the proof.

Notice that in the proof of uniqueness we used the function $E(t) = \frac{1}{2}m\dot{w}^2 + \frac{1}{2}kw^2$, which most readers will recognize as the energy—kinetic energy plus potential energy of the spring. Then $\dot{E}(t) = 0$ states the fact that energy is conserved. We did not have to know this to complete the proof. On the other hand, this illustrates how a knowledge of physical principles may aid the mathematician in his analysis of the mathematical model.

EXERCISES 5.2

1 Show that $y(t) = a\cos(\omega t + \phi)$ and $y(t) = A\sin\omega t + B\cos\omega t$ both satisfy $\ddot{y} + \omega^2 y = 0$ for arbitrary constants a, ϕ, A, and B. Does this contradict the uniqueness theorem? Explain.

2 Find the solution of the initial-value problem $\ddot{y} + \omega^2 y = 0$, $t \geq 0$, $y(0) = y_0$, $\dot{y}(0) = \dot{y}_0$ in the form $y(t) = A\sin\omega t + B\cos\omega t$.

3 Let $f(t)$ be a given function continuous for $t \geq 0$. Prove that if there exists a function $y(t)$ which is continuous and has continuous first and second derivatives for $t \geq 0$ satisfying $\ddot{y} + \omega^2 y = f(t)$, with $y(0) = y_0$, $\dot{y}(0) = \dot{y}_0$, then it is unique.

4 The differential equation satisfied by the angular displacement θ of the plane pendulum shown in the figure is $l\ddot{\theta} + g\sin\theta = 0$. Consider the total energy

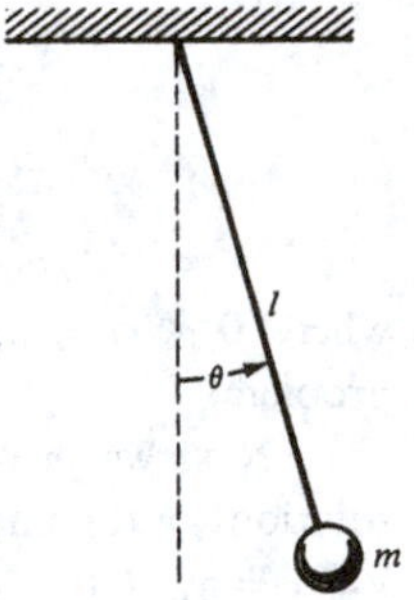

$E(t) = \frac{1}{2}ml^2\dot{\theta}^2 + mgl(1 - \cos\theta)$. Show that energy is conserved, that is, $\dot{E}(t) = 0$. If the pendulum has no initial displacement and no initial velocity, can it ever become displaced from equilibrium? Does this prove uniqueness of the solution of the differential equation subject to specified initial displacement and velocity in this case? Explain.

5 Assuming in Exercise 4 that $|\theta| \leq 5°$, so that $\sin\theta \approx \theta$, θ in radians, find an approximate solution to the initial-value problem. Find the approximate frequency of the pendulum, that is, the number of complete cycles per second.

6 Some people insist that an initial-value problem is not well formulated unless the solution is a continuous function of the initial data; that is, small changes in the initial data produce small changes in the solution. Prove that this is the case in the mass-spring problem of this section.

5.3 BASIC DEFINITIONS

Before beginning a systematic study of differential equations, we shall define some of the basic terms in the subject.

A differential equation is an equation which involves one or more independent variables, one or more dependent variables, and derivatives of the dependent variables with respect to some or all of the independent variables. If there is just one independent variable, then the derivatives are all ordinary derivatives, and the equation is an *ordinary differential equation.* We saw the ordinary differential equation

$$\frac{d^2y}{dt^2} + \omega^2 y = 0$$

in the last section. Another example is Bessel's equation

$$x^2 \frac{d^2y}{dx^2} + x \frac{dy}{dx} + (x^2 - n^2)y = 0$$

where x is the independent variable, y is the dependent variable, and n^2 is a constant. If there is more than one independent variable and partial derivatives appear in the equation, the equation is called a *partial differential equation.* Some common examples are the heat equation

$$k \frac{\partial u}{\partial t} = \frac{\partial^2 u}{\partial x^2}$$

the wave equation

$$\frac{1}{c^2} \frac{\partial^2 u}{\partial t^2} = \frac{\partial^2 u}{\partial x^2}$$

and Laplace's equation

$$\frac{\partial^2 u}{\partial x^2} + \frac{\partial^2 u}{\partial y^2} = 0$$

It is obvious in these cases which is the dependent variable and which are the independent variables.

The *order* of a differential equation is the order of the highest derivative which appears in the equation. Bessel's equation is a second order ordinary differential equation. The heat, wave, and Laplace equations are examples of second order partial differential equations. The general form of an nth order ordinary differential equation, in which y is the dependent variable and t is the independent variable, is

$$F(t, y, \dot{y}, \ldots, y^{(n)}) = 0$$

where the nth derivative must actually appear in this function. Lower order derivatives may be missing, however. If F is a linear function of the variables $y, \dot{y}, \ddot{y}, \ldots, y^{(n)}$, that is,

$$F(t, \dot{y}, \ddot{y}, \ldots, y^{(n)}) = a_0(t)y^{(n)} + a_1(t)y^{(n-1)} + \cdots + a_{n-1}(t)\dot{y} + a_n(t)y + f(t)$$

then the differential equation is said to be *linear*. The two ordinary differential equations cited above are linear. On the other hand, the equation

$$y\frac{dy}{dt} + y = f(t)$$

is nonlinear. There are corresponding definitions for nth order and linear nth order partial differential equations.

We have already indicated how a mathematical model may lead to a problem in differential equations. Clearly, to define the problem we must specify more than the differential equation. We must specify where the solution is to be found, what continuity conditions must be met by the solution and its derivatives, and also what values the solution and/or its derivatives must take on at certain points in its domain of definition. We say that a solution *exists* if there is at least one function which satisfies all these conditions. We say that the solution is *unique* if there is no more than one function which satisfies all the conditions. We usually say that the problem is overdetermined if there are no solutions, that is, too many conditions to be met, and underdetermined if there are solutions but the solution is not unique, that is, there are not enough conditions to single out a unique solution.

In speaking of ordinary differential equations, we say we have an *initial-value problem* if all the specified values of the solution and its derivatives are given at one point. The mass-spring problem of the previous section was an example. These problems are most frequently encountered in dynamical problems where time is the independent variable and the data are given at some initial time, say $t = 0$. For this reason, when we are dealing with initial-value problems we shall usually use t as the independent variable. On the other hand, if data are given for the solution at more than one point, we say we have a *boundary-value problem.* These problems occur most frequently when the independent variable is a space variable. For example, a beam is loaded with some static load (constant set of forces), and conditions are specified at the ends of the beam. For this reason, when we are dealing with boundary-value problems we shall generally use x as the dependent variable. Corresponding definitions of initial-value and boundary-value problems occur in partial differential equations, but descriptions become more complicated; we shall not study partial differential equations in this book.

EXERCISES 5.3

1 Classify each of the following ordinary differential equations as linear or nonlinear. Also determine the order of the equation.

(*a*) $t\dot{y} + y = e^t$

(*b*) $y\dot{y} + y = e^t$

(*c*) $\ddot{y} + \dot{y} + y = 0$

(*d*) $\dot{y}\ddot{y} = y$

(*e*) $t^3\dddot{y} + t^2\ddot{y} + t\dot{y} + y = t^4$

(*f*) $\sin y + x \cos y' = 0,\ y' = \dfrac{dy}{dx}$

(*g*) $(1 - x^2)y'' - 2xy' + n(n + 1)y = 0$ (n = constant)

(*h*) $e^y y''' + x^3 = 0$

2 Let $u(x,y)$ be a function of two independent variables. Describe the general nth order partial differential equation involving u.

3 Referring to Exercise 2, give the form of the general nth order linear partial differential equation involving u.

4 Classify each of the following partial differential equations as linear or nonlinear. Also determine the order of the equation.

(*a*) $\dfrac{\partial^2 u}{\partial x^2} + \dfrac{\partial^2 u}{\partial y^2} = 0$

(*b*) $\dfrac{\partial u}{\partial t} = \dfrac{\partial^2 u}{\partial x^2} + \dfrac{\partial^2 u}{\partial y^2}$

(*c*) $\dfrac{\partial^2 u}{\partial t^2} = \dfrac{\partial^2 u}{\partial x^2} + \dfrac{\partial^2 u}{\partial y^2} + \dfrac{\partial^2 u}{\partial z^2}$

(*d*) $u\dfrac{\partial u}{\partial x} + u\dfrac{\partial u}{\partial y} = 1$

(*e*) $\dfrac{\partial u}{\partial x} = e^u\dfrac{\partial u}{\partial y}$

(*f*) $\dfrac{\partial^3 u}{\partial x\,\partial y\,\partial z} = 0$

(*g*) $\dfrac{\partial^2 u}{\partial x\,\partial y} + u\dfrac{\partial u}{\partial x} = 0$

(*h*) $y\dfrac{\partial u}{\partial x} + x\dfrac{\partial u}{\partial y} = \sin xy$

5.4 FIRST ORDER LINEAR EQUATIONS

The general form of the first order linear ordinary differential equation is

$$a_1(t)\dot{y} + a_0(t)y = f(t)$$

We shall consider the initial-value problem based on this equation, which requires a solution continuous for $0 \leq t \leq b$ satisfying $y(0) = y_0$. If $a_1(0) = 0$, we say that $t = 0$ is a singularity of the differential equation. We shall consider singularities later, so for now we assume that $a_1(0) \neq 0$. We assume that $a_0(t)$, $a_1(t)$, and $f(t)$ are continuous for $0 \leq t \leq b$, and therefore if $a_1(0) \neq 0$, there is a $\beta > 0$ such that $a_1(t) \neq 0$ for $0 \leq t \leq \beta$. For simplicity let us assume that $\beta = b$. Dividing through by $a_1(t)$, we can write the equation as

$$\frac{dy}{dt} + q(t)y = r(t)$$

where $q(t)$ and $r(t)$ are continuous for $0 \leq t \leq b$.

It is convenient to introduce the operator L, which by definition is

$$Ly = \frac{dy}{dt} + q(t)y$$

We observe that L is linear, since

$$\begin{aligned} L(c_1y_1 + c_2y_2) &= \frac{d}{dt}(c_1y_1 + c_2y_2) + q(t)(c_1y_1 + c_2y_2) \\ &= c_1\left(\frac{dy_1}{dt} + qy_1\right) + c_2\left(\frac{dy_2}{dt} + qy_2\right) \\ &= c_1Ly_1 + c_2Ly_2 \end{aligned}$$

where c_1 and c_2 are arbitrary constants.

The equation $\dot{y} + qy = 0$ is called the *associated homogeneous equation*, where $\dot{y} + qy = r$ is considered the *nonhomogeneous equation* if $r \not\equiv 0$. We note that if y_1 is a solution of the associated homogeneous equation and y_2 is *any* solution of the nonhomogeneous equation, then $cy_1 + y_2$ is a solution of the nonhomogeneous equation for an arbitrary constant c. This is because

$$L(cy_1 + y_2) = cLy_1 + Ly_2 = r(t)$$

since $Ly_1 = 0$ and $Ly_2 = r$. The function $y = cy_1 + y_2$ is a one-parameter family of solutions. From this family we can select one which satisfies the initial condition $y(0) = y_0$; that is,

$$y(0) = cy_1(0) + y_2(0) = y_0$$

$$c = \frac{y_0 - y_2(0)}{y_1(0)}$$

provided $y_1(0) \neq 0$. To prove existence of a solution of the initial-value problem it then remains to show that y_1 exists with $y_1(0) \neq 0$ and y_2 exists.

Let us consider the homogeneous equation first. Since $\dot{y} = -qy$,

$$\frac{dy}{y} = -q(t)\,dt$$

Integrating, we have for $0 \leq t \leq b$

$$\ln|y| - \ln|y(0)| = -\int_0^t q(\tau)\,d\tau$$

$$\left|\frac{y}{y(0)}\right| = \exp\left[-\int_0^t q(\tau)\,d\tau\right]$$

It is not clear that we can remove the absolute-value sign. However, if we let

$$y_1 = \exp\left[-\int_0^t q(\tau)\,d\tau\right]$$

we see that this is a continuously differentiable solution of the homogeneous equation such that $y_1(0) = 1$. Hence, we have proved the existence of y_1.

Next we show that y_2, a solution of the nonhomogeneous equation, exists. Let†

$$Q(t) = \exp\left[\int_0^t q(\tau)\,d\tau\right]$$

Then multiplying $\dot{y}_2 + qy_2 = r$ by Q, we have

$$Q\dot{y}_2 + qQy_2 = rQ$$

$$\frac{d}{dt}(Qy_2) = rQ$$

$$Qy_2 = \int_0^t r(\tau)Q(\tau)\,d\tau$$

$$y_2 = \frac{1}{Q(t)}\int_0^t r(\tau)Q(\tau)\,d\tau$$

This function is a solution of the nonhomogeneous equation. In fact, since $y_1(0) = 1$ and $y_2(0) = 0$, a solution of the initial-value problem is

$$\begin{aligned} y &= y_0 y_1 + y_2 \\ &= y_0 \exp\left[-\int_0^t q(\tau)\,d\tau\right] + \exp\left[-\int_0^t q(\tau)\,d\tau\right]\int_0^t r(\tau)Q(\tau)\,d\tau \end{aligned}$$

† $Q(t)$ is called an *integrating factor* because by multiplying the differential equation through by Q we make the left-hand side of the equation an exact derivative.

To show that the solution to the initial-value problem is unique, we assume that there are two solutions $u(t)$ and $v(t)$. Then $\dot{u} + qu = r, u(0) = y_0$, and $\dot{v} + qv = r, v(0) = y_0$. Let $w = u - v$. Then $\dot{w} + qw = 0$ and $w(0) = 0$. The function $Q(t) > 0$, since it is an exponential function. Also

$$Q\dot{w} + qQw = 0$$

$$\frac{d}{dt}(Qw) = 0$$

This implies that $Q(t)w(t) = \text{const}$. But $Q(0)w(0) = 0$. Therefore, $Q(t)w(t) \equiv 0$, which implies that $w = u - v \equiv 0$ since $Q(t) > 0$. Therefore, $u \equiv v$. We have therefore proved the following theorem.

Theorem 5.4.1 Let $q(t)$ and $r(t)$ be continuous for $0 \le t \le b$. Then the initial-value problem $\dot{y} + qy = r$, $y(0) = y_0$, has a unique solution for $0 \le t \le b$, given by

$$y(t) = y_0 \exp\left[-\int_0^t q(\tau)\,d\tau\right] + \exp\left[-\int_0^t q(\tau)\,d\tau\right]\int_0^t r(\tau)Q(\tau)\,d\tau$$

where

$$Q(t) = \exp\left[\int_0^t q(\tau)\,d\tau\right]$$

EXAMPLE 5.4.1 Solve the initial-value problem $y(0) = 1$.

$$\dot{y} + \frac{1}{1-t}y = 1 - t \qquad 0 \le t \le b < 1$$

The integrating factor is

$$Q(t) = \exp\left[\int_0^t (1-\tau)^{-1}\,d\tau\right]$$

$$= \exp\left[-\ln(1-t)\right] = \frac{1}{1-t}$$

Multiplying by $Q(t)$, we have

$$\frac{1}{1-t}\dot{y} + \frac{1}{(1-t)^2}y = \frac{d}{dt}\left(\frac{y}{1-t}\right) = 1$$

Therefore,

$$y = t(1-t) + c(1-t)$$

where c is arbitrary. However, $y(0) = 1$ implies $c = 1$, and so the unique solution to the problem is

$$y(t) = 1 - t^2$$

EXAMPLE 5.4.2 Radioactive matter is known to decay (change its form) at a rate which is proportional to the amount present at any given time. Let A_0 be the amount of radioactive material present at the beginning of the process ($t = 0$). Then

$$\frac{dA}{dt} = -kA$$
$$A(0) = A_0$$

This is an initial-value problem for a first order linear differential equation. The solution is

$$A(t) = A_0 e^{-kt}$$

The *half-life* τ of the radioactive material is defined to be the time required for half the original material to decay. Hence,

$$\tfrac{1}{2}A_0 = A_0 e^{-k\tau}$$

Therefore, $\tau = (1/k) \ln 2$.

EXAMPLE 5.4.3 Initially tank I contains 100 gallons of salt brine with a concentration of 1 pound per gallon, and tank II contains 100 gallons of water. Liquid is pumped from tank I into tank II at a rate of 1 gallon per minute, and liquid is pumped from tank II into tank I at a rate of 2 gallons per minute. The tanks are kept well stirred. What is the concentration in tank I after 10 minutes? Let A_1 be the amount of salt in pounds in tank I and A_2 be the amount of salt in pounds in tank II. The concentration in tank I is

$$C_1 = \frac{A_1}{100 + t}$$

and the concentration in tank II is

$$C_2 = \frac{A_2}{100 - t}$$

Therefore, A_1 and A_2 satisfy

$$\frac{dA_1}{dt} = \frac{2A_2}{100 - t} - \frac{A_1}{100 + t}$$
$$\frac{dA_2}{dt} = \frac{A_1}{100 + t} - \frac{2A_2}{100 - t}$$
$$A_1 + A_2 = 100$$

Eliminating A_2, we have

$$\frac{dA_1}{dt} = \frac{2(100 - A_1)}{100 - t} - \frac{A_1}{100 + t}$$

$$\frac{dA_1}{dt} + \frac{(300 + t)A_1}{100^2 - t^2} = \frac{200}{100 - t}$$

This is a first order linear equation with $q(t) = (300 + t)/(100^2 - t^2)$ and $r(t) = 200/(100 - t)$. According to the above,

$$Q(t) = \exp\left[\int_0^t q(\tau)\,d\tau\right]$$

$$= \frac{100(100 + t)}{(100 - t)^2}$$

We multiply the equation by $(100 + t)/(100 - t)^2$, and we have

$$\frac{d}{dt}\left[\frac{(100 + t)A_1}{(100 - t)^2}\right] = \frac{200(100 + t)}{(100 - t)^3}$$

$$\frac{100 + t}{(100 - t)^2} A_1 = K + 200\left[\frac{100}{(100 - t)^2} - \frac{1}{(100 - t)}\right]$$

where K is an arbitrary constant. Therefore,

$$A_1 = \frac{K(100 - t)^2}{100 + t} + \frac{200t}{100 + t}$$

Also, since $A_1(0) = 100$, $K = 1$, and

$$A_1 = \frac{100^2 + t^2}{100 + t} \qquad 0 \le t \le 100$$

$$C_1 = \frac{100^2 + t^2}{(100 + t)^2} \qquad 0 \le t \le 100$$

When $t = 10$, $C_1 = \frac{101}{121}$ pounds per gallon.

EXAMPLE 5.4.4 Find a solution of $t\dot{y} + ay = 0$ (a constant), which is continuous for $t \ge 0$ and takes the value y_0 at $t = 0$. In this case, the equation is

$$\dot{y} + \frac{a}{t}y = 0$$

Hence, $q(t) = a/t$, which is not continuous at $t = 0$. Therefore, the existence and uniqueness Theorem 5.4.1 does not apply to this initial-value problem.

Nevertheless, we can find a solution, by multiplying by the integrating factor t^a,

$$y(t) = Kt^{-a}$$

where K is an arbitrary constant. If $a = 0$, then there is a unique solution $y(t) = y_0$. If $a < 0$, every solution goes to zero as $t \to 0$. Therefore, there is no solution unless $y_0 = 0$, but then the solution is not unique since K is arbitrary. If $a > 0$, then there are no solutions continuous at the origin unless $K = 0$. Hence, we can have any possibility if $q(t)$ is not continuous.

EXERCISES 5.4

1 Solve the following initial-value problems:

(*a*) $\dot{y} + ty = t$, $y(0) = y_0$.

(*b*) $t\dot{y} + y = t + 1$, $y(1) = 0$. *Hint*: Make the change of variable $\tau = t - 1$.

(*c*) $\dot{y} + e^t y = e^t$, $y(0) = y_0$.

(*d*) $(1 + t)\dot{y} + y = e^t$, $y(0) = 1$.

(*e*) $\dot{y} + (\tan t)y = \sec t$, $y(\pi/4) = 0$.

2 At a certain time a radioactive material is 90 percent pure (10 percent changed), and 1 hour later it is 75 percent pure. What is the half-life?

3 A certain radioactive material with a half-life of 50,000 years is 10 percent pure. When was the material created?

4 A 100-gallon tank holds salt brine with a concentration of 1 pound per gallon. Brine is drawn off at a rate of 2 gallons per minute while water is added at a rate of 1 gallon per minute. What is the concentration 10 minutes after the process is started?

5 An object of mass m falls from a great height under the influence of gravity (g = acceleration of gravity) with air resistance proportional to the velocity. What is its terminal velocity?

6 Solve the first order linear equation by the method of *variation of parameters*; that is, look for a solution of the form $y = A(t)y_1(t)$, where y_1 is a solution of the associated homogeneous equation. Solve part (*a*) of Exercise 1 by this method.

5.5 FIRST ORDER NONLINEAR EQUATIONS

A large class of first order equations can be written as

$$\frac{dy}{dt} = f(t,y)$$

or since $f(t,y)$ can always be written as the quotient of two functions, say

$$f(t,y) = -\frac{M(t,y)}{N(t,y)}$$

then

$$M(t,y)\,dt + N(t,y)\,dy = 0$$

In special cases, because of the forms of $M(t,y)$ and $N(t,y)$, we are able to find closed-form solutions. We shall say that we have a closed-form solution if we can find an implicit relation $F(t,y) = F(t_0,y_0)$ such that $dy/dt = -F_t/F_y = f(t,y)$. According to the implicit-function theorem,† if F_t and F_y are continuous within some circle with center at (t_0,y_0) and $F_y(t_0,y_0) \neq 0$, then provided (t,y) is sufficiently close to the point (t_0,y_0), we can solve $F(t,y) = F(t_0,y_0)$ explicitly for $y(t)$ such that $dy/dt = -F_t/F_y$ and $y(t_0) = y_0$.

In this section, we list a few of the cases where closed-form solutions of nonlinear equations can be found.

1 Reducible to Linear

Some equations, by a special transformation, can be reduced to a linear equation. For example, the *Bernoulli equation* $\dot{y} + q(t)y = r(t)y^n$ can be reduced to a linear equation by the substitution

$$w = y^{1-n}$$

We have

$$\frac{dw}{dt} = (1-n)y^{-n}\frac{dy}{dt}$$

$$\frac{dy}{dt} = \frac{1}{1-n}y^n\frac{dw}{dt}$$

Substituting in the differential equation, we have

$$\frac{1}{1-n}y^n\frac{dw}{dt} + q(t)y = r(t)y^n$$

$$\frac{dw}{dt} + (1-n)q(t)w = (1-n)r(t)$$

† If $F_y(t_0,y_0) = 0$ but $F_t(t_0,y_0) \neq 0$, we can reverse the roles of t and y and solve for t in terms of y.

EXAMPLE 5.5.1 Find a solution of the equation $\dot{y} - (1/t)y = -3ty^2$ such that $y(1) = a$. This is a Bernoulli equation. Therefore, let $w = y^{-1}$. The equation becomes

$$\dot{w} + \frac{1}{t} w = 3t$$

We multiply by

$$Q(t) = \exp\left[\int_1^t \tau^{-1}\, d\tau\right] = \exp(\ln t) = t$$

$$t\dot{w} + w = \frac{d}{dt}(tw) = 3t^2$$

$$w = t^2 + ct^{-1}$$

$$w(1) = \frac{1}{a} = 1 + c$$

$$y(t) = \frac{at}{at^3 - a + 1}$$

2 Separable

If $M(t,y)$ is a function of t only and $N(t,y)$ is a function of y only, then the variables are said to be *separated*. If the given equation can be written in the form

$$M(t)\, dt + N(y)\, dy = 0$$

then the equation is said to be *separable*. A solution satisfying $y(t_0) = y_0$ is then

$$\int_{t_0}^t M(\tau)\, d\tau + \int_{y_0}^y N(\eta)\, d\eta = 0$$

EXAMPLE 5.5.2 Find a solution of $dy/dt = ty^2$ satisfying $y(1) = a$. We can write

$$\frac{dy}{y^2} = t\, dt$$

$$\int_a^y \eta^{-2}\, d\eta = \int_1^t \tau\, d\tau$$

$$\frac{1}{a} - \frac{1}{y} = \frac{t^2}{2} - \frac{1}{2}$$

$$y = \frac{2a}{2 + a - at^2}$$

3 Reducible to Separable

Sometimes, by the proper change of variables, an equation can be changed so that it becomes separable. For example, if

$$M(t,y)\,dt + N(t,y)\,dy = 0$$

and M and N are both homogeneous† of degree k, then the change of variable

$$y = tu(t)$$

will produce a separable equation. This is seen as follows:

$$\begin{aligned} \dot{y} &= t\dot{u} + u \\ M(t,tu) + N(t,tu)(t\dot{u} + u) &= 0 \\ t^k M(1,u) + t^k N(1,u)(t\dot{u} + u) &= 0 \\ M(1,u) + uN(1,u) + tN(1,u)\dot{u} &= 0 \\ \frac{1}{t}\,dt + \frac{N(1,u)}{M(1,u) + uN(1,u)}\,du &= 0 \\ \ln|t| - \ln|t_0| + \int_{u_0}^{u} P(\mu)\,d\mu &= 0 \end{aligned}$$

where

$$P(u) = \frac{N(1,u)}{M(1,u) + uN(1,u)}$$

$$u_0 = \frac{y_0}{t_0}$$

EXAMPLE 5.5.3 Solve $(t^2 + y^2)\,dt + 2ty\,dy = 0$, $y(t_0) = y_0$, assuming t_0 and y_0 positive. Here $M = t^2 + y^2$, $N = 2ty$, and both are homogeneous of degree 2.

$$P(u) = \frac{2u}{1 + u^2 + 2u^2} = \frac{2u}{1 + 3u^2}$$

$$\ln t - \ln t_0 + \int_{u_0}^{u} \frac{2\mu}{1 + 3\mu^2}\,d\mu = 0$$

$$\ln t - \ln t_0 + \tfrac{1}{3}\ln(1 + 3u^2) - \tfrac{1}{3}\ln(1 + 3u_0^2) = 0$$

$$\ln \frac{t^3(1 + 3u^2)}{t_0^3(1 + 3u_0^2)} = 0$$

† A function $f(t,y)$ is said to be homogeneous of degree of k if $f(\lambda t,\lambda y) = \lambda^k f(t,y)$. For example, $f(t,y) = t^2 + y^2 + ty$ is homogeneous of degree 2 because

$$f(\lambda t,\lambda y) = \lambda^2 t^2 + \lambda^2 y^2 + \lambda^2 ty = \lambda^2 f(t,y)$$

$$1 + 3u^2 = \left(\frac{t_0}{t}\right)^3 (1 + 3u_0^2)$$

$$u = \left[\frac{t_0^3(1 + 3u_0^2) - t^3}{3t^3}\right]^{1/2}$$

$$y = \left[\frac{t_0(t_0^2 + 3y_0^2) - t^3}{3t}\right]^{1/2}$$

EXAMPLE 5.5.4 Show that the equation

$$(a_1t + b_1y + c_1)\,dt + (a_2t + b_2y + c_2)\,dy = 0$$

can always be solved by the method of the previous example. First, if $c_1 = c_2 = 0$, then M and N are homogeneous of degree 1. Second, if c_1 and c_2 are not both zero, then we make the substitutions

$$t = \tau + h$$
$$y = \eta + k$$

Then

$$\frac{dy}{dt} = \frac{d\eta}{d\tau} = -\frac{a_1t + b_1y + c_1}{a_2t + b_2y + c_2} = -\frac{a_1\tau + b_1\eta + (a_1h + b_1k + c_1)}{a_2\tau + b_2\eta + (a_2h + b_2k + c_2)}$$

Now the equations

$$a_1h + b_1k + c_1 = 0$$
$$a_2h + b_2k + c_2 = 0$$

have a unique solution for h and k if and only if $a_1b_2 - a_2b_1 \neq 0$. In this case, values of h and k can be found which reduce the equation to the case of homogeneous M and N. The third case is encountered when $a_1b_2 - a_2b_1 = 0$. In this case, we let

$$u = a_1t + b_1y + c_1$$

$$\frac{du}{dt} = a_1 + b_1\frac{dy}{dt}$$

$$\frac{dy}{dt} = \frac{1}{b_1}\left(\frac{du}{dt} - a_1\right)$$

$$= -\frac{u}{a_2t + b_2[(u - c_1 - a_1t)/b_1] + c_2}$$

$$= \frac{-b_1u}{b_2u - b_2c_1 + b_1c_2}$$

$$\frac{du}{dt} = \frac{-b_1^2u}{b_2u - b_2c_1 + b_1c_2} + a_1 = f(u)$$

$$\frac{du}{f(u)} = dt$$

and the variables are separated. If $b_2u - b_2c_1 + b_1c_2 \equiv 0$, then $a_2 = b_2 = c_2 = 0$ and this is not possible. If $b_1 = 0$, then this substitution will not work but the substitution $u = a_2t + b_2y + c_2$ will work, unless $b_2 = 0$. If both b_1 and b_2 are zero, then the equation is already separated.

4 Exact

If the form of the differential equation

$$M(t,y)\,dt + N(t,y)\,dy = 0$$

is an exact differential, that is,

$$M(t,y)\,dt + N(t,y)\,dy = \frac{\partial F}{\partial t}\,dt + \frac{\partial F}{\partial y}\,dy = dF = 0$$

for some function $F(t,y)$ which has continuous derivatives, then

$$F(t,y) = F(t_0,y_0)$$

gives a solution of the initial-value problem.

Theorem 5.5.1 Let $M(t,y)$ and $N(t,y)$ have continuous first partial derivatives in the rectangle $R = \{(t,y) \mid |t - t_0| \le a,\ |y - y_0| \le b\}$. Then there exists a function $F(t,y)$, defined on R, such that $dF = M(t,y)\,dt + N(t,y)\,dy = 0$ if and only if $M_y = N_t$ in R.

PROOF Suppose there is such a function $F(t,y)$ such that $F_t = M$ and $F_y = N$. Then $F_{ty} = M_y = F_{yt} = N_t$, since F has continuous second partial derivatives in R. Conversely, suppose $M_y = N_t$ in R; then define

$$F(t,y) = \int_{t_0}^{t} M(\tau,y_0)\,d\tau + \int_{y_0}^{y} N(t,\eta)\,d\eta$$

Clearly,

$$\begin{aligned}\frac{\partial F}{\partial t} &= M(t,y_0) + \int_{y_0}^{y} N_t(t,\eta)\,d\eta \\ &= M(t,y_0) + \int_{y_0}^{y} M_\eta(t,\eta)\,d\eta \\ &= M(t,y_0) + M(t,y) - M(t,y_0) \\ &= M(t,y) \\ \frac{\partial F}{\partial y} &= N(t,y)\end{aligned}$$

Also $F(t_0,y_0) = 0$, and so $F(t,y) = 0$ gives us a solution to the initial problem satisfying $y(t_0) = y_0$.

EXAMPLE 5.5.5 Solve $(y^2 - t^2)\,dt + 2ty\,dy = 0$ subject to $y(t_0) = y_0$. The equation is exact since, throughout the ty plane,

$$\frac{\partial M}{\partial y} = 2y = \frac{\partial N}{\partial t}$$

Then

$$\frac{\partial F}{\partial t} = y^2 - t^2$$

$$\frac{\partial F}{\partial y} = 2ty$$

Therefore,

$$F(t,y) = \int_{t_0}^{t} (y_0{}^2 - \tau^2)\,d\tau + \int_{y_0}^{y} 2t\eta\,d\eta$$

$$= ty^2 - \frac{t^3}{3} - t_0 y_0{}^2 + \frac{t_0{}^3}{3} = 0$$

gives us the desired solution. Solving for y, we have

$$y = \pm\left(\frac{t^3 - t_0{}^3 + 3t_0 y_0{}^2}{3t}\right)^{1/2}$$

The choice of sign will depend on the sign of y_0.

5 Reducible to Exact

If the equation $M(t,y)\,dt + N(t,y)\,dy = 0$ has a solution which can be written in the form

$$F(t,y) = F(t_0,y_0)$$

then

$$dF = \frac{\partial F}{\partial t}\,dt + \frac{\partial F}{\partial y}\,dy = 0$$

$$\frac{dy}{dt} = -\frac{F_t}{F_y} = -\frac{M(t,y)}{N(t,y)}$$

Therefore,

$$\frac{\partial F}{\partial t} = Q(t,y)M(t,y)$$

$$\frac{\partial F}{\partial y} = Q(t,y)N(t,y)$$

and hence multiplying $M\,dt + N\,dy = 0$ by $Q(t,y)$ makes the equation

$$QM\,dt + QN\,dy = \frac{\partial F}{\partial t}\,dt + \frac{\partial F}{\partial y}\,dy = 0$$

and it is exact. $Q(t,y)$ is called an *integrating factor*. In principle every equation has an integrating factor. However, it may not be easy to find an integrating factor in a given case.

Suppose $M(t,y)$ and $N(t,y)$ satisfy the condition of Theorem 5.5.1 but $M_y \neq N_t$. If there is an integrating factor $Q(t,y)$ with continuous first partial derivatives in R, then $(QM)_y = (QN)_t$ or

$$Q\left(\frac{\partial M}{\partial y} - \frac{\partial N}{\partial t}\right) = N\frac{\partial Q}{\partial t} - M\frac{\partial Q}{\partial y}$$

This equation is not easy to solve in general; however, if Q does not depend on y, then

$$\frac{1}{Q}\frac{dQ}{dt} = \frac{1}{N}\left(\frac{\partial M}{\partial y} - \frac{\partial N}{\partial y}\right)$$

and

$$Q = \exp\left[\int \frac{1}{N}\left(\frac{\partial M}{\partial y} - \frac{\partial N}{\partial t}\right) dt\right]$$

Conversely, if $(1/N)(\partial M/\partial y - \partial N/\partial t)$ does not depend on y, then the Q given by this formula is an integrating factor.

EXAMPLE 5.5.6 Find a solution of $(3t^4 - y)\,dt + t\,dy = 0$ satisfying $y(t_0) = y_0$. The equation is not exact since $M_y = -1 \neq N_t = 1$. However,

$$\frac{1}{N}\left(\frac{\partial M}{\partial y} - \frac{\partial N}{\partial t}\right) = \frac{-2}{t}$$

is independent of y. Therefore,

$$Q(t) = \exp\left(-\int \frac{2}{t}\,dt\right) = \frac{1}{t^2}$$

is an integrating factor. Multiplying by Q gives the equation

$$\left(3t^2 - \frac{y}{t^2}\right) dt + \frac{1}{t}\,dy = 0$$

which is exact, with solution

$$y = t\left(t_0^3 + \frac{y_0}{t_0}\right) - t^4$$

EXERCISES 5.5

1 Classify and solve the following equations:
(*a*) $\sin t \cos y \, dt + \cos t \sin y \, dy = 0$
(*b*) $(y^2 + 2ty) \, dt - t^2 \, dy = 0$
(*c*) $t^2\dot{y} + ty + y^3 = 0$
(*d*) $e^t \, dt + (y^2 + 1) \, dy = 0$
(*e*) $(2t - y - 2) \, dt + (4t + y - 4) \, dy = 0$

2 Show that, if $Q(t,y)$ is an integrating factor for $M \, dt + N \, dy = 0$, then $QG(F)$ is an integrating factor for any differentiable function G if the solution is expressible in the form $F(t,y) = F(t_0,y_0)$. Hence, there are an infinite number of integrating factors.

3 Prove that if $M \, dt + N \, dy = 0$ has an integrating factor $Q(y)$, which is a function of y only, then

$$Q(y) = e^{p(y)} \qquad \text{where } p(y) = \int \frac{1}{M}\left(\frac{\partial N}{\partial t} - \frac{\partial M}{\partial y}\right) dy$$

4 Find an integrating factor in each case and solve the equation.
(*a*) $ty \, dt + (t^2 + te^y) \, dy = 0$
(*b*) $3t^2y \, dt + (y^4 - t^3) \, dy = 0$
(*c*) $(y^2 - 2t^2y) \, dt + (2t^3 - ty) \, dy = 0$

5 If $M(t,y)$ and $N(t,y)$ are both homogeneous of the same degree, show that the equation $M \, dt + N \, dy = 0$ is reducible to separable by the substitution $u = t/y$.

5.6 APPLICATIONS OF FIRST ORDER EQUATIONS

In this section we shall give several examples of applications of first order differential equations.

EXAMPLE 5.6.1 Find a family of orthogonal trajectories for the system of curves $x^2 - y^2 = c$. Notice that for different values of $c \neq 0$, $x^2 - y^2 = c$ represents a family of hyperbolas. By an orthogonal trajectory we mean a curve which crosses each curve of the given family at right angles. Differentiating $x^2 - y^2 = c$, we have $x \, dx - y \, dy = 0$, or

$$\frac{dy}{dx} = \frac{x}{y}$$

A curve crossing these curves orthogonally should therefore have slope

$$\frac{dy}{dx} = \frac{-y}{x}$$

and hence we solve the differential equation

$$y \, dx + x \, dy = 0$$

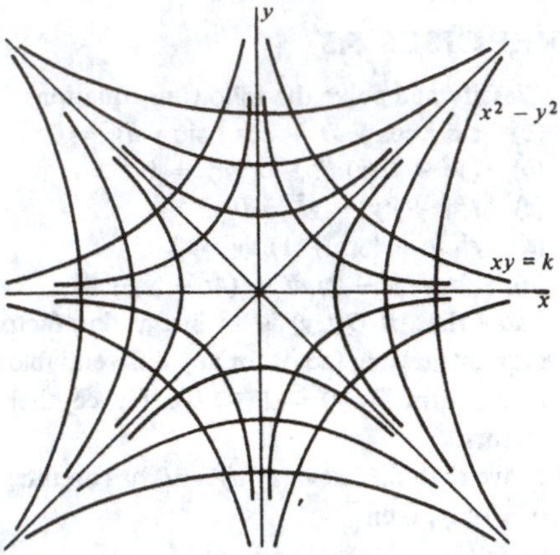

FIGURE 29

This differential equation is exact since $M_y = N_x = 1$. The solution is obviously $xy = k$, where k is an arbitrary constant. The trajectories form another family of hyperbolas. The original family and the family of orthogonal trajectories are indicated in Fig. 29. Note that $xy = 0$ is a member of the family of trajectories.

EXAMPLE 5.6.2 An object of mass m is dropped with zero initial velocity. The air resistance is assumed to be proportional to the square of the velocity. At what velocity will the object drop, and what distance will it travel in time t? The acceleration of the object is dv/dt, and m times this is to be put equal to the force of gravity mg minus kv^2, where k is a constant. Hence, if s is the distance from the starting point, the differential equation is

$$m\frac{d^2s}{dt^2} = m\frac{dv}{dt} = mg - kv^2$$

The initial conditions are at $t = 0$, $v = 0$, and $s = 0$. The differential equation is separable and we can write

$$\frac{dv}{a^2 - v^2} = b\,dt$$

where $a^2 = mg/k$ and $b = k/m$. Integrating, we have

$$\frac{1}{2a}[\ln(a + v) - \ln(a - v)] = bt + c$$

where c is arbitrary. When $t = 0$, $v = 0$ and this implies that $c = 0$. Therefore, we have

$$\ln \frac{a + v}{a - v} = 2abt$$

Solving for v we have

$$v = a \frac{e^{2abt} - 1}{e^{2abt} + 1} = a \tanh abt$$

Notice that as $t \to \infty$, $v \to a$, which is usually called the *terminal velocity.*

$$a = \sqrt{\frac{mg}{k}}$$

Finally,

$$v = \frac{ds}{dt} = a \tanh abt$$

$$s = \frac{1}{b} \ln \cosh abt$$

using the fact that $s = 0$ when $t = 0$.

EXAMPLE 5.6.3 Consider the following pursuit problem. An airplane is flying in a straight line with a constant speed of 200 miles per hour. A second plane is initially flying directly toward the first on a line perpendicular to its path. The second plane continues to pursue the first in such a way that the distance between the planes remains constant (5 miles) and the pursuing plane is always headed toward the other; that is, the tangent to the path of the pursuer passes through the other. Consider the problem in the xy plane (see Fig. 30). Let the coordinates of the pursuing plane be (x,y) and the coordinates of the other be $(s,0)$. The conditions of the problem can be stated in the following equations

$$(s - x)^2 + y^2 = 25$$

$$\frac{dy}{dx} = \frac{-y}{s - x}$$

$$s = 200t$$

We wish to find x and y as functions of t subject to the initial conditions at $t = 0$; $s = 0$, $x = 0$, $y = 5$. Differentiating the first equation, we have

$$(s - x)\left(200 - \frac{dx}{dt}\right) + y\frac{dy}{dt} = 0$$

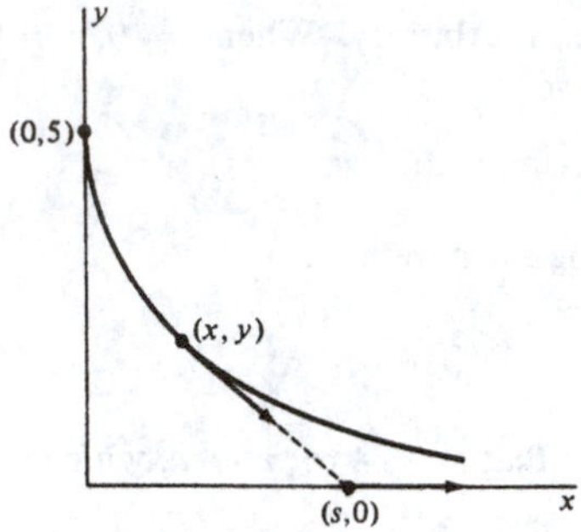

FIGURE 30

From the second equation we have

$$y\frac{dy}{dt} = \frac{-y^2}{s-x}\frac{dx}{dt} = \frac{(s-x)^2 - 25}{s-x}\frac{dx}{dt}$$

Eliminating y, we have

$$\frac{dx}{dt} = 8(s-x)^2 = 8(200t - x)^2$$

We introduce the new variable $u = 200t - x$. Then

$$\frac{dx}{dt} = 200 - \frac{du}{dt} = 8u^2$$

$$\frac{du}{dt} = 200 - 8u^2$$

This equation is separable. Hence,

$$\frac{du}{25 - u^2} = 8\,dt$$

$$\tfrac{1}{10}[\ln(5+u) - \ln(5-u)] = 8t + c$$

When $t = 0$, $u = 0$, which implies that $c = 0$. Solving for u, we have

$$x = 200t - u = 200t - 5\tanh 40t$$

Finally,

$$\frac{dy}{dt} = \frac{-y}{u}\frac{dx}{dt} = -8uy$$

$$\frac{dy}{y} = -8u\,dt = -40\tanh 40t\,dt$$

$$\ln y = -\ln\cosh 40t + k$$

When $t = 0$, $y = 5$, which gives us

$$y = 5 \operatorname{sech} 40t$$

EXERCISES 5.6

1 A ball is thrown straight up with initial velocity v_0. Neglecting air resistance, determine how high the ball will rise.

2 A ball is thrown straight up with initial velocity v_0. Suppose air resistance is proportional to the speed (magnitude of velocity). How high will the ball rise?

3 At a point 4,000 miles from the center of the earth a rocket has expended all its fuel and is moving radially outward with a velocity v_0. Let the force on the rocket due to the earth's gravitational attraction be $10^5 m/r^2$, where m is the mass of the rocket, r is the distance to the center of the earth measured in miles, and time is measured in seconds. Find the minimum v_0 such that the rocket will not return to the earth, sometimes called the *escape velocity*. If v_0 is less than the escape velocity, find the maximum altitude attained by the rocket.

4 Find the family of orthogonal trajectories for each of the given families:

(*a*) $y^2 = cx$ (*b*) $\dfrac{x^2}{4} + \dfrac{y^2}{9} = c^2$

(*c*) $x^2 - xy + y^2 = c^2$ (*d*) $x^2 - 2cx + y^2 = 1$

5 An airplane with a constant airspeed of 200 miles per hour starts out to a destination 300 miles due east. There is a wind out of the north of 25 miles per hour. The plane always flies so that it is headed directly at the destination. Find the path of the airplane.

6 Find the curve passing through the point (3,4) such that the tangent to the curve and the line to the origin are always perpendicular.

7 Find the equation of the curve passing through (2,4) such that the segment of the tangent line between the curve and the x axis is bisected by the y axis.

8 A ball is dropped from a great height. Assuming that the acceleration of gravity is constant and that air resistance is proportional to the square root of the speed, find the terminal velocity.

5.7 NUMERICAL METHODS

Out of all the nonlinear first order differential equations with solutions, only a relatively small number can be solved in closed form. Therefore, it is very important that there be numerical methods for solving differential equations. In this section, we give a very brief introduction to a very large subject, which is playing an increasingly important role in applied mathematics as nonlinear analysis becomes more and more inescapable in modern technology. Of course this development has been aided and abetted by the invention of large-scale

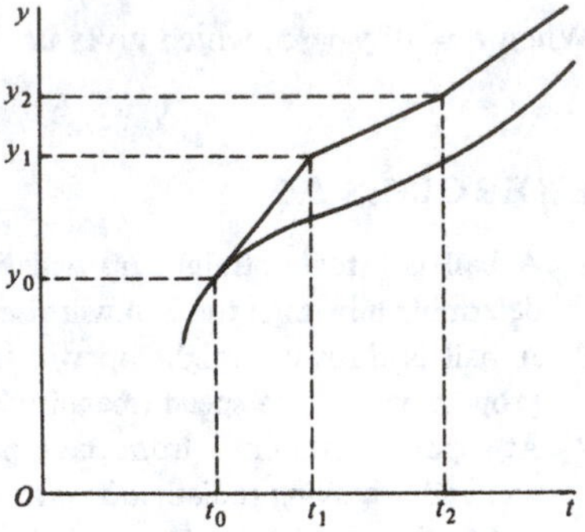

FIGURE 31

digital computers. It is not a coincidence that the development of numerical methods and very fast computers have gone hand in hand over the past 25 years.

We shall consider the numerical solution of the initial-value problem $\dot{y} = f(t,y)$, $t \geq 0$, $y(0) = y_0$. Perhaps the simplest method was invented by Euler. We observe that for small Δt, $\dot{y} \approx \Delta y/\Delta t$ and therefore

$$y_1 - y_0 = \Delta y \approx f(t_0,y_0)\,\Delta t = f(t_0,y_0)(t_1 - t_0)$$

In other words, we compute the *next value* of y, namely y_1, by the formula

$$y_1 = y_0 + f(t_0,y_0)(t_1 - t_0)$$

and, in general,

$$y_k = y_{k-1} + f(t_{k-1},y_{k-1})(t_k - t_{k-1})$$

$k = 1, 2, 3, \ldots$. This is sometimes called the *tangent-line method* because it gives the next approximation by moving along the tangent line from the given point (see Fig. 31).

EXAMPLE 5.7.1 Using the Euler method, find an approximate value of $y(t)$ at $t = 0.4$, where $y(t)$ is the solution of the initial-value problem $\dot{y} = ty^2$, $y(0) = 1$. We have intentionally picked an example where we could find a closed-form solution which could be checked against the numerical solution. The equation is separable, and we easily find the unique solution

$$y(t) = \frac{2}{2 - t^2}$$

which has the value $y(0.4) = 1.087$ correct to three decimal places. On the other hand, using equal $\Delta t = t_k - t_{k-1} = 0.1$, we compute $y_4 = 1.062$ from the

Euler method. Using a smaller $\Delta t = 0.05$ yields the value $y_8 = 1.074$, illustrating the fact that the accuracy of the Euler method tends to improve with smaller increments in the independent variable. The accompanying table summarizes the calculations.

t	Actual solution	$\Delta t = 0.1$	$\Delta t = 0.05$
0	1.000	1.000	1.000
0.05	1.001		1.000
0.10	1.005	1.000	1.003
0.15	1.011		1.008
0.20	1.020	1.010	1.015
0.25	1.032		1.025
0.30	1.047	1.030	1.038
0.35	1.065		1.055
0.40	1.087	1.062	1.074

If the solution $y(t)$ has a continuous second derivative,

$$y_1 = y_0 + \dot{y}_0(t_1 - t_0) + \frac{\ddot{y}(\eta)}{2}(t_1 - t_0)^2$$

where η is between t_0 and t_1. This would tend to show that the error in the Euler method is proportional to the square of Δt. This is not the whole story, however, for several reasons.

1 We may not have a good estimate of the second derivative, which may grow as we move away from the initial point.

2 The point (t_1, y_1) is not, in general, on the solution curve, so that the estimate $\dot{y}_1 = f(t_1, y_1)$ is in error compared with $\dot{y}(t_1)$. The same comment applies to the values $\dot{y}_k = f(t_k, y_k)$, $k > 1$.

3 Whatever computer we are using will have to round off the numbers in the computation, and this can introduce errors which will tend to build up as the number of steps in the calculation to the final result increases.

These comments have been made to indicate the complexity of error analysis. In general, it is very difficult to determine an upper bound on the error made in the numerical solution of a differential equation by a given method. Therefore, we shall have very little to say on the subject. Much research is still going on in this important matter.

EXAMPLE 5.7.2 Find a numerical solution at $t = 0.5$ by the Euler method of the second order equation

$$\ddot{y} + \frac{2t}{t^2 - 1}\dot{y} - \frac{6}{t^2 - 1}y = 0$$

subject to the initial conditions $y(0) = 1$, $\dot{y}(0) = 0$. We have not yet indicated that the Euler method is adaptable to second order equations. The idea is to convert the problem to one for a first order system by introducing $u = y$, $v = \dot{y}$. Then the problem can be stated as

$$\dot{u} = v$$
$$\dot{v} = \frac{6}{t^2 - 1} u - \frac{2t}{t^2 - 1} v$$

$u(0) = 1$, $v(0) = 0$. We shall use equal intervals $\Delta t = 0.1$. We shall proceed as follows:

$$\begin{aligned} u_1 &= u_0 + v_0 \, \Delta t \\ v_1 &= v_0 + \dot{v}_0 \, \Delta t \\ &= v_0 + \frac{6 \, \Delta t}{{t_0}^2 - 1} u_0 - \frac{2t_0 \, \Delta t}{{t_0}^2 - 1} v_0 \end{aligned}$$

and in general,

$$\begin{aligned} u_k &= u_{k-1} + v_{k-1} \, \Delta t \\ v_k &= v_{k-1} + \frac{6 \, \Delta t}{t_{k-1}^2 - 1} u_{k-1} - \frac{2t_{k-1} \, \Delta t}{t_{k-1}^2 - 1} v_{k-1} \end{aligned}$$

The calculation is summarized in the table. The solution, which we shall show later how to find, is $y = 1 - 3t^2$ and so $y(0.5) = 0.25$.

t	u	v
0	1.000	0.000
0.1	1.000	−0.600
0.2	0.940	−1.218
0.3	0.818	−1.835
0.4	0.634	−2.494
0.5	0.385	

The Euler method can be viewed from the following alternative point of view. If $y(t)$ is the solution of the initial-value problem $\dot{y} = f(t,y)$, $y(0) = y_0$, then

$$y(t) = y_0 + \int_{t_0}^{t} f(\tau, y(\tau)) \, d\tau$$

If we make the approximation

$$\begin{aligned} y(t_1) &= y_0 + \int_{t_0}^{t_1} f(\tau, y(\tau)) \, d\tau \\ &\approx y_0 + f(t_0, y_0)(t_1 - t_0) \end{aligned}$$

then we have, in effect, approximated the integral by

$$\int_{t_0}^{t_1} f(\tau, y(\tau)) \, d\tau \approx f(t_0, y_0)(t_1 - t_0)$$

A better approximation to the integral would be

$$\int_{t_0}^{t_1} f(\tau, y(\tau))\, d\tau \approx \frac{f(t_0, y_0) + f(t_1, y(t_1))}{2}(t_1 - t_0)$$

This would lead to the approximation

$$y_1 = y_0 + \frac{f(t_0, y_0) + f(t_1, y_1)}{2}(t_1 - t_0)$$

However, this requires the knowledge of y_1 to compute y_1, which is not acceptable. An alternative is to use the approximation to y_1 obtained by the Euler method. This leads to the formula

$$y_1 = y_0 + \frac{f(t_0, y_0) + f(t_1, y_0 + f(t_0, y_0)(t_1 - t_0))}{2}(t_1 - t_0)$$

or in general, with equal intervals $\Delta t = h$

$$y_k = y_{k-1} + \frac{f(t_{k-1}, y_{k-1}) + f(t_k, y_{k-1} + f(t_{k-1}, y_{k-1})h)}{2}h$$

This formula is called the *modified Euler formula.* Using this method on Example 5.7.1, with $h = 0.1$, we obtain the approximation $y_4 = 1.085$.

The *Runge-Kutta method* is based on an improved approximation of

$$\int_{t_0}^{t} f(\tau, y(\tau))\, d\tau$$

If three points are known

$$(t_0, f(t_0, y(t_0)))$$

$$\left(t_0 + \frac{h}{2}, f\left(t_0 + \frac{h}{2}, y\left(t_0 + \frac{h}{2}\right)\right)\right)$$

and

$$(t_0 + h, f(t_0 + h, y(t_0 + h)))$$

then by passing a parabola of the form $p(t) = at^2 + bt + c$ through the three points and integrating we obtain the following approximation:

$$\begin{aligned}\int_{t_0}^{t_0+h} f(\tau, y(\tau))\, d\tau &\approx \int_{t_0}^{t_0+h} (a\tau^2 + b\tau + c)\, d\tau \\ &= \frac{h}{6}\left[f(t_0, y(t_0)) + 4f\left(t_0 + \frac{h}{2}, y\left(t_0 + \frac{h}{2}\right)\right)\right. \\ &\qquad \left. + f(t_0 + h, y(t_0 + h))\right]\end{aligned}$$

This is known as *Simpson's rule.* In the Runge-Kutta method we attempt to solve the initial-value problem $\dot{y} = f(t,y)$, $y(t_0) = y_0$, by writing

$$y(t) = y_0 + \int_{t_0}^{t} f(\tau, y(\tau))\, d\tau$$

and approximating the integral by Simpson's rule. Unfortunately, to find

$$y_1 = y_0 + \int_{t_0}^{t_1} f(\tau, y(\tau))\, d\tau$$

we would need to know $y(t_0 + h/2)$ and $y(t_0 + h)$, where $h = t_1 - t_0$. Therefore, we need to approximate these values. We define the following quantities:

$$\begin{aligned} k_1 &= f(t_0, y_0) \\ k_2 &= f\left(t_0 + \frac{h}{2}, y_0 + \frac{h}{2} k_1\right) \\ k_3 &= f\left(t_0 + \frac{h}{2}, y_0 + \frac{h}{2} k_2\right) \\ k_4 &= f(t_0 + h, y_0 + hk_3) \end{aligned}$$

and make the following approximations:

$$4f\left(t_0 + \frac{h}{2}, y\left(t_0 + \frac{h}{2}\right)\right) \approx 2k_2 + 2k_3$$

$$f(t_0 + h, y(t_0 + h)) \approx k_4$$

The first iteration in the Runge-Kutta method is given by

$$y_1 = y_0 + \frac{h}{6}(k_1 + 2k_2 + 2k_3 + k_4)$$

Further iterations can be made by repeating the process for the initial-value problem $\dot{y} = f(t,y)$, $y(t_1) = y_1$, etc.

If we apply the Runge-Kutta method to the example of Exercise 5.7.1 using $h = 0.4$, we obtain the following data: $k_1 = 0$, $k_2 = 0.2$, $k_3 = 0.2163$, $k_4 = 0.4722$, and $y_1 = 1.087$. Hence, we obtain a result correct to three decimal places with but one iteration.

The *Milne method* again starts from the integral equation

$$y(t) = y_0 + \int_{t_0}^{t} f(\tau, y(\tau))\, d\tau$$

and uses a four-point numerical-integration formula for approximating the integral. Suppose we integrate from t_0 to t_4, where $\Delta t_i = t_i - t_{i-1} = h$, $i = 1, 2, 3, 4$. Then the appropriate integration formula is

$$\int_{t_0}^{t_4} \dot{y}(\tau)\, d\tau \approx \frac{4h}{3}(2\dot{y}_3 - \dot{y}_2 + 2\dot{y}_1)$$

Hence, the formula for approximating $y(t_4)$ is

$$y_4 = y_0 + \frac{4h}{3}(2\dot{y}_3 - \dot{y}_2 + 2\dot{y}_1)$$

where $\dot{y}_k = f(t_k, y_k)$, $k = 1, 2, 3$. To get the method started we need good approximations for y_1, y_2, and y_3. The usual procedure is to use the Runge-Kutta method to obtain y_1, y_2, and y_3. Then y_4 is obtained from the above formula, which is called the *predictor formula.* As a check we can then use the Simpson rule for recomputing y_4 using the formula

$$y_4 = y_2 + \frac{h}{3}(\dot{y}_2 + 4\dot{y}_3 + \dot{y}_4)$$

This is called the *corrector formula,* and the Milne method is known as a *predictor-corrector method.* Having obtained an approximation for y_4, we can then obtain y_5 from the predictor formula

$$y_5 = y_1 + \frac{4h}{3}(2\dot{y}_4 - \dot{y}_3 + 2\dot{y}_2)$$

This can be checked using the corrector formula

$$y_5 = y_3 + \frac{h}{3}(\dot{y}_3 + 4\dot{y}_4 + \dot{y}_5)$$

and so on. The corrector formula can be used over and over again until the change in the data is less than the error inherent in the Simpson integration formula, which is of order h^5.

EXERCISES 5.7

1 Consider the initial-value problem: $\dot{y} = y^2$, $y(0) = 1$. Find the solution and evaluate it correct to three decimal places at equal intervals of $\Delta t = 0.1$ up to and including $t = 0.5$.

2 Solve the problem of Exercise 1 approximately using the Euler method with equal intervals $\Delta t = 0.1$. Compare with the results of Exercise 1.

3 Solve the problem of Exercise 1 approximately using the modified Euler method with equal intervals of $\Delta t = 0.1$. Compare with the results of Exercise 1.

4 Solve the problem of Exercise 1 approximately at $t = 0.3$ using the Runge-Kutta method using the single interval $\Delta t = 0.3$. Compare with the results of Exercise 1.

5 Derive Simpson's rule

$$\int_{t_0}^{t_2} f(t)\,dt = \frac{h}{6}(f_0 + 4f_1 + f_2)$$

where $t_2 - t_1 = t_1 - t_0 = h/2$ and $f_0 = f(t_0)$, $f_1 = f(t_1)$, $f_2 = f(t_2)$. *Hint*: Fit a parabola $p(t) = at^2 + bt + c$ to the data and then approximate the integral by $\int_{t_0}^{t_2} p(t)\,dt$.

6 Consider the initial-value problem: $t^2\ddot{y} + t\dot{y} - y = 0$, $y(1) = 2$, $\dot{y}(1) = 0$. Show that $y = (t^2 + 1)/t$ is a solution. It can be shown that this solution is unique.

7 Using equal intervals of $\Delta t = 0.1$ find an approximate solution to the problem of Exercise 6 at $t = 0.4$ by the Euler method.

*5.8 EXISTENCE AND UNIQUENESS

The most general first order ordinary differential equation can be written in the form

$$F(t,y,\dot{y}) = 0$$

In most cases, although it is not guaranteed, we can solve this implicit relation for $\dot{y}$, giving

$$\frac{dy}{dt} = f(t,y)$$

A corresponding initial-value problem is to find a function $y(t)$ which is continuous and has a continuous derivative for $0 \le t \le a$ which satisfies $\dot{y} = f(t,y)$ and $y(0) = y_0$. We shall state and prove an existence and uniqueness theorem for this initial-value problem, but first we must define a new concept.

Definition 5.8.1 A function $f(t,y)$ satisfies a *Lipschitz condition* in $R = \{(t,y) \mid 0 \le t \le b, |y - y_0| \le c\}$ if there exists a constant $K > 0$ such that $|f(t,y_1) - f(t,y_2)| \le K|y_1 - y_2|$ for each (t,y_1) and (t,y_2) in R.

EXAMPLE 5.8.1 The function $f(t,y) = ty^2$ satisfies a Lipschitz condition for $0 \leq t \leq b < \infty$, $|y - y_0| \leq c < \infty$, since

$$|ty_1^2 - ty_2^2| = |t|\,|y_1 + y_2|\,|y_1 - y_2| \leq b(2c + 2|y_0|)|y_1 - y_2|$$

because

$$\begin{aligned} |y_1 + y_2| &= |y_1 - y_0 + y_2 - y_0 + 2y_0| \\ &\leq |y_1 - y_0| + |y_2 - y_0| + 2|y_0| \\ &\leq 2c + 2|y_0| \end{aligned}$$

Therefore, we can take $K = b(2c + 2|y_0|)$, and we have obtained the result.

EXAMPLE 5.8.2 The function $f(t,y) = \sqrt{y}$ does not satisfy a Lipschitz condition for $0 \leq t \leq b$, $|y - 1| \leq 1$ because

$$|\sqrt{y_1} - \sqrt{y_2}| = \frac{|y_1 - y_2|}{\sqrt{y_1} + \sqrt{y_2}}$$

and $1/(\sqrt{y_1} + \sqrt{y_2})$ is not bounded in the given region, since we can take y_1 and y_2 arbitrary close to zero. This example shows that the continuity in the region does *not* imply that the function satisfies a Lipschitz condition.

Theorem 5.8.1 If $f(t,y)$ has a continuous partial derivative f_y in $R = \{(t,y) \mid 0 \leq t \leq b < \infty,\ |y - y_0| \leq c < \infty\}$, then it satisfies a Lipschitz condition in R.

PROOF The mean-value theorem holds in R. Therefore,

$$f(t,y_1) - f(t,y_2) = f_y(t,\eta)(y_1 - y_2)$$

where η is between y_1 and y_2. Therefore, (t,η) is in R, and $f_y(t,\eta)$ is bounded in R. Hence,

$$|f(t,y_1) - f(t,y_2)| \leq K|y_1 - y_2|$$

where $K = \max |f_y|$ in R.

We now turn to the main theorem of this section.

Theorem 5.8.2 Let $f(t,y)$ be continuous and satisfy a Lipschitz condition in $R = \{(t,y) \mid 0 \leq t \leq b,\ |y - y_0| \leq c\}$. Let $M = \max |f(t,y)|$

in R. Then the initial-value problem $\dot{y} = f(t,y)$, $y(0) = y_0$ has a unique solution for $0 \leq t \leq \min [b,c/M]$.

PROOF We write the integral equation

$$y(t) = y_0 + \int_0^t f(\tau, y(\tau))\, d\tau$$

and see that $y(t)$ satisfies the initial-value problem if and only if it satisfies the integral equation. Next we propose to solve the integral equation by a method of successive approximation. As a first approximation we take

$$y_1(t) = y_0 + \int_0^t f(\tau, y_0)\, d\tau$$

Then

$$|y_1 - y_0| \leq \int_0^t |f(\tau, y_0)|\, d\tau \leq Mt \leq Ma$$

provided $0 \leq t \leq a \leq b$. But we must have that $Ma \leq c$. Therefore, we take

$$a = \min\left[b, \frac{c}{M}\right]$$

Next we take

$$y_2(t) = y_0 + \int_0^t f(\tau, y_1(\tau))\, d\tau$$

and, in general,

$$y_k(t) = y_0 + \int_0^t f(\tau, y_{k-1}(\tau))\, d\tau$$

We must show that $|y_k - y_0| \leq c$ for $k = 1, 2, 3, \ldots$, for all t such that $0 \leq t \leq a$. This we do by induction. We have already shown that this is true for $k = 1$. Now assume that it is true for $k = 1, 2, 3, \ldots, n - 1$. Then

$$|y_n - y_0| \leq \int_0^t |f(\tau, y_{n-1}(\tau))|\, d\tau \leq Ma \leq c$$

Therefore, by induction the inequality is correct for all k.

We shall show that the nth approximation converges to a solution of the integral equation. Observe that

$$y_n(t) = y_0 + \sum_{k=1}^{n} [y_k(t) - y_{k-1}(t)]$$

Now

$$|y_1 - y_0| = \left|\int_0^t f(\tau, y_0)\, d\tau\right| \le \int_0^t |f(\tau, y_0)|\, d\tau$$

$$\le \frac{M}{K} Kt \le \frac{M}{K} Ka$$

$$|y_2 - y_1| = \left|\int_0^t f(\tau, y_1) - f(\tau, y_0)\, d\tau\right|$$

$$\le \int_0^t |f(\tau, y_1) - f(\tau, y_0)|\, d\tau$$

$$\le K \int_0^t |y_1 - y_0|\, d\tau \le KM \int_0^t \tau\, d\tau$$

$$= \frac{KMt^2}{2} = \frac{M(Kt)^2}{K2!} \le \frac{M(Ka)^2}{K2!}$$

In general, we can prove that

$$|y_k - y_{k-1}| \le \frac{M(Kt)^k}{Kk!} \le \frac{M(Ka)^k}{Kk!}$$

We have already shown this for $k = 1, 2$. Now assume that it is true for $k = 1, 2, 3, \ldots, n - 1$. Then

$$|y_n - y_{n-1}| \le \int_0^t |f(\tau, y_{n-1}) - f(\tau, y_{n-2})|\, d\tau$$

$$\le K \int_0^t |y_{n-1} - y_{n-2}|\, d\tau$$

$$\le \frac{M}{K} \frac{K^n}{(n-1)!} \int_0^t \tau^{n-1}\, d\tau = \frac{M(Kt)^n}{Kn!} \le \frac{M(Ka)^n}{Kn!}$$

Therefore, the inequality is correct for all k.

From the definition of $y_n(t)$, we have that

$$\lim_{n\to\infty} y_n(t) = y_0 + \sum_{k=1}^{\infty} [y_k(t) - y_{k-1}(t)]$$

provided the series converges. Since

$$|y_k - y_{k-1}| \le \frac{M(Ka)^k}{Kk!}$$

we have, by comparison with the convergent series $\sum_{k=1}^{\infty} \frac{M(Ka)^k}{Kk!}$, that the

$\lim_{n\to\infty} y_n(t)$ exists uniformly in t. Therefore, since $y_k(t)$ is continuous for each k, for $0 \le t \le a$,

$$\lim_{n\to\infty} y_n(t) = y(t)$$

where $y(t)$ is a continuous function for $0 \le t \le a$.

The fact that for each t, $0 \le t \le a$, $|y_n(t) - y_0| \le c$ guarantees that in the limit $|y(t) - y_0| \le c$. Therefore, if given an $\varepsilon > 0$ there exists an $N(\varepsilon)$ such that

$$|y_n(t) - y(t)| < \varepsilon$$

for all $n > N(\varepsilon)$, it follows that

$$|f(t,y_n(t)) - f(t,y(t))| \le K|y_n(t) - y(t)| < K\varepsilon$$

which implies that $\lim_{n\to\infty} f(t,y_n(t)) = f(t,y(t))$ uniformly in t. This allows us to take the limit under the integral sign, and therefore

$$\begin{aligned} y(t) = \lim_{n\to\infty} y_n(t) &= \lim_{n\to\infty} \left[y_0 + \int_0^t f(\tau, y_{n-1}(\tau))\, d\tau \right] \\ &= y_0 + \int_0^t \lim_{n\to\infty} f(\tau, y_{n-1}(\tau))\, d\tau \\ &= y_0 + \int_0^t f(\tau, y(\tau))\, d\tau \end{aligned}$$

This proves that $y(t)$ satisfies the integral equation and hence is a solution of the initial-value problem.

Finally, we have to show that the solution is unique. Let $\tilde{y}(t)$ be any solution of the initial-value problem, from which it follows as before that $|\tilde{y}(t) - y_0| \le c$ for $0 \le t \le a$. Then

$$\begin{aligned} |\tilde{y} - y_1| &= \left| \int_0^t [f(\tau, \tilde{y}(\tau)) - f(\tau, y_0)]\, d\tau \right| \\ &\le K \int_0^t |\tilde{y} - y_0|\, d\tau \le Kct \le Kca \\ |\tilde{y} - y_2| &= \left| \int_0^t [f(\tau, \tilde{y}) - f(\tau, y_1)]\, d\tau \right| \\ &\le K \int_0^t |\tilde{y} - y_1|\, d\tau \le \frac{K^2ct^2}{2} \le \frac{c(Ka)^2}{2!} \end{aligned}$$

Inductively, we can prove that

$$|\tilde{y} - y_n| \leq \frac{c(Kt)^n}{n!} \leq \frac{c(Ka)^n}{n!} \to 0$$

as $n \to \infty$, since it is well known that the nth term of the convergent series $\sum_{k=1}^{\infty} \frac{(Ka)^k}{k!}$ must go to zero as $n \to \infty$. This proves that

$$\tilde{y}(t) = \lim_{n\to\infty} y_n(t) = y(t)$$

which completes the proof of the theorem.

Corollary 5.8.1 Let $f(t,y)$ be continuous and satisfy a Lipschitz condition in $R = \{(t,y) \mid -b \leq t \leq 0, |y - y_0| \leq c\}$. Let $M = \max |f(t,y)|$ in R. Then the initial-value problem $\dot{y} = f(t,y)$, $y(0) = y_0$ has a unique solution for $-\min [b,c/M] \leq t \leq 0$.

PROOF The proof follows, with very little change in detail, the proof of Theorem 5.8.2. The reader should check the details.

Corollary 5.8.2 Let $f(t,y)$ be continuous and satisfy a Lipschitz condition in $R = \{(t,y) \mid |t| \leq b, |y - y_0| \leq c\}$. Let $M = \max |f(t,y)|$ in R. Then the initial-value problem $\dot{y} = f(t,y)$, $y(0) = y_0$ has a unique solution for $|t| \leq \min [b,c/M]$.

PROOF The proof combines the results of Theorem 5.8.2 and Corollary 5.8.1. The fact that the solution is continuous and has a continuous derivative at $t = 0$ follows from

$$\lim_{t\to 0^+} y(t) = y_0 = \lim_{t\to 0^-} y(t)$$

$$\lim_{t\to 0^+} \dot{y}(t) = f(0, y_0) = \lim_{t\to 0^-} \dot{y}(t)$$

Corollary 5.8.3 Let $f(t,y)$ be continuous and satisfy a Lipschitz condition in $R = \{(t,y) \mid |t - t_0| \leq b, |y - y_0| \leq c\}$. Let $M = \max |f(t,y)|$ in R. Then the initial-value problem $\dot{y} = f(t,y)$, $y(t_0) = y_0$ has a unique solution for $|t - t_0| \leq \min [b,c/M]$.

PROOF This result is just Corollary 5.8.2 with the origin ($t = 0$) shifted to $t = t_0$.

The result of Corollary 5.8.3 is a *local result* in the following sense. We start out with an interval $\{t \mid |t - t_0| \leq b\}$ and a function $f(t,y)$ continuous

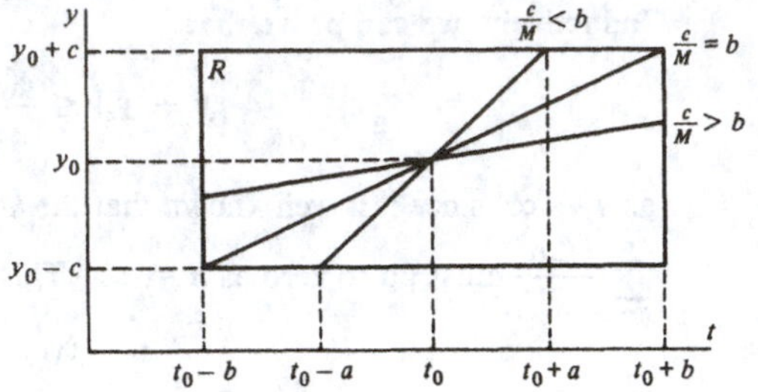

FIGURE 32

for t in the interval and $|y - y_0| \leq c$. But we end up proving only that the initial-value problem has a unique solution for $|t - t_0| \leq a$, where $a = \min[b, c/M]$ and $M = \max |f(t,y)|$ in R. It is clear why this is the case. Consider the utterly trivial case where $f(t,y) \equiv M$ in R. Then $\dot{y} = M$, $y(t_0) = y_0$ has the solution

$$y = y_0 + M(t - t_0)$$

If $M \leq c/b$, the solution reaches the end of the interval ($t = t_0 + b$) before $y = y_0 + c$. If $M > c/b$, the solution reaches the value $y_0 + c$ before it reaches the end of the interval. Therefore, existence in the interval $\{t \mid |t - t_0| \leq a\}$ is the best result we can hope for, and this will not give us the original interval unless $M \leq c/b$ (see Fig. 32).

Now consider the following situation. Suppose $c/M < b$. Then Corollary 5.8.3 does not establish existence of a solution throughout the interval given by $|t - t_0| \leq b$. However, if $y(t)$ is the solution for $|t - t_0| \leq a = c/M$, then the following left-hand limits exist:

$$\lim_{t \to (t_0 + a)^-} y(t) = y_a$$

$$\lim_{t \to (t_0 + a)^-} \dot{y}(t) = \lim_{t \to (t_0 + a)^-} f(t,y) = \dot{y}_a$$

Now imagine the following situation. A graph of $y = y(t)$, $t_0 - a \leq t \leq t_0 + a$, has been drawn. This is part of the solution curve through (t_0, y_0). It reaches the point $(t_0 + a, y_a)$. Suppose there is a rectangle

$$R' = \{(t,y) \mid |t - t_0 - a| \leq \beta, |y - y_a| \leq \gamma\}$$

in which $f(t,y)$ is defined, is continuous, and satisfies a Lipschitz condition. According to Corollary 5.8.3, if $\alpha = \min[\beta, \gamma/M']$, where $M' = \max |f(t,y)|$ in R', then there exists a unique solution of $\dot{y} = f(t,y)$, $y(t_0 + a) = y_a$, for $|t - t_0 - a| \leq \alpha$. Hence, we have *continued* the solution from $t_0 - a \leq t \leq$

$t_0 + a$ to $t_0 + a \le t \le t_0 + a + \alpha$. In a similar manner, if the point $(t_0 - a, y_{-a})$, where

$$y_{-a} = \lim_{t \to (t_0 - a)^+} y(t)$$

is not on the boundary of the original rectangle, the solution can be continued closer to $t_0 - b$. In fact, the process can be continued in both directions as long as the solution curve has not yet met the boundary of the original rectangle.† This does not imply, however, that the solution can always be continued to the ends of the interval given by $|t - t_0| \le b$, as will be seen in the next example.

EXAMPLE 5.8.3 Solve the initial-value problem $\dot{y} = y^2$, $y(0) = 1$. Here $f(t,y) = y^2$. Let $b = 1$ and $c = 1$. Then $|f(t,y)| = y^2 \le 4$, and

$$\begin{aligned} |f(t,y_1) - f(t,y_2)| &= |y_1{}^2 - y_2{}^2| \\ &= |y_1 - y_2|\,|y_1 + y_2| \\ &\le (|y_1| + |y_2|)|y_1 - y_2| \\ &\le 4|y_1 - y_2| \end{aligned}$$

Therefore, $f(t,y)$ satisfies a Lipschitz condition with $K = 4$, and there exists a unique solution for

$$|t| \le \min\,[1,\tfrac{1}{4}] = \tfrac{1}{4}$$

The integral equation is $y(t) = 1 + \int_0^t y^2\,d\tau$, and the method of successive approximations proceeds as follows:

$$y_1(t) = 1 + \int_0^t d\tau = 1 + t$$

$$y_2(t) = 1 + \int_0^t (1 + \tau)^2\,d\tau = 1 + t + t^2 + \frac{t^3}{3}$$

$$\begin{aligned} y_3(t) &= 1 + \int_0^t \left(1 + \tau + \tau^2 + \frac{\tau^3}{3}\right)^2 d\tau \\ &= 1 + t + t^2 + t^3 + \tfrac{2}{3}t^4 + \tfrac{1}{3}t^5 + \tfrac{1}{9}t^6 + \tfrac{1}{63}t^7 \end{aligned}$$

. . . .

Except for terms of order t^{k+1} the kth approximation is $1 + t + t^2 + \cdots + t^k$.

† The solution may be continued outside the original rectangle if there is a rectangle centered on a point on the solution curve and extending beyond the original rectangle where the hypotheses of Corollary 5.8.3 are satisfied.

We seem to be generating the series

$$1 + t + t^2 + \cdots = \frac{1}{1 - t} \qquad -1 < t < 1$$

Let $y(t) = 1/(1 - t)$. This clearly is a solution of the initial-value problem. Notice, however, that y does not exist for $t = 1$. This shows that the solution by successive approximations cannot be continued up to and including $t = 1$.

EXAMPLE 5.8.4 Let $f(t,y) = -q(t)y$, where $q(t)$ is continuous for $|t - t_0| \le b$. Let $K = \max |q(t)|$ in the interval. Then $|f(t,y_1) - f(t,y_2)| = |q(t)|\,|y_1 - y_2| \le K|y_1 - y_2|$, so that $f(t,y)$ satisfies a Lipschitz condition in $R = \{(t,y) \mid |t - t_0| \le b, |y - y_0| \le c\}$ for any y_0 and any c. Since $|y| = |y - y_0 + y_0| \le c + |y_0|$, and $M = \max |f(t,y)| \le K(c + |y_0|)$, $\min [b,c/M]$ is either b or is greater than $c/K(c + |y_0|)$. In the latter case, by taking c large c/M can be made larger than $1/2K$ whatever y_0. Hence, the solution can always be continued to the ends of the interval. Therefore, the initial-value problem $\dot{y} + q(t)y = 0$, $y(t_0) = y_0$, has a unique solution for $|t - t_0| \le b$. This is in complete agreement with Theorem 5.4.1.

EXAMPLE 5.8.5 Consider the initial-value problem $\dot{y} = \sqrt{y}$, $y(0) = 0$. We can write

$$\frac{dy}{\sqrt{y}} = dt$$

$$2\sqrt{y} = t + k$$

But $k = 0$ since $y(0) = 0$. Therefore, $y = t^2/4$ is a solution. However, $y \equiv 0$ is also a solution. In this case we have existence but not uniqueness of a solution. Here $f(t,y) = \sqrt{y}$ does not satisfy a Lipschitz condition in

$$R = \{(t,y) \mid 0 \le t \le b, 0 \le y \le c\}$$

(see Example 5.8.2). It can be shown that continuity of $f(t,y)$ in

$$R = \{(t,y) \mid |t - t_0| \le b, |y - y_0| \le c\}$$

is sufficient for existence of a solution of $\dot{y} = f(t,y)$, $y(t_0) = y_0$ but not for uniqueness.†

†See E. Coddington and N. Levinson, "Theory of Ordinary Differential Equations," McGraw-Hill, New York, 1955 (rpt. Krieger, Melbourne, Florida, 1984).

From the point of view of the physical applications, there is another requirement which we should impose on the mathematical model beside existence and uniqueness of the solution; that is, the solution should depend continuously on the initial data. The reason for this is that the data may be only approximate. They may be the result of some measurement which at best is somewhat inaccurate, or they may represent some starting configuration of a physical system which is not exactly reproducible. It would be comforting to know that for two slightly different starting conditions the solutions vary only slightly. Let us consider two solutions of the initial-value problem dealt with in Theorem 5.8.2, $u(t)$ and $v(t)$, such that $u(0) = u_0$ and $v(0) = v_0$, where $|u_0 - v_0| = \delta$. We shall assume that both solutions exist in R for $0 \leq t \leq a$. Then

$$u(t) = u_0 + \int_0^t f(\tau,u(\tau))\,d\tau$$

$$v(t) = v_0 + \int_0^t f(\tau,v(\tau))\,d\tau$$

$$u(t) - v(t) = u_0 - v_0 + \int_0^t [f(\tau,u(\tau)) - f(\tau,v(\tau))]\,d\tau$$

Using the Lipschitz condition, we have

$$|u(t) - v(t)| \leq |u_0 - v_0| + K \int_0^t |u(\tau) - v(\tau)|\,d\tau$$

Let $w(t) = |u(t) - v(t)|$ and $w_0 = |u_0 - v_0| = \delta$. Then

$$w(t) \leq \delta + K \int_0^t w(\tau)\,d\tau$$

If $W(t) = \int_0^t w(\tau)\,d\tau$, then $\dot{W}(t) = w(t)$ and $W(0) = 0$, and

$$\dot{W} - KW \leq \delta$$

Multiplying by e^{-Kt}, we have

$$\dot{W}e^{-Kt} - KWe^{-Kt} = \frac{d}{dt}(e^{-Kt}W) \leq \delta e^{-Kt}$$

and when we integrate from 0 to t,

$$e^{-Kt}W \leq \frac{\delta}{K}(1 - e^{-Kt})$$

which implies from the above that

$$|u(t) - v(t)| = w(t) \leq \delta e^{Kt} \leq \delta e^{Ka} = |u_0 - v_0|e^{Ka}$$

This result states that the absolute value of the difference between two solutions remains less than or equal to the absolute value of the difference between their initial values multiplied by the constant e^{Ka} for all t satisfying $0 \le t \le a$. Hence, if the initial values differ by very little, the solutions will be close to one another unless the interval considered is very large.

If in the above discussion $-a \le t \le 0$, then $w(t) \le \delta - K \int_0^t w(\tau)\, d\tau$ is the inequality which must be considered. But then we can prove that

$$-W = \int_t^0 w(\tau)\, d\tau \le \frac{\delta}{K}(e^{K|t|} - 1)$$

and it follows that

$$|u(t) - v(t)| \le \delta e^{K|t|} \le |u_0 - v_0| e^{Ka}$$

Combining these two results, we have the following inequality to handle the situation covered by Corollary 5.8.3.

$$|u(t) - v(t)| \le |u(t_0) - v(t_0)| e^{K|t-t_0|}$$

EXERCISES 5.8

1 Use Theorem 5.8.2 to show that the initial-value problem $\dot{y} = ty^2$, $y(0) = 1$, has a unique solution for $0 \le t \le \frac{1}{2}$.

2 Set up the appropriate integral equation for the successive-approximation solution of the problem in Exercise 1. Carry out the iteration for three steps.

3 Solve the problem of Exercise 1 by writing $dy/y^2 = t\, dt$ and integrating. Compare this solution with the successive approximations of Exercise 2 by expanding the solution in powers of t. Is it possible to continue the successive-approximation solution to the interval $0 \le t \le \sqrt{2}$?

4 Carry out one continuation of the solution of the initial-value problem of Exercise 5.8.1 into an interval of the type $\frac{1}{2} \le t \le b$. How far can you continue the solution at this stage? Why can you not reach $t = \sqrt{2}$ after a finite number of continuations?

5 Show that a solution of $\dot{y} = t\sqrt{y}$, $y(0) = 0$, exists but is not unique. Explain. Show that there are actually infinitely many solutions of this problem.

6 As a numerical method for solving nonlinear equations the method of successive approximations given in this section has a distinct advantage in that there is a bound on the error. Show that

$$|y(t) - y_n(t)| \le \frac{cK^n|t - t_0|^n}{n!}$$

7 Prove the inequality $|u(t) - v(t)| \le |u(t_0) - v(t_0)| e^{K|t-t_0|}$ under the hypotheses of Corollary 5.8.3.

8 Assuming the existence of solutions only, use the inequality in Exercise 7 to prove uniqueness.

9 Consider the linear equations $\dot{w} + p(t)w = q(t)$ and $\dot{W} + p(t)W = Q(t)$, for $0 \le t \le b$, where p, q, and Q are continuous. If $w(0) = W(0) = 0$ and $q(t) \le Q(t)$ in the interval, prove that $w(t) \le W(t)$.

10 Generalize Exercise 3 to the case where $w(0) = W(0) \neq 0$.

11 Consider the initial-value problem $\dot{y} + p(t)y = q(t)$, $y(t_0) = y_0$. Let $p(t)$ and $q(t)$ be continuous for $|t - t_0| \le b$. Prove that there exists a unique solution of the problem for $|t - t_0| \le b$ using Corollary 5.8.3.

6

LINEAR DIFFERENTIAL EQUATIONS

6.1 INTRODUCTION

This chapter deals with higher order linear differential equations. The second section deals with general theorems about solutions of the nonsingular initial-value problem (leading coefficient not zero). We state (without proof) the fundamental existence and uniqueness theorem and take up the general question of representation of solutions. This section depends heavily on the concept of basis from the linear algebra. The next section takes up the method of variation of parameters for constructing solutions of nonhomogeneous equations from fundamental systems of solutions of the corresponding homogeneous equation. The fourth section is a discussion of the very important class of equations with constant coefficients. Here the use of complex variables and, in particular, the exponential function are extremely important. The fifth section deals with the method of undetermined coefficients, a special method for determining solutions of nonhomogeneous equations of certain types. The sixth section takes up applications, and the last section (starred) develops the notion of the Green's function for linear initial-value and boundary-value problems and shows how these problems are related to the study of certain integral equations.

6.2 GENERAL THEOREMS

The most general nth order linear differential equation can be written as

$$a_n(t)y^{(n)} + a_{n-1}(t)y^{(n-1)} + \cdots + a_1(t)\dot{y} + a_0(t)y = f(t)$$

We shall discuss the initial-value problem for an interval of the form $\{t \mid a \le t \le b\}$, in which we shall seek an n-times continuously differentiable function $y(t)$ satisfying the differential equation and such that for some t_0 in the interval $y(t_0) = y_0, \dot{y}(t_0) = \dot{y}_0, \ldots, y^{(n-1)}(t_0) = y_0^{(n-1)}$, the constants $y_0, \dot{y}_0, \ldots, y_0^{(n-1)}$ being given. In order for the problem to be well set, we shall have to put some restrictions on the functions $a_0(t), a_1(t), \ldots, a_n(t), f(t)$. We shall, in fact, assume that they are all continuous on the interval $\{t \mid a \le t \le b\}$ and that $a_n(t)$ is not zero in the same interval.† We state, for the purposes of this chapter, the basic existence-uniqueness theorem for the initial-value problem. The proof will not be given until Chap. 9, when we discuss systems of differential equations (see Exercise 6.2.17).

Theorem 6.2.1 Let $a_0(t), a_1(t), \ldots, a_n(t), f(t)$ be continuous on the interval $\{t \mid a \le t \le b\}$, where $a_n(t)$ is never zero. Then there exists a unique n-times continuously differentiable function $y(t)$ satisfying

$$a_n(t)y^{(n)} + a_{n-1}(t)y^{(n-1)} + \cdots + a_1(t)\dot{y} + a_0(t)y = f(t)$$

and $y(t_0) = y_0, \dot{y}(t_0) = \dot{y}_0, \ldots, y^{(n-1)}(t_0) = y_0^{(n-1)}$, where $a \le t_0 \le b$ and $y_0, \dot{y}_0, \ldots, y_0^{(n-1)}$ are given constants.

The linear transformation

$$L(y) = a_n(t)y^{(n)} + a_{n-1}(t)y^{(n-1)} + \cdots + a_1(t)\dot{y} + a_0(t)y$$

is defined on the space of n-times continuously differentiable functions on $\{t \mid a \le t \le b\}$. The range space of the linear transformation is the space of continuous functions on the same interval. In terms of L we can write the differential equation as $L(y) = f(t)$. If $f(t) \not\equiv 0$, then we say that the differential equation is *nonhomogeneous*. The equation $L(y) = 0$ is the *associated homogeneous equation*. We note the following trivial but important fact. If y_1 is a solution of the nonhomogeneous equation $L(y) = f(t)$ and y_2 is a solution of the

† See Example 5.4.4, where we were treating the initial-value problem for a first order linear equation in which the leading coefficient was zero at $t = 0$. We failed to have existence or uniqueness of the solution in some cases. Actually, if we merely specify that $a_n(t_0) \ne 0$, then by the continuity of $a_n(t)$ there will exist an interval $\{t \mid \alpha \le t \le \beta\}$ containing t_0 where $a_n(t)$ is not zero.

associated homogeneous equation $L(y) = 0$, then $y_1 + y_2$ is a solution of the nonhomogeneous equation since

$$L(y_1 + y_2) = L(y_1) + L(y_2) = f(t)$$

Furthermore, if y is *any* solution of the nonhomogeneous equation and $\tilde{y}$ is a particular solution of the nonhomogeneous equation, then

$$L(y - \tilde{y}) = L(y) - L(\tilde{y}) = f(t) - f(t) = 0$$

so that $y - \tilde{y}$ is in the null space of the linear transformation L. Therefore, $y = \tilde{y} + w$, where w is in the null space of L. We shall show that the null space of L is n-dimensional by showing that every function in it can be expressed as a linear combination of n independent solutions of $L(y) = 0$.

Theorem 6.2.2 Let $y_k(t)$ be the solution of $L(y) = 0$ on $\{t \mid a \le t \le b\}$ such that for some t_0 in the interval

$$(y_k(t_0), \dot{y}_k(t_0), \ldots, y_k^{(n-1)}(t_0)) = \mathbf{e}_k \qquad k = 1, 2, \ldots, n$$

Then $y_1(t), y_2(t), \ldots, y_n(t)$ are independent on $\{t \mid a \le t \le b\}$. Furthermore, if $y(t)$ is the solution of $L(y) = 0$ satisfying $y(t_0) = c_1$, $\dot{y}(t_0) = c_2, \ldots, y^{(n-1)}(t_0) = c_n$, then

$$y(t) = c_1 y_1(t) + c_2 y_2(t) + \cdots + c_n y_n(t)$$

PROOF Theorem 6.2.1 guarantees that there are unique solutions $y_1, y_2, \ldots, y_n$ to the n problems set by the conditions

$$(y_k(t_0), \dot{y}_k(t_0), \ldots, y_k^{(n-1)}(t_0)) = \mathbf{e}_k$$

$k = 1, 2, \ldots, n$. Now we compute the Wronskian of the set of functions $y_1, y_2, \ldots, y_n$ at t_0.

$$W(t_0) = \begin{vmatrix} 1 & 0 & 0 & \cdots & 0 \\ 0 & 1 & 0 & \cdots & 0 \\ 0 & 0 & 1 & \cdots & 0 \\ \cdots & \cdots & \cdots & \cdots & \cdots \\ 0 & 0 & 0 & \cdots & 1 \end{vmatrix} = 1$$

Therefore, by Theorem 3.4.4, the set of functions is independent. Now consider the function

$$y(t) = c_1 y_1(t) + c_2 y_2(t) + \cdots + c_n y_n(t)$$

Clearly this is a solution of $L(y) = 0$. It satisfies the required conditions

$$y(t_0) = c_1 y_1(t_0) = c_1$$
$$\dot{y}(t_0) = c_2 \dot{y}_2(t_0) = c_2, \ldots, y^{(n-1)}(t_0) = c_n y_n^{(n-1)}(t_0) = c_n$$

and hence by uniqueness is the required solution. This completes the proof.

In Theorem 6.2.2, we have constructed a basis for the null space of L. This still leaves the question: Are there other bases? We know from Chap. 3 that there are sets of independent functions which have a Wronskian which is zero in the given interval. Therefore, if we compute the Wronskian of a set of functions at a point and find it is zero, this does not show that the set is dependent. However, if each of the n functions is a solution of $L(y) = 0$, then the situation is different, as indicated by the next theorem.

Theorem 6.2.3 Let $y_1(t), y_2(t), \ldots, y_n(t)$ be solutions of $L(y) = 0$ on the interval $\{t \mid a \le t \le b\}$. Then $y_1, y_2, \ldots, y_n$ are independent if and only if their Wronskian is never zero in the interval.

PROOF By Theorem 3.4.4 the set of functions is independent if their Wronskian is never zero. Conversely, suppose that $y_1, y_2, \ldots, y_n$ are independent. Let t_0 be *any* point in the given interval and assume that the Wronskian of $y_1, y_2, \ldots, y_n$ is zero at t_0. We form the linear combination

$$y(t) = c_1 y_1(t) + c_2 y_2(t) + \cdots + c_n y_n(t)$$

Clearly $L(y) = 0$. Now we choose a set of constants (not all zero) so that

$$\begin{aligned} y(t_0) &= c_1 y_1(t_0) + c_2 y_2(t_0) + \cdots + c_n y_n(t_0) = 0 \\ \dot{y}(t_0) &= c_1 \dot{y}_1(t_0) + c_2 \dot{y}_2(t_0) + \cdots + c_n \dot{y}_n(t_0) = 0 \\ &\cdots\cdots\cdots\cdots\cdots\cdots\cdots\cdots\cdots\cdots \\ y^{(n-1)}(t_0) &= c_1 y_1^{(n-1)}(t_0) + c_2 y_2^{(n-1)}(t_0) + \cdots + c_n y_n^{(n-1)}(t_0) = 0 \end{aligned}$$

This is possible because the determinant of the coefficient matrix of the system (c's treated as unknowns) $W(t_0) = 0$. However, $y(t) \equiv 0$ is a solution of the same initial-value problem and by uniqueness is the only solution. Therefore,

$$y(t) = c_1 y_1(t) + c_2 y_2(t) + \cdots + c_n y_n(t) = 0$$

for a set of constants not all zero. This is a contradiction because the set $y_1, y_2, \ldots, y_n$ is independent. We conclude that $W(t_0)$ could not have been zero. However, t_0 was any point in the interval, and therefore $W(t)$ can never be zero. This completes the proof.

EXAMPLE 6.2.1 Consider the differential equation $\dddot{y} - \dot{y} = 1$. Find a solution of this equation satisfying $y(0) = 1$, $\dot{y}(0) = -1$, $\ddot{y}(0) = 0$. In this case,

$L(y) = \dddot{y} - \dot{y}$, and the associated homogeneous equation is $\dddot{y} - \dot{y} = 0$. We shall see later that when L is an operator with constant coefficients, the homogeneous equation usually has exponential solutions. Therefore, let us substitute e^{mt} into the equation $L(y) = 0$:

$$L(e^{mt}) = m^3 e^{mt} - me^{mt} = m(m - 1)(m + 1)e^{mt}$$

Now if $m = 0$, 1, or -1, the equation $L(y) = 0$ is satisfied by $y = e^{mt}$. We shall then take $y_1 = 1$, $y_2 = e^t$, and $y_3 = e^{-t}$. Let us check to see if these three functions are independent by computing their Wronskian:

$$W(t) = \begin{vmatrix} 1 & e^t & e^{-t} \\ 0 & e^t & -e^{-t} \\ 0 & e^t & e^{-t} \end{vmatrix} = 2$$

Therefore, 1, e^t, e^{-t} are independent and form a basis for the null space of L. Any solution of the homogeneous equation can be written in the form $c_1 + c_2 e^t + c_3 e^{-t}$. Next we look for a particular solution of the nonhomogeneous equation $L(y) = 1$. In this case it is fairly easy to guess a solution since a constant times t will yield a constant upon one differentiation and will yield zero upon three differentiations. Therefore, $L(kt) = -k = 1$ if $k = -1$, and a solution is $-t$. We now know that the solution of the initial-value problem can be found among functions of the form

$$y(t) = c_1 + c_2 e^t + c_3 e^{-t} - t$$

To evaluate the constants c_1, c_2, and c_3 we must solve the system of equations

$$\begin{aligned} y(0) &= 1 = c_1 + c_2 + c_3 \\ \dot{y}(0) &= -1 = c_2 - c_3 - 1 \\ \ddot{y}(0) &= 0 = c_2 + c_3 \end{aligned}$$

The solution is $c_1 = 1$ and $c_2 = c_3 = 0$, and $y(t) = 1 - t$ is the unique solution to the initial-value problem. Note that the coefficient matrix of the above system is

$$\begin{pmatrix} 1 & 1 & 1 \\ 0 & 1 & -1 \\ 0 & 1 & 1 \end{pmatrix}$$

and its determinant is $W(0) = 2$. This guarantees that the equations determining the c's have a unique solution. This underlines the importance of having a system of solutions of the homogeneous equation with a Wronskian which is never zero in an interval where we wish to solve the initial-value problem and shows the importance of Theorem 6.2.3.

We conclude this section with some terminology which we shall use in what is to follow. It should be clear by now that a scheme for solving a given initial-value problem is the following:

1 Find n independent solutions of the homogeneous nth order equation $L(y) = 0$. We shall refer to such a set as a *fundamental system of solutions.*
2 Form an arbitrary solution of the homogeneous equation by forming a linear combination of a fundamental system with n arbitrary constants. We shall call this a *complementary solution* and denote it by $y_c(t)$.
3 Find any particular solution of the nonhomogeneous equation $L(y) = f(t)$. We shall call this a *particular solution* and denote it by $y_p(t)$.
4 Add $y_c(t)$ and $y_p(t)$ and call the sum the *general solution.*
5 Evaluate the n constants in $y_c(t)$ so that $y(t) = y_c(t) + y_p(t)$ satisfies the initial conditions.

EXERCISES 6.2

1 Show that $y_1 = e^t$ and $y_2 = e^{-t}$ form a basis for the null space of the operator $L(y) = \ddot{y} - y$.
2 Show that $y_1 = 1$, $y_2 = e^{2t}$, $y_3 = te^{2t}$ form a basis for the null space of the operator $L(y) = \dddot{y} + 4\ddot{y} + 4\dot{y}$.
3 Show that $y_1 = \sin \omega t$, $y_2 = \cos \omega t$ form a basis for the null space of the operator $L(y) = \ddot{y} + \omega^2 y$.
4 Show that $y_1 = t$, $y_2 = t^{-1}$ form a basis for the null space of the operator $L(y) = t^2\ddot{y} + t\dot{y} - y$ on the interval $\{t \mid 0 < a \le t \le b\}$.
5 Consider the differential equation $L(y) = \ddot{y} - 5\dot{y} + 6y = 0$. Look for solutions of the form $y = e^{mt}$. Find a basis for the null space of the operator L.
6 Find the general solution of $\ddot{y} - y = 1$ (see Exercise 1).
7 Find the general solution of $\ddot{y} + 4y = 1$ (see Exercise 3).
8 Find the general solution of $\dddot{y} + 4\ddot{y} + 4\dot{y} = 1$ (see Exercise 2).
9 Find the general solution of $t^2\ddot{y} + t\dot{y} - y = 1$ on the interval $\{t \mid 1 \le t \le 2\}$.
10 Find the solution of the initial-value problem $\ddot{y} - y = 1$, $y(0) = \dot{y}(0) = 1$ (see Exercise 6).
11 Find the solution of the initial-value problem $\ddot{y} + 4y = 1$, $y(0) = 1$, $\dot{y}(0) = 0$ (see Exercise 7).
12 Find the solution of the initial-value problem $\dddot{y} + 4\ddot{y} + 4\dot{y} = 1$, $y(0) = \dot{y}(0) = \ddot{y}(0) = 1$ (see Exercise 8).
13 Find the solution of the initial-value problem $t^2\ddot{y} + t\dot{y} - y = 1$, $y(1) = \dot{y}(1) = 1$ (see Exercise 9).
14 Consider two solutions y_1 and y_2 of the differential equation $\ddot{y} + p(t)\dot{y} + q(t)y = 0$ on the interval $\{t \mid a \le t \le b\}$, where p and q are continuous. Show that the Wronskian of y_1 and y_2 satisfies the differential equation $\dot{W} + pW = 0$.

Hence, show that the Wronskian of y_1 and y_2 is either identically zero or is never zero on the interval.

15 Consider the determinant

$$D = \begin{vmatrix} u_1 & u_2 & \cdots & u_n \\ v_1 & v_2 & \cdots & v_n \\ \cdots & \cdots & \cdots & \cdots \\ w_1 & w_2 & \cdots & w_n \end{vmatrix}$$

where the rows consist of differentiable functions of t. Prove, by induction, the formula

$$\dot{D} = \begin{vmatrix} \dot{u}_1 & \dot{u}_2 & \cdots & \dot{u}_n \\ v_1 & v_2 & \cdots & v_n \\ \cdots & \cdots & \cdots & \cdots \\ w_1 & w_2 & \cdots & w_n \end{vmatrix} + \begin{vmatrix} u_1 & u_2 & \cdots & u_n \\ \dot{v}_1 & \dot{v}_2 & \cdots & \dot{v}_n \\ \cdots & \cdots & \cdots & \cdots \\ w_1 & w_2 & \cdots & w_n \end{vmatrix} + \cdots + \begin{vmatrix} u_1 & u_2 & \cdots & u_n \\ v_1 & v_2 & \cdots & v_n \\ \cdots & \cdots & \cdots & \cdots \\ \dot{w}_1 & \dot{w}_2 & \cdots & \dot{w}_n \end{vmatrix}$$

16 Using the result of Exercise 15, show that the Wronskian of n solutions of the differential equation

$$y^{(n)} + p_1(t)y^{(n-1)} + \cdots + p_n(t)y = 0$$

on the interval $\{t \mid a \le t \le b\}$, where $p_1, p_2, \ldots, p_n$ are continuous, satisfies the differential equation $\dot{W} + p_1 W = 0$. Hence, show that the Wronskian of n solutions either is identically zero or is never zero on the interval.

17 Show that any solution $y(t)$ of the nth order linear differential equation

$$y^{(n)} + p_1(t)y^{(n-1)} + \cdots + p_n(t)y = f(t)$$

is a solution of the first order system $w_1 = y$, $w_2 = \dot{w}_1$, $w_3 = \dot{w}_2, \ldots, w_n = \dot{w}_{n-1}$, $\dot{w}_n = f(t) - p_n w_1 - p_{n-1}w_2 - \cdots - p_1 w_n$. Conversely, show that any solution of the first order system is a solution of the nth order linear equation.

6.3 VARIATION OF PARAMETERS

In the last section, we saw that we could find the general solution of the nth order nonhomogeneous linear differential equation if we could find a fundamental system of solutions of the associated homogeneous equation and any particular solution of the nonhomogeneous equation. In this section, we take up a method for finding a particular solution using a given system of fundamental solutions. Let us first illustrate the method with a simple example, and then we shall take up the general method.

EXAMPLE 6.3.1 Find a particular solution of the differential equation $\ddot{y} - y = e^t$. We know from Exercise 6.2.1 that a fundamental system of solutions is $y_1 = e^t$, $y_2 = e^{-t}$. To find a particular solution we seek a solution of the form $y_p(t) = A(t)y_1(t) + B(t)y_2(t) = A(t)e^t + B(t)e^{-t}$. Differentiating, we have

$$\dot{y}_p = A(t)e^t - B(t)e^{-t} + \dot{A}e^t + \dot{B}e^{-t}$$

At this point we assume that $\dot{A}e^t + \dot{B}e^{-t} = 0$. Then

$$\ddot{y}_p = A(t)e^t + B(t)e^{-t} + \dot{A}e^t - \dot{B}e^{-t}$$

Hence,

$$\ddot{y}_p - y_p = \dot{A}e^t - \dot{B}e^{-t} = e^t$$

We have to solve the system

$$\dot{A}e^t + \dot{B}e^{-t} = 0$$
$$\dot{A}e^t - \dot{B}e^{-t} = e^t$$

for $\dot{A}$ and $\dot{B}$. This is easily done, and we have $\dot{A} = \frac{1}{2}$ and $\dot{B} = -\frac{1}{2}e^{2t}$. We can take any integrals of $\dot{A}$ and $\dot{B}$; that is,

$$A = \tfrac{1}{2}t \qquad B = -\tfrac{1}{4}e^{2t}$$

Substituting, we find

$$y_p(t) = \tfrac{1}{2}te^t - \tfrac{1}{4}e^t$$

which is a particular solution.

The general method of variation of parameters is the following. We are seeking a particular solution of

$$a_n(t)y^{(n)} + a_{n-1}(t)y^{(n-1)} + \cdots + a_1\dot{y} + a_0y = f(t)$$

on $\{t \mid a \le t \le b\}$, where we assume that $a_n(t) \neq 0$. Let $y_1(t), y_2(t), \ldots, y_n(t)$ be a fundamental system of solutions of the associated homogeneous equation. We seek a particular solution in the form

$$y_p(t) = A_1(t)y_1(t) + A_2(t)y_2(t) + \cdots + A_n(t)y_n(t)$$

Differentiating, we have

$$\dot{y}_p = A_1\dot{y}_1 + A_2\dot{y}_2 + \cdots + A_n\dot{y}_n + \dot{A}_1y_1 + \dot{A}_2y_2 + \cdots + \dot{A}_ny_n$$

At this point we put

$$\dot{A}_1y_1 + \dot{A}_2y_2 + \cdots + \dot{A}_ny_n = 0$$

Differentiating again, we have

$$\ddot{y}_p = A_1\ddot{y}_1 + A_2\ddot{y}_2 + \cdots + A_n\ddot{y}_n + \dot{A}_1\dot{y}_1 + \dot{A}_2\dot{y}_2 + \cdots + \dot{A}_n\dot{y}_n$$

Again we put

$$\dot{A}_1\dot{y}_1 + \dot{A}_2\dot{y}_2 + \cdots + \dot{A}_n\dot{y}_n = 0$$

Continuing in this way, we finally have

$$y_p^{(n)} = A_1 y_1^{(n)} + A_2 y_2^{(n)} + \cdots + A_n y_n^{(n)} + \dot{A}_1 y_1^{(n-1)} + \dot{A}_2 y_2^{(n-1)} + \cdots + \dot{A}_n y_n^{(n-1)}$$

Substituting in the differential equation, we find that all the terms which involve the undifferentiated A's will drop out. This is because the functions $y_1, y_2, \ldots, y_n$ are all solutions of the associated homogeneous equation. This leaves us with the following system of equations to solve for $\dot{A}_1, \dot{A}_2, \ldots, \dot{A}_n$:

$$\begin{aligned}
\dot{A}_1 y_1 + \dot{A}_2 y_2 + \cdots + \dot{A}_n y_n &= 0 \\
\dot{A}_1 \dot{y}_1 + \dot{A}_2 \dot{y}_2 + \cdots + \dot{A}_n \dot{y}_n &= 0 \\
\cdots\cdots\cdots\cdots\cdots\cdots\cdots\cdots\cdots & \\
\dot{A}_1 y_1^{(n-2)} + \dot{A}_2 y_2^{(n-2)} + \cdots + \dot{A}_n y_n^{(n-2)} &= 0 \\
\dot{A}_1 y_1^{(n-1)} + \dot{A}_2 y_2^{(n-1)} + \cdots + \dot{A}_n y_n^{(n-1)} &= \frac{f(t)}{a_n(t)}
\end{aligned}$$

This is a system of n equations in n unknowns with the determinant of the coefficient matrix

$$W(t) = \begin{vmatrix} y_1 & y_2 & y_3 & \cdots & y_n \\ \dot{y}_1 & \dot{y}_2 & \dot{y}_3 & \cdots & \dot{y}_n \\ \cdots & \cdots & \cdots & \cdots & \cdots \\ y_1^{(n-1)} & y_2^{(n-1)} & y_3^{(n-1)} & \cdots & y_n^{(n-1)} \end{vmatrix}$$

which is the Wronskian of the fundamental system of solutions $y_1, y_2, \ldots, y_n$. This Wronskian is never zero. Therefore, we can always solve uniquely for $\dot{A}_1, \dot{A}_2, \ldots, \dot{A}_n$. In fact, for $k = 1, 2, \ldots, n$

$$\dot{A}_k = \frac{f(t)W_k(t)}{a_n(t)W(t)}$$

where $W_k(t)$ is the determinant obtained from $W(t)$ by replacing the kth column by $(0, 0, 0, \ldots, 0, 1)$. A particular solution of the nonhomogeneous equation is then

$$y_p(t) = \sum_{k=1}^{n} A_k(t) y_k(t) = \sum_{k=1}^{n} y_k(t) \int_a^t \frac{f(\tau)W_k(\tau)}{a_n(\tau)W(\tau)}\, d\tau$$

The lower limit of integration need not be a since any set of integrals of the $\dot{A}$'s will do.

EXAMPLE 6.3.2 Find the general solution of $\dddot{y} + 3\ddot{y} + 2\dot{y} = -e^{-t}$. To find solutions of the homogeneous equation $\dddot{y} + 3\ddot{y} + 2\dot{y} = 0$ we try $y = e^{mt}$.

Substituting, we have $e^{mt}(m^3 + 3m^2 + 2m) = 0$. We shall have solutions if $m = 0, -1$, or -2. The functions $y_1 = 1$, $y_2 = e^{-t}$, $y_3 = e^{-2t}$ are independent since their Wronskian

$$W(t) = \begin{vmatrix} 1 & e^{-t} & e^{-2t} \\ 0 & -e^{-t} & -2e^{-2t} \\ 0 & e^{-t} & 4e^{-2t} \end{vmatrix} = -2e^{-3t}$$

never vanishes. To find y_p we must evaluate

$$W_1(t) = \begin{vmatrix} 0 & e^{-t} & e^{-2t} \\ 0 & -e^{-t} & -2e^{-2t} \\ 1 & e^{-t} & 4e^{-2t} \end{vmatrix} = -e^{-3t}$$

$$W_2(t) = \begin{vmatrix} 1 & 0 & e^{-2t} \\ 0 & 0 & -2e^{-2t} \\ 0 & 1 & 4e^{-2t} \end{vmatrix} = 2e^{-2t}$$

$$W_3(t) = \begin{vmatrix} 1 & e^{-t} & 0 \\ 0 & -e^{-t} & 0 \\ 0 & e^{-t} & 1 \end{vmatrix} = -e^{-t}$$

Therefore,

$$\begin{aligned} y_p &= \frac{1}{2}\int_0^t (-e^{-\tau} + 2e^{-t} - e^{\tau}e^{-2t})\,d\tau \\ &= \tfrac{1}{2}(e^{-t} - 1 + 2te^{-t} - e^{-t} + e^{-2t}) \\ &= te^{-t} + \tfrac{1}{2}e^{-2t} - \tfrac{1}{2} \end{aligned}$$

The general solution is therefore

$$y = c_1 + c_2e^{-t} + c_3e^{-2t} + te^{-t}$$

We do not include terms in the particular solution if they already appear in the complementary solution.

EXERCISES 6.3

1 Find the general solution of $\ddot{y} - y = e^{-t}$.

2 Find the general solution of $\dddot{y} + 4\ddot{y} + 4\dot{y} = 2e^t$ (see Exercise 6.2.2).

3 Find the general solution of $\ddot{y} + \omega^2 y = 2\cos\omega t$ (see Exercise 6.2.3).

4 Find the general solution of $t^2\ddot{y} + t\dot{y} - y = t$ on the interval $\{t \mid 0 < a \le t \le b\}$ (see Exercise 6.2.4).

5 Find the general solution of $\ddot{y} - 5\dot{y} + 6y = 2t + 3$.

6 Find the solution of $\ddot{y} - y = e^{-t}$ satisfying $y(0) = 1$, $\dot{y}(0) = 2$.

7 Find the solution of $\dddot{y} + 4\ddot{y} + 4\dot{y} = 2e^t$ satisfying $\ddot{y}(0) = \dot{y}(0) = y(0) = 1$.

8 Find the general solution of $\ddot{y} + 9y = 9 \sec^2 3t$, $-\frac{\pi}{6} < t < \frac{\pi}{6}$.

9 Find the general solution of $\ddot{y} + y = \tan t$, $-\frac{\pi}{2} < t < \frac{\pi}{2}$.

6.4 EQUATIONS WITH CONSTANT COEFFICIENTS

In this section we shall be concerned with the problem of solving the general nth order linear equation with constant coefficients,

$$y^{(n)} + p_1 y^{(n-1)} + p_2 y^{(n-2)} + \cdots + p_{n-1}\dot{y} + p_n y = f(t)$$

where $p_1, p_2, \ldots, p_n$ are real constants. As we saw in Sec. 6.3, if we can find the general solution of the associated homogeneous equation (the complementary solution), then we can solve the nonhomogeneous equation by the method of variation of parameters. We shall later see two other methods of finding particular solutions, the method of underdetermined coefficients and the method of the Laplace transform.

It is convenient, in the present case, to identify differentiation with respect to t with the operator D. Hence, $Dy = \dot{y}$, $D^2 y = \ddot{y}, \ldots, D^n = y^{(n)}$. In terms of D, we can write the differential equation as

$$(D^n + p_1 D^{n-1} + p_2 D^{n-2} + \cdots + p_{n-1}D + p_n)y = f(t)$$

If $P(z) = z^n + p_1 z^{n-1} + \cdots + p_{n-1}z + p_n$ is a polynomial in z, then we can write the equation as

$$P(D)y = f(t)$$

The associated homogeneous equation is $P(D)y = 0$. We shall show that the polynomial $P(z)$ can be factored in a certain way and that the operator $P(D)$ can be factored the same way.

The fundamental theorem of algebra† tells us that if $n \geq 1$, $P(z) = 0$ has at least one complex solution r_1. Let r be any complex number. If we divide $P(z)$ by $z - r$, we obtain

$$\frac{P(z)}{z - r} = Q(z) + \frac{R}{z - r}$$

where $Q(z)$ is a polynomial of degree $n - 1$ and R is a constant. Therefore,

$$P(z) = (z - r)Q(z) + R$$

†See J. W. Dettman, "Applied Complex Variables," p. 116, Macmillan, New York, 1965 (rpt. Dover, New York, 1984).

If we now put $r = r_1$, $R = P(r_1) = 0$. Hence, we have shown that

$$P(z) = (z - r_1)Q(z)$$

and we have factored the polynomial. Similarly, if we take $P(D)$ and $Q(D)$ as operators obtained from $P(z)$ and $Q(z)$ by replacing z by D, we find the same factorization continues to hold, namely

$$P(D) = (D - r_1)Q(D)$$

This can be verified by performing the indicated operations and making use of the fact that the operations of multiplying by a constant and differentiating commute. It is also clear that the operators $D - r_1$ and $Q(D)$ commute, so that

$$P(D) = (D - r_1)Q(D) = Q(D)(D - r_1)$$

If the polynomial $Q(z)$ is of degree ≥ 1, then the equation $Q(z) = 0$ has at least one complex solution r_2 and we can factor $Q(z)$ as follows:

$$Q(z) = (z - r_2)S(z)$$

where $S(z)$ is a polynomial of degree $n - 2$. Hence, the operator $Q(D)$ factors into

$$Q(D) = (D - r_2)S(D) = S(D)(D - r_2)$$

and the operator $P(D)$ into

$$P(D) = (D - r_1)(D - r_2)S(D) = S(D)(D - r_2)(D - r_1)$$

Continuing in this way, we can factor $P(D)$ into

$$P(D) = (D - r_1)(D - r_2) \cdots (D - r_n)$$

where the complex numbers $r_1, r_2, \ldots, r_n$ are all solutions of $P(z) = 0$ but are not necessarily distinct. In general, we can write the operator $P(D)$ as

$$P(D) = (D - r_1)^{k_1}(D - r_2)^{k_2} \cdots (D - r_m)^{k_m}$$

where $r_1, r_2, \ldots, r_m$ are the m distinct solutions of $P(z) = 0$ and the integers $k_1, k_2, \ldots, k_m$ are the numbers of times the respective factors appear. Clearly,

$$k_1 + k_2 + \cdots + k_m = n$$

It should be reiterated that the factors in $P(D)$ can be written in any order.

Now suppose we wish to solve the homogeneous equation $P(D)y = 0$. We can write the equation as

$$(D - r_2)^{k_2}(D - r_3)^{k_3} \cdots (D - r_m)^{k_m}(D - r_1)^{k_1}y = 0$$

or

$$(D - r_2)^{k_2}(D - r_3)^{k_3} \cdots (D - r_m)^{k_m}w = 0$$

where $w = (D - r_1)^{k_1}y$. Any y which is a solution of

$$(D - r_1)^{k_1}y = 0$$

will be a solution of the full equation. For simplicity we start by considering equations of the form

$$(D - r)^k y = 0$$

where r is real and k is a positive integer.

The identity

$$(D - r)y = e^{rt}D(e^{-rt}y)$$

is easy to verify. Hence,

$$\begin{aligned}(D - r)^k y &= (D - r)^{k-1}(D - r)y \\ &= (D - r)^{k-1}e^{rt}D(e^{-rt}y) \\ &= (D - r)^{k-2}(D - r)e^{rt}D(e^{-rt}y) \\ &= (D - r)^{k-2}e^{rt}D^2(e^{-rt}y) \\ &= \cdots = e^{rt}D^k(e^{-rt}y)\end{aligned}$$

Therefore, since $e^{rt} \neq 0$, the equation $(D - r)^k y = 0$ is equivalent to

$$D^k(e^{-rt}y) = 0$$

which has as general solution

$$y = (c_1 + c_2 t + \cdots + c_k t^{k-1})e^{rt}$$

where $c_1, c_2, \ldots, c_k$ are arbitrary constants.

Now let us return to the full differential equation, but we shall assume for the moment that the numbers $r_1, r_2, \ldots, r_m$ are real (and of course distinct). Each factor $(D - r_j)^{k_j}$, $j = 1, 2, \ldots, m$, will contribute k_j solutions of the form

$$e^{r_j t}, te^{r_j t}, t^2 e^{r_j t}, \ldots, t^{k_j - 1}e^{r_j t}$$

Taking into account all the different factors, we can list the following solutions corresponding to the various r's:

$$\begin{aligned}r_1&: \quad e^{r_1 t}, te^{r_1 t}, t^2 e^{r_1 t}, \ldots, t^{k_1 - 1}e^{r_1 t} \\ r_2&: \quad e^{r_2 t}, te^{r_2 t}, t^2 e^{r_2 t}, \ldots, t^{k_2 - 1}e^{r_2 t} \\ r_m&: \quad e^{r_m t}, te^{r_m t}, t^2 e^{r_m t}, \ldots, t^{k_m - 1}e^{r_m t}\end{aligned}$$

There are $k_1 + k_2 + \cdots + k_m = n$ different functions in this list. It is not hard to show, although we shall not do it, that these n functions are independent. They therefore form a fundamental system of solutions of the homogeneous equation $P(D)y = 0$ in the case where the r's are real and distinct.

EXAMPLE 6.4.1 Find the general solution of $\ddddot{y} - 4\dddot{y} + 4\ddot{y} = 0$. The differential equation can be written as $(D^4 - 4D^3 + 4D^2)y = 0$, and the operator can be factored as follows:

$$D^2(D - 2)^2 y = 0$$

Therefore, $r_1 = 0$ and $r_2 = 2$, and from the above discussion the functions $y_1 = 1, y_2 = t, y_3 = e^{2t}, y_4 = te^{2t}$ are solutions. We shall show that they are independent by computing their Wronskian:

$$W(t) = \begin{vmatrix} 1 & t & e^{2t} & te^{2t} \\ 0 & 1 & 2e^{2t} & e^{2t} + 2te^{2t} \\ 0 & 0 & 4e^{2t} & 4e^{2t} + 4te^{2t} \\ 0 & 0 & 8e^{2t} & 12e^{2t} + 8te^{2t} \end{vmatrix} = 16e^{4t}$$

The general solution is therefore

$$y(t) = c_1 + c_2 t + c_3 e^{2t} + c_4 te^{2t}$$

We now return to the general case where $r_1, r_2, \ldots, r_m$ can be complex. Actually, the method used above for solving the equation $(D - r)^k y = 0$ is valid for r complex since it is based on the formula

$$\frac{d}{dt} e^{rt} = re^{rt}$$

which is valid for r complex (see Sec. 1.6). However, since we are assuming that the coefficients $p_1, p_2, \ldots, p_n$ are real, we expect that we shall be able to find a general solution containing only real-valued functions. Therefore, we shall take a different approach which will lead to a fundamental system of solutions consisting of real-valued functions.

We first note the following important fact. If $P(z)$ is a polynomial with real coefficients and r is a solution of $P(z) = 0$, then the conjugate $\bar{r}$ is also a solution. This is because

$$P(\bar{r}) = \overline{P(r)} = \bar{0} = 0$$

Therefore, the complex solutions of $P(z) = 0$ occur in conjugate pairs and $\bar{r} = r$ only if r is real. Suppose that $r = a + ib$, then $\bar{r} = a - ib$ and

$$\begin{aligned}(D - r)(D - \bar{r}) &= (D - a - ib)(D - a + ib) \\ &= D^2 - 2aD + a^2 + b^2 \\ &= (D - a)^2 + b^2\end{aligned}$$

and this operator has real coefficients. Let $r_1, r_2, \ldots, r_q$ be the distinct real solutions of $P(z) = 0$ and let $s_1, \bar{s}_1, s_2, \bar{s}_2, \ldots, s_p, \bar{s}_p$ be the distinct complex solutions. Then $P(D)$ can be factored as follows:

$$P(D) = (D - r_1)^{k_1} \cdots (D - r_q)^{k_q}[(D - a_1)^2 + b_1^2]^{l_1} \cdots [(D - a_p)^2 + b_p^2]^{l_p}$$

where $k_1 + k_2 + \cdots + k_q + 2l_1 + 2l_2 + \cdots + 2l_p = n$, and where $s_j = a_j + ib_j, j = 1, 2, \ldots, p$.

We shall solve the homogeneous equation $[(D - a)^2 + b^2]^l y = 0$, where a and b are real and l is a positive integer. The operator can be factored as follows:

$$[(D - a)^2 + b^2]^l = (D - a - ib)^l(D - a + ib)^l$$

Therefore, there are complex solutions of the form

$$e^{(a+ib)t}, te^{(a+ib)t}, t^2e^{(a+ib)t}, \ldots, t^{l-1}e^{(a+ib)t}$$
$$e^{(a-ib)t}, te^{(a-ib)t}, t^2e^{(a-ib)t}, \ldots, t^{l-1}e^{(a-ib)t}$$

By the linearity of the equation we can add and subtract solutions and obtain solutions. Hence,

$$\frac{e^{(a+ib)t} + e^{(a-ib)t}}{2} = e^{at} \cos bt$$
$$\frac{e^{(a+ib)t} - e^{(a-ib)t}}{2i} = e^{at} \sin bt$$

are both solutions. Similarly

$$te^{at} \cos bt, t^2e^{at} \cos bt, \ldots, t^{l-1}e^{at} \cos bt$$
$$te^{at} \sin bt, t^2e^{at} \sin bt, \ldots, t^{l-1}e^{at} \sin bt$$

are all solutions. It is possible to show that these $2l$ solutions are independent.

In the general case, where we have p operators of the form

$$[(D - a_j)^2 + b_j^2]^{l_j}$$

$j = 1, 2, \ldots, p$, corresponding to the distinct pairs of complex numbers $s_j = a_j + ib_j$ and $\bar{s}_j = a_j - ib_j$, we shall have $2l_j$ independent solutions for each operator as follows:

$$e^{a_1t} \cos b_1t, e^{a_1t} \sin b_1t, \ldots, t^{l_1-1}e^{a_1t} \cos b_1t, t^{l_1-1}e^{a_1t} \sin b_1t$$
$$e^{a_2t} \cos b_2t, e^{a_2t} \sin b_2t, \ldots, t^{l_2-1}e^{a_2t} \cos b_2t, t^{l_2-1}e^{a_2t} \sin b_2t$$
$$\cdots\cdots\cdots\cdots\cdots\cdots\cdots\cdots\cdots\cdots\cdots\cdots$$
$$e^{a_pt} \cos b_pt, e^{a_pt} \sin b_pt, \ldots, t^{l_p-1}e^{a_pt} \cos b_pt, t^{l_p-1}e^{a_pt} \sin b_pt$$

This accounts for $2l_1 + 2l_2 + \cdots + 2l_p$ solutions, and, of course, the real numbers $r_1, r_2, \ldots, r_q$ account for $k_1 + k_2 + \cdots + k_q$ solutions. As we have seen from above,

$$k_1 + k_2 + \cdots + k_q + 2l_1 + 2l_2 + \cdots + 2l_p = n$$

so we have achieved our goal of finding a fundamental system of n independent solutions of the homogeneous equation. We can complete the task of finding the general solution of the nonhomogeneous equation by using the method of variation of parameters. In the next section, we shall take up another method for finding particular solutions when the right-hand side of the equation has the form of a solution of some homogeneous linear differential equation with constant coefficients.

EXAMPLE 6.4.2 Find the general solution of

$$(D^5 + D^4 - D^3 - 3D^2 + 2)y = 0$$

To solve this equation we must find all the distinct solutions of $P(z) = z^5 + z^4 - z^3 - 3z^2 + 2 = 0$. There is a theorem† from algebra which says that if a polynomial equation with integer coefficients has a rational solution $r = p/q$, where p and q are integers, then p divides the constant term and q divides the coefficient of the highest power of z. In this case, the only possible rational roots are $-1, 1, -2, 2$. By substituting we find that $P(1) = P(-1) = 0$. Therefore,

$$P(z) = (z - 1)(z + 1)(z^3 + z^2 - 2)$$

Let $Q(z) = z^3 + z^2 - 2$. Then we find that $Q(1) = 0$. Hence,

$$P(z) = (z - 1)^2(z + 1)(z^2 + 2z + 2) = (z - 1)^2(z + 1)[(z + 1)^2 + 1]$$

and the operator $P(D)$ can be written

$$P(D) = (D - 1)^2(D + 1)[(D + 1)^2 + 1]$$

and the general solution is

$$y = c_1e^{-t} + c_2e^t + c_3te^t + c_4e^{-t}\cos t + c_5e^{-t}\sin t$$

EXERCISES 6.4

1 Let y be any twice-differentiable function. Show that $(D - a)(D - b)y = (D - b)(D - a)y = [D^2 - (a + b)D + ab]y$ where a and b are constants.

2 Let y be any three-times-differentiable function. Show that

$$(D - a)[(D - b)(D - c)]y = [(D - a)(D - b)](D - c)y$$

where a, b, and c are constants.

† The reader will be asked to verify this in Exercise 6.4.5.

3 Let $P(z) = z^n + p_1 z^{n-1} + p_2 z^{n-2} + \cdots + p_{n-1}z + p_n$. Let $P(z) = (z - r)Q(z) + R$, where $Q(z) = z^{n-1} + q_1 z^{n-2} + \cdots + q_{n-2}z + q_{n-1}$ and R is a constant. Show that the coefficients $q_1, q_2, \ldots, q_{n-1}, R$ are uniquely determined.

4 Let $P(z)$ and $Q(z)$ be as in Exercise 3 and suppose $P(r) = 0$. Show that for any n-times-differentiable function y, $P(D)y = (D - r)Q(D)y = Q(D)(D - r)y$.

5 Let $P(z) = a_n z^n + a_{n-1}z^{n-1} + \cdots + a_1 z + a_0$ be a polynomial with integer coefficients. Suppose $r = p/q$ is a rational solution of $P(z) = 0$, where p and q are integers with no common divisors other than ± 1. Show that p must divide a_0 and q must divide a_n. *Hint*: Show that

$$a_n p^n + a_{n-1}qp^{n-1} + a_{n-2}q^2p^{n-2} + \cdots + a_1 q^{n-1}p + a_0 q^n = 0$$

6 Using the result of Exercise 5, find all solutions of $P(z) = z^8 + 4z^7 - 2z^6 - 20z^5 + z^4 + 40z^3 - 8z^2 - 32z + 16 = 0$. Find the general solution of the differential equation $P(D)y = 0$.

7 Find the general solutions of each of the following differential equations:

(*a*) $(D^3 - 7D^2 + 16D - 12)y = 0$ (*b*) $(D^4 - 1)y = 0$
(*c*) $(D^4 + 1)y = 0$ (*d*) $(D^4 - D)y = 0$
(*e*) $(D^4 + 3D^2 - 4)y = 0$ (*f*) $(D^8 + 8D^4 + 16)y = 0$

6.5 METHOD OF UNDETERMINED COEFFICIENTS

For certain types of right-hand sides we can find a particular solution of the linear equation with constant coefficients without resorting to the method of variation of parameters. The idea is to "guess" the general form of the solution, leaving certain constants to be determined by substitution. We illustrate with the following example.

EXAMPLE 6.5.1 Find the general solution of $(D^3 - D)y = t + 1$. The operator can be factored as follows: $D^3 - D = D(D - 1)(D + 1)$. Therefore, the complementary solution is $c_1 + c_2e^t + c_3e^{-t}$. Since the right-hand side is a polynomial, we guess that a particular solution might be in the form of a polynomial. Therefore, we assume

$$y_p = at^2 + bt + c$$

Then $\dot{y}_p = 2at + b$, $\ddot{y}_p = 2a$, $\dddot{y}_p = 0$. Substituting in the equation we have $-2at - b = t + 1$. The choice of $b = -1$, $a = -\frac{1}{2}$, and $c = 0$ gives us a solution. The general solution is then

$$y = c_1 + c_2e^t + c_3e^{-t} - \tfrac{1}{2}t^2 - t$$

In this example c was superfluous, so we could have left the constant term out of y_p. Also, why did we stop with the t^2 term? The most efficient assumption would have been $y_p = at^2 + bt$, but how do we arrive at this choice? And what is the assumed form for other types of right-hand sides? To answer these questions we introduce the concept of an *annihilation operator*.

We consider only right-hand sides consisting of a finite number of terms each of which might appear in the complementary solution of some nth order linear differential equations with constant coefficients. In other words, we consider only terms of the form $t^k e^{rt}$, $t^k e^{\alpha t} \cos \beta t$, and $t^k e^{\alpha t} \sin \beta t$. Let us consider these one at a time. The function $t^k e^{rt}$ appears in the complementary solution of the equation $(D - r)^{k+1} y = 0$. Therefore, $(D - r)^{k+1} t^k e^{rt} = 0$, and we say that $(D - r)^{k+1}$ is an annihilation operator for $t^k e^{rt}$. Of course, $(D - r)^n$ for $n > k + 1$ would also annihilate $t^k e^{rt}$, but for simplicity we always take the operator of minimum order. Similarly, $[(D - \alpha)^2 + \beta^2]^{k+1}$ is an annihilation operator for both $t^k e^{\alpha t} \cos \beta t$ and $t^k e^{\alpha t} \sin \beta t$.

The procedure we follow is this. We write $f(t) = f_1(t) + f_2(t) + \cdots + f_m(t)$, where each f_j has a different annihilation operator. Because of the linearity, if we find m functions which are particular solutions of the differential equation with $f_1, f_2, \ldots, f_m$ on the right-hand side, then the sum of these functions will satisfy the equation with $f_1 + f_2 + \cdots + f_m$ on the right-hand side. Suppose the differential equation is $P(D)y = f(t)$. We consider $P(D)y_j = f_j$, $j = 1, 2, 3, \ldots, m$. Suppose $A_j(D)$ is an annihilation operator for f_j. Then

$$A_j(D)P(D)y_j = A_j(D)f_j = 0$$

Let $A_j(D)P(D) = Q_j(D)$. Then Q_j is a linear differential operator with constant coefficients, and the equation $Q_j(D)y = 0$ has a general solution containing n_j independent functions, where n_j is the order of Q_j. Now n, where n is the order of $P(D)$, of these functions are already in the complementary solution of $P(D)y = f(t)$. Clearly these n functions will be annihilated by $P(D)$, and there is no point in putting them in the assumed form for y_j. Therefore, since $Q_j(D)y_j = 0$, comparing the solutions of $Q_j(D)y = 0$ and $P(D)y = 0$ will give us a definite form to be assumed for y_j. This is the form, with undetermined coefficients, which is inserted in $P(D)y_j = f_j$ for the purpose of determining the coefficients. When this is done for each j and the results are summed, we have a particular solution of $P(D)y = f$.

Let us reconsider Example 6.5.1 in the light of this discussion. An annihilation operator for $t + 1$ is D^2. Therefore,

$$D^2(D^3 - D)y_p = D^2(t + 1) = 0$$
$$D^3(D - 1)(D + 1)y_p = 0$$

Hence, $y_p = c_1 + c_2e^t + c_3e^{-t} + bt + at^2$. Omitting the first three terms because they are already in the complementary solution, we have $y_p = at^2 + bt$, which is the form we know works.

EXAMPLE 6.5.2 Find the general solution of $(D^2 - 5D + 6)y = e^{2t} + \cos t$. Let $f_1 = e^{2t}$, and consider $(D^2 - 5D + 6)y_1 = e^{2t}$ or $(D - 2)(D - 3)y_1 = e^{2t}$. An annihilation operator for e^{2t} is $D - 2$. Hence

$$(D - 2)^2(D - 3)y_1 = (D - 2)e^{2t} = 0$$

$$y_1 = c_1e^{2t} + c_2e^{3t} + ate^{2t}$$

Omitting the first two terms because they are already in the complementary solution, we have $y_1 = ate^{2t}$. Then $Dy_1 = ae^{2t} + 2ate^{2t}$, $D^2y_1 = 4ae^{2t} + 4ate^{2t}$, and substituting gives

$$4ae^{2t} + 4ate^{2t} - 5ae^{2t} - 10ate^{2t} + 6ate^{2t} = -ae^{2t} = e^{2t}$$

and $a = -1$. Next let $f_2 = \cos t$, and consider $(D^2 - 5D + 6)y_2 = \cos t$. An annihilation operator for $\cos t$ is $D^2 + 1$. Hence,

$$(D^2 + 1)(D^2 - 5D + 6)y_2 = (D^2 + 1)\cos t = 0$$

$$y_2 = c_1e^{2t} + c_2e^{3t} + a\cos t + b\sin t$$

Omitting the first two terms because they are in the complementary solution of the equation, we have $y_2 = a\cos t + b\sin t$ and $Dy_2 = -a\sin t + b\cos t$, $D^2y_2 = -a\cos t - b\sin t$. Substituting, we have

$$-a\cos t - b\sin t + 5a\sin t - 5b\cos t + 6a\cos t + 6b\sin t = \cos t$$

$$(5a - 5b)\cos t + (5a + 5b)\sin t = \cos t$$

Therefore, $5a - 5b = 1$ and $5a + 5b = 0$, or $a = -b = \frac{1}{10}$. The general solution to the problem is then

$$y = c_1e^{2t} + c_2e^{3t} - te^{2t} + \tfrac{1}{10}\cos t - \tfrac{1}{10}\sin t$$

EXAMPLE 6.5.3 Find the general solution of $(D^2 + 2D + 2)y = t\cos 2t + \sin 2t$. This time the right-hand side is annihilated by $(D^2 + 4)^2$.

Therefore,

$$(D^2 + 4)^2[(D + 1)^2 + 1]y_p = (D^2 + 4)^2(t \cos 2t + \sin 2t) = 0$$

$$y_p = c_1e^{-t} \cos t + c_2e^{-t} \sin t + a \cos 2t + b \sin 2t + ct \cos 2t + dt \sin 2t$$

We omit the first two terms because they are in the complementary solution. Hence, we assume

$$y_p = a \cos 2t + b \sin 2t + ct \cos 2t + dt \sin 2t$$

and

$$Dy_p = (c + 2b) \cos 2t + (d - 2a) \sin 2t + 2dt \cos 2t - 2ct \sin 2t$$

$$D^2y_p = (4d - 4a) \cos 2t + (-4b - 4c) \sin 2t - 4ct \cos 2t - 4dt \sin 2t$$

$$\begin{aligned}(D^2 + 2D + 2)y_p = {} & (-2a + 4b + 2c + 4d) \cos 2t \\ & + (-4a - 2b - 4c + 2d) \sin 2t \\ & + (-2c + 4d)t \cos 2t + (-4c - 2d)t \sin 2t\end{aligned}$$

We must solve

$$\begin{aligned} -2a + 4b + 2c + 4d &= 0 \\ -4a - 2b - 4c + 2d &= 1 \\ -2c + 4d &= 1 \\ -4c - 2d &= 0 \end{aligned}$$

The solution is $a = \frac{1}{50}$, $b = -\frac{7}{50}$, $c = -\frac{1}{10}$, $d = \frac{1}{5}$, and the general solution is

$$y = c_1e^{-t} \cos t + c_2e^{-t} \sin t + \tfrac{1}{50} \cos 2t - \tfrac{7}{50} \sin 2t - \tfrac{1}{10}t \cos 2t + \tfrac{1}{5}t \sin 2t$$

EXERCISES 6.5

1 Find annihilation operators for each of the following functions:
(*a*) $2t^2 + 3t - 5$
(*b*) $(t^2 + 2t + 1)e^t$
(*c*) $te^{2t} \cos t + e^{2t} \sin t$
(*d*) $t^3e^{-t} \sin 3t + t^2e^{-t} \cos 3t$

2 Find the general solution of each of the following differential equations:
(*a*) $(D^2 + 2D + 1)y = 3e^t$
(*b*) $(D^2 - 5D + 6)y = 2e^{3t} + \cos t$
(*c*) $(D^3 - 1)y = te^t$
(*d*) $(D^2 + 4)y = t \cos 2t + \sin 2t$
(*e*) $(D^4 + 4D^2 + 4)y = \cos 2t - \sin 2t$
(*f*) $(D^2 - 2D + 5)y = te^t \sin 2t$

3 Solve the initial-value problem $(D^2 - 2D + 1)y = e^t$, $y(0) = \dot{y}(0) = 1$.

4 Solve the initial-value problem $(D^3 + D)y = te^t$, $y(0) = \dot{y}(0) = 0$, $\ddot{y}(0) = 1$.

5 If r is not a solution of $P(z) = 0$, show that $y_p = e^{rt}/P(r)$ is a particular solution of $P(D)y = e^{rt}$.

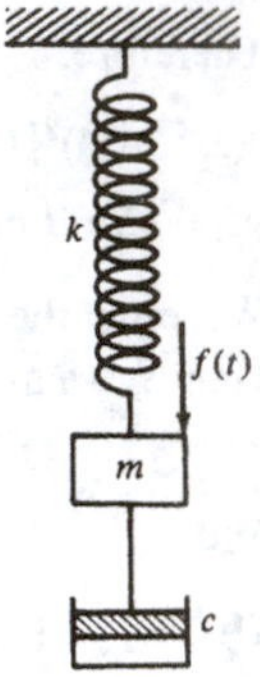

FIGURE 33

6.6 APPLICATIONS

There are many applications of linear differential equations. We shall illustrate with one from the theory of mechanical vibrations and one from the theory of electric networks. It will turn out that both problems lead to the same basic differential equation. This will illustrate the unification of two quite different fields of science through the study of a common differential equation.

Consider the following problem (see Fig. 33). A mass of m slugs is hanging on a spring with spring constant k pounds per foot. The motion of the spring is impeded by a dashpot which exerts a force counter to the motion and proportional to the velocity. The constant of proportionality is c pounds per foot per second. There is a variable force of $f(t)$ pounds driving the mass. If we pick a coordinate $Y(t)$ measured in feet from the position of natural length of the spring (unstretched), where $Y(t)$ is positive downward, then the forces on the mass are as follows:

$$\begin{aligned} mg &= \text{weight of mass} \\ -kY &= \text{restoring force of spring} \\ -c\dot{Y} &= \text{resistive force of dashpot} \\ f(t) &= \text{driving force} \end{aligned}$$

The sum of these forces gives the mass times the acceleration $\ddot{Y}$. Hence, the differential equation is

$$m\ddot{Y} = mg - kY - c\dot{Y} + f(t)$$

Let us introduce a new coordinate $y(t) = Y(t) - mg/k$, which is the displacement measured from the equilibrium position (recall Sec. 5.2). In terms of y the differential equation becomes

$$m\ddot{y} + c\dot{y} + ky = f(t)$$

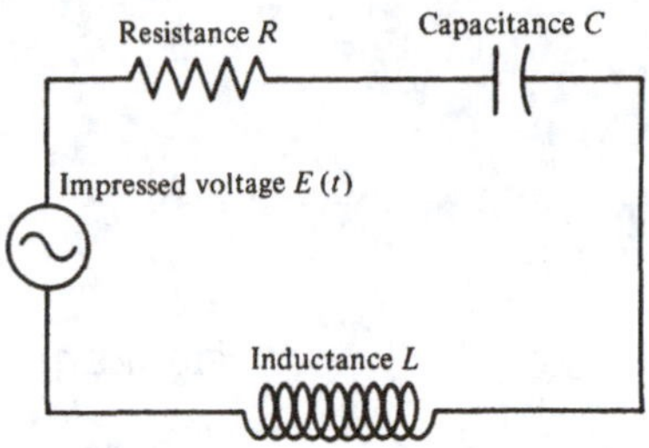

FIGURE 34

This is a nonhomogeneous second order differential equation with constant coefficients. An appropriate initial-value problem is to specify the initial displacement $y(0)$ and the initial velocity $\dot{y}(0)$.

Now let us consider the following electric network (see Fig. 34). Kirchhoff's law states that the impressed voltage is equal to the sum of the voltage drops around the circuit. If there is an instantaneous charge on the capacitor of Q coulombs, then the current flowing in the circuit is $I = \dot{Q}$ amperes and the voltage drops are as follows:

RI = voltage drop across resistance of R ohms
Q/C = voltage drop across capacitance of C farads
$L\dot{I}$ = voltage drop across inductance of L henrys

The appropriate equation is then

$$L\dot{I} + RI + \frac{Q}{C} = E(t)$$

In terms of Q this equation becomes

$$L\ddot{Q} + R\dot{Q} + \frac{Q}{C} = E(t)$$

and if we assume that R, C, and L do not change with time (a reasonable assumption in most cases), then we again have a linear second order equation with constant coefficients. If $E(t)$ is differentiable, then we can write an equation for the current I,

$$L\ddot{I} + R\dot{I} + \frac{I}{C} = \dot{E}(t)$$

which is again of the same type. An appropriate initial-value problem for the first equation is to specify the initial charge $Q(0)$ and initial current $I(0) = \dot{Q}(0)$. For the second equation we should specify the initial current $I(0)$ and the initial derivative $\dot{I}(0)$.

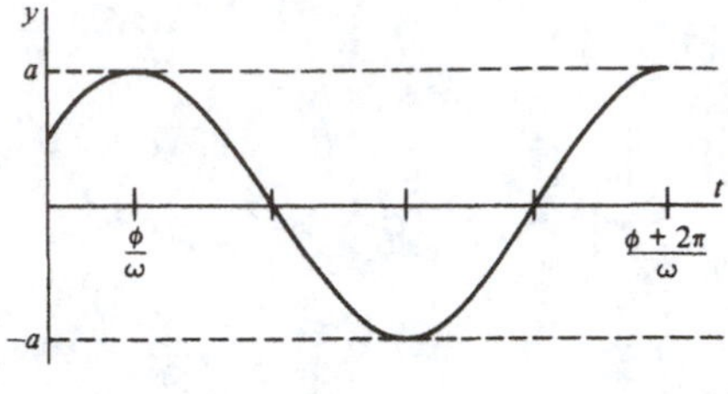

FIGURE 35

We shall study in detail the mechanical vibration problem, but the reader should keep in mind that the remarks apply equally well to the corresponding electrical problems. The cases are considered in order of increasing complexity.

1 Simple Harmonic Motion

Here we assume no damping ($c = 0$) and no forcing [$f(t) \equiv 0$]. The differential equation is $\ddot{y} + \omega^2 y = 0$, where $\omega^2 = k/m$. The general solution is

$$y = A \cos \omega t + B \sin \omega t$$

where A and B are arbitrary constants. Alternatively we can write

$$y = \sqrt{A^2 + B^2}\left(\frac{A}{\sqrt{A^2 + B^2}} \cos \omega t + \frac{B}{\sqrt{A^2 + B^2}} \sin \omega t\right) = a \cos (\omega t - \phi)$$

where $a = \sqrt{A^2 + B^2}$ is called the *amplitude* and $\phi = \tan^{-1} (B/A)$ is called the *phase angle*. This solution is plotted in Fig. 35. The *period*, which is the time required for the motion to go through one complete cycle, is

$$\tau = \frac{2\pi}{\omega}$$

Let us assume that the initial position $y(0)$ and initial velocity $\dot{y}(0)$ are given. Then $A = y(0)$ and $B = \dot{y}(0)/\omega$, and we have the following constants of the motion:

$$a = \sqrt{[y(0)]^2 + \left[\frac{\dot{y}(0)}{\omega}\right]^2}$$

$$\phi = \tan^{-1} \frac{\dot{y}(0)}{\omega y(0)}$$

$$\tau = \frac{2\pi\sqrt{m}}{\sqrt{k}}$$

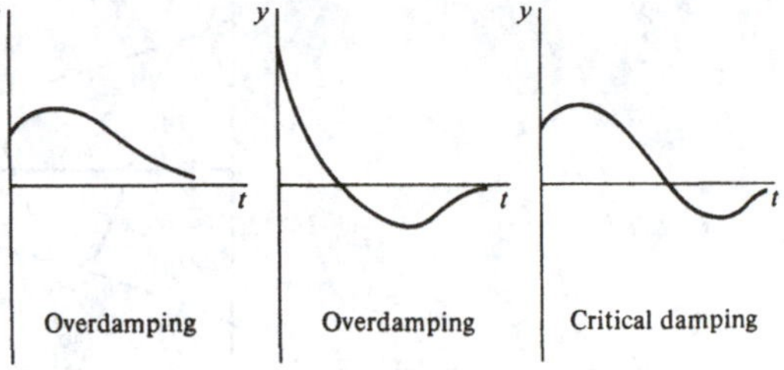

FIGURE 36

For ϕ we shall take the minimum nonnegative angle such that $\sin \phi = \dot{y}(0)/a\omega$ and $\cos \phi = y(0)/a$.

2 Free Vibration with Damping

In this case, we have $k > 0$ and $c > 0$ and $f(t) \equiv 0$. The differential equation is

$$(mD^2 + cD + k)y = 0$$

and the associated polynomial equation is $P(z) = mz^2 + cz + k = 0$. The solutions are

$$r_1, r_2 = \frac{-c \pm \sqrt{c^2 - 4km}}{2m}$$

The independent solutions of the differential equation depend on the value of $c^2 - 4km$ according to the following:

a *Overdamping:*

$$c^2 - 4km > 0 \qquad y = Ae^{r_1 t} + Be^{r_2 t} \qquad r_2 < r_1 < 0$$

b *Critical damping:*

$$c^2 - 4km = 0 \qquad y = (A + Bt)e^{r_1 t} \qquad r_1 = -\frac{c}{2m}$$

c *Underdamping:*

$$c^2 - 4km < 0 \qquad y = e^{-(c/2m)t}(A \cos \omega t + B \sin \omega t) \qquad \omega = \frac{\sqrt{4km - c^2}}{2m}$$

Some typical motions in cases *a* and *b* are illustrated in Fig. 36. In case *c*, we can write alternatively

$$y = ae^{-(c/2m)t} \cos (\omega t - \phi)$$

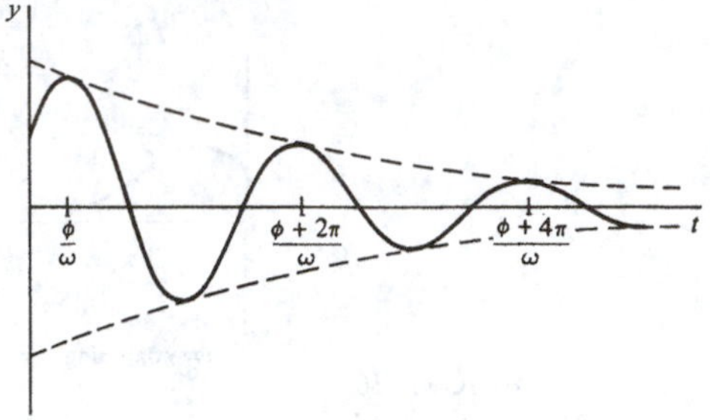

FIGURE 37

where $a = \sqrt{A^2 + B^2}$ and $\phi = \tan^{-1}(B/A)$. The motion is illustrated in Fig. 37. The motion is oscillatory, as in the case of simple harmonic motion, but the amplitude is diminishing according to the exponential factor $e^{-(c/2m)t}$. The motion is not periodic, but the time between successive peaks is $\tau = 2\pi/\omega$. If the damping constant c is very small compared with k and m, then $c/2m$ is very small and the exponential factor $e^{-(c/2m)t}$ is near 1 for reasonably small values of t. In this case the motion is very nearly simple harmonic motion, and τ is approximately a period.

For forcing we shall consider two cases.

3 Forced Vibrations without Damping

In this case $c = 0$, and for definiteness we shall take $f(t) = f_0 \cos \omega_0 t$, where f_0 is a constant and $\omega_0 \neq \omega = (k/m)^{1/2}$. The differential equation is

$$\ddot{y} + \omega^2 y = \frac{f_0}{m} \cos \omega_0 t$$

The complementary solution is

$$y_c = A \cos \omega t + B \sin \omega t$$

To find a particular solution we use the method of undetermined coefficients. We assume a solution of the form

$$y_p = C \cos \omega_0 t + D \sin \omega_0 t$$

Then

$$\ddot{y}_p + \omega^2 y_p = C(\omega^2 - \omega_0^2) \cos \omega_0 t + D(\omega^2 - \omega_0^2) \sin \omega_0 t = \frac{f_0}{m} \cos \omega_0 t$$

Therefore, $D = 0$ and

$$C = \frac{f_0}{m(\omega^2 - \omega_0^2)}$$

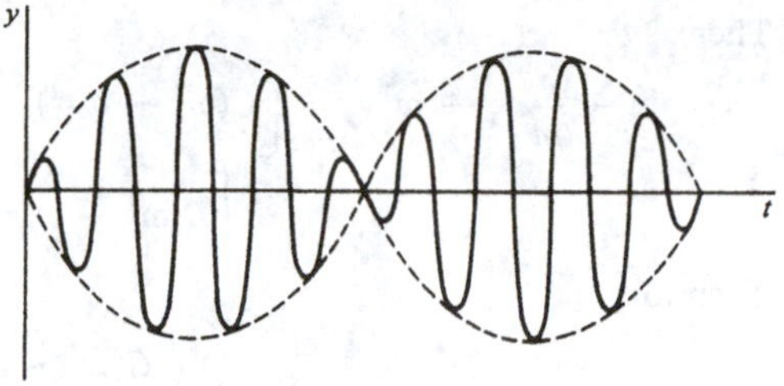

FIGURE 38

Suppose initially $y(0) = \dot{y}(0) = 0$. Then the solution is

$$y(t) = \frac{f_0}{m(\omega^2 - \omega_0{}^2)}(\cos \omega_0 t - \cos \omega t)$$

$$= \frac{2f_0}{m(\omega^2 - \omega_0{}^2)} \sin \frac{(\omega - \omega_0)t}{2} \sin \frac{(\omega + \omega_0)t}{2}$$

This motion is illustrated in Fig. 38. This phenomenon is known as *beating*. It is especially pronounced when ω_0 is approximately ω. Then one of the sine terms is slowly varying with a frequency of $(\omega - \omega_0)/4\pi$ while the other two terms are varying rapidly with a frequency of $(\omega + \omega_0)/4\pi$. This phenomenon is the basis for a technique known as *amplitude modulation* in electronics.

If the forcing term in this example had been $f_0 \cos \omega t$, then a different sort of solution would have been obtained. The reader will be asked to study this case in the exercises.

4 Forced Vibrations with Damping

In this case $k > 0$, $c > 0$, and for definiteness we shall again take $f(t) = f_0 \cos \omega_0 t$. The differential equation is

$$\ddot{y} + \frac{c}{m}\dot{y} + \omega^2 y = \frac{f_0}{m}\cos \omega_0 t$$

where $\omega^2 = k/m$. Depending on the value of $c^2 - 4km$, the complementary solution is one of three functions listed under case 2 above. To find a particular solution we assume

$$y_p = C \cos \omega_0 t + D \sin \omega_0 t$$

Then

$$\ddot{y}_p + \frac{c}{m}\dot{y}_p + \omega^2 y_p = \left[C(\omega^2 - \omega_0^2) + \frac{c}{m}\omega_0 D\right]\cos\omega_0 t + \left[D(\omega^2 - \omega_0^2) - \frac{c}{m}\omega_0 C\right]\sin\omega_0 t$$

Therefore,

$$C(\omega^2 - \omega_0^2) + \frac{c}{m}\omega_0 D = \frac{f_0}{m}$$

$$D(\omega^2 - \omega_0^2) - \frac{c}{m}\omega_0 C = 0$$

Solving, we have

$$C = \frac{m(\omega^2 - \omega_0^2)f_0}{m^2(\omega^2 - \omega_0^2) + c^2\omega_0^2}$$

$$D = \frac{c\omega_0 f_0}{m^2(\omega^2 - \omega_0^2) + c^2\omega_0^2}$$

We can write

$$y_p = \frac{f_0}{\sqrt{m^2(\omega^2 - \omega_0^2)^2 + c^2\omega_0^2}}\cos(\omega_0 t - \phi)$$

where

$$\phi = \tan^{-1}\frac{c\omega_0}{m(\omega^2 - \omega_0^2)}$$

It should be noted that for any one of the complementary solutions

$$\lim_{t\to\infty} y_c = 0$$

Therefore, after a long period of time the part of the solution which continues to contribute is y_p. For this reason y_p is called the *steady-state solution* in this problem. The other part is called the *transient solution.* The case $\omega_0 = \omega$ deserves special comment. The method of solution does not have to be changed, as was the case when there was no damping. However, the amplitude factor

$$\frac{f_0}{\sqrt{m^2(\omega^2 - \omega_0^2) + c^2\omega_0^2}}$$

is a maximum when $\omega_0 = \omega$. This phenomenon is known as *resonance.* A similar thing occurs in the electric-circuit case. If the values of L and C are adjusted so that $\omega = (LC)^{-1/2}$ is equal to the ω_0 for the impressed voltage, current flowing will be maximized. This is the basis for *tuning* a radio or TV.

EXERCISES 6.6

1 A mass weighing 1 pound hangs in equilibrium with a spring stretched 1 inch. The mass is pulled down 1 inch below the equilibrium position and released. What is the period and frequency of the motion? What is the amplitude?

2 If the mass of Exercise 1 is given an initial velocity of 1 foot per second along with the initial displacement of 1 inch, calculate the amplitude and phase angle.

3 Show that regardless of initial conditions, in the case of free vibrations with overdamping, the mass can go through the equilibrium position at most once.

4 Find the general solution in the case of forced vibrations without damping for $f(t) = f_0 \cos \omega t$, $\omega^2 = k/m$.

5 Consider a series electric circuit with no resistance and no impressed voltage. Show that the current is periodic with period $\tau = 2\pi\sqrt{LC}$, where L is the inductance and C is the capacitance. Find τ if $L = 1$ henry and $C = 4 \times 10^{-6}$ farad.

6 Find the steady-state current in a series circuit if the impressed voltage is $E(t) = 110 \sin 120\pi t$ volts and $L = 3$ henrys, $R = 500$ ohms, and $C = 5 \times 10^{-6}$ farad.

*6.7 GREEN'S FUNCTIONS

Green's functions are very useful in the treatment of linear differential equations, both for initial-value problems and boundary-value problems. They are also important in the study of linear problems in partial differential equations, and they supply an important link between differential equations and integral equations.

We shall introduce the ideas involved by finding the general solution of the simple problem $\ddot{y} = f(t)$, $y(0) = a$, $\dot{y}(0) = b$. The associated homogeneous equation is $\ddot{y} = 0$, and clearly the function $y_c = a + bt$ satisfies the homogeneous equation and the initial conditions. Therefore, we must find a particular solution y_p which satisfies $y_p(0) = \dot{y}_p(0) = 0$. We proceed by variation of parameters. Let $y_p = A + Bt$, where A and B are functions of t. Then

$$\dot{y}_p = B + \dot{A} + \dot{B}t$$

We put $\dot{A} + \dot{B}t = 0$, and then

$$\ddot{y}_p = \dot{B} = f(t)$$
$$\dot{A} = -tf(t)$$

Integrating, we find

$$y_p = \int_0^t (t - \tau)f(\tau)\, d\tau$$

It is easy to verify that $y_p(0) = \dot{y}_p(0) = 0$, and so the general solution is

$$y = a + bt + \int_0^t (t - \tau)f(\tau)\, d\tau$$

and this is a solution to the problem as long as $f(t)$ is integrable.

Let us define the following *Green's function*:

$$G(t,\tau) = \begin{cases} t - \tau & 0 \le \tau \le t \\ 0 & t \le \tau \end{cases}$$

We think of t as a parameter and τ as the variable in the function. If the dot refers to differentiation with respect to τ, we have that G satisfies the following conditions:

1 $G(t,\tau)$ is continuous for $0 \le \tau < \infty$ and $G(t,\tau) \equiv 0$ for $t \le \tau$.
2 $\ddot{G} = 0$ for $0 < \tau < t$.
3 There is a jump of 1 in the derivative $\dot{G}$ at $\tau = t$.

These conditions determine G uniquely. Condition 2 requires that $G(t,\tau) = \alpha + \beta\tau$. Condition 1 requires that $\alpha + \beta t = 0$, and condition 3 requires that

$$\lim_{\tau \to t^-} \dot{G}(t,\tau) = \beta = -1$$

Hence, $\alpha = t$ and $G(t,\tau) = t - \tau$ for $0 \le \tau \le t$.

Let us take another point of view of the example. Let us seek a function $H(t,\tau)$ which will produce a y_p satisfying $0 = y_p(0) = \dot{y}_p(0)$ upon multiplying by $f(\tau)$ and integrating; that is, $y_p = \int_0^t H(t,\tau)f(\tau)\, d\tau$. If we let†

$$\int_0^{t^-} F(\tau)\, d\tau = \lim_{\varepsilon \to 0^+} \int_0^{t-\varepsilon} F(\tau)\, d\tau$$

then

$$\begin{aligned}
\int_0^t H(t,\tau)f(\tau)\, d\tau &= \int_0^t H(t,\tau)\ddot{y}_p(\tau)\, d\tau \\
&= \dot{y}_p(\tau)H(t,\tau)\Big|_0^t - \int_0^{t^-} \dot{H}(t,\tau)\dot{y}_p\, d\tau \\
&= \dot{y}_p H(t,t) - \dot{H}y_p\Big|_0^{t^-} + \int_0^{t^-} \ddot{H}(t,\tau)y_p\, d\tau \\
&= \dot{y}_p H(t,t) - \dot{H}(t,t^-)y_p(t) + \int_0^{t^-} \ddot{H}y_p\, d\tau
\end{aligned}$$

† This is necessary because H may not have derivatives at $\tau = t$.

This last expression will be equal to $y_p(t)$ if and only if $H(t,t) = 0$, $\dot{H}(t,t^-) = -1$, and $\ddot{H} = 0$ for $0 < \tau < t$. But these are the conditions which determined $G(t,\tau)$. Hence, $H(t,\tau) = G(t,\tau)$.

EXAMPLE 6.7.1 Let us consider another initial-value problem: $\ddot{y} + q(t)\dot{y} = f(t)$, $y(0)$ and $\dot{y}(0)$ given. We shall assume that $q(t)$ and $\dot{q}(t)$ are continuous. There exists a complementary solution $y_c(t)$ satisfying $y_c(0) = y(0)$ and $\dot{y}_c(0) = \dot{y}(0)$. What remains is to determine a particular solution $y_p(t)$ satisfying $y_p(0) = \dot{y}_p(0) = 0$. By analogy we seek a solution in the form $\int_0^t G(t,\tau)f(\tau)\,d\tau$. Then

$$\begin{aligned}\int_0^t G(t,\tau)f(\tau)\,d\tau &= \int_0^t G(t,\tau)(\ddot{y}_p + q\dot{y}_p)\,d\tau \\ &= \dot{y}_p(\tau)G(t,\tau)\Big|_0^t - \int_0^{t^-} \dot{y}_p(\dot{G} - qG)\,d\tau \\ &= \dot{y}_p(t)G(t,t) - y_p(\dot{G} - qG)\Big|_0^{t^-} \\ &\qquad + \int_0^{t^-} y_p(\ddot{G} - q\dot{G} - \dot{q}G)\,d\tau\end{aligned}$$

We attempt to determine the Green's function $G(t,\tau)$ by the following conditions:

1 $G(t,\tau)$ is continuous for $0 \le \tau \le t$, and $G(t,\tau) \equiv 0$ for $t \le \tau$.
2 $\ddot{G} - q\dot{G} - \dot{q}G = 0$ for $0 < \tau < t$.
3 There is a jump of 1 in the derivative $\dot{G}$ at $\tau = t$.

Under these conditions $G(t,t) = 0$ and $\dot{G}(t,t^-) = -1$ and hence from above $\int_0^t G(t,\tau)f(\tau)\,d\tau = y_p$.

We shall now show that conditions 1, 2, and 3 determine G uniquely. The equation in condition 2 can be written as

$$\frac{d}{d\tau}(\dot{G} - qG) = 0$$

Therefore, $\dot{G} - qG$ is constant for $0 < \tau < t$. But if $q(t)$ is continuous and $G(t,t) = 0$, then $\dot{G} - qG = \dot{G}(t,t^-) = -1$. Therefore, our conditions reduce to $\dot{G} - qG = -1$, $G(t,t) = 0$. The integrating factor is

$$Q(\tau) = \exp\left[-\int_0^\tau q(\xi)\,d\xi\right]$$

Then $(d/d\tau)(QG) = -Q$, and

$$G = -\frac{1}{Q}\int_0^\tau Q(\eta)\,d\eta + \frac{c}{Q}$$

However, $G(t,t) = 0$, so

$$G(t,t) = \frac{-1}{Q(t)}\int_0^t Q(\eta)\,d\eta + \frac{c}{Q(t)} = 0$$

$$c = \int_0^t Q(\eta)\,d\eta$$

$$G(t,\tau) = \frac{1}{Q(\tau)}\int_\tau^t Q(\eta)\,d\eta$$

EXAMPLE 6.7.2 Find the solution of $\ddot{y} - \dot{y}/(t+1) = f(t)$ for $0 \le t < \infty$, $y(0) = \dot{y}(0) = 0$. The solution is

$$y(t) = \int_0^t G(t,\tau)f(\tau)\,d\tau$$

with

$$G(t,\tau) = \frac{1}{Q(\tau)}\int_\tau^t Q(\eta)\,d\eta$$

where

$$Q(\tau) = \exp\left[\int_0^\tau (\xi+1)^{-1}\,d\xi\right] = \tau + 1$$

Therefore,

$$G(t,\tau) = \frac{1}{\tau+1}\int_\tau^t (\eta+1)\,d\eta = \frac{(t-\tau)(t+\tau+2)}{2(\tau+1)}$$

EXAMPLE 6.7.3 Find the general solution of the problem $\ddot{y} + y = f(t)$, $y(0) = y_0$, $\dot{y}(0) = \dot{y}_0$. Again we shall find a Green's function. The complementary solution

$$y_c = y_0 \cos t + \dot{y}_0 \sin t$$

satisfies the associated homogeneous equation $\ddot{y} + y = 0$ and the initial conditions. Hence, we must find a particular solution y_p satisfying $y_p(0) = \dot{y}_p(0) = 0$. We look for a solution in the form $y_p(t) = \int_0^t G(t,\tau)f(\tau)\,d\tau$. Then

$$\begin{aligned}\int_0^t G(t,\tau)f(\tau)\,d\tau &= \int_0^t G(t,\tau)(\ddot{y}_p + y_p)\,d\tau \\ &= \dot{y}_p(\tau)G(t,\tau)\Big|_0^t - \int_0^{t-} (\dot{G}\dot{y}_p - Gy_p)\,d\tau \\ &= \dot{y}_p(t)G(t,t) - \dot{G}(t,\tau)\,y_p(\tau)\Big|_0^{t-} + \int_0^{t-} y_p(\ddot{G} + G)\,d\tau\end{aligned}$$

We shall look for a Green's function $G(t,\tau)$ satisfying $\ddot{G} + G = 0$, $0 < \tau < t$, $G(t,t) = 0$, and $\dot{G}(t,t^-) = -1$. From the differential equation we have

$$G(t,\tau) = A\cos\tau + B\sin\tau$$

Then $G(t,t) = A\cos t + B\sin t = 0$ and $\dot{G}(t,t^-) = -A\sin t + B\cos t = -1$. Solving for A and B, we have $A = \sin t$ and $B = -\cos t$. Therefore, $G(t,\tau) = \sin(t-\tau)$. The general solution to the problem is

$$y = y_0\cos t + \dot{y}_0\sin t + \int_0^t \sin(t-\tau)f(\tau)\,d\tau$$

For the general nonsingular second order linear equation $\ddot{y} + q\dot{y} + ry = f(t)$, where we assume that $f(t)$, $r(t)$, and $\dot{q}(t)$ are continuous, we proceed as follows to obtain a Green's function:

$$\begin{aligned}\int_0^t G(t,\tau)f(\tau)\,d\tau &= \int_0^t G(t,\tau)(\ddot{y}_p + q\dot{y}_p + ry_p)\,d\tau \\ &= \dot{y}_p(\tau)G(t,\tau)\Big|_0^t - \int_0^{t^-}[\dot{y}_p(\dot{G} - qG) - rGy_p]\,d\tau \\ &= \dot{y}_p(t)G(t,t) - y_p(\dot{G} - qG)\Big|_0^{t^-} \\ &\quad + \int_0^{t^-} y_p(\ddot{G} - q\dot{G} - \dot{q}G + rG)\,d\tau\end{aligned}$$

We take the following three properties for $G(t,\tau)$:

1. $G(t,\tau)$ is continuous for $0 \le \tau \le t$, and $G(t,\tau) \equiv 0$ for $t \le \tau$.
2. $\ddot{G} - q\dot{G} - \dot{q}G + rG = 0$ for $0 < \tau < t$.
3. There is a jump of 1 in the derivative $\dot{G}$ at $\tau = t$.

Satisfying these conditions is equivalent to solving the following initial-value problem for the interval $\{\tau \mid 0 \le \tau \le t\}$: $\ddot{G} - q\dot{G} - \dot{q}G + rG = 0$, $G(t,t) = 0$, $\dot{G}(t,t^-) = -1$. By Theorem 6.2.1, this problem has a unique solution. The differential equation can be written

$$\frac{d^2G}{d\tau^2} - \frac{d}{d\tau}(qG) + rG = 0$$

This is called the *adjoint differential equation*. The reader will be asked to investigate the form of the adjoint equation associated with higher order differential equations.

Let us reexamine the problem $\ddot{y} + q\dot{y} + ry = f(t)$, $y(0) = \dot{y}(0) = 0$, from a slightly different point of view. Let

$$G(t,\tau) = \frac{1}{Q(\tau)}\int_\tau^t Q(\eta)\,d\eta$$

where $Q(\tau) = \exp\left[-\int_0^\tau q(\xi)\, d\xi\right]$. Recall that $G(t,\tau)$ is the Green's function associated with Example 6.7.1. Let $y(t)$ be the solution to the problem, and consider the function

$$\begin{aligned} z(t) &= \int_0^t [f(\tau) - r(\tau)y(\tau)]G(t,\tau)\, d\tau \\ &= \int_0^t (\ddot{y} + q\dot{y})G(t,\tau)\, d\tau \\ &= \dot{y}(\tau)G(t,\tau)\Big|_0^t - \int_0^{t^-} (\dot{G} - qG)\dot{y}\, d\tau \\ &= -(\dot{G} - qG)y(\tau)\Big|_0^{t^-} + \int_0^{t^-} (\ddot{G} - q\dot{G} - \dot{q}G)y\, d\tau \\ &= y(t) \end{aligned}$$

Hence, if y is the solution of the problem, it is also a solution of the following *Volterra integral equation*:

$$y(t) = F(t) - \int_0^t r(\tau)G(t,\tau)y(\tau)\, d\tau$$

where $F(t) = \int_0^t G(t,\tau)f(\tau)\, d\tau$. It is not hard to show that any solution of the integral equation is also a solution of the differential equation, thus showing that the two problems are equivalent. In fact,

$$\dot{y} = \int_0^t [f(\tau) - r(\tau)y(\tau)]\frac{\partial G}{\partial t}\, d\tau$$

$$\dot{y} = \int_0^t [f(\tau) - r(\tau)y(\tau)]\frac{Q(t)}{Q(\tau)}\, d\tau$$

$$\ddot{y} = f(t) - r(t)y + \int_0^t [f(\tau) - r(\tau)y(\tau)]\frac{\dot{Q}(t)}{Q(\tau)}\, d\tau$$

where $\dot{Q}(t) = -q(t)Q(t)$. Hence, $\ddot{y} = f(t) - r(t)y - q(t)\dot{y}$. Obviously, $y(0) = y(0) = 0$. There are iterative methods for solving the integral equation even when the differential equation may not be easy to solve.

We conclude this section with a brief discussion of boundary-value problems. Throughout this book we use the independent variable t when we discuss initial-value problems and the dot to denote differentiation with respect to t. On the other hand, we shall use the variable x when we discuss boundary-value problems and a prime to denote differentiation with respect to x.

Consider the following boundary-value problem $y'' = f(x)$, $0 < x < 1$,

$y(0) = y(1) = 0$. According to what we did earlier in this section, the general solution is

$$y = a + bx + \int_0^x (x - \xi)f(\xi)\,d\xi$$

Then $y(0) = a = 0$, and

$$y(1) = b + \int_0^1 (1 - \xi)f(\xi)\,d\xi = 0$$

Hence,

$$b = \int_0^1 (\xi - 1)f(\xi)\,d\xi$$

and

$$\begin{aligned} y &= \int_0^1 x(\xi - 1)f(\xi)\,d\xi + \int_0^x (x - \xi)f(\xi)\,d\xi \\ &= \int_0^x \xi(x - 1)f(\xi)\,d\xi + \int_x^1 x(\xi - 1)f(\xi)\,d\xi \\ &= \int_0^1 G(x,\xi)f(\xi)\,d\xi \end{aligned}$$

where we have let

$$G(x,\xi) = \begin{cases} \xi(x - 1) & 0 \le \xi \le x \\ x(\xi - 1) & x \le \xi \le 1 \end{cases}$$

$G(x,\xi)$ is a Green's function associated with the boundary-value problem. It has the following properties:

1 $G(x,\xi)$ is continuous for $0 \le \xi \le 1$.
2 $G'' = 0$ for $0 < \xi < x$ and $x < \xi < 1$.
3 $G(0) = G(1) = 0$.
4 There is a jump of 1 in the derivative G' at $\xi = x$.

EXAMPLE 6.7.4 Find the solution of the boundary-value problem $y'' + y = f(x)$, $0 \le x \le L$, $y(0) = y(L) = 0$. We discovered earlier that the general solution is

$$y(x) = a\cos x + b\sin x + \int_0^x \sin(x - \xi)f(\xi)\,d\xi$$

Then $y(0) = a = 0$, and

$$y(L) = b \sin L + \int_0^L \sin (L - \xi) f(\xi)\, d\xi = 0$$

Provided $\sin L \neq 0$,

$$b = \frac{1}{\sin L} \int_0^L \sin (\xi - L) f(\xi)\, d\xi$$

and

$$\begin{aligned} y(x) &= \frac{\sin x}{\sin L} \int_0^L \sin (\xi - L) f(\xi)\, d\xi + \int_0^x \sin (x - \xi) f(\xi)\, d\xi \\ &= \int_0^x \frac{\sin x \sin (\xi - L) + \sin L \sin (x - \xi)}{\sin L} f(\xi)\, d\xi \\ &\qquad + \int_x^L \frac{\sin x \sin (\xi - L)}{\sin L} f(\xi)\, d\xi \\ &= \int_0^x \frac{\sin \xi \sin (x - L)}{\sin L} f(\xi)\, d\xi + \int_x^L \frac{\sin x \sin (\xi - L)}{\sin L} f(\xi)\, d\xi \\ &= \int_0^L G(x,\xi) f(\xi)\, d\xi \end{aligned}$$

where the Green's function is

$$G(x,\xi) = \begin{cases} \dfrac{\sin \xi \sin (x - L)}{\sin L} & 0 \le \xi \le x \\[2ex] \dfrac{\sin x \sin (\xi - L)}{\sin L} & x \le \xi \le L \end{cases}$$

This Green's function satisfies the conditions:

1. $G(x,\xi)$ is continuous for $0 \le \xi \le L$.
2. $G'' + G = 0$ for $0 < \xi < x$ and $x < \xi < L$.
3. $G(0) = G(L) = 0$.
4. There is a jump of 1 in the derivative G' at $\xi = x$.

These conditions determine the Green's function uniquely, provided $\sin L \neq 0$.

It is of interest to show how these four conditions can be used directly to show the representation of the solution of the boundary-value problem.

Consider

$$\begin{aligned}
\int_0^L G(x,\xi)f(\xi)\,d\xi &= \int_0^L G(x,\xi)(y'' + y)\,d\xi \\
&= \int_0^x G(x,\xi)(y'' + y)\,d\xi + \int_x^L G(x,\xi)(y'' + y)\,d\xi \\
&= y'G(x,\xi)|_0^x - \int_0^{x^-} (G'y' - Gy)\,d\xi \\
&\qquad + y'G(x,\xi)|_x^L - \int_{x^+}^L (G'y' - Gy)\,d\xi \\
&= -G'y|_0^{x^-} - G'y|_{x^+}^L + \int_0^{x^-} (G'' + G)y\,d\xi \\
&\qquad + \int_{x^+}^L (G'' + G)y\,d\xi
\end{aligned}$$

where we have already used the boundary conditions on G and the continuity of G at $\xi = x$ to drop certain terms. Finally, we use the boundary conditions on y, the jump of 1 in G' at $\xi = x$, and condition 2 to show that

$$\int_0^L G(x,\xi)f(\xi)\,d\xi = y(x)[G'(x,x^+) - G'(x,x^-)] = y(x)$$

Finally, we consider the boundary-value problem $y'' + q(x)y = f(x)$, $y(0) = y(1) = 0$, where q and f are continuous on $\{x \mid 0 \le x \le 1\}$. Let $G(x,\xi)$ be the Green's function considered earlier

$$G(x,\xi) = \begin{cases} \xi(x - 1) & 0 \le \xi \le x \\ x(\xi - 1) & x \le \xi \le 1 \end{cases}$$

and consider

$$\begin{aligned}
\int_0^L [f(\xi) &- q(\xi)y]G(x,\xi)\,d\xi \\
&= \int_0^L G(x,\xi)y''\,d\xi \\
&= \int_0^x G(x,\xi)y''\,d\xi + \int_x^L G(x,\xi)y''\,d\xi \\
&= y'G(x,\xi)|_0^{x^-} - \int_0^{x^-} G'y'\,d\xi + y'G(x,\xi)|_{x^+}^L - \int_{x^+}^L G'y'\,d\xi \\
&= -G'y|_0^{x^-} - G'y|_{x^+}^L + \int_0^{x^-} yG''\,d\xi + \int_{x^+}^L yG''\,d\xi \\
&= y(x)
\end{aligned}$$

using the properties of the Green's function listed above. Hence, we see that y satisfies the *Fredholm integral equation*

$$y(x) = F(x) - \int_0^L q(\xi)y(\xi)G(x,\xi)\,d\xi$$

where $F(x)$ is the known function $\int_0^L G(x,\xi)f(\xi)\,d\xi$. There are ways of solving the integral equation available even when the differential equation itself is not easy to solve.

EXERCISES 6.7

1 Find the Green's function associated with the initial-value problem $\ddot{y} + \dot{y} = f(t)$, $y(0) = y_0$, $\dot{y}(0) = \dot{y}_0$, by means of which the solution can be written as

$$y(t) = a + be^{-t} + \int_0^t G(t,\tau)f(\tau)\,d\tau$$

where a and b are constants depending on y_0 and $\dot{y}_0$.

2 Find the Green's function associated with the initial-value problem $\ddot{y} + \omega^2 y = f(t)$, $y(0) = y_0$, $\dot{y}(0) = \dot{y}_0$, by means of which the solution can be written

$$y(t) = y_0 \cos \omega t + \frac{\dot{y}_0}{\omega} \sin \omega t + \int_0^t G(t,\tau)f(\tau)\,d\tau$$

3 Find the Green's function associated with the initial-value problem $\ddot{y} - 3\dot{y} + 2y = f(t)$, $y(0) = \dot{y}(0) = 0$, by means of which the solution can be written

$$y(t) = \int_0^t G(t,\tau)f(\tau)\,d\tau$$

4 Find the Green's function associated with the initial-value problem $\ddot{y} - 2\dot{y} + y = f(t)$, $y(0) = \dot{y}(0) = 0$, by means of which the solution can be written

$$y(t) = \int_0^t G(t,\tau)f(\tau)\,d\tau$$

5 Find a set of conditions which will uniquely determine a Green's function associated with the initial-value problem $\dddot{y} + p(t)\ddot{y} + q(t)\dot{y} + r(t)y = f(t)$, $y(0) = \dot{y}(0) = \ddot{y}(0) = 0$, by means of which the solution can be written

$$y(t) = \int_0^t G(t,\tau)f(\tau)\,d\tau$$

Assume that $\ddot{p}$, $\dot{q}$, r, f are all continuous for $0 \le t < \infty$. What is the *adjoint differential equation* associated with $\dddot{y} + p\ddot{y} + q\dot{y} + ry = f(t)$? Can you give a general characterization of the adjoint differential equation for higher order problems?

6 Find a Volterra integral equation equivalent to the initial-value problem of Exercise 3.

7 Show that the linear second order initial-value problem, $\ddot{y} + p\dot{y} + qy = f(t)$, $y(0) = y_0$, $\dot{y}(0) = \dot{y}_0$, can always be modified to one in which the initial values are zero by subtracting from the solution the function $y_0 + \dot{y}_0 t$.

8 Referring to Sec. 5.8, see if you can devise an iteration method for solving Volterra integral equations of the type introduced in this section. Can you prove that the method converges to a solution?

9 Show that the four conditions listed for the Green's functions associated with the boundary-value problems of this section determine the Green's functions uniquely.

10 Find a Green's function associated with the boundary-value problem $y'' + \omega^2 y = f(x)$, $y(0) = y(L) = 0$, by means of which the solution can be written

$$y(x) = \int_0^L G(x,\xi)f(\xi)\,d\xi$$

provided $\sin \omega L \neq 0$ (see Exercise 2).

11 Show that $\sin \omega_k x$, $\omega_k = k\pi/L$, $k = 1, 2, 3, \ldots$, are solutions of $y'' + \omega_k^2 y = 0$, $y(0) = y(L) = 0$. These functions are called *characteristic solutions* of $y'' + \omega^2 y = 0$, $y(0) = y(L) = 0$, associated with *characteristic values* ω_k^2. Relate this to the solution of the boundary-value problem of Exercise 10.

12 Find a Green's function associated with the boundary-value problem $y'' + p(x)y' = f(x)$, $y(0) = y(1) = 0$, by means of which the solution can be written

$$y(x) = \int_0^1 G(x,\xi)f(\xi)\,d\xi$$

13 Find a Green's function associated with the boundary-value problem $y'' + \omega^2 y = f(x)$, $y(0) = y'(L) = 0$, by means of which the solution can be written

$$y(x) = \int_0^L G(x,\xi)f(\xi)\,d\xi$$

14 Show that the linear second order boundary-value problem $y'' + py' + qy = f(x)$, $y(0) = a$, $y(L) = b$ can always be modified to one in which the boundary values are zero by subtracting from the solution the function $a + (b - a)x/L$.

15 Show that the Fredholm integral equation introduced in this section is equivalent to the corresponding boundary-value problem by showing that any solution of the integral equation is also a solution of the boundary-value problem.

16 Find a Fredholm integral equation equivalent to the boundary-value problem $y'' + py' + qy = f(x)$, $0 \leq x \leq 1$, $y(0) = y(1) = 0$.

17 Referring to Sec. 5.8, see if you can devise an iteration method for solving Fredholm integral equations of the type introduced in this section. Can you prove that the method converges to a solution?

7

LAPLACE TRANSFORMS

7.1 INTRODUCTION

This chapter deals with the use of the Laplace transform in the solution of initial-value problems for linear ordinary differential equations. It begins with a definition of the transform and some sufficient conditions for its existence. We study some of the general properties of the transform and develop the transforms of some of the functions most often encountered in the solution of linear differential equations. We do not treat the general problem of inversion of the transform but study in some detail the method based on the partial-fraction expansion of the transform. This is followed by the method of Laplace transformation in the solution of linear differential equations, especially those with constant coefficients. The last unstarred section takes up some applications by this method. The starred section in this chapter discusses the question of uniqueness of the transform. This is extremely important since the use of the Laplace transform in the solution of differential equations depends so heavily on the ability to invert the transform uniquely.

7.2 EXISTENCE OF THE TRANSFORM

Let $f(t)$ be Riemann-integrable for $0 \leq t \leq T$ for any finite T and suppose that for real s

$$\lim_{T\to\infty} \int_0^T e^{-st} f(t)\, dt$$

exists. Then we define the Laplace transform of $f(t)$ as

$$\mathscr{L}[f(t)] = \lim_{T\to\infty} \int_0^T e^{-st} f(t)\, dt = \int_0^\infty e^{-st} f(t)\, dt$$

Generally the transform will exist for more than one value of s, and hence $\mathscr{L}[f(t)]$ defines a function of s when it exists.

If $\lim\limits_{T\to\infty} \int_0^T e^{-st}|f(t)|\, dt$ exists, we say that the improper integral *converges absolutely*. If the integral converges absolutely, it also converges in the ordinary sense. This can be seen as follows. Since

$$0 \leq |f(t)| + f(t) \leq 2|f(t)|$$

$$0 \leq \int_0^T e^{-st}|f(t)|\, dt + \int_0^T e^{-st} f(t)\, dt \leq 2 \int_0^T e^{-st}|f(t)|\, dt$$

and the existence of

$$\lim_{T\to\infty} \int_0^T e^{-st}|f(t)|\, dt$$

implies the existence of

$$\lim_{T\to\infty} \int_0^T e^{-st} f(t)\, dt$$

Theorem 7.2.1 Let $f(t)$ be Riemann-integrable for $0 \leq t \leq T$, for any finite T, and of exponential order; that is, there exist constants M and a such that $|f(t)| \leq Me^{at}$ for all positive t. Then $f(t)$ has a Laplace transform for $s > a$, and $\lim\limits_{s\to\infty} \mathscr{L}[f(t)] = 0$.

PROOF From the inequality we have

$$\left| \int_0^T e^{-st} f(t)\, dt \right| \leq \int_0^T e^{-st}|f(t)|\, dt$$

$$\leq M \int_0^T e^{-st} e^{at}\, dt = \frac{M}{s-a} - \frac{M}{s-a} e^{(a-s)T}$$

If $s > a$, $e^{(a-s)T} \to 0$ as $T \to \infty$. This establishes the absolute convergence of the integral and the inequality

$$|\mathscr{L}[f(t)]| \leq \frac{M}{s - a}$$

from which it follows that $\lim_{s\to\infty} \mathscr{L}[f(t)] = 0$.

EXAMPLE 7.2.1 Compute the Laplace transform of $f(t) = 1$.

$$\begin{aligned}\mathscr{L}[1] &= \int_0^\infty e^{-st}\,dt = \lim_{T\to\infty} \int_0^T e^{-st}\,dt \\ &= \lim_{T\to\infty} \frac{1 - e^{-sT}}{s} = \frac{1}{s}\end{aligned}$$

provided $s > 0$.

EXAMPLE 7.2.2 Compute the Laplace transform of $f(t) = e^{at}$.

$$\begin{aligned}\mathscr{L}[e^{at}] &= \int_0^\infty e^{(a-s)t}\,dt = \lim_{T\to\infty} \int_0^T e^{(a-s)t}\,dt \\ &= \lim_{T\to\infty} \frac{1 - e^{(a-s)T}}{s - a} = \frac{1}{s - a}\end{aligned}$$

provided $s > a$.

EXAMPLE 7.2.3 Compute the Laplace transform of $f(t) = \sin \omega t$.

$$\mathscr{L}[\sin \omega t] = \int_0^\infty e^{-st} \sin \omega t\,dt = \lim_{T\to\infty} \int_0^T e^{-st} \sin \omega t\,dt$$

Integrating by parts, we have

$$\begin{aligned}\int_0^T & e^{-st} \sin \omega t\,dt \\ &= -\frac{e^{-st}}{s} \sin \omega t\Big|_0^T + \frac{\omega}{s} \int_0^T e^{-st} \cos \omega t\,dt \\ &= -\frac{e^{-sT}}{s} \sin \omega T - \frac{\omega}{s^2} e^{-st} \cos \omega t\Big|_0^T - \frac{\omega^2}{s^2} \int_0^T e^{-st} \sin \omega t\,dt\end{aligned}$$

$$\left(1 + \frac{\omega^2}{s^2}\right) \int_0^T e^{-st} \sin \omega t\,dt = -\frac{e^{-sT}}{s} \sin \omega T - \frac{\omega}{s^2} e^{-sT} \cos \omega T + \frac{\omega}{s^2}$$

If $s > 0$, $\lim_{T\to\infty} e^{-sT} \sin \omega T = \lim_{T\to\infty} e^{-sT} \cos \omega T = 0$. Therefore,

$$\mathscr{L}[\sin \omega t] = \frac{\omega}{s^2 + \omega^2}$$

It is not essential that $f(t)$ be continuous at $t = 0$. If it is not, we treat the Laplace transform as an improper integral at $t = 0$ as well as $t = \infty$. Hence,

$$\mathscr{L}[f(t)] = \lim_{\substack{\varepsilon\to 0^+ \\ T\to\infty}} \int_\varepsilon^T e^{-st} f(t)\, dt$$

provided the limit exists. For example,

$$\mathscr{L}[t^{-1/2}] = \lim_{\substack{\varepsilon\to 0^+ \\ T\to\infty}} \int_\varepsilon^T e^{-st} t^{-1/2}\, dt$$

Near $t = 0$ the integrand behaves like $t^{-1/2}$, which is integrable, and so the transform exists if $s > 0$. We make the change of variable $\eta = st$, and

$$\int_0^\infty e^{-st} t^{-1/2}\, dt = \frac{1}{s^{1/2}} \int_0^\infty e^{-\eta} \eta^{-1/2}\, d\eta$$

The integral exists and is equal to $\sqrt{\pi}$. Therefore,

$$\mathscr{L}[t^{-1/2}] = \frac{\sqrt{\pi}}{s^{1/2}}$$

One of the most important properties of the Laplace transform is that it is a linear transformation. Let a and b be any scalars, and let $f(t)$ and $g(t)$ be any two functions with Laplace transforms. Then

$$\begin{aligned}\mathscr{L}[af(t) + bg(t)] &= \int_0^\infty ae^{-st} f(t)\, dt + \int_0^\infty be^{-st} g(t)\, dt \\ &= a \int_0^\infty e^{-st} f(t)\, dt + b \int_0^\infty e^{-st} g(t)\, dt \\ &= a\mathscr{L}[f(t)] + b\mathscr{L}[g(t)]\end{aligned}$$

There are various operations on Laplace transforms which lead to useful formulas for transforms. One of these is change of variable. Suppose $f(t)$ satisfies the hypotheses of Theorem 7.2.1. Then its transform

$$\phi(s) = \int_0^\infty e^{-st} f(t)\, dt$$

exists for $s > a$. If $s - b > a$, then we can obtain

$$\phi(s - b) = \int_0^\infty e^{-st} e^{bt} f(t)\, dt$$

which gives us the formula

$$\mathscr{L}[e^{bt} f(t)] = \mathscr{L}[f(t)]_{s \to s-b}$$

Another operation is differentiation of the transform with respect to s. If $f(t)$ satisfies the hypotheses of Theorem 7.2.1, then

$$\phi(s) = \int_0^\infty e^{-st} f(t)\, dt$$

converges uniformly† for $s \geq s_0 > a$. This allows us to differentiate under the integral sign. Hence,

$$\phi'(s) = -\int_0^\infty e^{-st} t f(t)\, dt$$

which leads to the formula

$$\frac{d}{ds} \mathscr{L}[f(t)] = -\mathscr{L}[tf(t)]$$

Finally, we shall consider what happens when we take the product of two transforms. It is *not* the case that the transform of a product is the product of the transforms. On the other hand, let $\phi(s) = \mathscr{L}[f(t)]$ and $\psi(s) = \mathscr{L}[g(t)]$ be the transforms of two functions satisfying the hypotheses of Theorem 7.2.1. Then

$$\begin{aligned} \phi(s)\psi(s) &= \lim_{T \to \infty} \int_0^T e^{-st} f(t)\, dt \int_0^T e^{-st} g(t)\, dt \\ &= \lim_{T \to \infty} \int_0^T\!\!\int_0^T e^{-s(t+\tau)} f(t) g(\tau)\, dt\, d\tau \end{aligned}$$

The last integral is over a square $\{(t,\tau) \mid 0 \leq t \leq T, 0 \leq \tau \leq T\}$. We make a change of variables $\eta = t + \tau$, $\tau = \tau$, and we have

$$\int_0^T\!\!\int_0^T e^{-s(t+\tau)} f(t) g(\tau)\, dt\, d\tau = \int_0^T g(\tau) \int_\tau^{T+\tau} e^{-s\eta} f(\eta - \tau)\, d\eta\, d\tau$$

where the integration is now over the region indicated in Fig. 39. Assuming

†See J. W. Dettman, "Applied Complex Variables," pp. 186–191, Macmillan, New York, 1965 (rpt. Dover, New York, 1984).

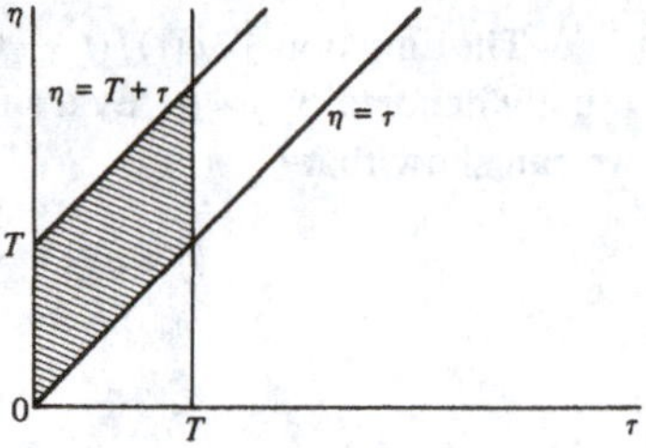

FIGURE 39

that we can interchange the order of integration (this is certainly the case if f and g are continuous), we have

$$\int_0^T g(\tau) \int_\tau^{T+\tau} e^{-s\eta} f(\eta - \tau)\, d\eta\, d\tau$$

$$= \int_0^T e^{-s\eta} \int_0^\eta g(\tau) f(\eta - \tau)\, d\tau\, d\eta + \int_T^{2T} e^{-s\eta} \int_{\eta - T}^T g(\tau) f(\eta - \tau)\, d\tau\, d\eta$$

Taking the limit as $T \to \infty$, the first integral becomes the Laplace transform of the function

$$h(t) = \int_0^t g(\tau) f(t - \tau)\, d\tau$$

and we can show that the second integral goes to zero as $T \to \infty$. In fact, if $|f(t)| \leq Me^{at}$ and $|g(t)| \leq Ne^{bt}$, then both f and g are of order e^{ct}, where $c = \max\,[a,b]$. Then

$$\left| \int_{\eta - T}^T g(\tau) f(\eta - \tau)\, d\tau \right| \leq MN \int_{\eta - T}^T e^{c\tau} e^{c(\eta - \tau)}\, d\tau = MNe^{c\eta}(2T - \eta)$$

But $\int_0^\infty \eta e^{-s\eta} e^{c\eta}\, d\eta$ converges provided $s > c$, which shows that

$$\lim_{T \to \infty} \int_T^{2T} e^{-s\eta} \eta e^{c\eta}\, d\eta \leq \lim_{T \to \infty} \int_T^\infty e^{-s\eta} \eta e^{c\eta}\, d\eta = 0$$

Also

$$\lim_{T \to \infty} 2T \int_T^{2T} e^{(c-s)\eta}\, d\eta = \lim_{T \to \infty} \frac{2T}{c - s} \left(e^{2(c-s)T} - e^{(c-s)T}\right) = 0$$

We have shown that

$$\mathscr{L}[f(t)]\mathscr{L}[g(t)] = \mathscr{L}\left[\int_0^t g(\tau) f(t - \tau)\, d\tau\right]$$

The function $\int_0^t g(\tau)f(t - \tau)\,d\tau$ is called the *convolution integral* and is usually denoted by $f * g$. By a simple change of variable from τ to $\eta = t - \tau$, we can show that

$$(f * g)(t) = \int_0^t g(\tau)f(t - \tau)\,d\tau$$
$$= \int_0^t f(\eta)g(t - \eta)\,d\eta = (g * f)(t)$$

Some of the other properties of the convolution integral are

$$f * (g + h) = f * g + f * h$$
$$(f * g) * h = f * (g * h)$$
$$0 * f = 0$$

However, $(1 * f)(t) = \int_0^t f(\tau)\,d\tau$, which is not in general equal to $f(t)$.

EXERCISES 7.2

1 Find the Laplace transform of t^n, n a positive integer.

2 Find the Laplace transform of $\cos \omega t$.

3 Show that if $f(t)$ is Riemann-integrable for every interval of the form $\{t \mid 0 < \varepsilon \leq t \leq T\}$, is of exponential order e^{at} as $t \to \infty$, and of order t^{-p}, $p < 1$, as $t \to 0^+$, then it has a Laplace transform for $s > a$.

4 Find the Laplace transform of t^α, $\alpha > -1$. *Hint:* The integral $\int_0^\infty \eta^{x-1}e^{-\eta}\,d\eta$ exists for $x > 0$ and is called the *gamma function*† $\Gamma(x)$.

5 Find the Laplace transform of $\cosh \omega t$ and $\sinh \omega t$.

6 By differentiating under the integral sign, find the Laplace transform of $t^n e^{at}$, n a positive integer.

7 By differentiating under the integral sign, find the Laplace transform of $t \sin \omega t$ and $t \cos \omega t$.

8 Find the Laplace transforms of $e^{bt} \sin \omega t$ and $e^{bt} \cos \omega t$.

9 Find the Laplace transforms of $te^{bt} \sin \omega t$ and $te^{bt} \cos \omega t$.

10 Find the Laplace transform of $\int_0^t (t - \tau) \sin \omega\tau\,d\tau$.

11 Assuming that $f(t)$ is a Riemann-integrable function of exponential order, find $\mathscr{L}\left[\int_0^t f(\tau)\,d\tau\right]$. *Hint:* Consider the integral as a convolution.

12 Assuming that $f(t)$ is continuous for $t \geq 0$, $f(0) = 0$, and f has a derivative satisfying the hypotheses of Theorem 7.2.1, prove that $\lim_{s \to \infty} s\mathscr{L}[f(t)] = 0$.

† See ibid., pp. 191–199.

13 Find a function whose transform is $1/s(s^2 + \omega^2)$. *Hint:* Consider the transform as the product of two transforms.

14 Find a function whose transform is $(s - a)^{-1}(s - b)^{-1}$, $a \neq b$.

15 If $f(t)$ and $g(t)$ are continuous for $0 \leq t$, show that $(f * g)(t)$ is continuous for $0 \leq t$.

16 If f and g are Riemann-integrable and of exponential order, show that $(f * g)(t)$ is of exponential order.

17 Prove:

(*a*) $f * (g + h) = f * g + f * h$.

(*b*) $(f * g) * h = f * (g * h)$.

(*c*) $0 * f = 0$.

(*d*) $1 * f \neq f$ (in general). As a matter of fact there is no Riemann-integrable function f such that $1 * f = f$.

7.3 TRANSFORMS OF CERTAIN FUNCTIONS

We already know how to compute the Laplace transforms of many functions commonly encountered in the solution of differential equations, such as t^n, $\cos \omega t$, $\sin \omega t$, e^{at}, $t^n e^{at}$, $t^n \cos \omega t$, $t^n \sin \omega t$, $t^n e^{at} \cos \omega t$, $t^n e^{at} \sin \omega t$. For example, if we start with

$$\mathscr{L}[\sin \omega t] = \frac{\omega}{s^2 + \omega^2}$$

then we have, by differentiating twice with respect to s,

$$\mathscr{L}[t^2 \sin \omega t] = \frac{d^2}{ds^2}\left(\frac{\omega}{s^2 + \omega^2}\right) = \frac{6\omega s^2 - 2\omega^3}{(s^2 + \omega^2)^3}$$

Then by making a change of variable we obtain

$$\mathscr{L}[t^2 e^{at} \sin \omega t] = \frac{6\omega(s - a)^2 - 2\omega^3}{[(s - a)^2 + \omega^2]^3}$$

We could obtain transforms of many more functions by taking convolutions of functions whose transforms are already known. However, given a function, it is not easy to determine if it is the convolution of two functions of the right type.

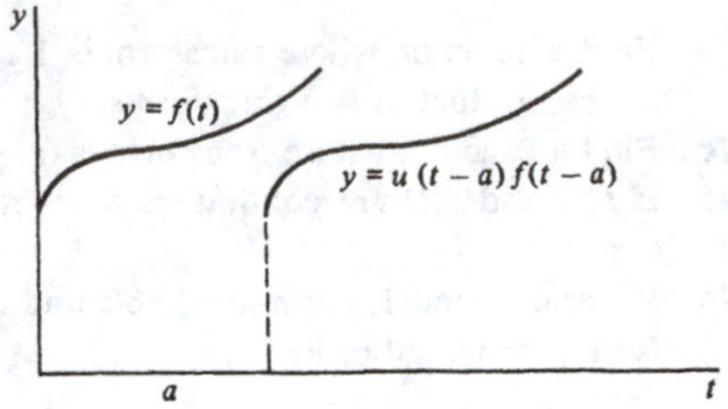

FIGURE 40

We can add to our list of functions with known transforms by considering operations on the transformed function. Let $c > 0$, and consider

$$\begin{aligned}\mathscr{L}[f(ct)] &= \int_0^\infty e^{-st} f(ct)\,dt \\ &= \frac{1}{c}\int_0^\infty e^{-s\tau/c} f(\tau)\,d\tau \\ &= \frac{1}{c}\mathscr{L}[f(t)]_{s\to s/c}\end{aligned}$$

Next we consider a shift in the independent variable. We shall find the transform of $u(t - a)f(t - a)$, where $a > 0$ and $u(t)$ is the unit step function

$$u(t) = \begin{cases} 1 & t \geq 0 \\ 0 & t < 0 \end{cases}$$

The function $u(t - a)f(t - a)$ is graphed in Fig. 40. We compute the transform as follows

$$\begin{aligned}\mathscr{L}[u(t - a)f(t - a)] &= \int_a^\infty e^{-st} f(t - a)\,dt \\ &= \int_0^\infty e^{-s(\tau + a)} f(\tau)\,d\tau \\ &= e^{-as}\int_0^\infty e^{-s\tau} f(\tau)\,d\tau \\ &= e^{-as}\mathscr{L}[f(t)]\end{aligned}$$

EXAMPLE 7.3.1 Find the Laplace transform of the function graphed in Fig. 41, consisting of one positive half-cycle of $\sin \omega t$ shifted by $t = a$. The function can be represented by the formula $f(t) = u(t - a)\sin \omega(t - a) +$

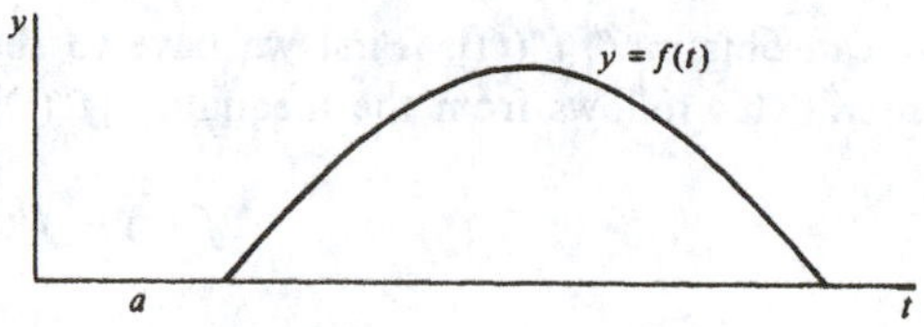

FIGURE 41

$u(t - a - \pi/\omega) \sin \omega(t - a - \pi/\omega)$. Therefore, using the result just derived, we have

$$\mathscr{L}[f(t)] = \frac{\omega e^{-as}}{s^2 + \omega^2} + \frac{\omega e^{-as} e^{-\pi s/\omega}}{s^2 + \omega^2}$$

$$= \frac{\omega e^{-as}(1 + e^{-\pi s/\omega})}{s^2 + \omega^2}$$

Next let $f(t)$ be continuous for $t \geq 0$ and have a derivative $f'(t)$ which is Riemann-integrable and of exponential order. Then

$$f(t) = f(0) + \int_0^t f'(\tau)\, d\tau$$

We consider the integral as a convolution $1 * f'$. Then according to the previous section,

$$\mathscr{L}[f(t)] = \frac{f(0)}{s} + \frac{\mathscr{L}[f'(t)]}{s}$$

and

$$\mathscr{L}[f'(t)] = s\mathscr{L}[f(t)] - f(0)$$

EXAMPLE 7.3.2 Find the Laplace transform of $\cos \omega t$. We can use the result just derived, for the derivative of $\cos \omega t$ is $-\omega \sin \omega t$, and

$$\mathscr{L}[-\omega \sin \omega t] = \frac{-\omega^2}{s^2 + \omega^2}$$

$$= s\mathscr{L}[\cos \omega t] - 1$$

Solving for $\mathscr{L}[\cos \omega t]$, we have

$$\mathscr{L}[\cos \omega t] = \frac{s}{s^2 + \omega^2}$$

If $f(t)$ and $f'(t)$ are continuous for $t \geq 0$ and $f''(t)$ is Riemann-integrable and of exponential order, then by repeated application of the above result we can obtain $\mathscr{L}[f''(t)]$. First we have to show that $f'(t)$ is of exponential order. This follows from the inequality $|f''(t)| \le Me^{at}$, where $a > 0$, and

$$\begin{aligned} f'(t) &= f'(0) + \int_0^t f''(\tau)\,d\tau \\ |f'(t)| &\le |f'(0)| + \int_0^t |f''(\tau)|\,d\tau \\ &\le |f'(0)| + M\int_0^t e^{a\tau}\,d\tau \\ &\le |f'(0)| + \frac{Me^{at}}{a} - \frac{M}{a} \\ &\le \left(|f'(0)| + \frac{M}{a}\right)e^{at} \end{aligned}$$

Applying the above result, we have

$$\begin{aligned} \mathscr{L}[f'(t)] &= s\mathscr{L}[f(t)] - f(0) \\ \mathscr{L}[f''(t)] &= s\mathscr{L}[f'(t)] - f'(0) \\ &= s^2\mathscr{L}[f(t)] - sf(0) - f'(0) \end{aligned}$$

Repeated applications of these principles give us the following theorem.

Theorem 7.3.1 Let $f, f', \ldots, f^{(n-1)}$ be continuous for $t \ge 0$ and let $f^{(n)}$ be Riemann-integrable and of exponential order. Then

$$\mathscr{L}[f^{(n)}(t)] = s^n\mathscr{L}[f(t)] - s^{n-1}f(0) - s^{n-2}f'(0) - \cdots - f^{(n-1)}(0)$$

EXAMPLE 7.3.3 Find the Laplace transform of $t\cos\omega t$. Here we use a technique based on the fact that the given function is the unique solution of an initial-value problem in ordinary differential equations. An annihilation operator for the function is $(D^2 + \omega^2)^2$. Therefore, the function satisfies the differential equation $(D^2 + \omega^2)^2 y = 0$. It also satisfies the initial conditions $y(0) = 0$, $\dot{y}(0) = 1$, $\ddot{y}(0) = 0$, $\dddot{y}(0) = -3\omega^2$. The differential equation is $(D^4 + 2\omega^2 D^2 + \omega^4)y = 0$. We take transforms using Theorem 7.3.1. If $\phi(s) = \mathscr{L}[t\cos\omega t]$, then

$$\begin{aligned} s^4\phi - s^2 + 3\omega^2 + 2\omega^2(s^2\phi - 1) + \omega^4\phi &= 0 \\ (s^4 + 2\omega^2 s^2 + \omega^4)\phi &= s^2 - \omega^2 \\ \phi &= \frac{s^2 - \omega^2}{(s^2 + \omega^2)^2} \end{aligned}$$

This example represents the inverse of the problem we are really interested in, namely the problem of finding the solution of an initial-value problem by finding its transform from the differential equation and the initial conditions. This will be the subject of Sec. 7.5.

We conclude this section with a short table of Laplace transforms, which also summarizes the general properties we have studied so far.†

$f(t)$	$\mathscr{L}[f(t)]$
$ag(t) + bh(t)$	$a\mathscr{L}[g(t)] + b\mathscr{L}[h(t)]$
$g(ct) \quad c > 0$	$\frac{1}{c}\mathscr{L}[g(t)]_{s \to s/c}$
$e^{at}g(t)$	$\mathscr{L}[g(t)]_{s \to s-a}$
$tg(t)$	$\frac{-d}{ds}\mathscr{L}[g(t)]$
$\int_0^t g(\tau)h(t-\tau)\,d\tau$	$\mathscr{L}[g(t)]\mathscr{L}[h(t)]$
$\int_0^t g(\tau)\,d\tau$	$\frac{1}{s}\mathscr{L}[g(t)]$
$u(t-a)g(t-a) \quad a > 0$	$e^{-as}\mathscr{L}[g(t)]$
$g'(t)$	$s\mathscr{L}[g(t)] - g(0)$
$g^{(n)}(t) \quad n = 1, 2, 3, \ldots$	$s^n\mathscr{L}[g(t)] - s^{n-1}g(0) - \cdots - g^{(n-1)}(0)$
1	s^{-1}
$t^n \quad n = 1, 2, 3, \ldots$	$n!/s^{n+1}$
$t^\alpha \quad \alpha > -1$	$\frac{\Gamma(\alpha+1)}{s^{\alpha+1}}$
e^{at}	$(s-a)^{-1}$
$t^n e^{at} \quad n = 1, 2, 3, \ldots$	$\frac{n!}{(s-a)^{n+1}}$
$t^\alpha e^{at} \quad \alpha > -1$	$\frac{\Gamma(\alpha+1)}{(s-a)^{\alpha+1}}$
$\sin \omega t$	$\frac{\omega}{s^2+\omega^2}$

(Continued overleaf)

†Much more extensive tables can be found in books such as R. V. Churchill, "Operational Mathematics," 3d ed., McGraw-Hill, New York, 1971.

$f(t)$	$\mathscr{L}[f(t)]$
$\cos \omega t$	$\dfrac{s}{s^2 + \omega^2}$
$e^{at} \sin \omega t$	$\dfrac{\omega}{(s - a)^2 + \omega^2}$
$e^{at} \cos \omega t$	$\dfrac{s - a}{(s - a)^2 + \omega^2}$
$t \sin \omega t$	$\dfrac{2\omega s}{(s^2 + \omega^2)^2}$
$t \cos \omega t$	$\dfrac{s^2 - \omega^2}{(s^2 + \omega^2)^2}$
$te^{at} \sin \omega t$	$\dfrac{2\omega(s - a)}{[(s - a)^2 + \omega^2]^2}$
$te^{at} \cos \omega t$	$\dfrac{(s - a)^2 + \omega^2}{[(s - a)^2 + \omega^2]^2}$
$\sinh \omega t$	$\dfrac{\omega}{s^2 - \omega^2}$
$\cosh \omega t$	$\dfrac{s}{s^2 - \omega^2}$

EXERCISES 7.3

1 Find the Laplace transform of $te^{at} \cos \omega t$.

2 Find the Laplace transform of $\sinh \omega t = \frac{1}{2}(e^{\omega t} - e^{-\omega t})$.

3 Find the Laplace transform of $\cosh \omega t = \frac{1}{2}(e^{\omega t} + e^{-\omega t})$.

4 Find the Laplace transforms of $t \sinh \omega t$ and $t \cosh \omega t$.

5 Find the Laplace transforms of $te^{at} \sinh \omega t$ and $te^{at} \cosh \omega t$.

6 Find the Laplace transform of $f(t)$ if $f(t) = 1$, $0 \le a \le t \le b$, and $f(t) = 0$ for all other values of t.

7 Find the Laplace transform of $f(t)$ if $f(t) = 1$, $0 \le t < c$, $f(t) = -1$, $c \le t < 2c$, and repeats periodically with period $2c$.

8 Find the Laplace transform of $f(t)$ if $f(t) = t$, $0 \le t < c$, and repeats periodically with period c.

9 Find a differential equation and a set of initial conditions satisfied by $f(t) = e^{at} \sin \omega t$, and use the initial-value problem to determine the Laplace transform of $f(t)$.

10 Find a differential equation and a set of initial conditions satisfied by $f(t) = te^{at} \sin \omega t$, and use the initial-value problem to determine the Laplace transform of $f(t)$.

7.4 INVERSION OF THE TRANSFORM

There are general formulas for finding a function whose Laplace transform is a given function.† However, these formulas are not easily applied, and in many cases there are simpler techniques for inverting the transform. We shall take up a few of these in this section. In the starred section of this chapter, we shall prove a uniqueness theorem for the Laplace transform which will assure us that once we have found a function with the given transform, there are essentially no others with the desired properties.

EXAMPLE 7.4.1 Find a function whose Laplace transform is

$$(s - a)^{-1}(s - b)^{-1} \qquad \text{where } a \neq b$$

We know that $\mathscr{L}[e^{at}] = (s - a)^{-1}$ and $\mathscr{L}[e^{bt}] = (s - b)^{-1}$. Hence, using the convolution integral, we have that $(s - a)^{-1}(s - b)^{-1}$ is the Laplace transform of

$$\begin{aligned}\int_0^t e^{a\tau}e^{b(t-\tau)}\,d\tau &= e^{bt}\int_0^t e^{(a-b)\tau}\,d\tau \\ &= \frac{1}{a-b}e^{at} - \frac{1}{a-b}e^{bt}\end{aligned}$$

The form of the answer suggests that the transform should be expressed as

$$\frac{1}{(s-a)(s-b)} = \frac{1}{a-b}\frac{1}{s-a} + \frac{1}{b-a}\frac{1}{s-b}$$

which suggests a partial-fraction expansion. Indeed, if we write

$$\begin{aligned}\frac{1}{(s-a)(s-b)} &= \frac{A}{s-a} + \frac{B}{s-b} \\ &= \frac{(A+B)s - Ab - Ba}{(s-a)(s-b)}\end{aligned}$$

this must be an identity in s. For this to be so we must have

$$\begin{aligned}A + B &= 0 \\ Ab + Ba &= -1\end{aligned}$$

Solving these equations, we have $A = (a - b)^{-1}$ and $B = (b - a)^{-1}$.

† See Dettman op. cit., p. 400.

The method based on the convolution integral is generally unwieldy if the given transform is the product of more than two recognizable transforms, and even then the integral may not be easy to evaluate. Generally, the method of the partial-fraction expansion is preferred when it is applicable, and, as we shall see in the next section, it will be applicable in problems involving linear differential equations with constant coefficients. We shall now review some of the methods for finding partial-fraction expansions.

Suppose we are asked to find a function whose Laplace transform is

$$\phi(s) = \frac{p(s)}{(s - a_1)(s - a_2)\cdots(s - a_n)}$$

where $p(s)$ is a polynomial† of degree less than n and $a_1, a_2, \ldots, a_n$ are all different. We seek a partial-fraction expansion in the form

$$\frac{p(s)}{(s - a_1)(s - a_2)\cdots(s - a_n)} = \frac{A_1}{s - a_1} + \frac{A_2}{s - a_2} + \cdots + \frac{A_n}{s - a_n}$$

Suppose we multiply through by $s - a_k$, $1 \leq k \leq n$, an operation which is valid for $s \neq a_k$. Then

$$\begin{aligned}\frac{p(s)}{(s - a_1)\cdots(s - a_{k-1})(s - a_{k+1})\cdots(s - a_n)} &= \frac{A_1(s - a_k)}{s - a_1} + \cdots + \frac{A_{k-1}(s - a_k)}{s - a_{k-1}} \\ &\quad + A_k + \frac{A_{k+1}(s - a_k)}{s - a_{k+1}} + \cdots + \frac{A_n(s - a_k)}{s - a_n}\end{aligned}$$

The equality holds for all s sufficiently close to a_k, where both sides are continuous functions of s. Therefore, we may take the limit as s approaches a_k, and we have

$$A_k = \frac{p(a_k)}{(a_k - a_1)\cdots(a_k - a_{k-1})(a_k - a_{k+1})\cdots(a_k - a_n)}$$

If $D(s) = (s - a_1)(s - a_2)\cdots(s - a_n)$ is the denominator in $\phi(s)$ and $D_k(s) = D(s)/(s - a_k)$, then

$$A_k = \frac{p(a_k)}{D_k(a_k)}$$

† For $\phi(s)$ to be the transform of a function satisfying the hypotheses of Theorem 7.2.1, $\lim_{s\to\infty} \phi(s) = 0$.

EXAMPLE 7.4.2 Find a function whose Laplace transform is

$$\phi(s) = \frac{2s^2 - s + 1}{(s - 1)(s - 2)(s - 3)}$$

We look for a partial-fraction expansion in the form

$$\phi(s) = \frac{A_1}{s - 1} + \frac{A_2}{s - 2} + \frac{A_3}{s - 3}$$

and if $D = (s - 1)(s - 2)(s - 3)$, then $D_1 = (s - 2)(s - 3)$, $D_2 = (s - 1)(s - 3)$, $D_3 = (s - 1)(s - 2)$, and $p(s) = 2s^2 - s + 1$. Therefore,

$$A_1 = \frac{p(1)}{D_1(1)} = 1$$

$$A_2 = \frac{p(2)}{D_2(2)} = -7$$

$$A_3 = \frac{p(3)}{D_3(3)} = 8$$

$$\phi(s) = \frac{1}{s - 1} - \frac{7}{s - 2} + \frac{8}{s - 3}$$

$$= \mathscr{L}[e^t - 7e^{2t} + 8e^{3t}]$$

We now consider the case where one of the factors in the denominator is repeated. Suppose

$$\phi(s) = \frac{p(s)}{(s - a)^m q(s)}$$

where $q(s)$ is a polynomial of degree n such that $q(a) \neq 0$ and $p(s)$ is a polynomial of degree less than $n + m$. We shall assume an expansion of the form

$$\frac{p(s)}{(s - a)^m q(s)} = \frac{A_1}{s - a} + \frac{A_2}{(s - a)^2} + \cdots + \frac{A_m}{(s - a)^m} + h(s)$$

where $h(s)$ represents the sum of all the terms due to $q(s)$. We multiply through by $(s - a)^m$, and then

$$\frac{p(s)}{q(s)} = A_1(s - a)^{m-1} + A_2(s - a)^{m-2} + \cdots + A_m + h(s)(s - a)^m$$

where the equality holds up to and including $s = a$ by the continuity of both sides at $s = a$. If we let $s = a$, we have $A_m = p(a)/q(a)$. Assuming that $h(s)$ is differentiable at $s = a$, we differentiate and

$$\frac{d}{ds}\left[\frac{p(s)}{q(s)}\right] = (m-1)A_1(s-a)^{m-2} + (m-2)A_2(s-a)^{m-3} + \cdots + A_{m-1} + h'(s)(s-a)^m + mh(s)(s-a)^{m-1}$$

If $m \geq 2$ and $r(s) = p(s)/q(s)$, then $A_{m-1} = r'(a)$. Differentiating again, we have

$$\begin{aligned} r''(s) = {} & (m-1)(m-2)A_1(s-a)^{m-3} \\ & + (m-2)(m-3)A_2(s-a)^{m-3} + \cdots + 2A_{m-2} \\ & + h''(s)(s-a)^m + 2mh'(s)(s-a)^{m-1} \\ & + m(m-1)h(s)(s-a)^{m-2} \end{aligned}$$

If $m \geq 3$, then $A_{m-2} = \frac{1}{2}r''(a)$. Continuing in this way, we obtain the general formula

$$A_{m-j} = \frac{1}{j!}r^{(j)}(a)$$

$j = 0, 1, 2, \ldots, m-1$, where $0! = 1$ and $r^{(0)}(a) = r(a)$. If $q(s)$ has a factor $(s-b)^k$, then of course $h(s)$ will contain terms of the form $B_1(s-b)^{-1}$, $B_2(s-b)^{-2}, \ldots, B_k(s-b)^{-k}$, but these can be handled in a similar way.

EXAMPLE 7.4.3 Find a function whose Laplace transform is

$$\phi(s) = \frac{s}{(s-1)^3(s-2)}$$

We look for a partial-fraction expansion in the form

$$\phi(s) = \frac{A_1}{s-1} + \frac{A_2}{(s-1)^2} + \frac{A_3}{(s-1)^3} + \frac{B}{s-2}$$

Then $B = \lim_{s \to 2}(s-2)\phi(s) = 2$. Let $r(s) = (s-1)^3\phi(s) = s/(s-2) = 1 + 2(s-2)^{-1}$. Then $A_3 = r(1) = -1$, $A_2 = r'(1) = -2$, $A_1 = \frac{1}{2}r''(1) = -2$. Therefore,

$$\begin{aligned} \phi(s) &= \frac{-2}{s-1} + \frac{-2}{(s-1)^2} + \frac{-1}{(s-1)^3} + \frac{2}{s-2} \\ &= \mathscr{L}[-2e^t - 2te^t - \tfrac{1}{2}t^2e^t + 2e^{2t}] \end{aligned}$$

Now suppose $\phi(s) = p(s)/q(s)$, where $q(s)$ is a polynomial of degree n and $p(s)$ is a polynomial of degree less than n (both with real coefficients). Suppose further that $q(s)$ has a nonrepeated complex root $a + ib$. Then, of course, $q(s)$ has a nonrepeated complex root $a - ib$ (see Sec. 6.4), and $q(s)$ can be written as

$$q(s) = (s - a - ib)(s - a + ib)r(s) = [(s - a)^2 + b^2]r(s)$$

Notice that the factor $(s - a)^2 + b^2$ has real coefficients. This factor is called an ***irreducible quadratic factor*** because it cannot be factored further without introducing complex coefficients. It can be shown that every polynomial in the real variable s of degree $n \geq 1$ with real coefficients can be factored into linear factors with real coefficients and irreducible quadratic factors, some possibly repeated. We now consider a partial-fraction expansion involving a nonrepeated irreducible quadratic factor.

Let

$$\phi(s) = \frac{p(s)}{[(s - a)^2 + b^2]r(s)}$$

where $r(a + ib) \neq 0$. We look for a partial-fraction expansion of the form

$$\frac{p(s)}{[(s - a)^2 + b]r(s)} = \frac{As + B}{(s - a)^2 + b^2} + h(s)$$

where $h(s)$ accounts for all the terms in the expansion due to the polynomial $r(s)$. We multiply through by $(s - a)^2 + b^2$. Then

$$\frac{p(s)}{r(s)} = As + B + h(s)[(s - a)^2 + b^2]$$

Let $C = p(a + ib)/r(a + ib)$. Then

$$\frac{p(a - ib)}{r(a - ib)} = \frac{\bar{p}(a + ib)}{\bar{r}(a + ib)} = \bar{C}$$

and we have the equations

$$A(a + ib) + B = C$$

$$A(a - ib) + B = \bar{C}$$

Solving for A and B, we obtain

$$A = \frac{C - \bar{C}}{2ib} = \frac{\text{Im}(C)}{b}$$

$$B = \frac{C + \bar{C}}{2} - aA = \text{Re}(C) - \frac{a}{b}\text{Im}(C)$$

EXAMPLE 7.4.4 Find a function whose Laplace transform is

$$\phi(s) = \frac{1}{(s-1)^2[(s+1)^2+4]}$$

We look for a partial-fraction expansion of the form

$$\phi(s) = \frac{A_1 s + B_1}{(s+1)^2+4} + \frac{A_2}{s-1} + \frac{A_3}{(s-1)^2}$$

According to the above discussion, if we let

$$C_1 = \frac{1}{(-2+2i)^2} = \tfrac{1}{8}i$$

then $A_1 = \frac{1}{16}$ and $B_1 = \frac{1}{16}$. Also, $A_3 = \frac{1}{8}$ and $A_2 = -\frac{1}{16}$. Therefore,

$$\begin{aligned}\phi(s) &= \frac{1}{16}\frac{s+1}{(s+1)^2+4} - \frac{1}{16}\frac{1}{s-1} + \frac{1}{8}\frac{1}{(s-1)^2} \\ &= \mathscr{L}[\tfrac{1}{16}e^{-t}\cos 2t - \tfrac{1}{16}e^t + \tfrac{1}{8}te^t]\end{aligned}$$

There are also methods for handling partial-fraction expansions with repeated irreducible quadratic factors. The reader will be asked to work one out in Exercise 7.4.3.

EXERCISES 7.4

1 Show that

$$\frac{As+B}{(s-a)^2+b^2} = \mathscr{L}\left[Ae^{at}\cos bt + \frac{B+aA}{b}e^{at}\sin bt\right]$$

2 Show that

$$\frac{As+B}{[(s-a)^2+b^2]^2} = \mathscr{L}[h(t)]$$

where

$$h(t) = \frac{e^{at}}{b}\int_0^t \left[A\cos b\tau \sin b(t-\tau) + \frac{B+aA}{b}\sin b\tau \sin b(t-\tau)\right] d\tau$$

Evaluate the integral.

3 Let $\phi(s) = g(s)/[(s-a)^2+b^2]^2$, where $g(s) = p(s)/r(s)$, $p(s)$ is a polynomial of degree less than $n+4$, and $r(s)$ is a polynomial of degree n, $p(a+ib) \neq 0$, $r(a+ib) \neq 0$. If

$$\phi(s) = \frac{A_1 s + B_1}{(s-a)^2+b^2} + \frac{A_2 s + B_2}{[(s-a)^2+b^2]^2} + h(s)$$

where $h(s)$ is the part of the partial-fraction expansion which does not involve $(s - a)^2 + b^2$, show that

$$A_1 = \frac{1}{b} \operatorname{Im}(C_1)$$

$$B_1 = \operatorname{Re}(C_1) - \frac{a}{b} \operatorname{Im}(C_1)$$

$$A_2 = \frac{1}{b} \operatorname{Im}(C_2)$$

$$B_2 = \operatorname{Re}(C_2) - \frac{a}{b} \operatorname{Im}(C_2)$$

where

$$C_1 = \frac{g'(a + ib) - A_2}{2ib} \qquad \text{and} \qquad C_2 = g(a + ib)$$

4 Find a function whose Laplace transform is

$$\frac{2s^3 - s^2 + s - 3}{s(s + 1)(s - 1)(s - 3)(s - 4)}$$

5 Find a function whose Laplace transform is

$$\frac{2s^2 + 3s + 5}{(s - 2)^3(s - 3)}$$

6 Find a function whose Laplace transform is

$$\frac{s^4 - s + 7}{s(s + 1)^2(s^2 + 2s + 2)}$$

7 Find a function whose Laplace transform is

$$\frac{s^2 - 2s + 1}{s^2(s^2 - 2s + 2)(s^2 + 2s + 5)}$$

8 Find a function whose Laplace transform is

$$\frac{s + 2}{(s + 1)^2(s^2 + 2s + 2)^2}$$

9 Find a function whose Laplace transform is

$$\frac{e^{-2s}}{(s + 1)(s - 1)^2(s^2 + 2s + 5)}$$

7.5 SOLUTION OF DIFFERENTIAL EQUATIONS

In this section we come to the main concern of this chapter, the solution of linear differential equations using the Laplace transform. We shall illustrate this technique in a series of examples. The first of these will involve the solution of an initial-value problem for a nonhomogeneous equation with constant coefficients, which could be solved readily using the method of undetermined coefficients. However, the next three examples will deal with problems which cannot be solved easily by other techniques. This should show that the Laplace transform method is quite versatile.

EXAMPLE 7.5.1 Solve the initial-value problem $\ddot{y} + 5\dot{y} + 6y = e^{-2t} + e^{-t}\cos 2t$, $y(0) = y_0$, $\dot{y}(0) = \dot{y}_0$. We take the Laplace transform and let $Y(s) = \mathscr{L}[y(t)]$. Then

$$s^2Y(s) - sy_0 - \dot{y}_0 + 5sY(s) - 5y_0 + 6Y(s) = \frac{1}{s+2} + \frac{s+1}{(s+1)^2+4}$$

Solving for $Y(s)$, we have

$$Y(s) = \frac{(s+5)y_0 + \dot{y}_0}{s^2+5s+6} + \frac{1}{(s+2)(s^2+5s+6)} + \frac{s+1}{[(s+1)^2+4](s^2+5s+6)}$$

Using partial-fraction expansions as in the last section, we have

$$Y(s) = \frac{3y_0 + \dot{y}_0}{s+2} - \frac{2y_0 + \dot{y}_0}{s+3} - \frac{1}{s+2} + \frac{1}{(s+2)^2} + \frac{1}{s+3} - \frac{1}{5}\frac{1}{s+2} + \frac{1}{4}\frac{1}{s+3} - \frac{1}{20}\frac{s-5}{(s+1)^2+4}$$

Inverting the transform, we have the solution

$$y(t) = (3y_0 + \dot{y}_0 - \tfrac{6}{5})e^{-2t} + (-2y_0 - \dot{y}_0 + \tfrac{5}{4})e^{-3t} + te^{-2t} - \tfrac{1}{20}e^{-t}\cos 2t + \tfrac{3}{20}e^{-t}\sin 2t$$

One advantage of the method is that it automatically incorporates the initial conditions and there is no need to solve a system of linear equations for the values of the constants in the complementary solution.

EXAMPLE 7.5.2 Solve the initial-value problem $\ddot{y} + 2\dot{y} + y = f(t)$, $y(0) = y_0$, $\dot{y}(0) = \dot{y}_0$, where

$$f(t) = \begin{cases} \sin(t-a) & a \le t \le a + \pi \\ 0 & \text{elsewhere} \end{cases}$$

We can write $f(t) = u(t - a)\sin(t - a) + u(t - a - \pi)\sin(t - a - \pi)$ where $u(t)$ is the unit step function. The Laplace transform of $f(t)$ is

$$\mathscr{L}[f(t)] = \frac{e^{-as}}{s^2 + 1} + \frac{e^{-(a+\pi)s}}{s^2 + 1}$$

Letting $Y(s) = \mathscr{L}[y(t)]$, we have

$$s^2 Y(s) - sy_0 - \dot{y}_0 + 2sY(s) - 2y_0 + Y(s) = \frac{e^{-as} + e^{-(a+\pi)s}}{s^2 + 1}$$

Solving for $Y(s)$ gives

$$Y(s) = \frac{(s + 2)y_0 + \dot{y}_0}{(s + 1)^2} + \frac{e^{-as} + e^{-(a+\pi)s}}{(s + 1)^2(s^2 + 1)}$$

Using partial-fraction expansions, we have

$$\frac{(s + 2)y_0 + \dot{y}_0}{(s + 1)^2} = \frac{y_0}{s + 1} + \frac{y_0 + \dot{y}_0}{(s + 1)^2}$$

$$\frac{1}{(s + 1)^2(s^2 + 1)} = \frac{1}{2(s + 1)} + \frac{1}{2(s + 1)^2} - \frac{s}{2(s^2 + 1)}$$

Inverting the transform, we obtain the solution

$$\begin{aligned} y(t) = {} & y_0 e^{-t} + (y_0 + \dot{y}_0)te^{-t} + \tfrac{1}{2}u(t - a)e^{-(t-a)} \\ & + \tfrac{1}{2}(t - a)u(t - a)e^{-(t-a)} - \tfrac{1}{2}u(t - a)\cos(t - a) \\ & + \tfrac{1}{2}u(t - a - \pi)e^{-(t-a-\pi)} \\ & + \tfrac{1}{2}(t - a - \pi)u(t - a - \pi)e^{-(t-a-\pi)} \\ & - \tfrac{1}{2}u(t - a - \pi)\cos(t - a - \pi) \end{aligned}$$

EXAMPLE 7.5.3 Solve the initial-value problem $\ddot{y} + 2\dot{y} + y = |\sin t|$, $y(0) = 0$, $\dot{y}(0) = 0$. Comparing this problem with the previous one, we can obtain the contribution to the solution of one positive half-cycle of $\sin t$ by putting $a = n\pi$. This contribution will be

$$\begin{aligned} y_n(t) = {} & \tfrac{1}{2}u(t - n\pi)e^{-(t-n\pi)} + \tfrac{1}{2}(t - n\pi)u(t - n\pi)e^{-(t-n\pi)} \\ & - \tfrac{1}{2}u(t - n\pi)\cos(t - n\pi) + \tfrac{1}{2}u(t - n\pi - \pi)e^{-(t-n\pi-\pi)} \\ & + \tfrac{1}{2}(t - n\pi - \pi)u(t - n\pi - \pi)e^{-(t-n\pi-\pi)} \\ & - \tfrac{1}{2}u(t - n\pi - \pi)\cos(t - n\pi - \pi) \end{aligned}$$

By the linearity of the equation the solution is the sum

$$y(t) = \sum_{n=0}^{\infty} y_n(t)$$

There is no problem with convergence of this series since for any finite t there are only a finite number of nonzero terms.

EXAMPLE 7.5.4 Solve the differential equation† $t\ddot{y} + \dot{y} + ty = 0$ subject to the initial conditions $y(0) = 1$, $\dot{y}(0) = 0$. Using the fact that $\mathscr{L}[tf(t)] = -(d/ds)\mathscr{L}[f(t)]$, we have $Y(s) = \mathscr{L}[y(t)]$,

$$-\frac{d}{ds}[s^2Y(s) - s] + sY(s) - 1 - \frac{d}{ds}Y(s) = 0$$

or

$$\frac{dY}{ds} + \frac{s}{1 + s^2}Y = 0$$

This time, instead of obtaining an algebraic equation for $Y(s)$ we obtain a differential equation; however, the equation is first order linear. Solving this equation, we have

$$Y(s) = \frac{c}{\sqrt{1 + s^2}}$$

where c is an arbitrary constant. We can evaluate c by the following device. We shall show in Sec. 8.4 that the derivative of the solution of our problem is bounded for all t. Therefore, assuming that $|\dot{y}(t)| \leq M$, we have

$$\mathscr{L}[\dot{y}(t)] = sY(s) - 1 = \frac{cs}{\sqrt{1 + s^2}} - 1$$

But

$$\begin{aligned} |\mathscr{L}[\dot{y}(t)]| &\leq \int_0^\infty |\dot{y}(t)|e^{-st}\,dt \\ &\leq \frac{M}{s} \end{aligned}$$

Therefore,

$$\lim_{s\to\infty} \mathscr{L}[\dot{y}(t)] = \lim_{s\to\infty}\left(\frac{cs}{\sqrt{1 + s^2}} - 1\right) = c - 1 = 0$$

† Note that this equation does not have constant coefficients.

Hence, $c = 1$ and $Y(s) = (1 + s^2)^{-1/2}$. Using the binomial expansion, valid for $|s| > 1$, we have

$$Y(s) = \frac{1}{s} - \frac{1}{2}\frac{1}{s^3} + \frac{1 \cdot 3}{2^2 \cdot 2!}\frac{1}{s^5} - \frac{1 \cdot 3 \cdot 5}{2^3 \cdot 3!}\frac{1}{s^7} + \cdots$$

$$= \sum_{n=0}^{\infty} \frac{(-1)^n (2n)!}{2^{2n}(n!)^2 s^{2n+1}}$$

Assuming† that we can invert the transform term by term, we obtain

$$y(t) = \sum_{n=0}^{\infty} \frac{(-1)^n t^{2n}}{2^{2n}(n!)^2}$$

This series converges for all t and hence may be differentiated as many times as we please. It is now a matter of showing that y satisfies the differential equation. It clearly satisfies the initial conditions.

EXERCISES 7.5

1 Solve the following initial-value problems using the Laplace transform:

(*a*) $\ddot{y} + 5\dot{y} + 4y = e^t + 2\cos 2t$, $y(0) = y_0$, $\dot{y}(0) = \dot{y}_0$
(*b*) $\ddot{y} + 4\dot{y} + 4y = 3e^{-2t}$, $y(0) = y_0$, $\dot{y}(0) = \dot{y}_0$
(*c*) $\ddot{y} + 2\dot{y} + 5y = \sin 3t$, $y(0) = y_0$, $\dot{y}(0) = \dot{y}_0$
(*d*) $\dddot{y} + 3\ddot{y} + 4\dot{y} + 2y = e^{-t}\cos t$, $y(0) = y_0$, $\dot{y}(0) = \ddot{y}(0) = 0$

2 Solve the following initial-value problem using the Laplace transform:

$$\ddot{y} + 3\dot{y} + 2y = f(t)$$

where $f(t) = t$, $0 \le t \le a$, $f(t) = 2a - t$, $a \le t \le 2a$, $f(t) = 0$, elsewhere; $y(0) = y_0$, $\dot{y}(0) = \dot{y}_0$.

3 Solve the following initial-value problem using the Laplace transform: $\ddot{y} + 3\dot{y} + 2y = f(t)$, $y(0) = \dot{y}(0) = 0$, where $f(t)$ is periodic with period $2a$ and is given for $0 \le t \le 2a$, as in Exercise 2.

4 Solve the following system of differential equations for $x(t)$ and $y(t)$ subject to the initial conditions $x(0) = x_0$, $y(0) = y_0$:

$$\dot{x} + \dot{y} - x + 3y = e^{-t}$$
$$\dot{x} - \dot{y} + 2x - 2y = e^{-2t}$$

Hint: Transform each equation and then solve for the Laplace transforms of $x(t)$ and $y(t)$.

† All of this can be justified by showing that the solution we finally obtain satisfies the differential equation and the initial conditions.

5 Solve the following initial-value problem: $\ddot{y} + ty = 0$, $y(0) = 1$, $\dot{y}(0) = 0$.
6 Solve the following initial-value problem: $t^2\ddot{y} + t\dot{y} + (t^2 - 1)y = 0$, $y(0) = 0$, $\dot{y}(0) = 1$.

7.6 APPLICATIONS

In this section, we shall consider some further applications to electric-circuit problems using the Laplace transform method.

EXAMPLE 7.6.1 Find the steady-state current in the circuit shown in Fig. 42 if the input voltage is $E_0 \sin \omega t$. The equation for the current is

$$H\frac{dI}{dt} + RI + \frac{1}{C}\left[q_0 + \int_0^t I(\tau)\,d\tau\right] = E_0 \sin \omega t$$

Let $I(0) = I_0$ and $q(0) = q_0$. Then if $\phi(s)$ is the Laplace transform of $I(t)$, we have

$$H(s\phi - I_0) + R\phi + \frac{1}{Cs}(q_0 + \phi) = \frac{E_0\omega}{s^2 + \omega^2}$$

and

$$\phi(s) = \frac{E_0 s\omega}{(s^2 + \omega^2)(Hs^2 + Rs + 1/C)} + \frac{HI_0 s}{Hs^2 + Rs + 1/C} - \frac{q_0}{C(Hs^2 + Rs + 1/C)}$$

It is clear that the second and third terms will lead to terms in the solution which decrease in amplitude exponentially. Therefore, we need only concern ourselves with the inversion of the first term. In fact, for the steady-state solution we only have to obtain the term in the partial-fraction expansion of the form $(As + B)/(s^2 + \omega^2)$. Using the technique of Sec. 7.4, we can find this term

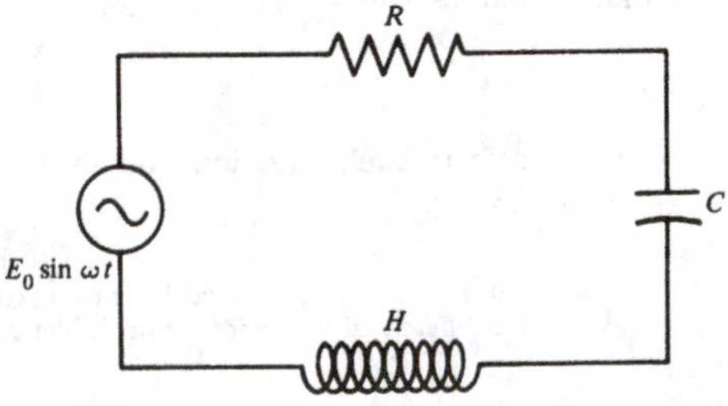

FIGURE 42

without bothering about the rest of the expansion. Let K be the value of $E_0\omega s/(Hs^2 + Rs + 1/C)$ at $s = i\omega$. Then

$$K = \frac{(iE_0\omega^2)[(1/C - H\omega^2) - iR\omega]}{(H\omega^2 - 1/C)^2 + R^2\omega^2}$$

and $A = \text{Im }(K)/\omega$ and $B = \text{Re }(K)$. Therefore,

$$A = \frac{-E_0(\omega H - 1/\omega C)}{R^2 + (\omega H - 1/\omega C)^2}$$

$$B = \frac{E_0 R\omega}{(\omega H - 1/\omega C)^2 + R^2}$$

Let $X = \omega H - 1/\omega C$, $Z = R + iX$. We call X the *reactance*, Z the *complex impedance*, and $\theta = \arg Z$ the *phase*. We have

$$A = \frac{-E_0 \sin\theta}{|Z|}$$

$$B = \frac{\omega E_0 \cos\theta}{|Z|}$$

and

$$\frac{As + B}{s^2 + \omega^2} = \frac{E_0}{|Z|}\,\frac{\omega\cos\theta - s\sin\theta}{s^2 + \omega^2}$$

and the steady-state solution is

$$\frac{E_0}{|Z|}\sin(\omega t - \theta)$$

EXAMPLE 7.6.2 Consider the electric circuit shown in Fig. 43. Assuming that the initial currents and charges are zero, find E_{out}. I_1 and I_2 must satisfy the following equations

$$E_{\text{in}} = RI_1 + \frac{1}{C}\int_0^t I_1(\tau)\,d\tau + H\frac{dI_1}{dt} - H\frac{dI_2}{dt}$$

$$0 = RI_2 + H\frac{dI_2}{dt} - H\frac{dI_1}{dt} + \frac{1}{C}\int_0^t I_2(\tau)\,d\tau$$

Taking the Laplace transforms, we have

$$E_{\text{in}} = RI_1 + \frac{1}{Cs}I_1 + HsI_1 - HsI_2$$

$$0 = RI_2 + HsI_2 - HsI_1 + \frac{1}{Cs}I_2$$

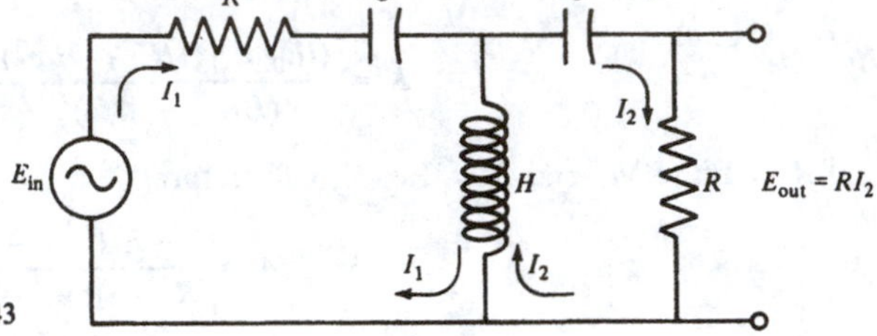

FIGURE 43

where, for simplicity, we have let I_1, I_2, E_{in} stand for the Laplace transform of the corresponding function. Solving for I_1 and I_2, we obtain

$$I_1 = \frac{E_{in}(Hs + R + 1/Cs)}{(Hs + R + 1/Cs)^2 - H^2s^2}$$

$$I_2 = \frac{E_{in}sH}{(Hs + R + 1/Cs)^2 - H^2s^2}$$

The Laplace transform of E_{out} is

$$\begin{aligned} E_{out} = RI_2 &= \frac{E_{in}sHR}{(Hs + R + 1/Cs)^2 - H^2s^2} \\ &= \frac{E_{in}s^3HR}{2HRs^3 + (2H/C + R^2)s^2 + (2R/C)s + 1/C^2} \\ &= (sE_{in})T(s) \end{aligned}$$

where

$$T(s) = \frac{s^2HR}{2HRs^3 + (2H/C + R^2)s^2 + (2R/C)s + 1/C^2}$$

Actually, $T(s)$ is the transform of E_{out} if E_{in} is a constant unit voltage. Let $A(t)$ be the inverse transform of $T(s)$, and assume that $\dot{E}_{in}(0) = 0$. Then sE_{in} is the transform of $\dot{E}_{in}(t)$, and by the convolution theorem

$$E_{out}(t) = \int_0^t \dot{E}_{in}(\tau)A(t - \tau)\, d\tau$$

EXAMPLE 7.6.3 Consider the electric circuit of Fig. 44. We are assuming that the two circuits are coupled by mutual inductance M and that the switch is

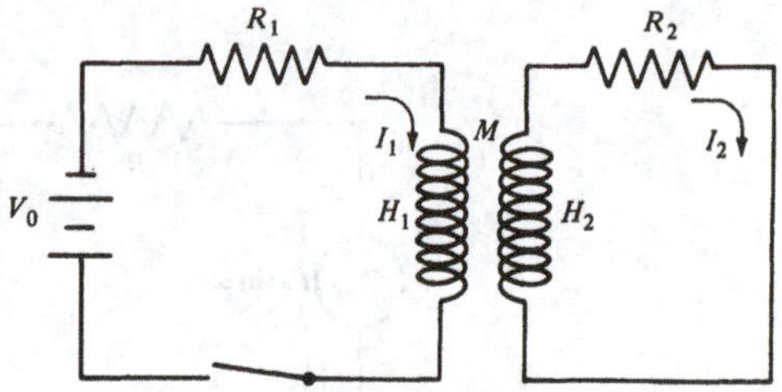

FIGURE 44

closed at $t = 0$ when I_1 and I_2 are zero. The equations for I_1 and I_2 are

$$H_1\dot{I}_1 + R_1 I_1 = M\dot{I}_2 + V_0$$
$$H_2\dot{I}_2 + R_2 I_2 = M\dot{I}_1$$

Taking Laplace transforms and letting I_1 and I_2 stand for the transforms gives

$$H_1 s I_1 + R_1 I_1 = M s I_2 + \frac{V_0}{s}$$
$$H_2 s I_2 + R_2 I_2 = M s I_1$$

Solving for I_1 and I_2, we have

$$I_1 = \frac{(V_0/s)(H_2 s + R_2)}{(H_1 s + R_1)(H_2 s + R_2) - M^2 s^2}$$

$$I_2 = \frac{V_0 M}{(H_1 s + R_1)(H_2 s + R_2) - M^2 s^2}$$

The denominators are of the form $as^2 + bs + c$. Let

$$\frac{1}{as^2 + bs + c} = \mathscr{L}[A(t)]$$

Then

$$I_1(t) = V_0 H_2 A(t) + V_0 R_2 \int_0^t A(\tau)\, d\tau$$
$$I_2(t) = V_0 M A(t)$$

EXERCISES 7.6

1 In a series circuit as in Example 7.6.1, let $R = 600$ ohms, $H = 0.5$ henry, and $C = 0.2$ microfarad. If the frequency of the input voltage is 400 cycles, compute the reactance, the impedance, and the phase angle. If $E_0 = 100$ volts, compute the steady-state current.

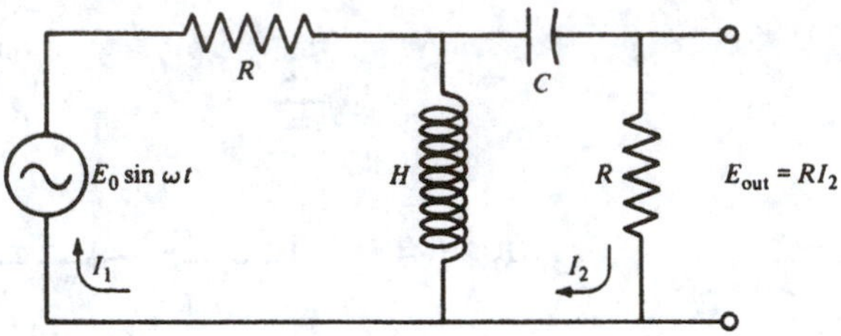

FIGURE 45

2 Repeat Exercise 1 with an input-voltage frequency of 600 cycles.
3 Find the steady-state output voltage, $E_{out} = RI_2$, in the circuit shown in Fig. 45.
4 Find the steady-state output voltage, $E_{out} = RI_2$, in the circuit shown in Fig. 46. M is the mutual inductance.

*7.7 UNIQUENESS OF THE TRANSFORM

From time to time we have commented on the importance of being able to invert the Laplace transform uniquely. This is especially important in solving differential equations where we have first found the transform of the unknown function and then inverted to find the solution. Of course, once we have a possible solution to the differential equation satisfying certain initial conditions, we can check it directly and then use a uniqueness theorem of the differential equation to guarantee that it is the only solution. Nevertheless, it is still of interest in analysis to know that in some sense the transform is unique, and in the proof we shall introduce some mathematical techniques which are of independent interest.

We cannot of course prove that the Laplace transform is unique. Suppose, for example, that $f(t)$ and $g(t)$ are both Riemann-integrable for all t such that $0 \le t < \infty$ and that $f(t) = g(t)$ except at $t = t_0$. Then clearly, if f has a Laplace transform,

$$\mathscr{L}[f(t)] = \int_0^\infty f(t)e^{-st}\,dt = \int_0^\infty g(t)e^{-st}\,dt = \mathscr{L}[g(t)]$$

In other words, the values of $f(t)$ and $g(t)$ at the one point t_0 have no effect on the value of the integral. We could in fact have f and g differing at a countable set of points and they would still have the same Laplace transform. To prove that two functions which have the same Laplace transform are "equal" we must have a more general definition of equal. If they differ, they must differ only at points which do not affect the Riemann integral. Therefore, we shall

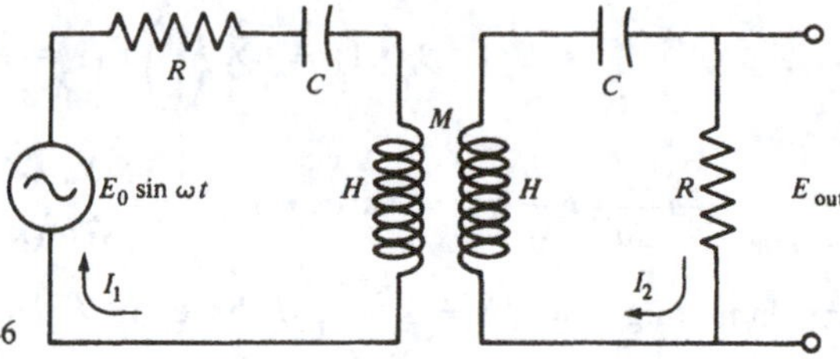

FIGURE 46

say, for the purpose of our uniqueness theorem, that two Riemann-integrable functions f and g are equal if for all positive t

$$\int_0^t f(\tau)\,d\tau = \int_0^t g(\tau)\,d\tau$$

If two functions have the same Laplace transform, then by the linearity of the transform their difference will have a zero transform. We shall therefore prove the following uniqueness theorem.

Theorem 7.7.1 If $f(t)$ is of exponential order, $\phi(s) = \int_0^\infty f(t)e^{-st}\,dt = 0$ for $s > a$, then $F(t) = \int_0^t f(\tau)\,d\tau \equiv 0$.

Before we prove this theorem, we shall need some preliminary results. One of these, known as the *Weierstrass approximation theorem*, states that any continuous function on $I = \{t \mid 0 \le t \le 1\}$ can be uniformly approximated by a polynomial.

Theorem 7.7.2 Let $f(t)$ be continuous on $I = \{t \mid 0 \le t \le 1\}$. Then given any $\varepsilon > 0$, there is a polynomial $p(t)$ such that $|f(t) - p(t)| < \varepsilon$ for all t in I.

PROOF We shall prove this theorem using the Bernstein polynomials

$$B_n(t;f) = \sum_{k=0}^{n} f\left(\frac{k}{n}\right)\binom{n}{k} t^k(1-t)^{n-k}$$

where $\binom{n}{k} = \dfrac{n!}{k!\,(n-k)!}$ are the binomial coefficients. We shall need to evaluate $B_n(t;1)$, $B_n(t;t)$, and $B_n(t;t^2)$. By the binomial theorem, we have

$$(a+b)^n = \sum_{k=0}^{n}\binom{n}{k} a^k b^{n-k}$$

Therefore, putting $a = t$, $b = 1 - t$, we have

$$B_n(t;1) = \sum_{k=0}^{n} \binom{n}{k} t^k(1-t)^{n-k} = 1$$

Next

$$a\frac{d}{da}(a+b)^n = na(a+b)^{n-1} = \sum_{k=0}^{n} \binom{n}{k} ka^k b^{n-k}$$

Again putting $a = t$, $b = 1 - t$, we have

$$B_n(t;t) = \sum_{k=0}^{n} \left(\frac{k}{n}\right)\binom{n}{k} t^k(1-t)^{n-k} = t$$

Differentiating again, we have

$$a^2 n(n-1)(a+b)^{n-2} = \sum_{k=0}^{n} \binom{n}{k} k(k-1)a^k b^{n-k}$$

$$\begin{aligned} B_n(t;t^2) &= \sum_{k=0}^{n} \binom{n}{k} \frac{k^2}{n^2} t^k(1-t)^{n-k} \\ &= \frac{a^2 n(n-1)(a+b)^{n-2} + an(a+b)^{n-1}}{n^2} \\ &= t^2 + \frac{t(1-t)}{n} \end{aligned}$$

when $a = t$, $b = 1 - t$.

We shall show that the polynomial $B_n(t;f)$ can be made arbitrarily close to $f(t)$ (uniformly in t) for n sufficiently large. We can write

$$B_n(t;f) - f(t) = \sum_{k=0}^{n} \left[f\left(\frac{k}{n}\right) - f(t)\right]\binom{n}{k} t^k(1-t)^{n-k}$$

We sum the series in two parts, S_1 the terms where $|t - k/n| < n^{-1/4}$ and S_2 the terms where $|t - k/n| \geq n^{-1/4}$. Since f is uniformly continuous on I, given $\varepsilon > 0$, there is an N_1 such that

$$\left|f(t) - f\left(\frac{k}{n}\right)\right| < \frac{\varepsilon}{2}$$

for $|t - k/n| < n^{-1/4}$ for all $n > N_1$. Therefore,

$$|S_1| < \frac{\varepsilon}{2}\sum_{k=0}^{n} \binom{n}{k} t^k(1-t)^{n-k} = \frac{\varepsilon}{2}$$

Since f is continuous, $|f(t)| \leq M$ on I for some positive M and

$$|S_2| \leq 2M \sum{}' \binom{n}{k} t^k(1-t)^{n-k}$$

where the sum is taken over those k for which $|t - k/n| \geq n^{-1/4}$. We have from the above calculations

$$\sum_{k=0}^{n} \binom{n}{k}\left(t - \frac{k}{n}\right)^2 t^k(1 - t)^{n-k}$$
$$= t^2 B_n(t;1) - 2tB_n(t;t) + B_n(t;t^2) = \frac{1}{n}\, t(1 - t)$$

The function $t(1 - t)/n$ has its maximum of $\frac{1}{4}n$ at $t = \frac{1}{2}$. Therefore, summing over the same k's as in S_2 gives

$$n^{1/2} \sum{}' \binom{n}{k}\left(t - \frac{k}{n}\right)^2 t^k(1 - t)^{n-k} < \frac{1}{4n^{1/2}}$$
$$\sum{}' \binom{n}{k} t^k(1 - t)^{n-k} < \frac{1}{4n^{1/2}}$$

Then

$$|B_n(t;f) - f(t)| \leq |S_1| + |S_2| < \frac{\varepsilon}{2} + \frac{M}{2n^{1/2}} < \varepsilon$$

if we pick $n > \max\,[N_1,N_2]$, where $N_2 > M^2/\varepsilon^2$. This completes the proof of Theorem 7.7.2.

The next result we need asserts that if all the moments of a function continuous on I are zero, then the function is identically zero.

Theorem 7.7.3 Let $f(t)$ be continuous on $I = \{t \mid 0 \leq t \leq 1\}$. If $\int_0^1 t^n f(t)\, dt = 0$ for $n = 0, 1, 2, \ldots$, then $f(t) \equiv 0$.

PROOF The proof is based on the Weierstrass approximation theorem. Given any $\varepsilon > 0$, there is a polynomial $p(t)$ such that $|f(t) - p(t)| < \varepsilon$ for all t in I. However,

$$\int_0^1 f(t)p(t)\, dt = 0$$

Therefore, $\int_0^1 [f(t)]^2\, dt = 0$ because

$$\int_0^1 [f(t)]^2\, dt = \int_0^1 f(t)[f(t) - p(t)]\, dt < M\varepsilon$$

where $M = \max |f(t)|$ and ε is arbitrary. Therefore, $f(t) \equiv 0$.

Proof of Theorem 7.7.1 Since $\phi(s) = \int_0^\infty e^{-st} f(t)\, dt \equiv 0$ for $s > a$, we know that $\phi(a + nb) = 0$ for $n = 1, 2, 3, \ldots$ with $b > 0$.

Now $F(t) = \int_0^t f(\tau)\,d\tau$ is continuous for $t \geq 0$ and of exponential order. Therefore, for some suitable α,

$$\lim_{t\to\infty} e^{-\alpha t}F(t) = 0$$

Integrating by parts, we can show that for $s > \alpha$

$$\phi(s) = s\int_0^\infty e^{-st}F(t)\,dt = s\psi(s)$$

and we know that $\psi(a + nb) = 0$ for $n = 1, 2, 3, \ldots$. We let $u = e^{-bt}$, and then $t = b^{-1}\ln(1/u)$ and

$$\psi(s) = \frac{1}{b}\int_0^1 u^{(a/b)+m-1}F(b^{-1}\ln u^{-1})u^{-m+(s-a)/b}\,du$$

where m is an integer so large that $(a/b) + m - 1 > (\alpha/b)$. Then

$$u^{(a/b)+m-1}F(b^{-1}\ln u^{-1})$$

is continuous for $u \in I$, and all its moments are zero. Therefore, $F(t) \equiv 0$, as we wished to prove.

EXERCISES 7.7

1 Show that if $f(t)$ and $g(t)$ are continuous for $t \geq 0$ and $\int_0^t f(\tau)\,d\tau = \int_0^t g(\tau)\,d\tau$ for all $t \geq 0$, then $f(t) \equiv g(t)$.

2 Compute $B_n(t;t^3)$.

3 Consider $f(t)$ defined on $I = \{t \mid a \leq t \leq b\}$. Show that $f[(b - a)x + a]$ is defined on $I^* = \{x \mid 0 \leq x \leq 1\}$. Use this to define Bernstein polynomials on I.

4 Find $B_n(t;|t|)$ for $-1 \leq x \leq 1$.

5 Let $\phi(s) = \mathscr{L}[f(t)]$, and suppose $\lim_{t\to 0^+} f(t)/t$ exists. Show that $\mathscr{L}[f(t)/t] = \int_s^\infty \phi(\sigma)\,d\sigma$.

6 Use the result of Exercise 5 to find the Laplace transform of:

(a) $\dfrac{\sin \omega t}{t}$ (b) $\dfrac{e^{at} - 1}{t}$ (c) $\dfrac{\sinh \omega t}{t}$

8

POWER-SERIES METHODS

8.1 INTRODUCTION

This chapter takes up the method of power-series solution of linear second order equations with nonconstant coefficients. The second section deals with power-series solutions near an ordinary point of the differential equation. The third section discusses the Frobenius method, which uses a modified power series for a solution near a regular singular point. This is followed by a complete discussion of the Bessel differential equation, partly because it is important in its own right and partly because it illustrates the earlier discussion so well. The next section takes up boundary-value problems for second order equations. Here again we encounter the notion of characteristic values and characteristic functions (characteristic vectors). The starred section in this chapter is devoted to the proofs of convergence which have been glossed over in the earlier discussion of power-series solutions.

8.2 SOLUTION NEAR ORDINARY POINTS

We shall be dealing exclusively with second order linear equations with nonconstant coefficients. The most general equation† of this type is

$$y''(x) + p(x)y'(x) + q(x)y(x) = r(x)$$

If p, q, and r are continuous for x in some interval $I = \{x \mid a \leq x \leq b\}$ and x_0 is in I, then there is a unique solution in I satisfying given conditions $y(x_0) = y_0$, $y'(x_0) = y'_0$. In Chap. 6 we gave the general theory of this initial-value problem. If we can find a fundamental system of independent solutions $y_1(x)$ and $y_2(x)$, then the complementary solution of the homogeneous equation is

$$y_c(x) = c_1 y_1(x) + c_2 y_2(x)$$

and the general solution of the nonhomogeneous equation is

$$y(x) = c_1 y_1(x) + c_2 y_2(x) + y_p(x)$$

where $y_p(x)$ is a particular solution, which can always be determined by the method of variation of parameters.

In the present case, with nonconstant coefficients, we have yet to determine general methods for finding fundamental systems of solutions of the associated homogeneous equation. With this in mind, we concentrate on the homogeneous equation

$$y'' + p(x)y' + q(x)y = 0$$

If $p(x)$ and $q(x)$ have Taylor series representations‡ converging for $|x - x_0| < R$, then we say that x_0 is an *ordinary point* of the differential equation. Before we begin discussing solutions near an ordinary point, we first show that by a simple change of variable, we can always consider an equation where the ordinary point is at $x = 0$. Let $\xi = x - x_0$. Then $x = x_0$ implies that $\xi = 0$, and the differential equation becomes

$$\frac{d^2y}{d\xi^2} + p(\xi + x_0)\frac{dy}{d\xi} + q(\xi + x_0)y = 0$$

Also $p(\xi + x_0)$ and $q(\xi + x_0)$ will have Taylor series representations for $|\xi| < R$. Hence, the new equation has an ordinary point at $\xi = 0$.

In Sec. 8.6 we shall prove the following theorem.

† Since these equations occur most frequently in boundary-value problems, we have switched to x as the independent variable. The prime denotes differentiation with respect to x.

‡ If p and q have different radii of convergence R_1 and R_2, then $R = \min [R_1, R_2]$.

Theorem 8.2.1 If the equation $y'' + p(x)y' + q(x)y = 0$ has an ordinary point at $x = 0$, due to the fact that p and q have power-series representations converging for $|x| < R$, then there are independent solutions $y_1(x)$ and $y_2(x)$ satisfying $y_1(0) = 1$, $y_1'(0) = 0$, $y_2(0) = 0$, $y_2'(0) = 1$ and each having a power-series representation converging for $|x| < R$.

Assuming the validity of this theorem, let us determine the power-series solutions $y_1(x)$ and $y_2(x)$. Let

$$y(x) = c_0 + c_1x + c_2x^2 + \cdots = \sum_{k=0}^{\infty} c_kx^k$$

be a solution where the series is known to converge for $|x| < R$. Differentiating twice, we have

$$y'(x) = \sum_{k=0}^{\infty} kc_kx^{k-1}$$

$$y''(x) = \sum_{k=0}^{\infty} k(k-1)c_kx^{k-2}$$

Assuming that

$$p(x) = p_0 + p_1x + p_2x^2 + \cdots = \sum_{k=0}^{\infty} p_kx^k$$

and

$$q(x) = q_0 + q_1x + q_2x^2 + \cdots = \sum_{k=0}^{\infty} q_kx^k$$

we have

$$\begin{aligned} &2c_2 + 6c_3x + 12c_4x^2 + \cdots \\ &\qquad + (p_0 + p_1x + p_2x^2 + \cdots)(c_1 + 2c_2x + 3c_3x^2 + \cdots) \\ &\qquad + (q_0 + q_1x + q_2x^2 + \cdots)(c_0 + c_1x + c_2x^2 + \cdots) = 0 \end{aligned}$$

Equating the coefficients of various powers of x to zero, we have

$$\begin{aligned} 2c_2 + p_0c_1 + q_0c_0 &= 0 \\ 6c_3 + p_1c_1 + 2p_0c_2 + q_1c_0 + q_0c_1 &= 0 \\ 12c_4 + p_2c_1 + 2p_1c_2 + 3p_0c_3 + q_2c_0 + q_1c_1 + q_0c_2 &= 0 \end{aligned}$$

etc. We can assign arbitrary values to c_0 and c_1 and solve the first equation for c_2. Then we can solve the second equation for c_3, the third equation for c_4, etc. To find y_1 we let $c_0 = 1$, $c_1 = 0$, and we have

$$y_1(x) = 1 - \frac{q_0}{2}x^2 + \frac{p_0q_0 - q_1}{6}x^3 + \cdots$$

To find y_2 we let $c_0 = 0$, $c_1 = 1$, and we have

$$y_2(x) = x - \frac{p_0}{2}x^2 + \frac{p_0^2 - p_1 - q_0}{6}x^3 + \cdots$$

These two solutions are independent because their Wronskian is nonzero at the origin.

EXAMPLE 8.2.1 Find the general solution of the equation $y'' + xy' + 2y = 0$. Here $p = x$, so that $p_0 = 0$, $p_1 = 1$, and $p_k = 0$, $k = 2, 3, \ldots$. Also $q = 2$, so that $q_0 = 2$, $q_k = 0$, $k = 1, 2, 3, \ldots$. Therefore, $x = 0$ is an ordinary point of the equation. Assuming $y(x) = \sum_{k=0}^{\infty} c_k x^k$ and substituting in the equation, we have

$$\sum_{k=2}^{\infty} k(k-1)c_k x^{k-2} + \sum_{k=0}^{\infty} kc_k x^k + \sum_{k=0}^{\infty} 2c_k x^k = 0$$

In the first summation we let $m = k - 2$, while in the second and third summations we let $m = k$. Therefore,

$$\sum_{m=0}^{\infty} (m+1)(m+2)c_{m+2}x^m + \sum_{m=0}^{\infty} mc_m x^m + \sum_{m=0}^{\infty} 2c_m x^m = 0$$

or

$$\sum_{m=0}^{\infty} [(m+1)(m+2)c_{m+2} + (m+2)c_m]x^m = 0$$

Equating the coefficients of x^m to zero, we have the recurrence relation for $m = 0, 1, 2, \ldots$

$$c_{m+2} = \frac{-c_m}{m+1}$$

To determine the first solution $y_1(x)$ described in Theorem 8.2.1, we put $c_0 = 1$, $c_1 = 0$. Then $c_2 = -c_0 = -1$, $c_4 = -c_2/3 = \frac{1}{3}$, $c_6 = -c_4/5 = -\frac{1}{15}$, etc. Therefore,

$$y_1(x) = 1 - x^2 + \frac{x^4}{3} - \frac{x^6}{3 \cdot 5} + \cdots$$
$$= \sum_{k=0}^{\infty} \frac{(-2)^k k!\, x^{2k}}{(2k)!}$$

It is easy to show by the ratio test that this series converges for all x. To determine the second solution $y_2(x)$, we put $c_0 = 0$, $c_1 = 1$. Then $c_3 = -c_1/2 = -\frac{1}{2}$, $c_5 = -c_3/4 = \frac{1}{8}$, $c_7 = -c_5/6 = -\frac{1}{48}$, etc. Therefore,

$$\begin{aligned} y_2(x) &= x - \frac{x^3}{2} + \frac{x^5}{2\cdot 4} - \frac{x^7}{2\cdot 4\cdot 6} + \cdots \\ &= x\sum_{k=0}^{\infty} \frac{(-1)^k x^{2k}}{2^k k!} \\ &= xe^{-x^2/2} \end{aligned}$$

It is again clear that the series converges for all x. The general solution of the equation is

$$y(x) = c_1 y_1(x) + c_2 y_2(x)$$

It is sometimes convenient to use the following device to find a second independent solution of a second order linear equation after a first solution has been obtained by another method. Suppose the equation is $y'' + py' + qy = 0$ and $u(x)$ is a solution. We attempt to find a second solution in the form $v(x) = A(x)u(x)$. We have

$$v'(x) = Au'(x) + uA'(x)$$

$$v''(x) = Au''(x) + 2u'(x)A'(x) + uA''(x)$$

Substituting, we have

$$A(u'' + pu' + qu) + uA'' + (2u' + pu)A' = 0$$

The first term vanishes because u is a solution of the equation. Therefore, we are left with the equation

$$A'' + \left(\frac{2u'}{u} + p\right)A' = 0$$

This equation is first order linear in A'. The integrating factor is

$$\exp\left(2\int \frac{u'}{u}\,dx\right)\exp\left[\int p(x)\,dx\right] = u^2 \exp\left[\int_{x_0}^{x} p(t)\,dt\right]$$

We have

$$\frac{d}{dx}\left\{A'u^2 \exp\left[\int_{x_0}^{x} p(t)\,dt\right]\right\} = 0$$

which implies that

$$A' = \frac{c}{u^2}\exp\left[-\int_{x_0}^{x} p(t)\,dt\right]$$

where c is an arbitrary constant. We let $c = 1$, and then

$$A = \int_{x_0}^{x} u^{-2} \exp\left[-\int_{x_0}^{t} p(s)\,ds\right] dt$$

and

$$v(x) = u(x)\int_{x_0}^{x} [u(t)]^{-2} \exp\left[-\int_{x_0}^{t} p(s)\,ds\right] dt$$

If $u(t)$ is continuous and does not vanish in the interval $I = \{t \mid x_0 \le t \le x\}$, then the integration can always be carried out and we do get a solution. The functions $u(x)$ and $v(x)$ are independent, because if not, one would be a multiple of the other, which is clearly not the case. If $u(x_0) = 1$ and $u'(x_0) = 0$, then $v(x_0) = 0$ and

$$v'(x_0) = [u(x_0)]^{-1} \exp\left[-\int_{x_0}^{x_0} p(s)\,ds\right]$$

leading to the conclusion that $v'(x_0) = 1$.

EXAMPLE 8.2.2 Find the general solution of $y'' - xy' - y = 0$. Since $p(x) = -x$ and $q(x) = -1$ have Taylor expansions about $x = 0$, zero is an ordinary point of the differential equation. Assuming a solution of the form

$$y(x) = \sum_{k=0}^{\infty} c_k x^k$$

we have, upon substituting in the differential equation,

$$\sum_{k=2}^{\infty} k(k-1)c_k x^k - \sum_{k=0}^{\infty} (k+1)c_k x^k = 0$$

In the first summation we let $k = m + 2$, and in the second summation we let $k = m$. Then we have

$$\sum_{m=0}^{\infty} [(m+1)(m+2)c_{m+2} - (m+1)c_m]x^m = 0$$

Equating coefficients to zero, we obtain for $m = 0, 1, 2, \ldots$

$$c_{m+2} = \frac{c_m}{m+2}$$

If we put $c_0 = 1$ and $c_1 = 0$, we obtain the solution

$$y_1(x) = 1 + \frac{x^2}{2} + \frac{x^4}{2 \cdot 4} + \frac{x^6}{2 \cdot 4 \cdot 6} + \cdots$$

$$= \sum_{k=0}^{\infty} \frac{1}{k!}\left(\frac{x^2}{2}\right)^k = e^{x^2/2}$$

We can get another series solution by putting $c_0 = 0, c_1 = 1$, but this time let us look for a solution of the form $y_2(x) = A(x)e^{x^2/2}$. According to the above, the differential equation satisfied by A is

$$A'' + xA' = 0$$

which has an integrating factor $e^{x^2/2}$. Then

$$\frac{d}{dx}(A'e^{x^2/2}) = 0$$

$$A' = e^{-x^2/2}$$

$$y_2(x) = e^{x^2/2} \int_0^x e^{-t^2/2}\, dt$$

Notice that $y_2(0) = 0$, $y_2'(0) = 1$, so y_1 and y_2 are clearly independent.

EXERCISES 8.2

1 Show that each of the following equations has an ordinary point at $x = 0$. In each case determine the minimum radius† of convergence of the power-series solution:

(*a*) $y'' + x^2y' + xy = 0$

(*b*) $y'' + (1 - x)y' + x^2y = 0$

(*c*) $(1 - x)y'' + y' + xy = 0$

(*d*) $(1 - x^2)y'' + xy' + \dfrac{1}{2 + x}\, y = 0$

(*e*) $y'' - 2xy' + \lambda y = 0$ (Hermite equation), λ constant

(*f*) $y'' = xy$ (Airy equation)

(*g*) $(1 - x^2)y'' - 2xy' + \lambda y = 0$ (Legendre equation), λ constant

(*h*) $(1 - x^2)y'' - xy' + \lambda y = 0$ (Tchebysheff equation), λ constant

2 Determine two independent power-series solutions of the Airy differential equation, $y'' - xy = 0$. Where do they converge?

3 Determine two independent power-series solutions of the Hermite equation, $y'' - 2xy' + \lambda y = 0$. Where do they converge? Show that if $\lambda = 2n$, $n = 0, 1, 2, 3, \ldots$, then there is a polynomial solution. These polynomials are proportional to the Hermite polynomials.

4 Determine two independent power-series solutions of the Legendre equation, $(1 - x^2)y'' - 2xy' + \lambda y = 0$. Where do they converge? Show that if $\lambda = n(n + 1)$, $n = 0, 1, 2, \ldots$, then there is a polynomial solution. These polynomials are proportional to the Legendre polynomials.

5 Determine two independent power-series solutions of the Tchebysheff equation, $(1 - x^2)y'' - xy' + \lambda y$. Where do they converge? Show that if $\lambda = n^2$,

† Note that Theorem 8.2.1 does not preclude the possibility that the power-series solutions could converge for $|x| \geq R$.

$n = 0, 1, 2, \ldots$, then there is a polynomial solution. These polynomials are proportional to the Tchebysheff polynomials.

6 Suppose $y'' + p(x)y' + q(x)y = 0$ has an ordinary point at $x = 0$. If $y(0) = a$, $y'(0) = b$, show directly from the differential equation that $y''(0) = -p(0)b - q(0)a$. Similarly determine $y'''(0)$, $y^{(4)}(0)$, etc. Write down the first few terms of the power-series expansion for $y(x)$.

7 Find the general solution of $y'' + xy' + 2y = x$ in the form of a power series. *Hint:* Assume a power series for $y_p(x)$.

8 Suppose $y'' + p(x)y' + q(x)y = 0$ has an ordinary point at $x = 0$. Is it possible for the differential equation to have solutions x and x^2? Explain.

8.3 SOLUTION NEAR REGULAR SINGULAR POINTS

We again consider the equation

$$y'' + p(x)y' + q(x)y = r(x)$$

If any of p, q, or r fails to be continuous at $x = x_0$, then we say that x_0 is a *singular point* of the differential equation. We cannot, in general, solve the equation subject to given initial conditions $y(x_0) = y_0$, $y'(x_0) = y_0'$ in some interval $I = \{x \mid a \leq x \leq b\}$ containing the singular point x_0. There are, however, certain kinds of singular points where we can obtain special solutions. These come up frequently in the applications, and so they deserve special attention. As in the previous section, we note that we can, without loss of generality, assume that the singular point is at $x = 0$. Suppose we multiply the homogeneous equation through by x^2. Then

$$x^2y'' + x^2p(x)y' + x^2q(x)y = 0$$

We say that the equation has a *regular singular point* at $x = 0$ if $xp(x)$ and $x^2q(x)$ have Taylor series expansions† converging for $|x| < R$.

Let

$$xp(x) = a_0 + a_1x + a_2x^2 + \cdots = \sum_{k=0}^{\infty} a_kx^k$$

$$x^2q(x) = b_0 + b_1x + b_2x^2 + \cdots = \sum_{k=0}^{\infty} b_kx^k$$

For very small x, $xp(x) \approx a_0$ and $x^2q(x) \approx b_0$, and the differential equation is approximately

$$x^2y'' + a_0xy' + b_0y = 0$$

† In other words, we are assuming that

$$p(x) = \frac{1}{x}\text{ (power series)} \qquad \text{and} \qquad q(x) = \frac{1}{x^2}\text{ (power series)}$$

This is an Euler differential equation (see Sec. 6.2). If we assume a solution of the form $y = x^m$, we have

$$[m(m - 1) + a_0 m + b_0]x^m = 0$$

Setting $m(m - 1) + a_0 m + b_0 = 0$, we obtain at least one and possibly two distinct solutions† x^{m_1} and x^{m_2}. If m_1 and m_2 are distinct real roots, then we have two solutions. If $m_1 = m_2$, we have one solution. If $m_1 = \alpha + i\beta$, then $m_2 = \alpha - i\beta$ and we have two solutions, $x^\alpha \cos(\beta \ln x)$ and $x^\alpha \sin(\beta \ln x)$.

In the general case, we shall assume that our solutions are of the form

$$\begin{aligned} y(x) &= x^m(c_0 + c_1 x + c_2 x^2 + \cdots) \\ &= x^m \sum_{k=0} c_k x^k \end{aligned}$$

Differentiating, we have

$$y' = \sum_{k=0}^{\infty} (k + m)c_k x^{k+m-1}$$

$$y'' = \sum_{k=0}^{\infty} (k + m)(k + m - 1)c_k x^{k+m-2}$$

Substituting in the equation, we obtain

$$\sum_{k=0}^{\infty} (k + m)(k + m - 1)c_k x^{k+m} + \sum_{k=0}^{\infty} a_k x^k \sum_{k=0}^{\infty} (k + m)c_k x^{k+m} + \sum_{k=0}^{\infty} b_k x^k \sum_{k=0}^{\infty} c_k x^{k+m} = 0$$

There is a term in this equation containing x^m with coefficient

$$c_0[m(m - 1) + a_0 m + b_0]$$

If we set this equal to zero, we obtain the *indicial equation*‡

$$m(m - 1) + a_0 m + b_0 = 0$$

This equation has at most two distinct roots m_1 and m_2, which may be real or complex. We shall have to consider all possible cases, but before we do, let us consider the coefficients of other powers of x. For the power x^{m+1}, we have

$$c_1[m(m + 1) + a_0(m + 1) + b_0] + c_0(ma_1 + b_1) = 0$$

† We are assuming that $x > 0$. However, if x is replaced by $-x$ in the differential equation, the equation does not change. Therefore, we can write all our solutions with x replaced by $|x|$.

‡ If we set $c_0 = 0$, we shall obtain only the trivial solution $y \equiv 0$.

If we assign any value to c_0, we can solve this equation for c_1 provided $m(m+1) + a_0(m+1) + b_0 \neq 0$, assuming that we have determined m from the indicial equation. For the power x^{m+2}, we have

$$c_2[(m+1)(m+2) + a_0(m+2) + b_0] + c_0(ma_2 + b_2) + c_1[(m+1)a_1 + b_1] = 0$$

Having determined c_1 from the previous equation, we can solve this equation for c_2 provided $(m+1)(m+2) + a_0(m+2) + b_0 \neq 0$.

For the power x^{m+k}, $k = 1, 2, 3, \ldots$, we have

$$c_k[(m+k-1)(m+k) + a_0(m+k) + b_0] + \sum_{j=0}^{k-1} c_j[(m+j)a_{k-j} + b_{k-j}] = 0$$

which we can solve for c_k provided that

$$(m+k-1)(m+k) + a_0(m+k) + b_0 \neq 0$$

Let $I(m) = m(m-1) + a_0 m + b_0$, which is the left-hand side of the indicial equation. Then the critical factor is $I(m+k)$, and at each stage in solving for c_k we require that $I(m+k) \neq 0$, $k = 1, 2, 3, \ldots$. We are now prepared to consider the various cases in what we shall call the *Frobenius method.*

CASE 1 The roots of the indicial equation are real, distinct, and do not differ by an integer. Let m_1 and m_2 be the roots. Then $I(m_1) = 0$, but $I(m_1 + k) \neq 0$ because the only other root of $I(m)$ is m_2, which does not differ from m_1 by an integer. If we assign an arbitrary value to c_0 (other than zero), we obtain a solution of the form

$$y_1(x) = x^{m_1} \sum_{k=0}^{\infty} c_k x^k$$

Actually, it is clear from the above that every coefficient c_k is a multiple of c_0. Therefore,

$$y_1(x) = c_0 w_1(x)$$

where w_1 is a unique nontrivial solution of the form x^{m_1} times a power series. We shall show in Sec. 8.6 that this power series converges at least for $|x| < R$. Also $I(m_2) = 0$ but $I(m_2 + k) \neq 0$, so we can repeat the determination of the coefficients using m_2 and obtain a second solution†

$$y_2(x) = x^{m_2} \sum_{k=0}^{\infty} \gamma_k x^k = \gamma_0 w_2(x)$$

† We are calling the coefficients γ_k, $k = 0, 1, 2, \ldots$, to avoid double-subscript notation.

where w_2 is a unique solution. The solutions w_1 and w_2 are, in general, not solutions at $x = 0$ because of the factors x^{m_1} and x^{m_2}. However, for $0 < x < R$ they are solutions and are clearly independent since $m_1 \neq m_2$. Therefore, the general solution for $0 < x < R$ is

$$y(x) = c_0 w_1(x) + \gamma_0 w_2(x)$$

where c_0 and γ_0 are arbitrary constants.

CASE 2 The roots of the indicial equation are real and distinct but differ by an integer. Let m_1 and m_2 be the roots and $m_1 = m_2 + p$, where p is a positive integer. Then $I(m_1) = 0$, but $I(m_1 + k) = I(m_2 + p + k) \neq 0$, $k = 1, 2, 3, \ldots$. Therefore, as in Case 1, we can determine a solution $y_1(x) = x^{m_1} \sum_{k=0}^{\infty} c_k x^k = c_0 w_1(x)$ containing one arbitrary constant c_0. If we attempt to find a second solution using the root m_2, we may have trouble because $I(m_2 + p) = I(m_1) = 0$. The equation for γ_p is

$$\gamma_p I(m_2 + p) + \sum_{j=0}^{p-1} \gamma_j[(m_2 + j)a_{p-j} + b_{p-j}] = 0$$

This equation cannot be solved for γ_p. In fact there is no solution unless

$$\sum_{j=0}^{p-1} \gamma_j[(m_2 + j)a_{p-j} + b_{p-j}] = 0$$

In the latter case, γ_p is arbitrary, and we may continue to determine the coefficients in the series since $I(m_2 + k) \neq 0$ for $k = p + 1, p + 2, p + 3, \ldots$. In this case, the second solution obtained using the root m_2 contains arbitrary constants γ_0 and γ_p. This does not mean that the general solution contains three arbitrary constants. If we write the linear combination

$$y(x) = c_0 w_1(x) + \gamma_0 w_2(x) + \gamma_p w_3(x)$$

then w_1, w_2, w_3 are not independent. However, w_1 and w_2 must be independent because w_1 starts off with x^{m_1} and w_2 starts off with x^{m_2}, where $m_1 \neq m_2$. In summary, we always obtain one solution using the larger of the two roots. We may or may not get a second solution. If we do not, we can always look for a solution in the form $A(x)w_1(x)$, as in the previous section.

CASE 3 The roots of the indicial equation are real, but $m_1 = m_2$. Clearly the Frobenius method will give but one solution.

CASE 4 The roots of the indicial equation are complex. Let $m_1 = \alpha + i\beta$, $\beta \neq 0$. Then $m_2 = \overline{m}_1 = \alpha - i\beta$ and $m_1 - m_2 = 2i\beta$, which is not an integer. In this case, both series using m_1 and m_2 can be determined. However, the series will have complex coefficients. Let $w_1(x) = u_1(x) + iv_1(x)$ be the first solution, where u_1 and v_1 are real-valued. Then

$$w_1'' + p(x)w_1' + q(x)w_1 = 0$$

and since p and q are real-valued, we have, upon taking real and imaginary parts,

$$u_1'' + p(x)u_1' + q(x)u_1 = 0$$
$$v_1'' + p(x)v_1' + q(x)v_1 = 0$$

Hence, u_1 and v_1 are real-valued solutions. Similarly, if $w_2(x) = u_2(x) + iv_2(x)$ is the second series solution, then u_2 and v_2 are real-valued solutions. Clearly, u_1, v_1, u_2, and v_2 are not independent. In fact, no more than two of them can form an independent set of solutions. The situation here is analogous to the case of conjugate complex exponential solutions in the case of second order equations with constant coefficients.

EXAMPLE 8.3.1 Find two independent solutions of the equation $xy'' + (x - 1)y' - y = 0$. The origin is a singular point because of the coefficient x of y''. If we multiply by x, we have

$$x^2y'' + x(-1 + x)y' - xy = 0$$

where $xp(x) = -1 + x$ and $x^2q(x) = -x$. Therefore, $x = 0$ is a regular singular point. Now $p_0 = -1$ and $q_0 = 0$, so that the indicial equation is

$$m(m - 1) - m = m(m - 2) = 0$$

The roots are $m_1 = 2$ and $m_2 = 0$, and so we have Case 2. We know that we shall get a solution using m_1. We let

$$y_1(x) = x^2 \sum_{k=0}^{\infty} c_k x^k = \sum_{k=0}^{\infty} c_k x^{k+2}$$

Substituting in the differential equation, we have

$$\sum_{k=0}^{\infty} (k + 2)(k + 1)c_k x^{k+2} - \sum_{k=0}^{\infty} (k + 2)c_k x^{k+2} + \sum_{k=0}^{\infty} (k + 2)c_k x^{k+3} - \sum_{k=0}^{\infty} c_k x^{k+3} = 0$$

In the first and second summations we put $k = m$, while in the third and fourth summations we put $k = m - 1$. Then

$$\sum_{m=1}^{\infty} [m(m + 2)c_m + mc_{m-1}]x^{m+2} = 0$$

The first coefficient c_0 is arbitrary, while the remaining coefficients are determined by the recurrence relation

$$c_m = \frac{-c_{m-1}}{m+2}$$

$m = 1, 2, 3, \ldots$. We have recursively $c_1 = -c_0/3$, $c_2 = -c_1/4 = c_0/3 \cdot 4$, $c_3 = -c_2/5 = -c_0/3 \cdot 4 \cdot 5$, etc. If we let $c_0 = \alpha/2$, then

$$y_1(x) = \alpha\left(\frac{x^2}{2!} - \frac{x^3}{3!} + \frac{x^4}{4!} - \cdots\right) = \alpha(e^{-x} - 1 + x)$$

One can easily check that $e^{-x} - 1 + x$ is a solution of the equation.

Since the roots of the indicial equation differ by an integer, we cannot be sure that a series corresponding to $m_2 = 0$ can be determined. Nevertheless we try a series of the form

$$y_2(x) = \sum_{k=0}^{\infty} \gamma_k x^k$$

Substituting, we have

$$\sum_{k=1}^{\infty} k(k-1)\gamma_k x^{k-1} + \sum_{k=1}^{\infty} k\gamma_k x^k - \sum_{k=1}^{\infty} k\gamma_k x^{k-1} - \sum_{k=0}^{\infty} \gamma_k x^k = 0$$

The coefficient of x^0 is $-\gamma_1 - \gamma_0$. Therefore, we put $\gamma_1 = -\gamma_0$. The coefficient of x^1 is $2\gamma_2 + \gamma_1 - 2\gamma_2 - \gamma_1 = 0$. But this is satisfied for arbitrary γ_2. To determine the general recurrence relation we put $m = k - 1$ in the first and third summations and $k = m$ in the second and fourth summations. This leads to

$$\sum_{m=2}^{\infty} [(m-1)(m+1)\gamma_{m+1} + (m-1)\gamma_m]x^m = 0$$

or

$$\gamma_{m+1} = -\frac{\gamma_m}{m+1}$$

$m = 2, 3, 4, \ldots$. Letting $\gamma_2 = \beta/2$, we have

$$y_2(x) = \gamma_0(1 - x) + \beta\left(\frac{x^2}{2!} - \frac{x^3}{3!} + \frac{x^4}{4!} - \cdots\right)$$
$$= \gamma_0(1 - x) + \beta(e^{-x} - 1 + x)$$

Since γ_0 and β are arbitrary, we know that $1 - x$ and $e^{-x} - 1 + x$ are separately solutions, and since they are independent, $y_2(x)$ gives us the general solution.

EXERCISES 8.3

1 Show that the Bessel equation $x^2y'' + xy' + (x^2 - \nu^2)y = 0$ has a regular singular point at $x = 0$. ν^2 is a constant. Find the roots of the indicial equation.

2 Show that the Legendre equation $(1 - x^2)y'' - 2xy' + \lambda y = 0$ has regular singular points at $x = 1$ and $x = -1$. λ is a constant. Find the roots of the indicial equation at each point.

3 Show that the Laguerre differential equation $xy'' + (1 - x)y' + \lambda y = 0$ (λ constant) has a regular singular point at $x = 0$. Find the roots of the indicial equation. Find one series solution of the equation.

4 Show that $xy'' + 2y' + xy = 0$ has a regular singular point at $x = 0$. Find the roots of the indicial equation and at least one solution of the equation.

5 Show that $xy'' + y' - xy = 0$ has a regular singular point at $x = 0$. Find the roots of the indicial equation and at least one solution of the equation.

6 Show that $2x(1 - x)y'' + (1 - 4x)y' + y = 0$ has a regular singular point at $x = 0$. Find the roots of the indicial equation and two independent solutions of the equation.

7 Find a power-series solution of the zeroth order Bessel equation $xy'' + y' + xy = 0$. Determine a second solution by means of the substitution $A(x)y_1(x)$, where y_1 is the solution already found.

8 Show that the method of Frobenius fails to give a nontrivial solution of $x^3y'' + x^2y' + y = 0$ valid near $x = 0$. Note that the singularity at $x = 0$ is not a regular singularity.

8.4 BESSEL FUNCTIONS

In this section, we shall study solutions of Bessel's differential equation

$$x^2y'' + xy' + (x^2 - \nu^2)y = 0$$

which has a regular singular point at $x = 0$. This is a good example to study because the equation occurs frequently in the applications, its solutions (the Bessel functions) are very important, and depending on the constant ν, all the cases of the previous section are illustrated.

According to the notation of the last section, $p(x) = 1/x$ and $q(x) = (x^2 - \nu^2)/x^2$. Therefore, the indicial equation is

$$m^2 - \nu^2 = 0$$

with roots $m_1 = \nu$ and $m_2 = -\nu$. Depending on ν, we have four cases.

CASE 1 If ν is real and 2ν is not an integer, then the roots are real and distinct and do not differ by an integer. In this case, we can always find two independent solutions of the form

$$J_\nu(x) = x^\nu \sum_{k=0}^{\infty} c_k x^k$$

$$J_{-\nu}(x) = x^{-\nu} \sum_{k=0}^{\infty} \gamma_k x^k$$

For certain choices of c_0 and γ_0 these solutions are called *Bessel functions of the first kind.* The general solution, in this case, is

$$y(x) = AJ_\nu(x) + BJ_{-\nu}(x)$$

where A and B are arbitrary constants.

CASE 2 If ν is real and 2ν is a positive integer, then the roots of the indicial equation are real and distinct but differ by an integer. Assuming that ν is positive, the Frobenius method always gives a solution of the form

$$J_\nu(x) = x^\nu \sum_{k=0}^{\infty} c_k x^k$$

Again, for a certain choice of c_0, this is called a *Bessel function of the first kind.* In the case where 2ν is an odd integer, the method also yields a second independent solution

$$J_{-\nu}(x) = x^{-\nu} \sum_{k=0}^{\infty} \gamma_k x^k$$

and the general solution is as in Case 1. If 2ν is an even integer, the method fails to give a second solution, but by the substitution considered in Sec. 8.2, we can obtain an independent solution $Y_\nu(x)$, which is called a *Bessel function of the second kind.*

CASE 3 If $\nu = 0$, the roots of the indicial equation are equal. In this case, we get a power-series solution

$$J_0(x) = \sum_{k=0}^{\infty} c_k x^k$$

To obtain a second solution we must use the substitution considered in Sec. 8.2.

CASE 4 If ν^2 is negative, the roots are $m_1 = i\nu$ and $m_2 = -i\nu$. In this case, $m_1 \neq m_2$, they do not differ by an integer, and there are two independent solutions of the form $\text{Re}\left(x^{i\nu} \sum_{k=0}^{\infty} c_k x^k\right)$ and $\text{Im}\left(x^{i\nu} \sum_{k=0}^{\infty} c_k x^k\right)$. We shall not consider this case further here.

In Cases 1 to 3, we always get a solution of the form ($\nu \geq 0$)

$$y(x) = \sum_{k=0}^{\infty} c_k x^{\nu+k}$$

Differentiating, we have

$$y'(x) = \sum_{k=0}^{\infty} (\nu + k)c_k x^{\nu+k-1}$$

$$y''(x) = \sum_{k=0}^{\infty} (\nu + k)(\nu + k - 1)c_k x^{\nu+k-2}$$

We substitute in the differential equation

$$\sum_{k=0}^{\infty} [(\nu + k)(\nu + k - 1) + (\nu + k) - \nu^2]c_k x^{\nu+k} + \sum_{k=0}^{\infty} c_k x^{\nu+k+2} = 0$$

or

$$\sum_{k=1}^{\infty} k(k + 2\nu)c_k x^{\nu+k} + \sum_{k=0}^{\infty} c_k x^{\nu+k+2} = 0$$

Consider the term $k = 1$ in the first sum, which is $c_1(2\nu + 1)x^{\nu+1}$. Since there is no term in $x^{\nu+1}$ in the second sum and $2\nu + 1 > 0$, $c_1 = 0$. Therefore, in the first sum we can let $k = m + 2$, and we have

$$\sum_{m=0}^{\infty} [(m + 2)(m + 2 + 2\nu)c_{m+2} + c_m]x^{\nu+m+2} = 0$$

and the recurrence relation is

$$c_{m+2} = -\frac{c_m}{(m + 2)(m + 2 + 2\nu)}$$

$m = 0, 1, 2, \ldots$. However, since $c_1 = 0$, we have $c_m = 0$ for all odd m. So we let $m = 2n$, $n = 0, 1, 2, \ldots$, and then

$$c_{2n+2} = -\frac{c_{2n}}{2^2(n + 1)(n + \nu + 1)}$$

To begin with c_0 is arbitrary. Then

$$c_2 = \frac{-c_0}{2^2(\nu + 1)}$$

$$c_4 = \frac{-c_2}{2^2 2(\nu + 2)} = \frac{c_0}{2^4 2!\,(\nu + 2)(\nu + 1)}$$

$$c_6 = \frac{-c_4}{2^2 3(\nu + 3)} = \frac{-c_0}{2^6 3!\,(\nu + 3)(\nu + 2)(\nu + 1)}$$

etc. In general,

$$c_{2n} = \frac{(-1)^n c_0}{2^{2n} n!\,(\nu + n)(\nu + n - 1)\cdots(\nu + 2)(\nu + 1)}$$

There is a function, called the *gamma function*, defined by the improper integral

$$\Gamma(x) = \int_0^\infty t^{x-1} e^{-t}\,dt$$

the integral converging for all $x > 0$. Integrating by parts, we find that

$$\begin{aligned} \Gamma(x+1) &= \int_0^\infty t^x e^{-t}\,dt \\ &= -t^x e^{-t}\Big|_0^\infty + x \int_0^\infty t^{x-1} e^{-t}\,dt \\ &= x\Gamma(x) \end{aligned}$$

and

$$\begin{aligned} \Gamma(\nu + n + 1) &= (\nu + n)\Gamma(\nu + n) \\ &= (\nu + n)(\nu + n - 1)\Gamma(\nu + n - 1) \\ &= (\nu + n)(\nu + n - 1)\cdots(\nu + 1)\Gamma(\nu + 1) \end{aligned}$$

If we let $c_0 = [2^\nu \Gamma(\nu + 1)]^{-1}$, then

$$y(x) = \sum_{n=0}^\infty \frac{(-1)^n (x/2)^{2n+\nu}}{n!\,\Gamma(\nu + n + 1)} = J_\nu(x)$$

is a particular solution of Bessel's equation for $\nu \geq 0$. $J_\nu(x)$ is the Bessel function of the first kind of order ν. The series converges for all x, as one can easily see by comparison with

$$\sum_{n=0}^\infty \frac{(x^2/4)^n}{n!} = e^{x^2/4}$$

Some particular examples of Bessel functions are

$$J_0(x) = \sum_{n=0}^\infty \frac{(-1)^n (x/2)^{2n}}{(n!)^2}$$

$$J_1(x) = \sum_{n=0}^\infty \frac{(-1)^n (x/2)^{2n+1}}{n!\,(n+1)!}$$

$$\begin{aligned} J_{1/2}(x) &= \sum_{n=0}^\infty \frac{(-1)^n (x/2)^{2n+1/2}}{n!\,\Gamma(n + \frac{3}{2})} \\ &= \sqrt{\frac{2}{x}} \sum_{n=0}^\infty \frac{(-1)^n x^{2n+1}}{2^n n!\,(2n+1)(2n-1)\cdots 3 \cdot \Gamma(\frac{1}{2})} \end{aligned}$$

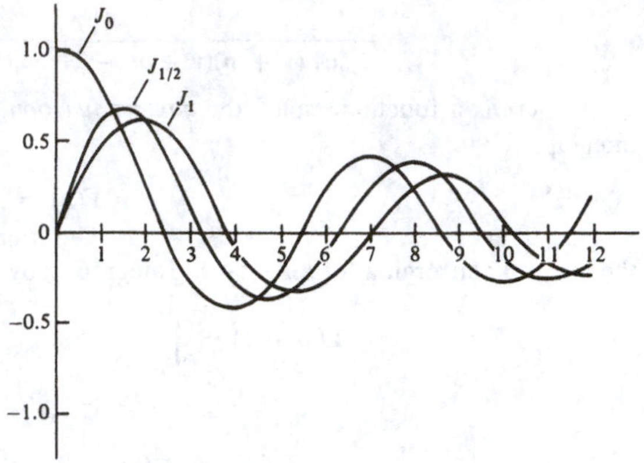

FIGURE 47

Since $\Gamma(\frac{1}{2}) = \sqrt{\pi}$,

$$J_{1/2}(x) = \sqrt{\frac{2}{\pi x}} \sum_{n=0}^{\infty} \frac{(-1)^n x^{2n+1}}{(2n+1)!}$$

$$= \sqrt{\frac{2}{\pi x}} \sin x$$

The graphs of $J_0(x)$, $J_{1/2}(x)$, and $J_1(x)$ are shown in Fig. 47.

We have shown that $J_\nu(x)$ is a solution of the Bessel equation in Cases 1 to 3. Now we have to show how to get a second solution in each case. In Case 1, the root $m_2 = -\nu$ should give us a solution. In fact the solution obtained above (up to the point of introducing the gamma function) is valid. Hence,

$$y(x) = c_0 \sum_{n=0}^{\infty} \frac{(-1)^n x^{2n-\nu}}{2^{2n} n! (-\nu + n)(-\nu + n + 1) \cdots (-\nu + 2)(-\nu + 1)}$$

is a solution for arbitrary c_0. The integral we used to define the gamma function does not converge for $x \leq 0$. We know that for $x > 0$

$$\Gamma(x + n) = (x + n - 1)(x + n - 2) \cdots x\Gamma(x)$$

For $x < 0$ there is some positive integer n such that $\Gamma(x + n)$ is defined by the integral. Now if x is not a negative integer, the quantity

$$\frac{\Gamma(x + n)}{(x + n - 1)(x + n - 2) \cdots (x + 1)x}$$

is defined, so we define

$$\Gamma(x) = \frac{\Gamma(x+n)}{(x+n-1)(x+n-2)\cdots(x+1)x}$$

for $x < 0$ and not a negative integer.

With this definition we can let $c_0 = [2^{-\nu}\Gamma(-\nu+1)]^{-1}$, and then

$$y(x) = \sum_{n=0}^{\infty} \frac{(-1)^n (x/2)^{2n-\nu}}{n!\,\Gamma(-\nu+n+1)}$$

is defined for all $\nu \neq 1, 2, 3, \ldots$. We define this series as $J_{-\nu}(x)$, the Bessel function of the first kind of order $-\nu$.

In Case 2, where 2ν is a positive integer, we have two different situations. The easier case is when 2ν is odd; that is, when ν is half an odd integer, $\frac{1}{2}, \frac{3}{2}, \frac{5}{2}, \ldots$. In this case, $J_{-\nu}(x)$ is defined and can be verified as a solution by direct substitution. Another way to see this is to proceed with the substitution of

$$y(x) = x^{-\nu} \sum_{k=0}^{\infty} \gamma_k x^k$$

As above, we arrive at the necessary condition $(1 - 2\nu)\gamma_1 = 0$. If $\nu = \frac{1}{2}$, then γ_1 is arbitrary and we obtain a series with odd powers of x and the recurrence relation is

$$\gamma_{2n+1} = -\frac{-\gamma_{2n-1}}{2n(2n+1)}$$

$n = 1, 2, 3, \ldots$. Then

$$\gamma_3 = \frac{-\gamma_1}{2 \cdot 3} = \frac{-\gamma_1}{2^2(\frac{1}{2}+1)}$$

$$\gamma_5 = \frac{-\gamma_3}{4 \cdot 5} = \frac{\gamma_1}{2^4 2!\,(\frac{1}{2}+2)(\frac{1}{2}+1)}$$

and, in general,

$$\gamma_{2n+1} = \frac{(-1)^n \gamma_1}{2^{2n} n!\,(\frac{1}{2}+1)(\frac{1}{2}+2)\cdots(\frac{1}{2}+n)}$$

If we let $\gamma_1 = [2^{1/2}\Gamma(\frac{1}{2}+1)]^{-1}$, then the series is

$$J_{1/2}(x) = \sum_{n=0}^{\infty} \frac{(-1)^n (x/2)^{2n+1/2}}{n!\,\Gamma(\frac{1}{2}+n+1)}$$

Now the other part of the solution is based on the recurrence relation

$$\gamma_{2n+2} = \frac{-\gamma_{2n}}{2^2(n+1)(n+\frac{1}{2})}$$

$n = 0, 1, 2, 3, \ldots$. Then

$$\gamma_2 = \frac{-\gamma_0}{2^2(-\frac{1}{2} + 1)}$$

$$\gamma_4 = \frac{-\gamma_2}{2^2 2(-\frac{1}{2} + 2)} = \frac{\gamma_0}{2^4 2!\,(-\frac{1}{2} + 1)(-\frac{1}{2} + 2)}$$

and, in general,

$$\gamma_{2n} = \frac{(-1)^n \gamma_0}{2^{2n} n!\,(-\frac{1}{2} + n)(-\frac{1}{2} + n - 1)\cdots(-\frac{1}{2} + 2)(-\frac{1}{2} + 1)}$$

This series gives us the solution, when $\gamma_0 = [2^{-1/2}\Gamma(-\frac{1}{2} + 1)]^{-1}$,

$$J_{-1/2}(x) = \sum_{n=0}^{\infty} \frac{(-1)^n (x/2)^{2n-1/2}}{n!\,\Gamma(-\frac{1}{2} + n + 1)}$$

Therefore, for the case $\nu = \frac{1}{2}$, the general solution is

$$y(x) = AJ_{1/2}(x) + BJ_{-1/2}(x)$$

where A and B are arbitrary.

If $2\nu = p$ is odd and larger than 1, we have a similar situation. To satisfy the necessary condition $\gamma_1(-2\nu + 1) = 0$, we must put $\gamma_1 = 0$. Then $\gamma_3 = \gamma_5 = \gamma_{p-2} = 0$. But the relation $p(p - 2\nu)\gamma_p = 0$ must be satisfied, so γ_p is arbitrary. The resulting series of odd powers starts out with a term in $x^{-\nu}x^p = x^\nu$. For the correct choice of γ_p this series will give $J_\nu(x)$, and the other part of the series with even powers will give $J_{-\nu}(x)$. So in the case when 2ν is odd the Frobenius method gives two independent solutions, $J_\nu(x)$ and $J_{-\nu}(x)$.

The situation is different when 2ν is even and not zero. The condition $(-2\nu + 1)\gamma_1 = 0$ requires that $\gamma_1 = 0$ and hence all the odd powers are zero. The recurrence relation

$$(2n + 2)(2n + 2 - 2\nu)\gamma_{2n+2} + \gamma_{2n} = 0$$

cannot be satisfied when $2n + 2 = 2\nu$ since $\gamma_{2n} \neq 0$. Therefore, the Frobenius method fails to give a second solution in the case where 2ν is a positive even integer. In this case, to get a second solution we must make a substitution $y(x) = A(x)J_\nu(x)$. This is also required in the case $\nu = 0$, when the roots of the indicial equation are equal. We start with this case since it is the easiest to consider.

We can write the Bessel equation of order zero in the following form:

$$xy'' + y' + xy = 0$$

We assume a solution of the form $y(x) = A(x)J_0(x)$. Then, upon substituting in the differential equations, we have

$$A(xJ_0'' + J_0' + xJ_0) + xA''J_0 + 2xA'J_0' + A'J_0 = 0$$

The first term is zero since J_0 is a solution of the differential equation. Therefore, solving for A''/A', we obtain†

$$\frac{A''}{A'} = -\frac{1}{x} - \frac{2J_0'}{J_0}$$

$$\ln A' = -\ln x - 2\ln J_0 = -\ln xJ_0^2$$

$$A' = \frac{1}{xJ_0^2}$$

The power series for J_0 starts with the constant 1, and therefore so does the power series for J_0^2. It can be shown that J_0^{-2} has a power-series representation starting with 1. This shows that

$$A' = \frac{1}{x} + \text{power series}$$

$$A = \ln x + \text{power series}$$

Since the product of two power series is a power series, our second solution is of the form

$$y(x) = J_0(x)\ln x + \sum_{k=0}^{\infty} c_k x^k$$

Substituting in the differential equation, we have

$$(xJ_0'' + J_0' + xJ_0)\ln x + 2J_0' + \sum_{k=1}^{\infty} k^2 c_k x^{k-1} + \sum_{k=0}^{\infty} c_k x^{k+1} = 0$$

The first term is zero since J_0 is a solution. Now

$$J_0'(x) = \sum_{k=1}^{\infty} \frac{(-1)^k (x/2)^{2k-1}}{(k-1)!\,k!}$$

so the only constant term in the equation is c_1. Therefore, $c_1 = 0$, and we can write

$$\sum_{k=1}^{\infty} \frac{(-1)^k x^{2k-1}}{2^{2k-2}(k-1)!\,k!} + \sum_{k=0}^{\infty} [(k+2)^2 c_{k+2} + c_k] x^{k+1} = 0$$

Since there are only odd powers in the series for J_0' the even-power terms in the other series must be zero. This means that $c_1 = c_3 = c_5 = \cdots = 0$. In the second sum we can replace $k + 1$ by $2k - 1$ if we sum from 1 to infinity. We have

$$\sum_{k=1}^{\infty} \left[\frac{(-1)^k}{2^{2k-2}(k-1)!\,k!} + (2k)^2 c_{2k} + c_{2k-2}\right] x^{2k-1} = 0$$

† We are assuming throughout that $x > 0$.

This gives us the recurrence relation

$$(2k)^2 c_{2k} + c_{2k-2} = \frac{(-1)^{k+1}}{2^{2k-2}(k-1)!\,k!}$$

for $k = 1, 2, 3, \ldots$. Starting with $k = 1$, we have

$$4c_2 + c_0 = 1$$
$$16c_4 + c_2 = -\tfrac{1}{8}$$
$$36c_6 + c_4 = \tfrac{1}{192}$$

etc. The constant c_0 is arbitrary. If we put† $c_0 = 0$, then $c_2 = \frac{1}{4}$, $c_4 = -\frac{3}{128}$, $c_6 = \frac{11}{13824}$, etc. The solution is

$$y(x) = J_0(x)\ln x + \tfrac{1}{4}x^2 - \tfrac{3}{128}x^4 + \tfrac{11}{13824}x^6 - \cdots$$

Obviously $y(x)$ and $J_0(x)$ are independent.

We have already observed that when $\nu = n$, a positive integer, the Frobenius method fails to give two independent solutions. So we try to find a solution of the equation

$$x^2y'' + xy' + (x^2 - n^2)y = 0$$

of the form $y(x) = A(x)J_n(x)$. Substituting, we have

$$A[x^2J_n'' + xJ_n' + (x^2 - n^2)J_n] + x^2(2A'J' + JA'') + xA'J = 0$$

The first term is zero since J_n is a solution of the differential equation. Therefore, we have

$$\frac{A''}{A'} = -\frac{1}{x} - \frac{2J_n'}{J_n}$$
$$\ln A' = -\ln x - 2\ln J_n = -\ln xJ_n^{\,2}$$
$$A' = \frac{1}{xJ_n^{\,2}}$$

The series for $J_n^{\,2}$ begins with a term of the form ax^{2n}. Therefore,

$$A' = \frac{1}{ax^{2n+1}}\text{(power series)}$$

Integrating, we find

$$A = \ln x + \frac{1}{x^{2n}}\text{(power series)}$$

† If $c_0 \neq 0$, this has the effect of adding a multiple of J_0 to the solution.

and so we look for a solution of the form

$$y(x) = J_n(x) \ln x + x^{-n} \sum_{k=0}^{\infty} c_k x^k$$

We shall leave the determination of the coefficients for the exercises.

EXERCISES 8.4

1 Consider the Bessel equation of order $\frac{1}{2}$, $x^2y'' + xy' + (x^2 - \frac{1}{4})y = 0$. Make the change of variable $y(x) = w(x)/\sqrt{x}$, and show that the general solution is of the form $y(x) = (A \sin x + B \cos x)/\sqrt{x}$. Reconcile this with the Bessel function solution.

2 Show that $J_0'(x) = -J_1(x)$.

3 Certain equations can be solved by reducing them to the Bessel equation. Show that the Airy equation $y'' + xy = 0$ can be solved by making the change of variables $y(x) = x^{1/2}w(t)$, where $t = \frac{2}{3}x^{3/2}$.

4 Show that the equation $x^2y'' + (\alpha^2\beta^2x^{2\beta} + \frac{1}{4} - \nu^2\beta^2)y = 0$ can be reduced to the Bessel equation by the change of variables $y(x) = x^{1/2}w(\alpha x^\beta)$.

5 Prove the following identities:

(*a*) $J_{\nu-1}(x) + J_{\nu+1}(x) = \dfrac{2\nu}{x} J_\nu(x)$ (*b*) $J_{\nu-1}(x) - J_{\nu+1}(x) = 2J_\nu'(x)$

(*c*) $\dfrac{\nu}{x} J_\nu(x) + J_\nu'(x) = J_{\nu-1}(x)$ (*d*) $\dfrac{\nu}{x} J_\nu(x) - J_\nu'(x) = J_{\nu+1}(x)$

6 Determine the coefficients in a solution of the form $y = J_n(x) \ln x + \sum_{k=0}^{\infty} c_k x^{k-n}$ of the equation $x^2y'' + xy' + (x^2 - n^2)y = 0$.

7 Consider the problem $xy'' + y' + \lambda xy = 0$, $0 \le x \le 1$, λ constant, $y(0)$ finite, $y(1) = 0$.

(*a*) Show that if the problem has a nontrivial solution ($y \not\equiv 0$), then λ is real. Evaluate $(\lambda - \bar{\lambda}) \int_0^1 xy\bar{y}\,dx$.

(*b*) Show that if the problem has a nontrivial solution, then λ is positive. Evaluate $\lambda \int_0^1 xy^2\,dx$.

(*c*) Show that any nontrivial solution will be of the form $J_0(\sqrt{\lambda}\,x)$, where $J_0(\sqrt{\lambda}) = 0$.

(*d*) It can be shown that $J_0(x)$ has a sequence of positive zeros x_i such that $\lim_{i\to\infty} x_i = \infty$. If $\lambda_i^2 = x_i$, show that $J_0(\sqrt{\lambda_i}\,x)$ and $J_0(\sqrt{\lambda_j}\,x)$, $i \ne j$, are orthogonal in the following sense:

$$\int_0^1 xJ_0(\sqrt{\lambda_i}x)J_0(\sqrt{\lambda_j}x)\,dx = 0$$

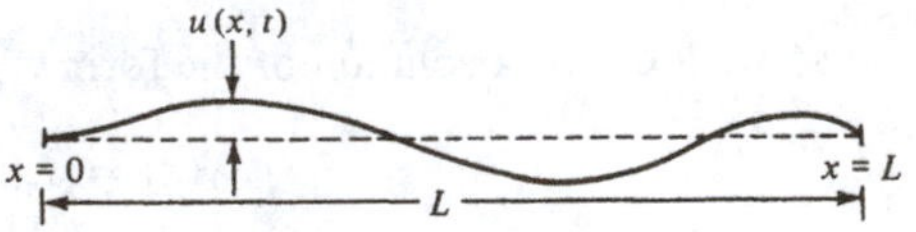

FIGURE 48

8.5 BOUNDARY-VALUE PROBLEMS

In order to introduce the subject of boundary-value problems, let us consider the following physical problem. A uniform elastic string is stretched between two fixed points a distance L apart (see Fig. 48). Assuming that the weight of the string can be neglected, that the tension is uniform along the string, and that the vertical displacement $u(x,t)$ is small, we can show that the displacement satisfies the following conditions:

$$\frac{\partial^2 u}{\partial x^2} = \frac{1}{c^2}\frac{\partial^2 u}{\partial t^2} \qquad t > 0 \qquad 0 < x < L$$

$$u(0,t) = u(L,t) = 0 \qquad t \geq 0$$

$$u(x,0) = f(x) \qquad 0 \leq x \leq L$$

$$u_t(x,0) = g(x) \qquad 0 \leq x \leq L$$

where c^2 is a positive constant depending on the density of the string and the tension. The given functions $f(x)$ and $g(x)$ are the prescribed initial displacement and velocity, respectively.†

We shall not attempt the complete solution of this problem, which is beyond the scope of this book, but shall indicate the method of *separation of variables.* We look for solutions of the form

$$u(x,t) = y(x)w(t)$$

Substituting in the differential equation, we have

$$w(t)y''(x) = \frac{1}{c^2}\, y(x)w''(t)$$

$$\frac{y''(x)}{y(x)} = \frac{1}{c^2}\frac{w''(t)}{w(t)}$$

The left-hand side is a function of x only, while the right-hand side is a function of t only. Therefore, for equality each side must be equal to the same constant,

† For the derivation of these conditions see J. W. Dettman, "Mathematical Methods in Physics and Engineering," 2d ed., McGraw-Hill, New York, 1969.

which for convenience we call $-\lambda^2$. Then y and w satisfy the following differential equations

$$y'' + \lambda^2 y = 0$$
$$w'' + c^2\lambda^2 w = 0$$

For the boundary conditions to be satisfied for all t, $y(0) = y(L) = 0$. We are led naturally to the boundary-value problem for y: $y'' + \lambda^2 y = 0$, $0 \le x \le L$, $y(0) = y(L) = 0$. Of course we must also determine for what constants λ^2 there are solutions.

We first show that λ^2 must be real. If λ^2 and y are complex, then $\bar{y}'' + \bar{\lambda}^2\bar{y} = 0$, $\bar{y}(0) = \bar{y}(L) = 0$. Therefore,

$$(\lambda^2 - \bar{\lambda}^2)\int_0^L y\bar{y}\,dx = \int_0^L (\bar{y}''y - y''\bar{y})\,dx = \bar{y}'y\big|_0^L - y'\bar{y}\big|_0^L = 0$$

Since $\int_0^L |y|^2\,dx > 0$ if $y \not\equiv 0$, $\lambda^2 = \bar{\lambda}^2$, which shows that λ^2 is real. Next we show that $\lambda^2 > 0$. In fact,

$$\begin{aligned}\lambda^2\int_0^L y^2\,dx &= -\int_0^L y''y\,dx \\ &= -y'y\big|_0^L + \int_0^L (y')^2\,dx \\ &= \int_0^L (y')^2\,dx > 0\end{aligned}$$

unless $y' \equiv 0$. However, if $y' \equiv 0$, then y is constant and this constant must be zero to satisfy the boundary conditions. Therefore, unless $y \equiv 0$ (the trivial solution), $\lambda^2 > 0$.

The general solution of $y'' + \lambda^2 y = 0$ is

$$y = A\cos\lambda x + B\sin\lambda x$$

But $y(0) = A = 0$. The other boundary condition requires that $B\sin\lambda L = 0$. This is possible if $B = 0$ or $\lambda L = n\pi$, $n = 1, 2, 3, \ldots$. If $B = 0$, we have the trivial solution, $y \equiv 0$, which is of no interest. The other possibility is that $\lambda = n\pi/L$ and B is arbitrary. These are the solutions which are of interest. The constants $\lambda_n^2 = n^2\pi^2/L^2$ are called *characteristic values*, and the functions $y_n = \sin(n\pi x/L)$ are called *characteristic functions*.

If $\lambda = n\pi/L$, $n = 1, 2, 3, \ldots$, then

$$w_n'' + \frac{c^2n^2\pi^2}{L^2}w_n = 0$$

and this has the general solution

$$w_n(t) = B_n\cos\frac{cn\pi t}{L} + C_n\sin\frac{cn\pi t}{L}$$

The functions

$$u_n(x,t) = \left(B_n \cos \frac{cn\pi t}{L} + C_n \sin \frac{cn\pi t}{L}\right) \sin \frac{n\pi x}{L}$$

are solutions of the partial differential equation and boundary conditions for $n = 1, 2, 3, \ldots$. The final step of the solution is to obtain a linear combination of these functions which satisfies the initial conditions. The idea is to seek a solution of the form

$$u(x,t) = \sum_{n=1}^{\infty} \left(B_n \cos \frac{cn\pi t}{L} + C_n \sin \frac{cn\pi t}{L}\right) \sin \frac{n\pi x}{L}$$

such that

$$u(x,0) = \sum_{n=1}^{\infty} B_n \sin \frac{n\pi x}{L} = f(x)$$

$$u_t(x,0) = \frac{c\pi}{L} \sum_{n=1}^{\infty} nC_n \sin \frac{n\pi x}{L} = g(x)$$

This involves the question of when a given function can be represented as a sum of trigonometric functions (Fourier series).† We shall not pursue the subject further here.

EXAMPLE 8.5.1 For what values of λ^2 does the boundary-value problem $y'' + \lambda^2 y = 0$, $0 \le x \le L$, $y(0) = y'(L) = 0$ have nontrivial solutions? We first show that λ^2 is real. If λ^2 and y are complex, then

$$\begin{aligned}(\lambda^2 - \bar{\lambda}^2) \int_0^L y\bar{y}\, dx &= \int_0^L (\bar{y}''y - y''\bar{y})\, dx \\ &= (\bar{y}'y - y'\bar{y})|_0^L = 0\end{aligned}$$

Therefore, since $\int_0^L |y|^2\, dx > 0$ unless $y \equiv 0$, $\lambda^2 = \bar{\lambda}^2$. Next we show that $\lambda^2 > 0$, since

$$\begin{aligned}\lambda^2 \int_0^L y^2\, dx &= -\int_0^L y''y\, dx \\ &= -y'y|_0^L + \int_0^L (y')^2\, dx \\ &= \int_0^L (y')^2\, dx \ge 0\end{aligned}$$

If $y' \equiv 0$, then $y \equiv 0$. Hence, for nontrivial solutions $\int_0^L y^2\, dx > 0$ and $\int_0^L (y')^2\, dx > 0$, and $\lambda^2 > 0$. The general solution is $y(x) = A \cos \lambda x + B \sin \lambda x$.

† See ibid.

However, $y(0) = A = 0$ and $y'(L) = \lambda B \cos \lambda L = 0$. If the solution is to be nontrivial, $\lambda L = (2n - 1)\pi/2$, $n = 1, 2, 3, \ldots$, and the characteristic values must be

$$\lambda_n^2 = \left[\frac{(2n - 1)\pi}{2L}\right]^2$$

and the characteristic functions must be

$$y_n(x) = \sin \frac{(2n - 1)\pi x}{2L}$$

We shall consider the following general homogeneous boundary-value problem: $y'' + p(x,\lambda)y' + q(x,\lambda)y = 0$, $0 \leq x \leq 1$, $a_0 y(0) + b_0 y'(0) = 0$, $a_1 y(1) + b_1 y'(1) = 0$, where the real-valued functions p and q are continuous for all x in the given interval and all real values of λ and the constants a_0, b_0, a_1, and b_1 are real-valued. The unit interval $\{x \mid 0 \leq x \leq 1\}$ is general because any finite interval $\{x \mid a \leq x \leq b\}$ can be reduced to it by the change of variable $\xi = (x - a)/(b - a)$. We seek real values λ for which the problem has nontrivial solutions. For a given value λ suppose that the general solution is

$$y(x,\lambda) = Au(x,\lambda) + Bv(x,\lambda)$$

The boundary conditions require that

$$A[a_0 u(0,\lambda) + b_0 u'(0,\lambda)] + B[a_0 v(0,\lambda) + b_0 v'(0,\lambda)] = 0$$
$$A[a_1 u(1,\lambda) + b_1 u'(1,\lambda)] + B[a_1 v(1,\lambda) + b_1 v'(1,\lambda)] = 0$$

Considering this as a system of homogeneous linear equations for A and B, we see that if the determinant of coefficients is not zero, then $A = B = 0$ and we have the trivial solution. Therefore, the condition for nontrivial solutions is

$$D(\lambda) = \begin{vmatrix} a_0 u(0,\lambda) + b_0 u'(0,\lambda) & a_0 v(0,\lambda) + b_0 v'(0,\lambda) \\ a_1 u(1,\lambda) + b_1 u'(1,\lambda) & a_1 v(1,\lambda) + b_1 v'(1,\lambda) \end{vmatrix} = 0$$

Any real value of λ for which $D(\lambda) = 0$ is a characteristic value for the problem.

EXAMPLE 8.5.2 Find all the characteristic values and characteristic functions of the boundary-value problem $y'' + \lambda^2 y = 0$, $0 \leq x \leq 1$, $y(0) + y'(0) = 0$, $y(1) + y'(1) = 0$. We first show that the characteristic values are real. If λ^2 and y are complex-valued, then

$$(\lambda^2 - \bar{\lambda}^2) \int_0^1 y\bar{y}\, dx = \int_0^1 (\bar{y}''y - y''\bar{y})\, dx$$
$$= (\bar{y}'y - y'\bar{y})\Big|_0^1 = 0$$

Therefore, for nontrivial solutions $\lambda^2 = \bar{\lambda}^2$. In this case, we cannot show that $\lambda^2 > 0$. Therefore, we consider three cases.

CASE 1 $\lambda^2 = 0$ In this case $y(x) = Ax + B$, and

$$D(0) = \begin{vmatrix} 1 & 1 \\ 2 & 1 \end{vmatrix} = -1$$

Therefore, $\lambda^2 = 0$ is not a characteristic value.

CASE 2 $\lambda^2 < 0$ Then if $\lambda^2 = -\mu^2$, $y(x) = Ae^{\mu x} + Be^{-\mu x}$, and

$$\begin{aligned} D(\lambda) &= \begin{vmatrix} 1 + \mu & 1 - \mu \\ (1 + \mu)e^{\mu} & (1 - \mu)e^{-\mu} \end{vmatrix} \\ &= 2(\mu^2 - 1) \sinh \mu \end{aligned}$$

Since $\sinh \mu \neq 0$, for $D(\lambda) = 0$, $\mu^2 - 1 = 0$, $\mu^2 = 1$. In this case, $\lambda^2 = -1$ is a characteristic value with a corresponding characteristic function e^{-x}.

CASE 3 $\lambda^2 > 0$ In this case, $y(x) = A \cos \lambda x + B \sin \lambda x$, and

$$\begin{aligned} D(\lambda) &= \begin{vmatrix} 1 & \lambda \\ \cos \lambda - \lambda \sin \lambda & \sin \lambda + \lambda \cos \lambda \end{vmatrix} \\ &= (1 + \lambda^2) \sin \lambda \end{aligned}$$

For $D(\lambda) = 0$, $\sin \lambda = 0$, and $\lambda = n\pi$, $n = 1, 2, 3, \ldots$. The characteristic values are $\lambda_n{}^2 = n^2\pi^2$ with corresponding characteristic functions $y_n(x) = n\pi \cos n\pi x - \sin n\pi x$.

It is possible to solve boundary-value problems where there is a singular point in the interval.

EXAMPLE 8.5.3 Find all characteristic values and characteristic functions for the boundary-value problem $xy'' + y' + \lambda xy = 0$, $0 \leq x \leq 1$, $y(0)$ finite, $y(1) = 0$. The differential equation can be written as $(xy')' + \lambda xy = 0$. If λ and y are complex-valued, then $(x\bar{y}')' + \bar{\lambda} x\bar{y} = 0$, $\bar{y}(0)$ finite, $\bar{y}(1) = 0$. Then

$$\begin{aligned} (\lambda - \bar{\lambda}) \int_0^1 xy\bar{y}\, dx &= \int_0^1 [(x\bar{y}')'y - (xy')'\bar{y}]\, dx \\ &= (x\bar{y}'y - xy'\bar{y})\big|_0^1 = 0 \end{aligned}$$

Since $|y|^2 > 0$ unless $y \equiv 0$, $\lambda = \bar{\lambda}$. Next we show that $\lambda > 0$, since

$$\lambda \int_0^1 xy^2\,dx = -\int_0^1 (xy')'y\,dx$$
$$= -xy'y|_0^1 + \int_0^1 x(y')^2\,dx$$
$$= \int_0^1 x(y')^2\,dx > 0$$

Now we make the change of variable $\xi = \sqrt{\lambda}x$. The differential equation becomes $\xi y'' + y' + \xi y = 0$, which is the Bessel equation of order zero. Since the solution must be finite at $\xi = 0$, the only possible solution is $AJ_0(\xi) = AJ_0(\sqrt{\lambda}x)$. To satisfy the other boundary condition $J_0(\sqrt{\lambda}) = 0$. It can be shown that J_0 has an infinite sequence of positive zeros $\mu_1, \mu_2, \mu_3, \ldots$, approaching infinity. Therefore, the characteristic values are $\lambda_n = \mu_n{}^2$, $n = 1, 2, 3, \ldots$, and the characteristic function are $y_n(x) = J_0(\sqrt{\lambda_n}x)$.

We conclude this section by considering some nonhomogeneous boundary-value problems.

EXAMPLE 8.5.4 Find all possible solutions of $y'' + \lambda^2 y = e^x$, $\lambda^2 > 0$, $0 \le x \le 1$, $y(0) = 0$, $y(1) = 1$. The general solution of the differential equation is

$$y(x) = A\cos\lambda x + B\sin\lambda x + \frac{1}{1+\lambda^2}e^x$$

To satisfy the boundary conditions we must have

$$y(0) = A + \frac{1}{1+\lambda^2} = 0$$
$$y(1) = A\cos\lambda + B\sin\lambda + \frac{1}{1+\lambda^2}e = 1$$

The determinant of the coefficients of A and B is

$$D(\lambda) = \begin{vmatrix} 1 & 0 \\ \cos\lambda & \sin\lambda \end{vmatrix} = \sin\lambda$$

If $\sin\lambda \neq 0$, or, in other words, if λ^2 is not a characteristic value of the homogeneous problem $y'' + \lambda^2 y = 0$, $y(0) = y(1) = 0$, then there is a unique solution for A and B and there is a solution of the nonhomogeneous boundary-value problem. In that case, $A = -1/(1 + \lambda^2)$, and

$$B = \left(1 - \frac{e}{1 + \lambda^2} + \frac{\cos \lambda}{1 + \lambda^2}\right) \csc \lambda$$

In this case, we can prove that the solution is unique. Suppose there were two solutions y_1 and y_2. Then if $w = y_1 - y_2$, $w'' + \lambda^2 w = 0$, $w(0) = w(1) = 0$. But $w \equiv 0$ if λ^2 is not a characteristic value. Therefore, $y_1 \equiv y_2$. If $\lambda = n\pi$, $n = 1, 2, 3, \ldots$, then we have a different situation in which $D(\lambda) = 0$. Now we know from linear algebra that there may still be solutions for A and B but they will not be unique. To determine this we could check the values of a couple of 2×2 determinants. However, there is an analytical way to proceed. We first convert the problem to one with homogeneous boundary conditions by subtracting† x from y. Let $v(x) = y(x) - x$. Then

$$v'' + \lambda^2 v = e^x + \lambda^2 x = g(x)$$

and $v(0) = v(1) = 0$. If there were a solution $v(x)$, then

$$\begin{aligned}\int_0^1 g(x) \sin \lambda x \, dx &= \int_0^1 (v'' + \lambda^2 v) \sin \lambda x \, dx \\ &= v'(x) \sin \lambda x\Big|_0^1 - \lambda \int_0^1 (v' \cos \lambda x - \lambda v \sin \lambda x) \, dx \\ &= -\lambda v(x) \cos \lambda x\Big|_0^1 = 0\end{aligned}$$

Therefore, if there is a solution of the nonhomogeneous boundary-value problem in the case where λ^2 is a characteristic value, then necessarily $\int_0^1 g(x) \sin \lambda x \, dx = 0$, where $\sin \lambda x$ is the corresponding characteristic function. It can be shown that this is also a sufficient condition for a solution, which, however, is not unique because if $v(x)$ exists, any multiple of $\sin \lambda x$ can be added to it. In the present example, $\int_0^1 g(x) \sin \lambda x \, dx \neq 0$, so there is no solution.

EXAMPLE 8.5.5 Find all possible solutions of the nonhomogeneous boundary-value problem $y'' + \pi^2 y = \cos \pi x$, $0 \leq x \leq 1$, $y(0) = y(1) = 0$. In this case, π^2 is a characteristic value of the homogeneous boundary-value problem $y'' + \lambda^2 y = 0$, $y(0) = y(1) = 0$. However,

$$\int_0^1 \sin \pi x \cos \pi x \, dx = \frac{1}{2} \int_0^1 \sin 2\pi x \, dx = 0$$

† The function $u(x) = x$ satisfies the nonhomogeneous boundary conditions but does not satisfy the differential equation.

and therefore there is a nonunique solution of the boundary-value problem. The general solution of the differential equation is

$$y(x) = A \cos \pi x + B \sin \pi x + \frac{1}{2\pi} x \sin \pi x$$

For the boundary conditions we have

$$y(0) = A = 0$$

$$y(1) = B \sin \pi + \frac{\sin \pi}{2\pi} = 0$$

and B is arbitrary. Therefore, all solutions are of the form $y(x) = B \sin \pi x + (1/2\pi)x \sin \pi x$.

EXERCISES 8.5

1 Find all the characteristic values and characteristic functions of the boundary-value problem $y'' + \lambda^2 y = 0$, $0 \le x \le L$, $y'(0) = y'(L) = 0$.

2 Find all the characteristic values and characteristic functions of the boundary-value problem $y'' + \lambda^2 y = 0$, $0 \le x \le L$, $y'(0) = y(L) = 0$.

3 Show that the characteristic functions $y_n(x)$ of the boundary-value problem $y'' + \lambda^2 y = 0$, $y(0) = y(L) = 0$ are orthogonal on the interval $\{x \mid 0 \le x \le L\}$; that is,

$$\int_0^L y_n(x) y_m(x)\, dx = 0 \text{ for } n \ne m$$

Hint: This can be shown directly from the differential equation and boundary conditions without explicitly knowing the characteristic functions.

4 Show that the characteristic functions $y_n(x)$ of the boundary-value problem $y'' + \lambda^2 y = 0$, $y'(0) = y'(L)$ are orthogonal on the interval $\{x \mid 0 \le x \le L\}$.

5 Find the characteristic values and characteristic functions of the boundary-value problem $y'' + \lambda^2 y = 0$, $0 \le x \le 1$, $y(0) = y(1) + y'(1) = 0$. Show that these characteristic functions are orthogonal on the interval $\{x \mid 0 \le x \le 1\}$.

6 Show that the characteristic functions of Example 8.5.3 are orthogonal on the interval $\{x \mid 0 \le x \le 1\}$; that is, show that

$$\int_0^1 x y_n(x) y_m(x)\, dx = 0 \text{ for } n \ne m$$

7 Find the characteristic values and characteristic functions of the boundary-value problem $(xy')' + (\lambda x - 1/x)y = 0$, $0 \le x \le 1$, $y(0)$ finite, $y(1) = 0$. Show that the characteristic functions are orthogonal on the interval $\{x \mid 0 \le x \le 1\}$; that is, show that

$$\int_0^1 x y_n(x) y_m(x)\, dx = 0 \text{ for } n \ne m$$

8 Show that the boundary-value problem $x^2y'' - \lambda xy' + \lambda y = 0$, $1 \le x \le 2$, $y(1) = y(2) = 0$, has no characteristic values. *Hint:* The differential equation is an Euler equation.

9 Find all possible solutions of $y'' + y = \sin x$, $0 \le x \le 1$, $y(0) = y'(1) = 1$.

10 Find all possible solutions of $y'' + \pi^2 y = \sin \pi x$, $0 \le x \le 1$, $y(0) = 0$, $y(1) = 1$.

11 Find all possible solutions of $y'' + \pi^2 y = \pi^2 x$, $0 \le x \le 1$, $y(0) = 0$, $y(1) = 1$.

*8.6 CONVERGENCE THEOREMS

In this section, we give the proofs of the two basic theorems of Secs. 8.2 and 8.3, dealing with the convergence of the series solutions near ordinary points and regular singular points. Before we begin these proofs, we recall the following important facts about real power series.

The power series $\sum_{k=0}^{\infty} a_k x^k$, with real coefficients, (1) converges for $x = 0$ and for no other x, or (2) it converges absolutely for all x, or (3) there exists an $R > 0$ such that the series converges absolutely for all $|x| < R$ and diverges for all $|x| > R$. In case 3 we call R the radius of convergence.

If $\sum_{k=0}^{\infty} a_k x^k = f(x)$ for $|x| < R$, then $f(x)$ is continuous for $|x| < R$.

If the series $\sum_{k=0}^{\infty} a_k x^k$ converges for $|x| < R$, then the series can be differentiated term by term as many times as we wish and all the differentiated series converge for $|x| < R$. If $f(x) = \sum_{k=0}^{\infty} a_k x^k$, then

$$f^{(n)}(x) = \sum_{k=n}^{\infty} k(k-1)\cdots(k-n+1)a_k x^{k-n}$$

$n = 1, 2, 3, \ldots$.

If the series $\sum_{k=0}^{\infty} a_k x^k$ and $\sum_{k=0}^{\infty} b_k x^k$ both converge for $|x| < R$, then $\sum_{k=0}^{\infty} a_k x^k + \sum_{k=0}^{\infty} b_k x^k = \sum_{k=0}^{\infty} (a_k + b_k)x^k$ and the latter series converges for $|x| < R$.

If the series $\sum_{k=0}^{\infty} a_k x^k$ and $\sum_{k=0}^{\infty} b_k x^k$ both converge for $|x| < R$, then the Cauchy product $\sum_{k=0}^{\infty} \left(\sum_{j=0}^{k} a_j b_{k-j}\right) x^k$ converges for $|x| < R$ and

$$\left(\sum_{k=0}^{\infty} a_k x^k\right)\left(\sum_{k=0}^{\infty} b_k x^k\right) = \sum_{k=0}^{\infty} \left(\sum_{j=0}^{k} a_j b_{k-j}\right) x^k$$

If the series $\sum_{k=0}^{\infty} a_k x^k$ converges to zero for $|x| < R$, then $a_k = 0$ for all $k = 0, 1, 2, 3, \ldots$.

If the series $\sum_{k=0}^{\infty} a_k x^k$ converges absolutely for $|x| \leq r$, then there exists a positive constant M such that $|a_k| \leq Mr^{-k}$, $k = 0, 1, 2, 3, \ldots$.

We now restate and prove Theorem 8.2.1 (p. 317), dealing with power-series solutions near ordinary points.

Theorem 8.2.1 If the equation $y'' + p(x)y' + q(x)y = 0$ has an ordinary point at $x = 0$, due to the fact that $p(x)$ and $q(x)$ have power-series representations converging for $|x| < R$, then there are independent solutions $y_1(x)$ and $y_2(x)$ satisfying $y_1(0) = 1$, $y_1'(0) = 0$, $y_2(0) = 0$, $y_2'(0) = 1$ and each having a power-series representation converging for $|x| < R$.

PROOF As in Sec. 8.2, if we assume

$$p(x) = p_0 + p_1 x + p_2 x^2 + \cdots = \sum_{k=0}^{\infty} p_k x^k$$

$$q(x) = q_0 + q_1 x + q_2 x^2 + \cdots = \sum_{k=0}^{\infty} q_k x^k$$

then we define recursively the coefficients for a possible power-series solution

$$y(x) = c_0 + c_1 x + c_2 x^2 + \cdots = \sum_{k=0}^{\infty} c_k x^k$$

by the equations

$$\begin{aligned} 2c_2 + p_0 c_1 + q_0 c_0 &= 0 \\ 6c_3 + 2p_0 c_2 + p_1 c_1 + q_0 c_1 + q_1 c_0 &= 0 \\ 12c_4 + 3p_0 c_3 + 2p_1 c_2 + p_2 c_1 + q_0 c_2 + q_1 c_1 + q_2 c_0 &= 0 \end{aligned}$$

etc. The first equation defines c_2 in terms of c_0 and c_1, which are arbitrary; the second equation defines c_3; the third equation defines c_4; etc. Since the series for $p(x)$ and $q(x)$ converge absolutely for $|x| \leq r < R$, there are constants M and N such that $|p_k| \leq Mr^{-k}$ and $|q_k| \leq Nr^{-k}$. Let K be the larger of M and Nr; then $|p_k| \leq Kr^{-k}$ and $|q_k| \leq Kr^{-k-1}$. If $|c_0| = a_0$ and $|c_1| = a_1$, then

$$\begin{aligned} 2|c_2| &\leq a_1|p_0| + a_0|q_0| \\ &\leq 2Ka_1 + Ka_0 r^{-1} \end{aligned}$$

so that $|c_2| \leq a_2$, where $2a_2 = K(2a_1 + a_0 r^{-1})$. Furthermore,

$$\begin{aligned} 6|c_3| &\leq 2a_2|p_0| + a_1|p_1| + a_1|q_0| + a_0|q_1| \\ &\leq 3a_2K + 2a_1Kr^{-1} + a_0Kr^{-2} \end{aligned}$$

and

$$\begin{aligned} 12|c_4| &\leq 3a_3|p_0| + 2a_2|p_1| + a_1|p_2| + a_2|q_0| + a_1|q_1| + a_0|q_2| \\ &\leq 4a_3K + 3a_2Kr^{-1} + 2a_1Kr^{-2} + a_0Kr^{-3} \end{aligned}$$

Therefore, $|c_3| \leq a_3$ and $|c_4| \leq a_4$, where

$$\begin{aligned} 6a_3 &= K(3a_2 + 2a_1r^{-1} + a_0r^{-2}) \\ 12a_4 &= K(4a_3 + 3a_2r^{-1} + 2a_1r^{-2} + a_0r^{-3}) \end{aligned}$$

Continuing in this way, we have $|c_k| \leq a_k$, where $(k-1)ka_k = K[ka_{k-1} + (k-1)a_{k-2}r^{-1} + \cdots + 2a_1r^{-k+2} + a_0r^{-k+1}]$ for $k = 2, 3, 4, \ldots$. If $k \geq 3$, we can write

$$\begin{aligned} &(k-2)(k-1)a_{k-1}r^{-1} \\ &\qquad = K[(k-1)a_{k-2}r^{-1} + \cdots + 2a_1r^{-k+2} + a_0r^{-k+1}] \end{aligned}$$

Subtracting, we have

$$(k-1)ka_k - (k-2)(k-1)a_{k-1}r^{-1} = Kka_{k-1}$$

from which it follows that

$$\frac{a_k}{a_{k-1}} = \frac{k-2}{kr} + \frac{K}{k-1}$$

Now consider the series $\sum_{k=0}^{\infty} a_k x^k$. Applying the ratio test, we have

$$\lim_{k\to\infty} \frac{a_k|x|}{a_{k-1}} = \frac{|x|}{r} < 1$$

for $|x| < r$. Therefore, $\sum_{k=0}^{\infty} a_k x^k$ converges absolutely for $|x| < r$, where the series $\sum_{k=0}^{\infty} c_k x^k$ also converges absolutely by comparison. However, r is any positive number less than R, so that $\sum_{k=0}^{\infty} c_k x^k$ converges absolutely for $|x| < R$. It is now a simple matter to substitute the series into the differential equation and, using the known properties of power series, to verify that it is a solution. Finally, since c_0 and c_1 are arbitrary, we can find two independent solutions $y_1(x)$ and $y_2(x)$ by assuming $c_0 = 1$, $c_1 = 0$, for y_1 and $c_0 = 0$, $c_1 = 1$, for y_2. This completes the proof.

We now turn to the question of series representation of solutions near regular singular points. Recall that the differential equation $y'' + p(x)y' + q(x)y = 0$ has a regular singular point at $x = 0$ if $xp(x)$ and $x^2q(x)$ have power-series representations converging for $|x| < R$. We assume that

$$xp(x) = a_0 + a_1x + a_2x^2 + \cdots = \sum_{k=0}^{\infty} a_kx^k$$

$$x^2q(x) = b_0 + b_1x + b_2x^2 + \cdots = \sum_{k=0}^{\infty} b_kx^k$$

where the coefficients a_k and b_k are real. The indicial equation is

$$I(m) = m(m-1) + a_0m + b_0 = 0$$

We assume that the roots of the indicial equation, m_1 and m_2, are both real† and that $m_1 \geq m_2$.

Theorem 8.6.2 If the equation $y'' + p(x)y' + q(x)y = 0$ has a regular singular point at $x = 0$ due to the fact that $xp(x) = \sum_{k=0}^{\infty} a_kx^k$ and $x^2q(x) = \sum_{k=0}^{\infty} b_kx^k$ converge for $|x| < R$, and if the roots m_1 and m_2 of the indicial equation are both real, $m_1 \geq m_2$, then for $x > 0$ there is a solution of the form $y(x) = x^{m_1} \sum_{k=0}^{\infty} c_kx^k$ where the series converges for $|x| < R$.

PROOF We begin by assigning an arbitrary nonzero value to c_0. Then the other coefficients are determined by

$$c_kI(m_1 + k) = -\sum_{j=0}^{k-1} c_j[(m_1 + j)a_{k-j} + b_{k-j}]$$

where $I(m) = m(m-1) + a_0m + b_0 = (m - m_1)(m - m_2)$. Therefore, $I(m_1 + k) = k(k + m_1 - m_2)$. For all $k > m_1 - m_2$ we can write the inequality

$$\begin{aligned} k(k - m_1 + m_2)|c_k| &\leq |I(m_1 + k)|\,|c_k| \\ &\leq \sum_{j=0}^{k-1} |c_j|[(|m_1| + j)|a_{k-j}| + |b_{k-j}|] \end{aligned}$$

Let $|c_j| = C_j$ for $j < n$, where n is some integer greater than $m_1 - m_2$.

†A similar theorem holds if m_1 and m_2 are complex, but the proof involves complex-variable techniques beyond the scope of this book. See J. W. Dettman, "Applied Complex Variables," Macmillan, New York, 1965 (rpt. Dover, New York, 1984).

Then

$$n(n - m_1 + m_2)|c_n| \leq \sum_{j=0}^{n-1} C_j[(|m_1| + j)|a_{n-j}| + |b_{n-j}|]$$

There exists a positive constant K such that $|a_k| \leq Kr^{-k}$ and $|b_k| \leq Kr^{-k}$ for $0 < r < R$. Then

$$n(n - m_1 + m_2)|c_n| \leq K \sum_{j=0}^{n-1} C_j(|m_1| + j + 1)r^{-n+j}$$

and $|c_n| \leq C_n$, where

$$n(n - m_1 + m_2)C_n = K \sum_{j=0}^{n-1} C_j(|m_1| + j + 1)r^{-n+j}$$

Furthermore, for $k > n$

$$k(k - m_1 + m_2)|c_k| \leq K \sum_{j=0}^{k-1} C_j(|m_1| + j + 1)r^{-k+j}$$

and $|c_k| \leq C_k$, where

$$k(k - m_1 + m_2)C_k = K \sum_{j=0}^{k-1} C_j(|m_1| + j + 1)r^{-k+j}$$

Replacing k by $k - 1$ and dividing by r, we have

$$(k - 1)(k - 1 - m_1 + m_2)C_{k-1}r^{-1} = K \sum_{j=0}^{k-2} C_j(|m_1| + j + 1)r^{-k+j}$$

Subtracting, we obtain

$$k(k - m_1 + m_2)C_k - (k - 1)(k - 1 - m_1 + m_2)C_{k-1}r^{-1} = KC_{k-1}(|m_1| + k)r^{-1}$$

or

$$\frac{C_k}{C_{k-1}} = \frac{(k - 1)(k - 1 - m_1 + m_2)}{k(k - m_1 + m_2)r} + \frac{K(|m_1| + k)}{k(k - m_1 + m_2)r}$$

We compare the series $\sum_{k=0}^{\infty} c_k x^k$ with $\sum_{k=0}^{\infty} C_k x^k$. The latter converges absolutely for $|x| < r$ by the ratio test, since

$$\lim_{k \to \infty} \frac{C_k|x|}{C_{k-1}} = \frac{|x|}{r} < 1$$

Therefore, the series $\sum_{k=0}^{\infty} c_k x^k$ converges absolutely for $|x| < R$, since r is any positive number less than R. It remains to substitute the proposed solution $y(x) = x^{m_1} \sum_{k=0}^{\infty} c_k x^k$ and use the properties of power series to show that it satisfies the differential equation. This completes the proof.

If m_2 does not differ from m_1 by a positive integer, then there is a solution of the differential equation of the form

$$y(x) = x^{m_2} \sum_{k=0}^{\infty} c_k x^k$$

where the series converges for $|x| < R$. This can be proved using the above proof with slight modifications. The important thing is that $I(m_2 + k) \neq 0$ for $k = 1, 2, 3, \ldots$.

EXERCISES 8.6

1 Referring to Theorem 8.6.1, show that if $m_1 - m_2$ is not a positive integer, there is a solution of the differential equation in the form

$$y(x) = x^{m_2} \sum_{k=0}^{\infty} c_k x$$

where the series converges for $|x| < R$.

2 Consider the differential equation $y'' + p(z)y' + q(z)y = 0$ where prime means derivative with respect to the complex variable z. Assume that

$$zp(z) = \sum_{k=0}^{\infty} a_k z^k, \; z^2 q(z) = \sum_{k=0}^{\infty} b_k z^k$$

where the series converge for $|z| < R$. Prove the complex-variable version of Theorem 8.6.1, where m_1 and m_2 need not be real. Does this take care of the real-variable case where m_1 and m_2 are complex?

9

SYSTEMS OF DIFFERENTIAL EQUATIONS

9.1 INTRODUCTION

The modern theory of ordinary differential equations is best discussed in terms of systems of equations, usually of the first order. Here the ability to manipulate with vector-valued functions is extremely important. In dealing with linear systems all the methods of the linear algebra discussed earlier play an important role. Hence, in this final chapter we come full circle and unite in an essential way the linear algebra and the differential equations. We treat first order systems, especially linear first order systems, and to a limited extent higher order systems. Finally, in the starred section we prove the basic existence and uniqueness theorem for first order systems which includes nonlinear as well as linear problems.

9.2 FIRST ORDER SYSTEMS

Systems of differential equations are quite basic in the applications. For example, we have already seen how systems arise in the study of electric networks. As another example, consider n particles moving in a force field which is a function of the positions of the n particles. If m_k is the mass of the kth particle with coordinates x_{1k}, x_{2k}, x_{3k}, which are functions of time t, then the differential equations satisfied, according to Newton's law, are

$$m_k \ddot{x}_{jk} = F_{jk}(x_{11}, x_{21}, x_{31}, \ldots, x_{1n}, x_{2n}, x_{3n})$$

$j = 1, 2, 3$; $k = 1, 2, 3, \ldots, n$, where F_{jk} is the jth coordinate of force on the kth particle. This is a second order system. However, it can easily be turned into a first order system by introducing the velocity variables $v_{jk} = \dot{x}_{jk}$. The system then becomes a first order system in the $6n$ variables x_{jk}, v_{jk},

$$\begin{aligned} m_k \dot{v}_{jk} &= F_{jk}(x_{11}, x_{21}, x_{31}, \ldots, x_{1n}, x_{2n}, x_{3n}) \\ \dot{x}_{jk} &= v_{jk} \end{aligned}$$

$j = 1, 2, 3$; $k = 1, 2, 3, \ldots, n$.

Consider the nth order linear differential equation

$$y^{(n)} + a_1 y^{(n-1)} + a_2 y^{(n-2)} + \cdots + a_{n-1}\dot{y} + a_n y = f(t)$$

If we introduce the variables $u_1 = y, u_2 = \dot{y}, u_3 = \ddot{y}, \ldots, u_n = y^{(n-1)}$, then the single nth order equation is equivalent to the first order system: $\dot{u}_1 = u_2, \dot{u}_2 = u_3, \ldots, \dot{u}_{n-1} = u_n$,

$$\dot{u}_n = f(t) - a_n u_1 - a_{n-1} u_2 - \cdots - a_1 u_n$$

In this section, we shall begin our study of first order systems of the form

$$\begin{aligned} \dot{y}_1(t) &= f_1(t, y_1, y_2, \ldots, y_n) \\ \dot{y}_2(t) &= f_2(t, y_1, y_2, \ldots, y_n) \\ &\cdots\cdots\cdots\cdots\cdots\cdots \\ \dot{y}_n(t) &= f_n(t, y_1, y_2, \ldots, y_n) \end{aligned}$$

We can simplify the notation if we introduce vector notation. Let $(y_1(t), y_2(t), \ldots, y_n(t)) = \mathbf{Y}(t)$ be the n-dimensional vector depending on t which denotes the n dependent variables. Let $(f_1, f_2, \ldots, f_n) = \mathbf{F}$ be the

n-dimensional vector of given functions on the right-hand side. The system can then be written as

$$\dot{\mathbf{Y}}(t) = \mathbf{F}(t,\mathbf{Y})$$

Except for the vector notation, this appears as a single first order equation. In fact, after one makes allowances for the vector notation, much of the analysis of the first order system follows the analysis of a single first order equation of Chap. 5. This is the principal reason for reducing many problems to first order systems.

We shall say that $\mathbf{Y}(t)$ is a solution of $\dot{\mathbf{Y}} = F(t,\mathbf{Y})$ on the interval $\{t \mid a < t < b\}$ if there exists a vector $\mathbf{Y}(t) = (y_1(t), y_2(t), \ldots, y_n(t))$ differentiable in the interval and satisfying, for each t, the differential equation. If the interval includes either or both of its end points, then we shall expect the obvious one-sided derivative to exist at the end point included. We shall consider the following initial-value problem:

$$\dot{\mathbf{Y}} = \mathbf{F}(t,\mathbf{Y}) \qquad a < t < b$$

$$\mathbf{Y}(t_0) = \mathbf{A} \qquad a < t_0 < b$$

where t_0 is given and $\mathbf{A} = (a_1, a_2, \ldots, a_n)$ is a given vector. Consider the $(n + 1)$-dimensional rectangle

$$R_{n+1} = \{(t, y_1, y_2, \ldots, y_n) \mid |t - t_0| \le r_0, |y_1 - a_1| \le r_1, \ldots, |y_n - a_n| \le r_n\}$$

where $r_0, r_1, \ldots, r_n$ are all positive. If $\mathbf{F}(t,\mathbf{Y})$ is continuous in R_{n+1}, then it is possible to show that there is a positive h such that the initial-value problem has a solution for $|t - t_0| < h$. However, this condition is not sufficient to prove that the solution is unique. In Sec. 9.6, we shall prove an existence-uniqueness theorem for the initial-value problem based on a Lipschitz condition which we shall define.

Definition 9.2.1 The vector-valued function $\mathbf{F}(t,\mathbf{Y})$ satisfies a Lipschitz condition with Lipschitz constant K in the "rectangle"

$$R_{n+1} = \{(t, y_1, y_2, \ldots, y_n) \mid |t - t_0| \le r_0, |y_1 - a_1| \le r_1, \ldots, |y_n - a_n| \le r_n\}$$

if for every† $(t,\mathbf{Y}_1)$ and $(t,\mathbf{Y}_2)$ in R_{n+1}

$$\|F(t,\mathbf{Y}_1) - F(t,\mathbf{Y}_2)\| \le K\|\mathbf{Y}_1 - \mathbf{Y}_2\|$$

† The reader is reminded that $\|\mathbf{Y}\|^2 = y_1{}^2 + y_2{}^2 + \cdots + y_n{}^2$.

EXAMPLE 9.2.1 Let $\mathbf{F} = (y_1 + y_2, y_1^2 + y_2^2)$. Show that $\mathbf{F}$ satisfies a Lipschitz condition in $R_3 = \{(t, y_1, y_2) \mid |t| \leq 1, |y_1| \leq 1, |y_2| \leq 1\}$. Let $\mathbf{Y}_1 = (y_{11}, y_{21})$, $\mathbf{Y}_2 = (y_{12}, y_{22})$. Then

$$
\begin{aligned}
&\|\mathbf{F}(\mathbf{Y}_1) - \mathbf{F}(\mathbf{Y}_2)\| \\
&\quad = \|(y_{11} - y_{12} + y_{21} - y_{22}, y_{11}^2 + y_{21}^2 - y_{12}^2 - y_{22}^2)\| \\
&\quad \leq |y_{11} - y_{12}| + |y_{21} - y_{22}| + |y_{11}^2 - y_{12}^2| + |y_{21}^2 - y_{22}^2| \\
&\quad \leq |y_{11} - y_{12}| + |y_{21} - y_{22}| + (|y_{11}| + |y_{12}|)|y_{11} - y_{12}| \\
&\qquad + (|y_{21}| + |y_{22}|)|y_{21} - y_{22}| \\
&\quad \leq 3|y_{11} - y_{12}| + 3|y_{21} - y_{22}| \\
&\quad \leq 6 \max [|y_{11} - y_{12}|, |y_{21} - y_{22}|] \\
&\quad \leq 6[(y_{11} - y_{12})^2 + (y_{21} - y_{22})^2]^{1/2} = 6\|\mathbf{Y}_1 - \mathbf{Y}_2\|
\end{aligned}
$$

This establishes the Lipschitz condition with Lipschitz constant 6.

EXAMPLE 9.2.2 Show that $\mathbf{F} = (y_1, \sqrt{y_2})$ does not satisfy a Lipschitz condition in $R_3 = \{(t, y_1, y_2) \mid |t| \leq 1, |y_1| \leq 1, 0 \leq y_2 \leq 1\}$. Let $\mathbf{Y}_1 = (y_{11}, y_{21})$, $\mathbf{Y}_2 = (y_{12}, y_{22})$. Then

$$\|\mathbf{F}(\mathbf{Y}_1) - \mathbf{F}(\mathbf{Y}_2)\| = \|(y_{11} - y_{12}, \sqrt{y_{21}} - \sqrt{y_{22}})\|$$

If we let $y_{11} = y_{12} = 0$, then

$$\|\mathbf{F}(\mathbf{Y}_1) - \mathbf{F}(\mathbf{Y}_2)\| = \frac{|y_{21} - y_{22}|}{\sqrt{y_{21}} + \sqrt{y_{22}}} = \frac{\|\mathbf{Y}_1 - \mathbf{Y}_2\|}{\sqrt{y_{21}} + \sqrt{y_{22}}}$$

But $(\sqrt{y_{21}} + \sqrt{y_{22}})^{-1}$ can be made larger than any constant by taking y_{21} and y_{22} close enough to zero. This example shows that continuity in R_{n+1} does not imply a Lipschitz condition.

Theorem 9.2.1 If $\mathbf{F}(t, y_1, y_2, \ldots, y_n)$ has a continuous partial derivative with respect to each of the variables $y_1, y_2, \ldots, y_n$ in

$$R_{n+1} = \{(t, y_1, y_2, \ldots, y_n) \mid |t - t_0| \leq r_0, |y_1 - a_1| \leq r_1, \ldots, |y_n - a_n| \leq r_n\}$$

then $\mathbf{F}$ satisfies a Lipschitz condition in R_{n+1}.

PROOF For simplicity consider the case where **F** and **Y** have two coordinates $f_1(t, y_1, y_2)$ and $f_2(t, y_1, y_2)$. Let $\mathbf{Y}_1 = (y_{11}, y_{12})$ and $\mathbf{Y}_2 = (y_{21}, y_{22})$ be in R_3. Then

$$\|\mathbf{F}(t,\mathbf{Y}_1) - \mathbf{F}(t,\mathbf{Y}_2)\| \le |f_1(t,y_{11},y_{12}) - f_1(t,y_{21},y_{22})| + |f_2(t,y_{11},y_{12}) - f_2(t,y_{21},y_{22})|$$

By the mean-value theorem

$$f_1(t,y_{11},y_{12}) - f_1(t,y_{21},y_{22}) = \frac{\partial f_1}{\partial y_1}(y_{11} - y_{12}) + \frac{\partial f_1}{\partial y_2}(y_{21} - y_{22})$$

$$f_2(t,y_{11},y_{12}) - f_2(t,y_{21},y_{22}) = \frac{\partial f_2}{\partial y_1}(y_{11} - y_{12}) + \frac{\partial f_2}{\partial y_2}(y_{21} - y_{22})$$

where the partial derivatives are evaluated, in each case, at a point between $\mathbf{Y}_1$ and $\mathbf{Y}_2$ which is in R_3. Since the partial derivatives are all continuous in R_3, let M stand for the maximum of the absolute values of the four first partial derivatives in R_3. Then

$$\begin{aligned}\|\mathbf{F}(t,\mathbf{Y}_1) - \mathbf{F}(t,\mathbf{Y}_2)\| &\le 2M|y_{11} - y_{22}| + 2M|y_{21} - y_{22}| \\ &\le 4M \max\,[|y_{11} - y_{12}|, |y_{21} - y_{22}|] \\ &\le 4M\,\|\mathbf{Y}_1 - \mathbf{Y}_2\|\end{aligned}$$

This completes the proof in the simple case. The general case follows by an obvious extension of these ideas.

In Sec. 9.6, we shall prove the following theorem.

Theorem 9.2.2 If $\mathbf{F}(t,\mathbf{Y})$ is continuous and satisfies a Lipschitz condition in

$$R_{n+1} = \{(t, y_1, y_2, \ldots, y_n) \mid |t - t_0| \le r_0, |y_1 - a_1| \le r_1, \ldots, |y_n - a_n| \le r_n\}$$

$r_0, r_1, r_2, \ldots, r_n$ all positive, then there exists a unique solution of the initial-value problem $\dot{\mathbf{Y}} = \mathbf{F}(t,\mathbf{Y})$, $\mathbf{Y}(t_0) = \mathbf{A} = (a_1, a_2, \ldots, a_n)$ for $|t - t_0| \le h = \min\,[r_0, r_1/M, r_2/M, \ldots, r_n/M]$ where $M = \max \|\mathbf{F}(t,\mathbf{Y})\|$ in R_{n+1}.

EXAMPLE 9.2.3 Show that there exists a unique solution of the initial-value problem $\dot{y}_1 = y_1 + y_2$, $\dot{y}_2 = y_1{}^2 + y_2{}^2$, $y_1(0) = 0$, $y_2(0) = 0$. According to Example 9.2.1, $\mathbf{F} = (y_1 + y_2, y_1{}^2 + y_2{}^2)$ satisfies a Lipschitz condition in

$R_3 = \{(t,y_1,y_2) \mid |t| \leq 1, |y_1| \leq 1, |y_2| \leq 1\}$ which contains the point (0,0,0). According to Theorem 9.2.2, there exists a unique solution to the initial-value problem. Clearly the constant function $\mathbf{Y} = (0,0)$ satisfies all the conditions. Therefore, this is the unique solution.

EXAMPLE 9.2.4 Show that the initial-value problem $\dot{y}_1 = y_1$, $\dot{y}_2 = \sqrt{y_2}$, $y_1(0) = 0$, $y_2(0) = 0$, has a nonunique solution. Clearly $y_1 = 0$, $y_2 = 0$ satisfies all the conditions. However, so does the solution $y_1 = 0$, $y_2 = t^2/4$. Notice that $\mathbf{F} = (y_1,\sqrt{y_2})$ does not satisfy a Lipschitz condition in

$$R_3 = \{(t,y_1,y_2) \mid |t| \leq 1, |y_1| \leq 1, 0 \leq y_2 \leq 1\}$$

EXAMPLE 9.2.5 Show that the initial-value problem $\dot{y}_1 = a_{11}(t)y_1 + a_{12}(t)y_2 + b_1(t)$, $\dot{y}_2 = a_{21}(t)y_1 + a_{22}(t)y_2 + b_2(t)$, $y_1(t_0) = c_1$, $y_2(t_0) = c_2$, where a_{11}, a_{12}, a_{21}, a_{22}, b_1, and b_2 are all continuous for $|t - t_0| \leq r_0$, has a unique solution in the entire interval $I = \{t \mid |t - t_0| \leq r_0\}$. Consider the "rectangle" $R_3 = \{(t,y_1,y_2) \mid |t - t_0| \leq r_0, |y_1 - c_1| \leq r_1, |y_2 - c_2| \leq r_2\}$. Let $f_1 = a_{11}y_1 + a_{12}y_2 + b_1$, $f_2 = a_{21}y_1 + a_{22}y_2 + b_2$. Since $\partial f_1/\partial y_1 = a_{11}(t)$, $\partial f_1/\partial y_2 = a_{12}(t)$, $\partial f_2/\partial y_1 = a_{21}(t)$, $\partial f_2/\partial y_2 = a_{22}(t)$ are all continuous for $|t - t_0| \leq r_0$ for all y_1 and y_2, $\mathbf{F} = (f_1,f_2)$ satisfies a Lipschitz condition for arbitrary positive r_1 and r_2. It can be shown (Exercise 9.2.1) that $M = \max \|\mathbf{F}\| \leq \alpha r_1 + \beta r_2 + \gamma$, where α, β, and γ are positive constants. Therefore, $\min [r_0,r_1/M,r_2/M]$ is either r_0 or some fixed positive number independent of where the initial conditions are taken in I. Hence, Theorem 9.2.2 gives us existence and uniqueness in the entire interval or gives us a minimum positive distance away from the initial point where a unique solution can be obtained. In the latter case, a finite sequence of initial-value problems can be solved to continue the solution uniquely throughout the interval I. The situation covered in this example is typical of the general linear first order system, which will be discussed in the next two sections.

EXERCISES 9.2

1 Referring to Example 9.2.5, show that $M = \max \|\mathbf{F}\| \leq \alpha r_1 + \beta r_2 + \gamma$, where α, β, and γ are positive constants. (α and β depend only on a_{11}, a_{12}, a_{21}, a_{22}.)

2 Find a first order system equivalent to the second order equation $\ddot{y} + 5\dot{y} + 6y = e^t$. Use Theorem 9.2.2 to prove that the initial-value problem $y(0) = y_0$, $\dot{y}(0) = \dot{y}_0$ has a unique solution.

3 Find a first order system equivalent to the second order equation $t^2\ddot{y} + 6t\dot{y} + 6y = 0$. Use Theorem 9.2.2 to prove that the initial-value problem $y(1) = 0$, $\dot{y}(1) = 1$ has a unique solution. Find the solution and state where it is valid.

4 Show that $\mathbf{F} = (t(y_1^2 + y_2^2),\ y_1^2 - y_2^2)$ satisfies a Lipschitz condition in $R_3 = \{(t,y_1,y_2) \mid |t - t_0| \le r_0,\ |y_1 - a_1| \le r_1,\ |y_2 - a_2| \le r_2\}$. Find a Lipschitz constant.

5 Show that the initial-value problem $\dot{y}_1 = t(y_1^2 + y_2^2)$, $\dot{y}_2 = 0$, $y_1(0) = 1$, $y_2(0) = 0$, has a unique solution. Can the solution be continued up to $t = \sqrt{2}$?

9.3 LINEAR FIRST ORDER SYSTEMS

A first order system is called linear if it can be written in the form

$$\dot{\mathbf{Y}} = M(t)\mathbf{Y} + \mathbf{B}(t)$$

where $M(t)$ is an $n \times n$ matrix and $\mathbf{B}(t)$ is an $n \times 1$ vector which do not depend on $\mathbf{Y}$. In this notation $\mathbf{Y}$ is considered an $n \times 1$ vector. Example 9.2.5 showed that for such a system, in which $M(t)$ and $\mathbf{B}(t)$ are continuous in some interval $\{t \mid |t - t_0| \le r_0\}$, the initial-value problem $\dot{\mathbf{Y}} = M(t)\mathbf{Y} + B(t)$, $\mathbf{Y}(t_0) = \mathbf{A}$ always has a unique solution in the entire interval. The proof of this general case is exactly like the example, taking into account the higher dimension. Therefore, we state without further comment the basic theorem of this section.

Theorem 9.3.1 The initial-value problem $\dot{\mathbf{Y}} = M(t)\mathbf{Y} + \mathbf{B}(t)$, $\mathbf{Y}(t_0) = \mathbf{A}$ has a unique solution in the interval $I = \{t \mid |t - t_0| \le r_0\}$ provided $M(t)$ and $\mathbf{B}(t)$ are continuous in I.

In the remainder of this section, we discuss the general question of finding solutions of the initial-value problem, and in the next section we deal with the special case where M is constant. It is convenient to write the linear system as $(D - M)\mathbf{Y} = \mathbf{B}$, where $D\mathbf{Y} = \dot{\mathbf{Y}}$. We first observe that if $\mathbf{Z}$ is a solution of the homogeneous equation $(D - M)\mathbf{Y} = \mathbf{0}$ and $\mathbf{W}$ is a solution of the nonhomogeneous equation $(D - M)\mathbf{Y} = \mathbf{B}$, then $\mathbf{Z} + \mathbf{W}$ is a solution of the nonhomogeneous equation since

$$(D - M)(\mathbf{Z} + \mathbf{W}) = (D - M)\mathbf{Z} + (D - M)\mathbf{W} = \mathbf{B}$$

Conversely, if $\mathbf{Y}_p$ is a particular solution of the nonhomogeneous equation and $\mathbf{Y}$ is *any* other solution, then $\mathbf{Y}^* = \mathbf{Y} - \mathbf{Y}_p$ is in the null space of the operator $D - M$; that is,

$$(D - M)\mathbf{Y}^* = (D - M)(\mathbf{Y} - \mathbf{Y}_p) = \mathbf{B} - \mathbf{B} = \mathbf{0}$$

Therefore, $\mathbf{Y} = \mathbf{Y}^* + \mathbf{Y}_p$ and the problem reduces to finding the general representation of vectors in the null space. We shall find that the null space is spanned by n independent vectors and is therefore n-dimensional.

Suppose that $\mathbf{Y}$ is any particular solution of the homogeneous equation. Then $\mathbf{Y}$ is the unique solution of some initial-value problem $\dot{\mathbf{Y}} = M\mathbf{Y}$, $\mathbf{Y}(t_0) = \mathbf{A}$, where t_0 is in the interval $\{t \mid a \leq t \leq b\}$, where M is continuous. Suppose $\mathbf{Y}_1, \mathbf{Y}_2, \ldots, \mathbf{Y}_n$ is a set of solutions of the homogeneous equation which is independent in the interval. Clearly,

$$\mathbf{Y}_c = c_1\mathbf{Y}_1 + c_2\mathbf{Y}_2 + \cdots + c_n\mathbf{Y}_n$$

is a solution of the homogeneous equation for arbitrary constants $c_1, c_2, \ldots, c_n$. It remains to show that $\mathbf{Y} = \mathbf{Y}_c$ for some particular choice of constants. We have to show that

$$\mathbf{A} = c_1\mathbf{Y}_1(t_0) + c_2\mathbf{Y}_2(t_0) + \cdots + c_n\mathbf{Y}_n(t_0)$$

has a unique solution for constants $c_1, c_2, \ldots, c_n$. But this is assured by the independence of $\mathbf{Y}_1, \mathbf{Y}_2, \ldots, \mathbf{Y}_n$, since the given vector $\mathbf{A}$ has a unique representation in terms of $\mathbf{Y}_1(t_0), \mathbf{Y}_2(t_0), \ldots, \mathbf{Y}_n(t_0)$. We have therefore shown that the general solution of the homogeneous equation can be written in the form

$$\mathbf{Y}_c = c_1\mathbf{Y}_1 + c_2\mathbf{Y}_2 + \cdots + c_n\mathbf{Y}_n$$

where $\mathbf{Y}_1, \mathbf{Y}_2, \ldots, \mathbf{Y}_n$ is a set of independent solutions in the interval where M is continuous. We call such a system of solutions a *fundamental system*. Finally, we have shown that the general solution of the nonhomogeneous equation can be written in the form

$$\mathbf{Y} = \mathbf{Y}_c + \mathbf{Y}_p$$

where $\mathbf{Y}_p$ is any particular solution of the nonhomogeneous equation. We call $\mathbf{Y}_c$ the *complementary solution*, and we can see that it contains n arbitrary constants.

If $\mathbf{Y}_1(t), \mathbf{Y}_2(t), \ldots, \mathbf{Y}_n(t)$ are n n-dimensional vector functions defined in the interval $I = \{t \mid a \leq t \leq b\}$, then we define the Wronskian of $\mathbf{Y}_1, \mathbf{Y}_2, \ldots, \mathbf{Y}_n$ as the determinant

$$W(t) = \begin{vmatrix} y_{11}(t) & y_{12}(t) & \cdots & y_{1n}(t) \\ y_{21}(t) & y_{22}(t) & \cdots & y_{2n}(t) \\ \cdots & \cdots & \cdots & \cdots \\ y_{n1}(t) & y_{n2}(t) & \cdots & y_{nn}(t) \end{vmatrix}$$

We shall now show that if $\mathbf{Y}_1, \mathbf{Y}_2, \ldots, \mathbf{Y}_n$ are solutions of the homogeneous equation $\dot{\mathbf{Y}} = M(t)\mathbf{Y}$ in I, where M is continuous, then $\mathbf{Y}_1, \mathbf{Y}_2, \ldots, \mathbf{Y}_n$ is a fundamental system if and only if their Wronskian never vanishes in I. First suppose that $W(t)$ is never zero, and consider the equation

$$c_1\mathbf{Y}_1(t_0) + c_1\mathbf{Y}_2(t_0) + \cdots + c_n\mathbf{Y}_n(t_0) = \mathbf{0}$$

as a homogeneous system of linear equations for the unknowns $c_1, c_2, \ldots, c_n$. The determinant of the system is $W(t_0)$, which is not zero. Hence, $c_1 = c_2 = \cdots = c_n = 0$. This shows that $\mathbf{Y}_1(t_0), \mathbf{Y}_2(t_0), \ldots, \mathbf{Y}_n(t_0)$ are independent. But t_0 is any point in I, which shows that $\mathbf{Y}_1, \mathbf{Y}_2, \ldots, \mathbf{Y}_n$ are independent on I. Conversely, suppose that $\mathbf{Y}_1, \mathbf{Y}_2, \ldots, \mathbf{Y}_n$ are independent in I and $W(t_0) = 0$ for some t_0 in I. Consider the initial-value problem $\dot{\mathbf{Y}} = M\mathbf{Y}$, $\mathbf{Y}(t_0) = \mathbf{0}$. This problem has the unique solution $\mathbf{Y} \equiv \mathbf{0}$. On the other hand, according to the above discussion, there is a unique set of constants $c_1, c_2, \ldots, c_n$ satisfying

$$\mathbf{0} \equiv c_1\mathbf{Y}_1 + c_2\mathbf{Y}_2 + \cdots + c_n\mathbf{Y}_n$$

However, $W(t_0) = 0$ implies that a set of constants, not all zero, can be found satisfying this equation. This contradicts the independence of $\mathbf{Y}_1, \mathbf{Y}_2, \ldots, \mathbf{Y}_n$ on I. Therefore, $W(t_0) \neq 0$, and since t_0 was any point in I, $W(t)$ never vanishes. This result simplifies the search for fundamental systems, since a check of the Wronskian will determine whether or not a given set of n solutions is fundamental.

We can now complete the discussion of the dimension of the null space of $D - M$ by showing the existence of a fundamental system. Consider the initial-value problems $\dot{\mathbf{Y}}_k = M\mathbf{Y}_k$, $\mathbf{Y}_k(t_0) = \mathbf{e}_k$, where $\mathbf{e}_1, \mathbf{e}_2, \ldots, \mathbf{e}_n$ is the standard basis. Each of these problems has a unique solution, by Theorem 9.3.1. Now we compute the Wronskian of $\mathbf{Y}_1, \mathbf{Y}_2, \ldots, \mathbf{Y}_n$ at t_0,

$$W(t_0) = \begin{vmatrix} 1 & 0 & 0 & \cdots & 0 \\ 0 & 1 & 0 & \cdots & 0 \\ 0 & 0 & 1 & \cdots & 0 \\ \cdots & \cdots & \cdots & \cdots & \cdots \\ 0 & 0 & 0 & \cdots & 1 \end{vmatrix} = 1$$

and since $W(t_0) \neq 0$, $W(t)$ will never vanish in I and hence $\mathbf{Y}_1, \mathbf{Y}_2, \ldots, \mathbf{Y}_n$ is a fundamental system.

EXAMPLE 9.3.1 Find the general solution of the system $\dot{y}_1 = y_1 + 2y_2$, $\dot{y}_2 = 3y_1 + 2y_2$. We can easily eliminate y_2 by the following computation:

$$\begin{aligned} \ddot{y}_1 = \dot{y}_1 + 2\dot{y}_2 &= \dot{y}_1 + 2(3y_1 + 2y_2) \\ &= \dot{y}_1 + 6y_1 + 2(\dot{y}_1 - y_1) \\ &= 3\dot{y}_1 + 4y_1 \end{aligned}$$

In other words, we must solve the equation

$$\ddot{y}_1 - 3\dot{y}_1 - 4y_1 = 0$$

The general solution of this equation is

$$y_1 = ae^{4t} + be^{-t}$$

Substituting into $\dot{y}_2 = 3y_1 + 2y_2$, we have

$$\dot{y}_2 - 2y_2 = 3ae^{4t} + 3be^{-t}$$

This is a linear first order equation with general solution

$$y_2 = \tfrac{3}{2}ae^{4t} - be^{-t} + ce^{2t}$$

However, y_1 and y_2 must satisfy the equation $\dot{y}_1 = y_1 + 2y_2$. Hence,

$$4ae^{4t} - be^{-t} = ae^{4t} + be^{-t} + 3ae^{4t} - 2be^{-t} + 2ce^{2t}$$

Therefore, $c = 0$, and the general solution of the system is

$$\begin{pmatrix} y_1 \\ y_2 \end{pmatrix} = \begin{pmatrix} ae^{4t} + be^{-t} \\ \tfrac{3}{2}ae^{4t} - be^{-t} \end{pmatrix} = a\begin{pmatrix} e^{4t} \\ \tfrac{3}{2}e^{4t} \end{pmatrix} + b\begin{pmatrix} e^{-t} \\ -e^{-t} \end{pmatrix}$$

Consider the vectors

$$\mathbf{Y}_1 = \begin{pmatrix} 1 \\ \tfrac{3}{2} \end{pmatrix} e^{4t} \qquad \mathbf{Y}_2 = \begin{pmatrix} 1 \\ -1 \end{pmatrix} e^{-t}$$

It is easy to show that $\mathbf{Y}_1$ and $\mathbf{Y}_2$ are independent solutions of the given system. In fact, their Wronskian is

$$W(t) = \begin{vmatrix} 1 & 1 \\ \tfrac{3}{2} & -1 \end{vmatrix} e^{3t} = -\tfrac{5}{2}e^{3t}$$

Furthermore, the numbers 4 and -1, appearing in the exponentials, are the characteristic values of the matrix

$$M = \begin{pmatrix} 1 & 2 \\ 3 & 2 \end{pmatrix}$$

of the system. This is seen from the characteristic equation

$$|M - \lambda I| = \begin{vmatrix} 1 - \lambda & 2 \\ 3 & 2 - \lambda \end{vmatrix} = \lambda^2 - 3\lambda - 4 = 0$$

Finally, the vectors $(1, \tfrac{3}{2})$ and $(1, -1)$ are corresponding characteristic vectors. We could have seen this if we had looked for exponential solutions in the form $\mathbf{Y} = \mathbf{X}e^{\lambda t}$, where $\mathbf{X}$ is constant. Then $\dot{\mathbf{Y}} = \lambda\mathbf{X}e^{\lambda t}$, and

$$\lambda\mathbf{X}e^{\lambda t} = M\mathbf{X}e^{\lambda t}$$

$$(M - \lambda I)\mathbf{X} = \mathbf{0}$$

which would require that $|M - \lambda I| = 0$ and that $\mathbf{X}$ be a characteristic vector. The condition that $W(t) \neq 0$ is exactly the condition that $\mathbf{X}_1$ and $\mathbf{X}_2$ be independent. So we come back to the same question raised in Chap. 4, namely when does an $n \times n$ matrix of constants have n independent characteristic vectors? We shall look at linear first order systems with constant coefficients from this point of view in the next section.

EXAMPLE 9.3.2 Find the general solution of the system

$$\begin{aligned} \dot{y}_1 &= 9y_1 - 3y_1 \\ \dot{y}_2 &= -3y_1 + 12y_2 - 3y_3 \\ \dot{y}_3 &= -3y_2 + 9y_3 \end{aligned}$$

We look for solutions of the form $\mathbf{Y} = \mathbf{X}e^{\lambda t}$. Substituting, we have

$$(\lambda\mathbf{X} - M\mathbf{X})e^{\lambda t} = \mathbf{0}$$

A necessary condition for nontrivial solutions is

$$|M - \lambda I| = \begin{vmatrix} 9-\lambda & -3 & 0 \\ -3 & 12-\lambda & -3 \\ 0 & -3 & 9-\lambda \end{vmatrix} = -(\lambda - 6)(\lambda - 9)(\lambda - 15)$$

The characteristic values are $\lambda_1 = 6$, $\lambda_2 = 9$, $\lambda_3 = 15$. The corresponding characteristic vectors are

$$\mathbf{X}_1 = \begin{pmatrix} 1 \\ 1 \\ 1 \end{pmatrix} \qquad \mathbf{X}_2 = \begin{pmatrix} 1 \\ 0 \\ -1 \end{pmatrix} \qquad \mathbf{X}_3 = \begin{pmatrix} 1 \\ -2 \\ 1 \end{pmatrix}$$

The general solution is therefore

$$\mathbf{Y} = c_1 \begin{pmatrix} 1 \\ 1 \\ 1 \end{pmatrix} e^{6t} + c_2 \begin{pmatrix} 1 \\ 0 \\ -1 \end{pmatrix} e^{9t} + c_3 \begin{pmatrix} 1 \\ -2 \\ 1 \end{pmatrix} e^{15t}$$

where c_1, c_2, and c_3 are arbitrary constants.

If the linear system is nonhomogeneous and we have a fundamental system of solutions $\mathbf{Y}_1, \mathbf{Y}_2, \ldots, \mathbf{Y}_n$ of the homogeneous system, then it is always possible to find a solution of the nonhomogeneous system by the method of *variation of parameters*. We write the system as

$$\dot{\mathbf{Y}} - M\mathbf{Y} = \mathbf{B}(t)$$

and look for a solution in the form

$$\mathbf{Y} = A_1(t)\mathbf{Y}_1 + A_2(t)\mathbf{Y}_2 + \cdots + A_n(t)\mathbf{Y}_n$$

Differentiating, we have

$$\dot{\mathbf{Y}} = A_1\dot{\mathbf{Y}}_1 + A_2\dot{\mathbf{Y}}_2 + \cdots + A_n\dot{\mathbf{Y}}_n + \dot{A}_1\mathbf{Y}_1 + \dot{A}_2\mathbf{Y}_2 + \cdots + \dot{A}_n\mathbf{Y}_n$$

Substituting and using the fact that $\dot{\mathbf{Y}}_k - M\mathbf{Y}_k = \mathbf{0}$, $k = 1, 2, \ldots, n$, gives

$$\dot{A}_1\mathbf{Y}_1 + \dot{A}_2\mathbf{Y}_2 + \cdots + \dot{A}_n\mathbf{Y}_n = \mathbf{B}(t)$$

This is a system of linear algebraic equations for the unknowns $\dot{A}_1, \dot{A}_2, \ldots, \dot{A}_n$. The determinant of the coefficient matrix is the Wronskian of $\mathbf{Y}_1, \mathbf{Y}_2, \ldots, \mathbf{Y}_n$, which is never zero. Hence, we can always solve for $\dot{A}_1, \dot{A}_2, \ldots, \dot{A}_n$. Integrating, we obtain $A_1, A_2, \ldots, A_n$, giving us the required solution.

EXAMPLE 9.3.3 Find the general solution of the system $\dot{y}_1 = y_1 + 2y_2 + e^t$, $\dot{y}_2 = 3y_1 + 2y_2 - e^{-t}$. This system is nonhomogeneous so we use the method of variation of parameters. We have already determined a fundamental system of solutions of the homogeneous equations, namely $\mathbf{Y}_1 = \begin{pmatrix} 2 \\ 3 \end{pmatrix} e^{4t}$ and $\mathbf{Y}_2 = \begin{pmatrix} 1 \\ -1 \end{pmatrix} e^{-t}$. We seek a solution of the nonhomogeneous system of the form

$$\mathbf{Y}(t) = A_1(t)\mathbf{Y}_1 + A_2(t)\mathbf{Y}_2$$

According to the above, we must solve

$$\dot{A}_1\mathbf{Y}_1 + \dot{A}_2\mathbf{Y}_2 = \mathbf{B}(t) = \begin{pmatrix} e^t \\ -e^{-t} \end{pmatrix}$$

or

$$2\dot{A}_1 e^{4t} + \dot{A}_2 e^{-t} = e^t$$
$$3\dot{A}_1 e^{4t} - \dot{A}_2 e^{-t} = -e^{-t}$$

Solving for $\dot{A}_1$ and $\dot{A}_2$, we have

$$\dot{A}_1 = \tfrac{1}{5}e^{-3t} - \tfrac{1}{5}e^{-5t}$$
$$\dot{A}_2 = \tfrac{3}{5}e^{2t} + \tfrac{2}{5}$$

or

$$A_1 = -\tfrac{1}{15}e^{-3t} + \tfrac{1}{25}e^{-5t}$$
$$A_2 = \tfrac{3}{10}e^{2t} + \tfrac{2}{5}t$$

and the general solution is

$$\mathbf{Y}(t) = c_1\mathbf{Y}_1 + c_2\mathbf{Y}_2 + (\tfrac{1}{25}e^{-5t} - \tfrac{1}{15}e^{-3t})\mathbf{Y}_1 + (\tfrac{3}{10}e^{2t} + \tfrac{2}{5}t)\mathbf{Y}_2$$

EXERCISES 9.3

1 Let M be a constant $n \times n$ matrix. Show that the system $\dot{\mathbf{Y}} = M\mathbf{Y}$ has fundamental system of solutions of the form $\mathbf{X}_1 e^{\lambda_1 t}, \mathbf{X}_2 e^{\lambda_2 t}, \ldots, \mathbf{X}_n e^{\lambda_n t}$ if M has n independent characteristic vectors $\mathbf{X}_1, \mathbf{X}_2, \ldots, \mathbf{X}_n$ with corresponding characteristic values $\lambda_1, \lambda_2, \ldots, \lambda_n$.

2 Let M be a constant $n \times n$ matrix. Show that the system $\dot{\mathbf{Y}} = M\mathbf{Y}$ has a fundamental system of solutions of the form $\mathbf{X}_1 e^{\lambda_1 t}, \mathbf{X}_2 e^{\lambda_2 t}, \ldots, \mathbf{X}_n e^{\lambda_n t}$, where $\mathbf{X}_1, \mathbf{X}_2, \ldots, \mathbf{X}_n$ are characteristic vectors of M, if the characteristic values $\lambda_1, \lambda_2, \ldots, \lambda_n$ are distinct.

3 Let M be a constant $n \times n$ matrix. Show that the system $\dot{\mathbf{Y}} = M\mathbf{Y}$ has a fundamental system of solutions of the form $\mathbf{X}_1 e^{\lambda_1 t}, \mathbf{X}_2 e^{\lambda_2 t}, \ldots, \mathbf{X}_n e^{\lambda_n t}$, where $\mathbf{X}_1, \mathbf{X}_2, \ldots, \mathbf{X}_n$ are characteristic vectors of M, if M is real and symmetric.

4 Find the general solution of the system $2\dot{y}_1 = y_1 + y_2$, $2\dot{y}_2 = y_1 + y_2$.

5 Find the particular solution of the system in Exercise 4 satisfying $y_1(0) = 1$, $y_2(0) = -1$.

6 Find the general solution of the system $2\dot{y}_1 = y_1 + y_2 + e^t$, $2\dot{y}_2 = y_1 + y_2 - t$.

7 Find the general solution of the system $\dot{y}_1 = 8y_1 + 9y_2 + 9y_3$, $\dot{y}_2 = 3y_1 + 2y_2 + 3y_3$, $\dot{y}_3 = -9y_1 - 9y_2 - 10y_3$.

8 Find the particular solution of the system in Exercise 7 satisfying $y_1(0) = 1$, $y_2(0) = 0$, $y_3(0) = -1$.

9 Find the general solution of the system $\dot{y}_1 = 8y_1 + 9y_2 + 9y_3 + e^{-t}$, $\dot{y}_2 = 3y_1 + 2y_2 + 3y_3 - t$, $\dot{y}_3 = -9y_1 - 9y_2 - 10y_3 + e^{2t}$.

10 Find the general solution of $\dot{y}_1 = y_1 + 2y_2 + 3y_3$, $\dot{y}_2 = 2y_2 + 3y_3$, $\dot{y}_3 = 2y_3$. *Hint:* Solve the third equation, then the second, then the first. Compare this solution with that obtained using characteristic vectors.

11 Find the general solution of the system $\dot{y}_1 = y_1 - 2y_2$, $\dot{y}_2 = y_1 - y_2$. Express the answer using real-valued functions only.

12 Consider the linear first order system $\dot{\mathbf{Y}} = M\mathbf{Y}$, with a fundamental system of solutions $\mathbf{Y}_1, \mathbf{Y}_2, \ldots, \mathbf{Y}_n$. Show that $\dot{W} = (m_{11} + m_{22} + \cdots + m_{nn})W$, where W is the Wronskian. This is another way to show that either $W \equiv 0$ or $W \neq 0$. Why?

9.4 LINEAR FIRST ORDER SYSTEMS WITH CONSTANT COEFFICIENTS

In this section, we consider only first order systems of the form $\dot{\mathbf{Y}} = M\mathbf{Y}$, where M is a constant $n \times n$ matrix. If M has n distinct characteristic values $\lambda_1, \lambda_2, \lambda_3, \ldots, \lambda_n$, then, according to Theorem 4.5.3, M has n independent characteristic vectors $\mathbf{X}_1, \mathbf{X}_2, \mathbf{X}_3, \ldots, \mathbf{X}_n$. In this case, a fundamental system of solutions of the differential equations $\dot{\mathbf{Y}} = M\mathbf{Y}$ is $\mathbf{Y}_1 = \mathbf{X}_1 e^{\lambda_1 t}$, $\mathbf{Y}_2 = \mathbf{X}_2 e^{\lambda_2 t}$, $\mathbf{Y}_3 = \mathbf{X}_3 e^{\lambda_3 t}, \ldots, \mathbf{Y}_n = \mathbf{X}_n e^{\lambda_n t}$. The Wronskian of $\mathbf{Y}_1, \mathbf{Y}_2, \ldots, \mathbf{Y}_n$ is

$$W(t) = |\mathbf{X}_1 \quad \mathbf{X}_2 \quad \cdots \quad \mathbf{X}_n| \exp(\lambda_1 + \lambda_2 + \cdots + \lambda_n)t$$

which is not zero by the independence of the characteristic vectors. The general solution of the system $\dot{\mathbf{Y}} = M\mathbf{Y}$ is therefore

$$\mathbf{Y}(t) = c_1\mathbf{X}_1 e^{\lambda_1 t} + c_2\mathbf{X}_2 e^{\lambda_2 t} + \cdots + c_n\mathbf{X}_n e^{\lambda_n t}$$

where $c_1, c_2, \ldots, c_n$ are arbitrary constants. There are cases where the characteristic values are not distinct but where we can still get a complete set of independent characteristic vectors.† The general solution is as shown, but $\lambda_1, \lambda_2, \ldots, \lambda_n$ are not distinct.

If M is real and the initial values $\mathbf{Y}(t_0)$ are real, then the solution should be real and it should be possible to express the general solution with real-valued functions and constants $c_1, c_2, \ldots, c_n$. However, one or more of the characteristic values may be complex. If $\lambda = \alpha + i\beta$ is a complex characteristic value, then $\bar{\lambda} = \alpha - i\beta$ is also a characteristic value. This is because the coefficients in the characteristic equation are real. Also, the equation for the characteristic vector $(M - \lambda)\mathbf{X} = \mathbf{0}$ implies that $(M - \bar{\lambda})\bar{\mathbf{X}} = \mathbf{0}$ and hence that $\bar{\mathbf{X}}$ is a characteristic vector. In this case, $\mathbf{X}$ and $\bar{\mathbf{X}}$ are independent because they correspond to different characteristic values since $\beta \neq 0$. Let $\varphi = \operatorname{Re}(\mathbf{X})$ and $\psi = \operatorname{Im}(\mathbf{X})$. Then

$$\operatorname{Re}(\mathbf{X}e^{\lambda t}) = e^{\alpha t}(\varphi \cos \beta t - \psi \sin \beta t)$$
$$\operatorname{Im}(\mathbf{X}e^{\lambda t}) = e^{\alpha t}(\varphi \sin \beta t + \psi \cos \beta t)$$

are real-valued solutions of the system. They are independent because if a linear combination

$$A \operatorname{Re}(\mathbf{X}e^{\lambda t}) + B \operatorname{Im}(\mathbf{X}e^{\lambda t}) = A\frac{\mathbf{X}e^{\lambda t} + \bar{\mathbf{X}}e^{\bar{\lambda}t}}{2} + B\frac{\mathbf{X}e^{\lambda t} - \bar{\mathbf{X}}e^{\bar{\lambda}t}}{2i} = 0$$

then

$$\frac{A - iB}{2}\mathbf{X}e^{\lambda t} + \frac{A + iB}{2}\bar{\mathbf{X}}e^{\bar{\lambda}t} = 0$$

implies that $A - iB = 0$ and $A + iB = 0$. But this implies that $A = B = 0$. Therefore, if there are n distinct characteristic values with some of them complex, we may express the general solution of the system in terms of real-valued functions.

† For example, if M is real and symmetric or M is hermitian.

EXAMPLE 9.4.1 Find the general solution of the system $\dot{y}_1 = y_1 - y_2$, $\dot{y}_2 = y_1 + y_2$. The matrix M in this case is

$$M = \begin{pmatrix} 1 & -1 \\ 1 & 1 \end{pmatrix}$$

and the characteristic equation is $|M - \lambda I| = \lambda^2 - 2\lambda + 2 = 0$. The characteristic values are $\lambda_1 = 1 + i$ and $\lambda_2 = 1 - i$. The corresponding characteristic vectors are

$$\mathbf{X}_1 = \begin{pmatrix} i \\ 1 \end{pmatrix} \qquad \mathbf{X}_2 = \begin{pmatrix} -i \\ 1 \end{pmatrix}$$

Therefore,

$$\text{Re}\,[\mathbf{X}_1 e^t (\cos t + i \sin t)] = \begin{pmatrix} -e^t \sin t \\ e^t \cos t \end{pmatrix}$$

$$\text{Im}\,[\mathbf{X}_1 e^t (\cos t + i \sin t)] = \begin{pmatrix} e^t \cos t \\ e^t \sin t \end{pmatrix}$$

are independent real-valued solutions. The general solution is then

$$\mathbf{Y}(t) = c_1 \begin{pmatrix} -e^t \sin t \\ e^t \cos t \end{pmatrix} + c_2 \begin{pmatrix} e^t \cos t \\ e^t \sin t \end{pmatrix}$$

Now we consider the nature of the general solution when some of the characteristic values of M are repeated roots of the characteristic equation and there are not enough independent solutions of the form $\mathbf{X}e^{\lambda t}$ to span the null space of $D - M$. For this to be the case there must be some characteristic value λ of multiplicity m with only $k < m$ corresponding independent characteristic vectors.† Let $\mathbf{X}_1, \mathbf{X}_2, \ldots, \mathbf{X}_k$ be a set of independent characteristic vectors corresponding to λ. Then the solutions $\mathbf{Y}_1 = \mathbf{X}_1 e^{\lambda t}, \mathbf{Y}_2 = \mathbf{X}_2 e^{\lambda t}, \ldots, \mathbf{Y}_k = \mathbf{X}_k e^{\lambda t}$ are independent fundamental solutions. However, corresponding to this characteristic value we are deficient by $m - k$ solutions. To generate more solutions we try something of the form‡

$$\mathbf{Y}(t) = \mathbf{Z}_1 e^{\lambda t} + \mathbf{X}_1 t e^{\lambda t}$$

Then

$$\dot{\mathbf{Y}} = \lambda \mathbf{Z}_1 e^{\lambda t} + \lambda \mathbf{X}_1 t e^{\lambda t} + \mathbf{X}_1 e^{\lambda t}$$

$$(D - M)\mathbf{Y} = [(\lambda I - M)\mathbf{Z}_1 + \mathbf{X}_1]e^{\lambda t} + (\lambda I - M)\mathbf{X}_1 t e^{\lambda t}$$

We already know that $(\lambda I - M)\mathbf{X}_1 = \mathbf{0}$, and so for $\mathbf{Y}$ to be a solution we must have

$$(M - \lambda I)\mathbf{Z}_1 = \mathbf{X}_1$$

† In other words, the dimension of the null space of $M - \lambda I$ is k.
‡ The rationale for this is given in Sec. 4.7.

This is a nonhomogeneous system of equations which may or may not have a solution $\mathbf{Z}_1$ since $|M - \lambda I| = 0$. If it does have a solution, then $\mathbf{X}_1$ and $\mathbf{Z}_1$ are independent, because if $\mathbf{X}_1 = \alpha\mathbf{Z}_1$, then

$$[M - (\lambda + \alpha)I]\mathbf{Z}_1 = \mathbf{0}$$

which implies that $\mathbf{Z}_1$ is a characteristic vector corresponding to $\lambda + \alpha$. But characteristic vectors corresponding to different characteristic values are independent. This presents a contradiction to $\mathbf{X}_1 = \alpha\mathbf{Z}_1$.

If $\mathbf{Z}_1$ can be found and $m - k > 1$, then next we try for a solution of the form

$$\mathbf{Y}(t) = \mathbf{W}_1 e^{\lambda t} + \mathbf{Z}_1 t e^{\lambda t} + \mathbf{X}_1 \frac{t^2}{2} e^{\lambda t}$$

Then

$$\dot{\mathbf{Y}} = \lambda\mathbf{W}_1 e^{\lambda t} + \lambda\mathbf{Z}_1 t e^{\lambda t} + \mathbf{Z}_1 e^{\lambda t} + \lambda\mathbf{X}_1 \frac{t^2}{2} e^{\lambda t} + \mathbf{X}_1 t e^{\lambda t}$$

$$\begin{aligned}(D - M)\mathbf{Y} &= [(\lambda I - M)\mathbf{W}_1 + \mathbf{Z}_1]e^{\lambda t} \\ &\quad + [(\lambda I - M)\mathbf{Z}_1 + \mathbf{X}_1]te^{\lambda t} + (\lambda I - M)\mathbf{X}_1 \frac{t^2}{2} e^{\lambda t}\end{aligned}$$

Since $(\lambda I - M)\mathbf{X}_1 = \mathbf{0}$ and $(\lambda I - M)\mathbf{Z}_1 + \mathbf{X}_1 = \mathbf{0}$, we seek $\mathbf{W}_1$ as a solution of

$$(M - \lambda I)\mathbf{W}_1 = \mathbf{Z}_1$$

If a $\mathbf{W}_1$ can be found and $m - k > 2$, then we look for another solution of the form

$$\mathbf{Y}(t) = \mathbf{V}_1 e^{\lambda t} + \mathbf{W}_1 t e^{\lambda t} + \mathbf{Z}_1 \frac{t^2}{2!} e^{\lambda t} + \mathbf{X}_1 \frac{t^3}{3!} e^{\lambda t}$$

etc. If this sequence of trials does not give the required number $(m - k)$ of missing solutions corresponding to λ, then we repeat the procedure with $\mathbf{X}_2, \mathbf{X}_3, \ldots, \mathbf{X}_k$. The theory of Jordan forms of Sec. 4.7 guarantees that the method will give all the required solutions.

EXAMPLE 9.4.2 Find the general solution of the system

$$\begin{aligned}\dot{y}_1 &= 2y_1 + y_2 + y_4 \\ \dot{y}_2 &= y_1 + 3y_2 - y_3 + 3y_4 \\ \dot{y}_3 &= y_2 + 2y_3 + y_4 \\ \dot{y}_4 &= y_1 - y_2 - y_3 - y_4\end{aligned}$$

The matrix M is

$$M = \begin{pmatrix} 2 & 1 & 0 & 1 \\ 1 & 3 & -1 & 3 \\ 0 & 1 & 2 & 1 \\ 1 & -1 & -1 & -1 \end{pmatrix}$$

and its characteristic equation is

$$|M - \lambda I| = \lambda(\lambda - 2)^3 = 0$$

The characteristic values are $\lambda_1 = 0$ with multiplicity 1, and $\lambda_2 = 2$ with multiplicity 3. There is a characteristic vector $\mathbf{X}_1 = (0,1,0,-1)$ corresponding to λ_1 and a characteristic vector $\mathbf{X}_2 = (1,0,1,0)$ corresponding to λ_2, but the null space of $M - \lambda_2 I$ is of dimension 1, so we cannot find more independent characteristic vectors. Instead, we look for a solution of the form

$$\mathbf{Y} = \mathbf{Z}_2 e^{2t} + \mathbf{X}_2 t e^{2t}$$

The equation for $\mathbf{Z}_2$ is $(M - 2I)\mathbf{Z}_2 = \mathbf{X}_2$. Solving, we have $\mathbf{Z}_2 = (0,\frac{3}{2},0,-\frac{1}{2})$. Next we look for a solution of the form

$$\mathbf{Y} = \mathbf{W}_2 e^{2t} + \mathbf{Z}_2 t e^{2t} + \mathbf{X}_2 \frac{t^2}{2} e^{2t}$$

The equation for $\mathbf{W}_2$ is $(M - 2I)\mathbf{W}_2 = \mathbf{Z}_2$. Solving, we have $\mathbf{W}_2 = (0,-\frac{1}{2},-\frac{1}{2},\frac{1}{2})$. The general solution is therefore

$$\begin{aligned} \mathbf{Y}(t) &= c_1\mathbf{X}_1 + c_2\mathbf{X}_2 e^{2t} + c_3(\mathbf{Z}_2 e^{2t} + \mathbf{X}_2 t e^{2t}) \\ &\quad + c_4\left(\mathbf{W}_2 e^{2t} + \mathbf{Z}_2 t e^{2t} + \mathbf{X}_2 \frac{t^2}{2} e^{2t}\right) \end{aligned}$$

If some of the repeated characteristic values are complex, then we can carry out the above procedure with these complex values, but we should expect the vectors **Z**, **W**, V, etc., to be complex even if M is real. In the end we can express the general solution in terms of real-valued functions. We should expect these solutions to involve functions like $te^{\alpha t}\cos\beta t$, $te^{\alpha t}\sin\beta t$, $t^2e^{\alpha t}\cos\beta t$, $t^2e^{\alpha t}\sin\beta t$, etc.

We conclude this section by considering the solution of first order systems with constant coefficients by the use of Laplace transforms. The idea here is that in transforming the differential equations we change the system into a system of algebraic equations. We solve these algebraic equations for the transforms of the unknowns and then invert to find the solution.

EXAMPLE 9.4.3 Find the solution of the system $\dot{x} = 5x - y$, $\dot{y} = 3x + y$ satisfying $x(0) = x_0$, $y(0) = y_0$. Transforming each equation we have, letting $X(s) = \mathscr{L}[x(t)]$, $Y(s) = \mathscr{L}[y(t)]$,

$$sX - x_0 = 5X - Y$$
$$sY - y_0 = 3X + Y$$

Solving for X and Y, we have

$$X = \frac{x_0 s - x_0 - y_0}{s^2 - 6s + 8} = \frac{1}{2}\frac{y_0 - x_0}{s - 2} + \frac{1}{2}\frac{3x_0 - y_0}{s - 4}$$

$$Y = \frac{y_0 s - 5y_0 + 3x_0}{s^2 - 6s + 8} = \frac{3}{2}\frac{y_0 - x_0}{s - 2} + \frac{1}{2}\frac{3x_0 - y_0}{s - 4}$$

Inverting the transforms, we obtain

$$x(t) = \tfrac{1}{2}(y_0 - x_0)e^{2t} + \tfrac{1}{2}(3x_0 - y_0)e^{4t}$$
$$y(t) = \tfrac{3}{2}(y_0 - x_0)e^{2t} + \tfrac{1}{2}(3x_0 - y_0)e^{4t}$$

EXERCISES 9.4

1 Find the general solution of each of the following systems:

(*a*) $\dot{y}_1 = 2y_1 + y_2$
$\dot{y}_2 = y_1 + 2y_2$

(*b*) $\dot{y}_1 = y_1 - 2y_2$
$\dot{y}_2 = y_1 - y_2$

(*c*) $\dot{y}_1 = y_1 + y_2$
$\dot{y}_2 = -4y_1 + y_2$

(*d*) $2\dot{y}_1 = 3y_1 + y_2$
$2\dot{y}_2 = -y_1 + y_2$

2 Find the general solution of each of the following systems:

(*a*) $\dot{y}_1 = 3y_1 + y_2$
$\dot{y}_2 = y_1 + 3y_2$
$\dot{y}_3 = 2y_3$

(*b*) $\dot{y}_1 = 5y_1 + y_2 + y_3$
$\dot{y}_2 = -3y_1 + y_2 - 3y_3$
$\dot{y}_3 = -2y_1 - 2y_2 + 2y_3$

(*c*) $\dot{y}_1 = -8y_1 + 5y_2 + 4y_3$
$\dot{y}_2 = 5y_1 + 3y_2 + y_3$
$\dot{y}_3 = 4y_1 + y_2$

(*d*) $\dot{y}_1 = y_1 + y_2 + y_3$
$\dot{y}_2 = 2y_1 + y_2 - y_3$
$\dot{y}_3 = -y_2 + y_3$

3 Find the general solution of $\dot{y}_1 = y_1 - 2y_2 - \sin t$, $\dot{y}_2 = y_1 - y_2 + \cos t$.

4 Find the general solution of $\dot{y}_1 = y_1 + y_2 + y_3 - 3e^{-t}$, $\dot{y}_2 = 2y_1 + y_2 - y_3 + 6e^{-t}$, $\dot{y}_3 = -y_2 + y_3$.

5 Find the general solution of the system:

$$\dot{y}_1 = -7y_1 - 4y_4$$
$$\dot{y}_2 = -13y_1 - 2y_2 - y_3 - 8y_4$$
$$\dot{y}_3 = 6y_1 + y_2 + 4y_4$$
$$\dot{y}_4 = 15y_1 + y_2 + 9y_4$$

Express the solution in terms of real-valued functions.

6 Solve part (*a*) of Exercise 1 using the Laplace transform.
7 Solve part (*d*) of Exercise 1 using the Laplace transform.
8 Solve part (*c*) of Exercise 2 using the Laplace transform.
9 Solve Exercise 3 using the Laplace transform.
10 Solve Exercise 4 using the Laplace transform.

9.5 HIGHER ORDER LINEAR SYSTEMS

In this section we consider some additional techniques for solving higher order linear systems. In most cases, these can be reduced to first order systems, but it is not necessary to do so to solve them. We begin with an example.

EXAMPLE 9.5.1 Find the general solution of the system

$$\ddot{y}_1 - 3\dot{y}_1 + \dot{y}_2 = -2y_1 + y_2$$
$$\dot{y}_1 + \dot{y}_2 = 2y_1 - y_2$$

If we let D stand for the derivative with respect to t, then we can write the system as

$$(D^2 - 3D + 2)y_1 + (D - 1)y_2 = 0$$
$$(D - 2)y_1 + (D + 1)y_2 = 0$$

To eliminate y_2, we operate with $D + 1$ on the first equation and with $D - 1$ on the second equation and subtract. The result is

$$[(D + 1)(D^2 - 3D + 2) - (D - 1)(D - 2)]y_1 = 0$$
$$D(D - 1)(D - 2)y_1 = 0$$

The general solution of this last equation is

$$y_1 = c_1 + c_2e^t + c_3e^{2t}$$

Substituting in the second equation, we have

$$(D + 1)y_2 = -(D - 2)y_1 = 2c_1 + c_2e^t$$

Solving for y_2 gives

$$y_2 = 2c_1 + \frac{c_2}{2}e^t + c_4e^{-t}$$

The y_1 and y_2 found above satisfy the second equation, but do they satisfy the first equation? Substituting, we have

$$(D^2 - 3D + 2)y_1 + (D - 1)y_2 = -2c_4e^{-t} = 0$$

In order for this equation to be satisfied $c_4 = 0$. Therefore, the solution contains three arbitrary constants, and it is

$$y_1 = c_1 + c_2e^t + c_3e^{2t}$$

$$y_2 = 2c_1 + \frac{c_2}{2}e^t$$

This example raises a couple of interesting questions. Do we have the general solution? How many arbitrary constants should the solution contain? The answer to both questions can be obtained by finding an equivalent first order system. Let $z_1 = y_1$, $z_2 = \dot{y}_1$, $z_3 = y_2$. Then

$$\dot{z}_1 = z_2$$

$$\dot{z}_2 + \dot{z}_3 = -2z_1 + 3z_2 + z_3$$

$$\dot{z}_3 = 2z_1 - z_2 - z_3$$

or

$$\dot{z}_1 = z_2$$

$$\dot{z}_2 = -4z_1 + 4z_2 + 2z_3$$

$$\dot{z}_3 = 2z_1 - z_2 - z_3$$

Therefore, since the given system can be written as a first order linear system with three unknowns, the general solution exists with three arbitrary constants. These constants could be taken as initial values $z_1(0) = y_1(0)$, $z_2(0) = \dot{y}_1(0)$, and $z_3(0) = y_2(0)$. As we have displayed the solution,

$$y_1(0) = c_1 + c_2 + c_3$$

$$\dot{y}_1(0) = c_2 + 2c_3$$

$$y_2(0) = 2c_1 + \frac{c_2}{2}$$

Since the determinant of the coefficients of c_1, c_2, c_3 is not zero, our solution is equivalent to the general solution of the first order system containing z_1, z_2, z_3.

Example 9.5.1 illustrates what one should do to solve higher order linear systems. If the highest order of the derivatives of y_1 in the system is k_1, then introduce k_1 new variables $z_1 = y_1$, $z_2 = \dot{y}_1, \ldots, z_{k_1} = y_1^{(k_1-1)}$. Similarly introduce new variables for $y_2, \dot{y}_2, \ldots, y_2^{(k_2-1)}$, $y_3, \dot{y}_3, \ldots, y_3^{(k_3-1)}, \ldots, y_n$, $\dot{y}_n, \ldots, y_n^{(k_n-1)}$, where $k_2, k_3, \ldots, k_n$ are the highest orders of the derivatives of $y_2, y_3, \ldots, y_n$, respectively. Let p be the number of new variables introduced; that is,

$$p = k_1 + k_2 + \cdots + k_n$$

We then write the system as

$$a_{11}\dot{z}_1 + a_{12}\dot{z}_2 + \cdots + a_{1p}\dot{z}_p = b_{11}z_1 + b_{12}z_2 + \cdots + b_{1p}z_p + f_1$$
$$a_{21}\dot{z}_1 + a_{22}\dot{z}_2 + \cdots + a_{2p}\dot{z}_p = b_{21}z_1 + b_{22}z_2 + \cdots + b_{2p}z_p + f_2$$
$$\cdots\cdots\cdots\cdots\cdots\cdots\cdots\cdots\cdots\cdots\cdots\cdots\cdots$$
$$a_{p1}\dot{z}_1 + a_{p2}\dot{z}_2 + \cdots + a_{pp}\dot{z}_p = b_{p1}z_1 + b_{p2}z_2 + \cdots + b_{pp}z_p + f_p$$

If the determinant of the coefficients of $\dot{z}_1, \dot{z}_2, \ldots, \dot{z}_p$ is different from zero on some interval, then we can write a linear first order system of the type we studied in Sec. 9.3 equivalent to the original system in that interval. In this case, the general solution of the system will contain p arbitrary constants. If the determinant of the coefficients of $\dot{z}_1, \dot{z}_2, \ldots, \dot{z}_p$ is zero, then we say that the system is *degenerate*. In this case, the original system may or may not have solutions, depending on other considerations. We shall illustrate with some examples.

EXAMPLE 9.5.2 Find all possible solutions of the system

$$\ddot{y}_1 - \dot{y}_1 + \dot{y}_2 - y_2 = 0$$
$$\ddot{y}_1 + \dot{y}_1 + \dot{y}_2 + 2y_2 = 0$$

This system is degenerate because if we introduce the variables $z_1 = y_1$, $z_2 = \dot{y}_1$, $z_3 = y_2$, the system becomes

$$\dot{z}_1 = z_2$$
$$\dot{z}_2 + \dot{z}_3 = z_2 + z_3$$
$$\dot{z}_2 + \dot{z}_3 = -z_2 - 2z_3$$

and the determinant of the coefficient matrix of $\dot{z}_1, \dot{z}_2, \dot{z}_3$ is zero. However, subtracting the first equation from the second, we can show that $y_2 = -\frac{2}{3}\dot{y}_1$, from which it follows that $\ddot{y}_1 - \dot{y}_1 = 0$. The general solution is

$$y_1 = c_1 + c_2e^t$$
$$y_2 = -\tfrac{2}{3}c_2e^t$$

It only contains two arbitrary constants, not the expected three.

EXAMPLE 9.5.3 Find all possible solutions of

$$\ddot{y}_1 - 4y_1 + \dot{y}_2 - 2y_2 = t$$
$$\dot{y}_1 + 2y_1 + y_2 = e^t$$

The system is degenerate because if we introduce the variables $z_1 = y_1$, $z_2 = \dot{y}_1$, $z_3 = y_2$ the system becomes

$$\dot{z}_1 = z_2$$

$$\dot{z}_2 + \dot{z}_3 = 4z_1 + 2z_3 + t$$

$$0 = -2z_1 - z_2 - z_3 + e^t$$

and the determinant of the coefficient matrix of $\dot{z}_1, \dot{z}_2, \dot{z}_3$ is zero. If we operate with $D - 2$ on the second equation, we have

$$\ddot{y}_1 - 4y_1 + \dot{y}_2 - 2y_2 = (D - 2)e^t = -e^t$$

But this is clearly inconsistent with the first equation, so we have no solutions in this case.

If a linear system is nondegenerate and has constant coefficients, except for the nonhomogeneous terms, then it can be solved using the Laplace transform with the initial values of $z_1, z_2, \ldots, z_p$ serving as arbitrary parameters.

EXAMPLE 9.5.4 Solve the following system

$$(D^2 - 3D + 2)x(t) + (D - 1)y(t) = e^{3t}$$

$$(D - 2)x(t) + (D + 1)y(t) = 0$$

subject to $x(0) = 1$, $\dot{x}(0) = 0$, $y(0) = -1$. Let $X(s) = \mathscr{L}[x(t)]$ and $Y(s) = \mathscr{L}[y(t)]$. Then taking Laplace transforms we have,

$$(s^2 - 3s + 2)X + (s - 1)Y = \frac{1}{s - 3} + s - 4$$

$$(s - 2)X + (s + 1)Y = 0$$

Solving for $X(s)$ and $Y(s)$ gives

$$X(s) = \frac{s + 1}{s(s - 1)(s - 2)(s - 3)} + \frac{(s + 1)(s - 4)}{s(s - 1)(s - 2)}$$

$$Y(s) = \frac{-1}{s(s - 1)(s - 3)} - \frac{s - 4}{s(s - 1)}$$

They can be written, using partial fraction expansions, as

$$X(s) = -\frac{13}{6}\frac{1}{s} + \frac{7}{s-1} - \frac{9}{2}\frac{1}{s-2} + \frac{2}{3}\frac{1}{s-3}$$

$$Y(s) = -\frac{13}{3}\frac{1}{s} + \frac{7}{2}\frac{1}{s-1} - \frac{1}{6}\frac{1}{s-3}$$

Inverting, we obtain

$$x(t) = -\tfrac{13}{6} + 7e^t - \tfrac{9}{2}e^{2t} + \tfrac{2}{3}e^{3t}$$
$$y(t) = -\tfrac{13}{3} + \tfrac{7}{2}e^t - \tfrac{1}{6}e^{3t}$$

EXERCISES 9.5

1 Show that each of the following systems is nondegenerate and find the general solution.

(*a*) $2\ddot{y}_1 + 2\dot{y}_1 + \ddot{y}_2 - 3\dot{y}_2 = y_1 - 2y_2$
$\quad \dot{y}_1 + \dot{y}_2 = -2y_1 + 2y_2$

(*b*) $3\ddot{y}_1 + y_1 + \ddot{y}_2 + 3y_2 = 0$
$\quad 2\ddot{y}_1 + y_1 + \ddot{y}_2 + 2y_2 = 0$

(*c*) $\dot{y}_1 + \dot{y}_3 = 8y_1 + 4y_2 + 12y_3 + \sin 3t$
$\quad \dot{y}_1 + \dot{y}_2 + 2\dot{y}_3 = -y_1 - y_3$
$\quad 3\dot{y}_1 + 2\dot{y}_2 + \ddot{y}_3 + 5\dot{y}_3 = 6y_1 + 4y_2 + 11y_3$

(*d*) $\ddot{y}_1 - 4\dot{y}_1 + 3\dot{y}_2 = -4y_1 + 1$
$\quad \dot{y}_1 + \dot{y}_2 = 2y_1 - 2y_2 + t$

2 Show that each of the following systems is degenerate and find the general solution if possible.

(*a*) $\ddot{y}_1 + \ddot{y}_2 = y_1 - y_2$
$\quad \dot{y}_1 + \dot{y}_2 = -2y_1 + y_2$

(*b*) $\dot{y}_1 + \dot{y}_2 = -2y_1 - 2y_2 + e^{2t}$
$\quad \dot{y}_1 + \dot{y}_2 = 2y_1 + 2y_2 - e^{-2t}$

(c) $\dot{y}_1 + \dot{y}_2 + 2\dot{y}_3 = y_1 + 1$
$\quad \dot{y}_1 - \dot{y}_2 + \dot{y}_3 = -y_3 + t$
$\quad \dot{y}_1 + 5\dot{y}_2 + 4\dot{y}_3 = 3y_1 + 2y_3 + 4 - 2t$

3 Solve Exercise 1 using the Laplace transform.

*9.6 EXISTENCE AND UNIQUENESS THEOREM

The main purpose of this section is to prove Theorem 9.2.2, which is an existence and uniqueness theorem for solutions of linear or nonlinear first order systems of differential equations. The method of proof is based on the famous Picard iteration method, which was also used in the corresponding proof for a single first order equation in Sec. 5.8.

Before proceeding with the proof, the reader should be reminded of the following simple facts about n-dimensional real vectors. Let $\mathbf{U} = (u_1, u_2, \ldots, u_n)$.

Then

$$\|\mathbf{U}\| = (u_1^2 + u_2^2 + \cdots + u_n^2)^{1/2}$$

$$|u_k| = \sqrt{u_k^2} \leq (u_1^2 + u_2^2 + \cdots + u_n^2)^{1/2} = \|\mathbf{U}\|$$

If $\mu = \max [|u_1|, |u_2|, \ldots, |u_n|]$, then

$$\|\mathbf{U}\| = (u_1^2 + u_2^2 + \cdots + u_n^2)^{1/2} \leq \sqrt{n\mu^2} = \sqrt{n}\,\mu$$

In fact, $\mu \leq \|\mathbf{U}\| \leq \sqrt{n}\,\mu$ and $n^{-1/2}\|\mathbf{U}\| \leq \mu \leq \|\mathbf{U}\|$.

Theorem 9.2.2 If $\mathbf{F}(t,\mathbf{Y})$ is continuous and satisfies a Lipschitz condition in

$$R_{n+1} = \{(t, y_1, y_2, \ldots, y_n) \mid |t - t_0| \leq r_0, |y_1 - a_1| \leq r_1, \ldots, |y_n - a_n| \leq r_n\}$$

$r_0, r_1, r_2, \ldots, r_n$ all positive, then there exists a unique solution of the initial-value problem $\dot{\mathbf{Y}} = \mathbf{F}(t,\mathbf{Y})$, $\mathbf{Y}(t_0) = \mathbf{A} = (a_1, a_2, \ldots, a_n)$ for $|t - t_0| \leq h = \min [r_0, r_1/M, r_2/M, \ldots, r_n/M]$, where $M = \max \|F(t,\mathbf{Y})\|$ in R_{n+1}.

PROOF We begin by writing the equivalent integral equation

$$\mathbf{Y}(t) = \mathbf{Y}(t_0) + \int_{t_0}^{t} \mathbf{F}[\tau,\mathbf{Y}(\tau)]\,d\tau$$

It is clear that the system $\dot{\mathbf{Y}} = \mathbf{F}(t,\mathbf{Y})$ will have a unique solution if and only if the integral equation has a unique solution. We shall assume that $t \geq t_0$. Otherwise, the proof can be repeated with minor modifications. We define the following iterates:

$$\mathbf{Y}_0 = \mathbf{Y}(t_0)$$

$$\mathbf{Y}_1 = \mathbf{Y}_0 + \int_{t_0}^{t} \mathbf{F}(\tau,\mathbf{Y}_0)\,d\tau$$

$$\mathbf{Y}_2 = \mathbf{Y}_0 + \int_{t_0}^{t} \mathbf{F}[\tau,\mathbf{Y}_1(\tau)]\,d\tau$$

$$\cdots\cdots\cdots\cdots\cdots\cdots\cdots\cdots$$

$$\mathbf{Y}_k = \mathbf{Y}_0 + \int_{t_0}^{t} \mathbf{F}[\tau,\mathbf{Y}_{k-1}(\tau)]\,d\tau$$

$$\cdots\cdots\cdots\cdots\cdots\cdots\cdots\cdots$$

We must first show that for all $k \geq 1$ the iterates are in the "rectangle" R_{n+1} if $t - t_0$ is not too large. We begin with $\mathbf{Y}_1$. Let y_{1j} be the jth coordinate of $\mathbf{Y}_1$ and F_j be the jth coordinate of $\mathbf{F}$. Then

$$|y_{1j} - a_j| = \left| \int_{t_0}^{t} F_j(\tau,\mathbf{Y}_0)\, d\tau \right| \leq \int_{t_0}^{t} |F_j(\tau,\mathbf{Y}_0)|\, d\tau$$
$$\leq \int_{t_0}^{t} \|\mathbf{F}(\tau,\mathbf{Y}_0)\|\, d\tau \leq M(t - t_0)$$

We want $t - t_0 \leq r_0$ and $|y_{1j} - a_j| \leq r_j$ for $j = 1, 2, \ldots, n$. Therefore, we take

$$t - t_0 \leq \min\left(r_0, \frac{r_1}{M}, \frac{r_2}{M}, \frac{r_3}{M}, \ldots, \frac{r_n}{M}\right) = h$$

Assuming that $|y_{kj} - a_j| \leq r_j$ for $j = 1, 2, \ldots, n$, we have

$$|y_{k+1j} - a_j| = \left| \int_{t_0}^{t} F_j(\tau,\mathbf{Y}_k)\, d\tau \right|$$
$$\leq \int_{t_0}^{t} |F_j(\tau,\mathbf{Y}_k)|\, d\tau$$
$$\leq \int_{t_0}^{t} \|\mathbf{F}(\tau,\mathbf{Y}_k)\|\, d\tau \leq M(t - t_0) \leq Mh \leq r_j$$

This shows by induction that $(t,\mathbf{Y}_k)$ is in R_{n+1} for all k and $t - t_0 \leq h$.

The next part of the proof is to show that the iterates converge to a vector $\mathbf{Y}(t)$ which is a solution of the integral equation. This will use the Lipschitz condition

$$\|\mathbf{F}(t,\mathbf{Y}) - \mathbf{F}(t,\mathbf{Y}^*)\| \leq K\|\mathbf{Y} - \mathbf{Y}^*\|$$

for each $(t,\mathbf{Y})$ and $(t,\mathbf{Y}^*)$ in R_{n+1}. Let y_{kj} be the jth coordinate of $\mathbf{Y}_k$, and consider the series

$$\mathbf{Y}_k = \mathbf{Y}_0 + \sum_{i=0}^{k-1} (\mathbf{Y}_{i+1} - \mathbf{Y}_i)$$

with jth coordinate

$$y_{kj} = a_j + \sum_{i=0}^{k-1} (y_{i+1j} - y_{ij})$$

Beginning with $i = 0$, we have

$$
\begin{aligned}
|y_{1j} - y_{0j}| &= \left| \int_{t_0}^{t} F_j(\tau, \mathbf{Y}_0)\, d\tau \right| \\
&\le \int_{t_0}^{t} |F_j(\tau, \mathbf{Y}_0)|\, d\tau \\
&\le \int_{t_0}^{t} \|\mathbf{F}(\tau, \mathbf{Y}_0)\|\, d\tau \le M(t - t_0) \le Mh \\
|y_{2j} - y_{1j}| &= \left| \int_{t_0}^{t} [F_j(\tau, \mathbf{Y}_1) - F_j(\tau, \mathbf{Y}_0)]\, d\tau \right| \\
&\le \int_{t_0}^{t} \|\mathbf{F}(\tau, \mathbf{Y}_1) - \mathbf{F}(\tau, \mathbf{Y}_0)\|\, d\tau \\
&\le K \int_{t_0}^{t} \|\mathbf{Y}_1 - \mathbf{Y}_0\|\, d\tau \\
&\le \sqrt{n}\, KM \int_{t_0}^{t} (\tau - t_0)\, d\tau = \sqrt{n}\, KM \frac{(t - t_0)^2}{2} \\
|y_{3j} - y_{2j}| &= \left| \int_{t_0}^{t} [F_j(\tau, \mathbf{Y}_2) - F_j(\tau, \mathbf{Y}_1)]\, d\tau \right| \\
&\le \int_{t_0}^{t} \|\mathbf{F}(\tau, \mathbf{Y}_2) - \mathbf{F}(\tau, \mathbf{Y}_1)\|\, d\tau \\
&\le K \int_{t_0}^{t} \|\mathbf{Y}_2 - \mathbf{Y}_1\|\, d\tau \\
&\le nK^2M \int_{t_0}^{t} \frac{(\tau - t_0)^2}{2}\, d\tau = nK^2M \frac{(t - t_0)^3}{3!}
\end{aligned}
$$

By an obvious induction, we have for $i = 0, 1, 2, \ldots$

$$
|y_{i+1\,j} - y_{ij}| \le \frac{M(\sqrt{n}\, K)^i (t - t_0)^{i+1}}{(i + 1)!} = \frac{M(\sqrt{n}\, K)^i h^{i+1}}{(i + 1)!}
$$

Therefore, the series

$$
a_j + \sum_{i=0}^{\infty} (y_{i+1\,j} - y_{ij})
$$

converges absolutely and uniformly for $t - t_0 \le h$. This means that

$$
\lim_{k \to \infty} y_{kj}(t) = y_j(t)
$$

exists and is continuous for $t - t_0 \leq h$. We know that

$$y_{kj}(t) = a_j + \int_{t_0}^{t} F_j[\tau,\mathbf{Y}_{k-1}(\tau)]\, d\tau$$

and taking the limit of both sides, we have

$$y_j(t) = a_j + \int_{t_0}^{t} F_j[\tau,\mathbf{Y}(\tau)]\, d\tau$$

This shows that $\mathbf{Y}(t) = (y_1(t), y_2(t), \ldots, y_n(t))$ is a solution of the integral equation and therefore a solution of the system of differential equations.

To prove uniqueness assume that there is another solution $\mathbf{Y}^*$ such that

$$\mathbf{Y}^*(t) = \mathbf{Y}_0 + \int_{t_0}^{t} \mathbf{F}[\tau,\mathbf{Y}^*(\tau)]\, d\tau$$

Then†

$$\begin{aligned}
\|\mathbf{Y}^* - \mathbf{Y}_0\| &= \left\| \int_{t_0}^{t} \mathbf{F}(\tau,\mathbf{Y}^*)\, d\tau \right\| \\
&\leq \int_{t_0}^{t} \|\mathbf{F}(\tau,\mathbf{Y}^*)\|\, d\tau \leq M(t - t_0) \leq Mh \\
\|\mathbf{Y}^* - \mathbf{Y}_1\| &= \left\| \int_{t_0}^{t} [\mathbf{F}(\tau,\mathbf{Y}^*) - \mathbf{F}(\tau,\mathbf{Y}_0)]\, d\tau \right\| \\
&\leq \int_{t_0}^{t} \|\mathbf{F}(\tau,\mathbf{Y}^*) - \mathbf{F}(\tau,\mathbf{Y}_0)\|\, d\tau \\
&\leq K \int_{t_0}^{t} \|\mathbf{Y}^* - \mathbf{Y}_0\|\, d\tau \\
&\leq KM \int_{t_0}^{t} (\tau - t_0)\, d\tau \\
&= KM \frac{(t - t_0)^2}{2} \leq \frac{KMh^2}{2}
\end{aligned}$$

Again we can show by induction that

$$\|\mathbf{Y}^* - \mathbf{Y}_k\| \leq \frac{MK^k(t - t_0)^{k+1}}{(k + 1)!} \leq \frac{MK^k h^{k+1}}{(k + 1)!}$$

† The inequality used here can be proved by approximating the integral with Riemann sums and then using the triangle inequality. Alternately, one can work with coordinates in this part of the proof as we did in the first part.

As $k \to \infty$, $\|\mathbf{Y}^* - \mathbf{Y}_k\| \to 0$ uniformly for $t - t_0 \le h$. Therefore,

$$\begin{aligned}\|\mathbf{Y}^* - \mathbf{Y}\| &= \|\mathbf{Y}^* - \mathbf{Y}_k + \mathbf{Y}_k - \mathbf{Y}\| \\ &\le \|\mathbf{Y}^* - \mathbf{Y}_k\| + \|\mathbf{Y}_k - \mathbf{Y}\| \to 0\end{aligned}$$

as $k \to \infty$. Hence $\mathbf{Y}^* = \mathbf{Y}$, which proves uniqueness. Finally we notice that the inequality

$$\|\mathbf{Y} - \mathbf{Y}_k\| \le \frac{MK^k(t - t_0)^{k+1}}{(k+1)!} \le \frac{MK^k h^{k+1}}{(k+1)!}$$

gives an upper bound on the error made by approximating the solution $\mathbf{Y}$ by the kth iterate. This completes the proof of Theorem 9.2.2.

Theorem 9.2.2, is a local result in that it only guarantees *some* positive h such that a unique solution exists for $|t - t_0| \le h$. However, having found such a solution, we reach a point $t = t_0 + h$ where the solution has values $\mathbf{Y}(t_0 + h)$. We can then pose another initial-value problem with initial values $\mathbf{Y}(t_0 + h)$. If there is an appropriate rectangle centered on $(t_0 + h, \mathbf{Y}(t_0 + h))$ for the application of Theorem 9.2.2, then the solution can be uniquely continued beyond the point $t_0 + h$. This process can be repeated indefinitely as long as the theorem continues to apply. However, there is no guarantee in general that a solution can be found up to and including the point $t = t_0 + r_0$. We can also consider the possibility of continuing the solution beyond the point $t = t_0 - h$. In the linear case, we have already shown that we can obtain a unique solution throughout the interval given by $|t - t_0| \le r_0$. This is because at each stage h can be found independently of the initial value of t (provided it satisfies $|t - t_0| \le r_0$). Therefore, the continuation proceeds a minimum distance at each stage, and so after some finite number of steps we must reach the end of the original interval. This does not always happen in the nonlinear case, as we saw in Example 5.8.3.

Finally, we remark that existence alone can be proved under the hypotheses of Theorem 9.2.2 without the Lipschitz condition.† However, the methods used are more advanced than we care to use in this book, so we shall not give the proof.

EXERCISES 9.6

1 Prove the inequalities involving $\|\mathbf{U}\|$ and $\mu = \max\,[|u_1|, |u_2|, \ldots, |u_n|]$ for complex vectors.

†This is done in E. Coddington and N. Levinson, "Theory of Ordinary Differential Equations," McGraw-Hill, New York, 1955 (rpt. Krieger, Melbourne, Florida, 1984).

2 Prove that $\mu(\mathbf{U}) = \max\,[|u_1|, |u_2|, \ldots, |u_n|]$ is a norm; that is, show that:
(*a*) $\mu(\mathbf{U}) \geq 0$.
(*b*) $\mu(\mathbf{U}) = 0$ implies $\mathbf{U} = \mathbf{0}$.
(*c*) $\mu(a\mathbf{U}) = |a|\mu(\mathbf{U})$ and $\mu(\mathbf{U}_1 + \mathbf{U}_2) \leq \mu(\mathbf{U}_1) + \mu(\mathbf{U}_2)$.

3 Complete the proof of Theorem 9.2.2 for the case $-r_0 \leq t - t_0 \leq 0$.

4 Consider two solutions $\mathbf{U}(t)$ and $\mathbf{V}(t)$ of the system $\dot{\mathbf{Y}} = \mathbf{F}(t,\mathbf{Y})$ where $\mathbf{U}(t_0) = \mathbf{U}_0$ and $\mathbf{V}(t_0) = \mathbf{V}_0$ and $\mathbf{F}(t,\mathbf{Y})$ satisfies a Lipschitz condition with Lipschitz constant K. Show that $\|\mathbf{U}(t) - \mathbf{V}(t)\| \leq \|\mathbf{U}_0 - \mathbf{V}_0\| e^{K|t-t_0|}$. How does this show that the solution depends continuously on the initial data?

ANSWERS AND HINTS FOR SELECTED EXERCISES

EXERCISES 1.2

1 $z_1 + z_2 = -1 + 6i$, $z_1 - z_2 = 5 - 4i$, $z_1z_2 = -11 + 7i$, $z_1/z_2 = -\frac{1}{34} - 13i/34$, $\bar{z}_1 = 2 - i$, $\bar{z}_2 = -3 - 5i$, $|z_1| = \sqrt{5}$, $|z_2| = \sqrt{34}$.

2 $z_1 + z_2 = 1 - i$, $z_1 - z_2 = -3 + 7i$, $z_1z_2 = 10 + 14i$, $z_1/z_2 = -\frac{7}{10} + i/10$, $\bar{z}_1 = -1 - 3i$, $\bar{z}_2 = 2 + 4i$, $|z_1| = \sqrt{10}$, $|z_2| = 2\sqrt{5}$.

6 $a + 0i - (b + 0i) = a - b + 0i$
$(a + 0i)/(b + 0i) = (ab + 0i)/b^2 = (a/b) + 0i$, $b \neq 0$

7 Substitute $u = x/(x^2 + y^2)$, $v = -y/(x^2 + y^2)$. Conversely, solve for u and v by elimination, given that $x^2 + y^2 \neq 0$.

9 $x^2 + y^2 \geq 0$, and $x^2 + y^2 = 0$ if and only if $x = y = 0$.

12 $|x + 0i| = (x^2 + 0^2)^{1/2} = (x^2)^{1/2} = |x|$

13 If a and b are real, $0 \leq (a - b)^2 = a^2 + b^2 - 2ab$. Hence, $a^2 + b^2 \geq 2ab$.

14
$$\begin{aligned} |z + w|^2 &= (x + u)^2 + (y + v)^2 \\ &= x^2 + y^2 + u^2 + v^2 + 2xu + 2yv \\ &\leq |z|^2 + |w|^2 + 2|xu + yv| \end{aligned}$$

16 $|z| = |z - w + w| \leq |z - w| + |w|$
$|w| = |w - z + z| \leq |w - z| + |z|$

EXERCISES 1.3

1 $\arg z_1 = 3\pi/4$, $\arg z_2 = \pi/6$, $\arg z_1z_2 = 11\pi/12$, $\arg z_1/z_2 = 7\pi/12$.

3 The arrow of z is rotated through the angle α.

5 The arrow of z is reversed.
6 The arrow of z is reflected in the x axis.
8 If $\alpha = p/q$ then there are q distinct powers if p is even, and $2q$ distinct powers if p is odd. If α is irrational there are infinitely many distinct powers.
9 $z = -2, 1 + i\sqrt{3}, 1 - i\sqrt{3}$.
12 $z = \cos(2\pi k/n) + i \sin(2\pi k/n), k = 0,1,2,\ldots,n-1$;
$z^n - 1 = (z-1)(z^{n-1} + z^{n-2} + \cdots + z + 1)$.
14 The triangle inequality is an equality if and only if $z_1 z_2 = 0$ or the arrows of z_1 and z_2 point in the same direction.
16 The circle with center at z_0 and radius r.
17 The bisector of the line segment joining z_1 and z_2.
19 A circle and its exterior. The center of the circle is on the line through z_1 and z_2. Can you find the center and radius?

EXERCISES 1.4

1 $\mathbf{v}_1 + \mathbf{v}_2 = (-2,3)$, $\mathbf{v}_1 - \mathbf{v}_2 = (4,-7)$, $2\mathbf{v}_1 + \mathbf{v}_2 = (-1,1)$, $\frac{1}{2}(\mathbf{v}_2 - \mathbf{v}_1) = (-2,\frac{7}{2})$.
2 The wind is from the northwest at $2\sqrt{2}$ miles per hour.
5 $(x,y) = (-1,2) + t(5,2), -\infty < t < \infty$. The vector $(2,-5)$ is perpendicular to the line.
6 $(x,y) = (1,3) + 5(\cos\theta, \sin\theta), 0 \le \theta < 2\pi$. Tangent line: $(x,y) = (4,7) + t(-4,3)$, $-\infty < t < \infty$.
7 $(x'(2), y'(2)) = (11,4)$.
11 $|\mathbf{T}(t)|^2 = 1$ implies $\mathbf{T}' \cdot \mathbf{T} = 0$.
12 $\mathbf{v}(t) = s(t)\mathbf{T}$, $\mathbf{a}(t) = s'(t)\mathbf{T} + s(t)|\mathbf{T}'|\mathbf{T}'/|\mathbf{T}'|$.

EXERCISES 1.5

1 The domain is all z. $z^3 = x^3 - 3xy^2 + i(3x^2y - y^3)$ is differentiable everywhere. $f'(z) = 3x^2 - 3y^2 + i(6xy) = 3z^2$.
2 $\partial x/\partial x = 1 \neq \partial 0/\partial y = 0$; $-\partial y/\partial y = -1 \neq \partial 0/\partial x = 0$.
4 $dz/dz = 1z^0$, $dz^{n+1}/dz = z^n(dz/dz) + z(dz^n/dz) = (n+1)z^n$.
6 The domain is all z. $f(z)$ is differentiable everywhere. $f'(z) = e^x \cos y + i\, e^x \sin y = f(z)$.
8 The domain is all z except $z = 0$. $f(z)$ is differentiable everywhere except at $z = 0$. $f'(z) = -1/z^2$.
10 The function is not continuous on the positive real axis. The function is differentiable everywhere except at $z = 0$ and on the positive real axis. $f'(z) = 1/2f(z)$.
11 The domain is all z except $z = 0$. $f(z)$ is differentiable everywhere except at $z = 0$ and on the positive real axis. $f'(z) = 1/z$.

EXERCISES 1.6

2 $\partial u/\partial x = e^x \cos y = \partial v/\partial y$, $\partial u/\partial y = -e^x \sin y = -\partial v/\partial x$.
4 $\overline{e^z} = e^x \cos y - ie^x \sin y = e^x \cos(-y) + ie^x \sin(-y) = e^{\bar{z}}$
6 $e^{z+2k\pi i} = e^x \cos(y + 2\pi k) + ie^x \sin(y + 2\pi k) = e^z$
9 $e^{2\pi ki/n} = \cos(2\pi k/n) + i \sin(2\pi k/n)$
10 $(d/dz)\cos z = -(\sin x \cosh y + i \cos x \sinh y) = -\sin z$
$(d/dz)\sin z = \cos x \cosh y - i \sin x \sinh y = \cos z$
12 $|\cos z|^2 = \cos^2 x \cosh^2 y + \sin^2 x \sinh^2 y$
$|\sin z|^2 = \sin^2 x \cosh^2 y + \cos^2 x \sinh^2 y$
14 $z = (2n+1)\pi i/2$; $n = 0, \pm 1, \pm 2, \ldots$.
15 $z = n\pi i$; $n = 0, \pm 1, \pm 2, \ldots$.

17 Everywhere except where $\sinh z = 0$.

20 $(\partial/\partial x)\frac{1}{2}\ln(x^2+y^2) = x/(x^2+y^2) = (\partial/\partial y)\tan^{-1}(y/x)$
$(\partial/\partial y)\frac{1}{2}\ln(x^2+y^2) = y/(x^2+y^2) = -(\partial/\partial x)\tan^{-1}(y/x)$
$\log z$ is not differentiable at $z = 0$ and on the positive real axis.

22 a^z is analytic everywhere. $(d/dz)a^z = a^z \log a$.

EXERCISES 1.7

2 $|k^{-z}| = k^{-x}$, and $\sum_{k=1}^{\infty} k^{-x}$ converges for $x > 1$.

3 $\lim_{n\to\infty} [n^2/(n+1)^2] = 1$. The series converges absolutely for $|z| \leq 1$.

4 $\lim_{n\to\infty} |(a_n+1)/a_n| = |a|/|a| = 1$

5 The series converges absolutely for $|w| = |z|^2/2 < 1$ and diverges for $|w| \geq 1$. The radius of convergence is $\sqrt{2}$.

7 If α is a nonnegative integer, the series is a polynomial and $R = \infty$. Otherwise $R = 1$.

10 $f^{(n)}(0) = n!a_n = n!b_n$

EXERCISES 2.2

1 Coefficient matrix $\begin{pmatrix} 1 & -2 & 3 \\ 2 & 1 & 5 \\ 1 & -1 & 1 \end{pmatrix}$

Augmented matrix $\begin{pmatrix} 1 & -2 & 3 & 7 \\ 2 & 1 & 5 & -6 \\ 1 & -1 & 1 & 0 \end{pmatrix}$

2 $A - B = \begin{pmatrix} -1 & 6 & -2 & 7 \\ 2 & -1 & 4 & 9 \\ 3 & -2 & -4 & 1 \end{pmatrix}$ $\quad 3A = \begin{pmatrix} 3 & 0 & -9 & 6 \\ 0 & -3 & 21 & 15 \\ 6 & 9 & -12 & 0 \end{pmatrix}$

$A - B = \begin{pmatrix} 3 & -6 & -4 & -3 \\ -2 & -1 & 10 & 1 \\ 1 & 8 & -4 & -1 \end{pmatrix}$ $\quad -2B = \begin{pmatrix} 4 & -12 & -2 & -10 \\ -4 & 0 & 6 & -8 \\ -2 & 10 & 0 & -2 \end{pmatrix}$

$5A - 7B = \begin{pmatrix} 19 & -42 & -22 & -25 \\ -14 & -5 & 56 & -3 \\ 3 & 50 & -20 & -7 \end{pmatrix}$

3 $AC = \begin{pmatrix} -8 & 18 \\ 22 & 11 \\ -13 & 12 \end{pmatrix}$ $\quad BC = \begin{pmatrix} -5 & 19 \\ -7 & 30 \\ 6 & 7 \end{pmatrix}$

8 $\begin{pmatrix} 1 & 1 \\ -1 & 1 \end{pmatrix}\begin{pmatrix} a & b \\ c & d \end{pmatrix} = \begin{pmatrix} a+c & b+d \\ -a+c & -b+d \end{pmatrix} = \begin{pmatrix} 1 & 0 \\ 0 & 1 \end{pmatrix}$

$a = \frac{1}{2}, b = -\frac{1}{2}, c = \frac{1}{2}, d = \frac{1}{2}$.

9 $E_1A = \begin{pmatrix} a_{11} & a_{12} & a_{13} & a_{14} \\ ka_{21} & ka_{22} & ka_{23} & ka_{24} \\ a_{31} & a_{32} & a_{33} & a_{34} \end{pmatrix}$. $\quad E_2A = \begin{pmatrix} a_{31} & a_{32} & a_{33} & a_{34} \\ a_{21} & a_{22} & a_{23} & a_{24} \\ a_{11} & a_{12} & a_{13} & a_{14} \end{pmatrix}$

$E_3A = \begin{pmatrix} a_{11}+a_{21} & a_{12}+a_{22} & a_{13}+a_{23} & a_{14}+a_{24} \\ a_{21} & a_{22} & a_{23} & a_{24} \\ a_{31} & a_{32} & a_{33} & a_{34} \end{pmatrix}$

10
$$\begin{aligned}(A - I)(A + I) &= A(A + I) - I(A + I)\\ &= A^2 + A - A - I = A^2 - I\end{aligned}$$
$$\begin{aligned}(A + I)(A - I) &= A(A - I) + I(A - I)\\ &= A^2 - A + A - I = A^2 - I\end{aligned}$$
$$\begin{aligned}(A - I)(A^2 + A + I) &= A(A^2 + A + I) - I(A^2 + A + I)\\ &= A^3 + A^2 + A - A^2 - A - I = A^3 - I\end{aligned}$$
$$\begin{aligned}(A^2 + A + I)(A - I) &= (A^2 + A + I)A - (A^2 + A + I)I\\ &= A^3 + A^2 + A - A^2 - A - I = A^3 - I\end{aligned}$$

EXERCISES 2.3

1 (*a*), (*c*), (*f*).

2 (*a*)
$$\begin{pmatrix} 1 & 2 & 3 & 4 & 5\\ -2 & 0 & 1 & 2 & 3\\ 0 & 1 & 5 & -2 & 4 \end{pmatrix} \to \begin{pmatrix} 1 & 2 & 3 & 4 & 5\\ 0 & 1 & 5 & -2 & 4\\ 0 & 0 & 1 & -\frac{18}{13} & \frac{3}{13} \end{pmatrix}$$

(*c*)
$$\begin{pmatrix} 1 & 2 & 3 & 4\\ 5 & 6 & 7 & 8\\ 9 & 10 & 11 & 12\\ 13 & 14 & 15 & 16 \end{pmatrix} \to \begin{pmatrix} 1 & 2 & 3 & 4\\ 0 & 1 & 2 & 3\\ 0 & 0 & 0 & 0\\ 0 & 0 & 0 & 0 \end{pmatrix}$$

3 $a = 5, b = -1$.

6 (*a*) $x_1 = a, x_2 = x_3 = 0, x_4 = -a$, a arbitrary.

(*c*) $x_1 = -\frac{21}{26}, x_2 = \frac{137}{26}, x_3 = -\frac{35}{26}, x_4 = -\frac{3}{2}$.

(*e*) $x_1 = -3a + 3b, x_2 = 3 - a + 3b, x_3 = a, x_4 = 1 - b, x_5 = b$, a and b arbitrary.

7 (*d*)
$$\begin{pmatrix} x_1\\ x_2\\ x_3\\ x_4\\ x_5 \end{pmatrix} = \begin{pmatrix} 3\\ -\frac{6}{5}\\ -\frac{22}{5}\\ 0\\ 0 \end{pmatrix} + a\begin{pmatrix} -2\\ 0\\ 3\\ 1\\ 0 \end{pmatrix} + b\begin{pmatrix} -1\\ 1\\ 2\\ 0\\ 1 \end{pmatrix}$$

8
$$\begin{pmatrix} x_1\\ x_2\\ x_3\\ x_4\\ x_5 \end{pmatrix} = a\begin{pmatrix} 1\\ -2\\ 1\\ 0\\ 0 \end{pmatrix} + b\begin{pmatrix} -1\\ 2\\ 0\\ -2\\ 1 \end{pmatrix} \qquad a \text{ and } b \text{ arbitrary}$$

10 $x_1 = -10 - 5x_3, x_2 = 2 + x_3, x_4 = 3, x_5 = -1$, x_3 arbitrary.

11 (*a*) Has a unique solution.

(*b*) If solution exists it is not unique.

EXERCISES 2.4

6 (*a*) -2 (*b*) 0 (*c*) 118

7 3,431

10 **Theorem 2.3.3** A homogeneous system of m linear algebraic equations in m unknowns has no nontrivial solutions if and only if the determinant of the coefficient matrix is not zero.

Theorem 2.3.5 A nonhomogeneous system of m linear algebraic equations in m unknowns has a unique solution if and only if the determinant of the coefficient matrix is not zero.

11 $$\begin{vmatrix} 2 & 1 & -1 & 1 \\ 1 & -1 & -1 & 1 \\ 1 & -4 & -2 & 2 \\ 4 & 1 & -3 & 3 \end{vmatrix} = \begin{vmatrix} 2 & 1 & -1 & 0 \\ 1 & -1 & -1 & 0 \\ 1 & -4 & -2 & 0 \\ 4 & 1 & -3 & 0 \end{vmatrix} = 0$$

The system does not have a unique solution.

12 $$\begin{vmatrix} 9-\lambda & -3 & 0 \\ -3 & 12-\lambda & -3 \\ 0 & -3 & 9-\lambda \end{vmatrix} = (6-\lambda)(9-\lambda)(15-\lambda) = 0$$

System has nontrivial solutions for $\lambda = 6, 9$, or 15.

13 $$\begin{vmatrix} 1-\lambda & 1 \\ -1 & 1-\lambda \end{vmatrix} = \lambda^2 - 2\lambda + 2 = 0, \lambda = 1 \pm i$$

There are no real nontrivial solutions.

EXERCISES 2.5

1 (*a*),(*c*),(*d*),(*f*).

2 (*a*) $\begin{pmatrix} -\frac{1}{3} & \frac{2}{3} \\ \frac{2}{3} & -\frac{1}{3} \end{pmatrix}$ (*c*) $\begin{pmatrix} \frac{1}{2} & 0 & \frac{1}{2} \\ 0 & -1 & 0 \\ \frac{1}{2} & 0 & -\frac{1}{2} \end{pmatrix}$

(*d*) $\begin{pmatrix} \frac{1}{4} & \frac{1}{2} & \frac{3}{4} \\ 0 & -1 & 0 \\ \frac{1}{4} & \frac{1}{2} & -\frac{1}{4} \end{pmatrix}$ (*f*) $\begin{pmatrix} 1 & -2 & 1 & 0 \\ 0 & 1 & -2 & 1 \\ 0 & 0 & 1 & -2 \\ 0 & 0 & 0 & 1 \end{pmatrix}$

4 Use the formula given in Theorem 2.5.2.

5 $x_1 = 4\frac{2}{3}, x_2 = 1\frac{1}{3}, x_3 = -1$.

6 $$A^{-1} = \begin{pmatrix} \frac{7}{15} & -\frac{2}{15} & \frac{1}{5} \\ \frac{1}{3} & \frac{1}{3} & 0 \\ -\frac{1}{5} & \frac{1}{5} & \frac{1}{5} \end{pmatrix} \qquad B = A^{-1}C = \begin{pmatrix} \frac{11}{15} & \frac{17}{15} & -\frac{4}{15} \\ \frac{5}{3} & -\frac{1}{3} & \frac{5}{3} \\ \frac{2}{5} & \frac{4}{5} & \frac{2}{5} \end{pmatrix}$$

7 If a solution X exists, $\tilde{C}AX = \tilde{C}B = 0$.

9 $|AB| = |A|\,|B| \neq 0$. $(B^{-1}A^{-1})(AB) = B^{-1}(A^{-1}A)B = I$.

10 $|\tilde{A}| = |A| \neq 0$. $I = \tilde{I} = \widetilde{AA^{-1}} = \widetilde{A^{-1}}\tilde{A}$.

12 If $|A| \neq 0$, then $A^{-1}(AB) = B = A^{-1}0 = 0$.

14 If $\tilde{A}A = I$ then $|A|^2 = 1$, $|A| = \pm 1$.

15 $\widetilde{A^{-1}} = (\tilde{A})^{-1} = (A^{-1})^{-1}$.

18 If $|A| = 0$ then $|C| = 0$. If $|A| \neq 0$ then $|C| = |A|^{n-1}$.

EXERCISES 2.6

3 A homogeneous system of m linear algebraic equations in m unknowns has no nontrivial solution if and only if the rank of the coefficient matrix is m.

4 A nonhomogeneous system of m linear algebraic equations in m unknowns has a unique solution if and only if the rank of the coefficient matrix is m.

5 (*a*) The rank of the coefficient matrix is 2. The rank of the augmented matrix is 3. No solutions.

(*b*) The ranks of the coefficient matrix and the augmented matrix are 4. There is a unique solution.

(*c*) The rank of the coefficient matrix is 3. The rank of the augmented matrix is 4. No solutions.

(*d*) The ranks of the coefficient matrix and the augmented matrix are 4. There are solutions but they are not unique because $4 < 5$, the number of unknowns.

6 The system has a solution if and only if the rank of the augmented matrix is $m - 1$. For this to happen all of the determinants containing the column B must be zero.

EXERCISES 3.2

1 $\mathbf{u} + \mathbf{v} = (4,-1,-3)$, $\mathbf{u} - \mathbf{v} = (-2,-3,5)$, $2\mathbf{u} = (2,-4,2)$, $-\frac{1}{2}\mathbf{v} = (-\frac{3}{2},-\frac{1}{2},2)$, $-\mathbf{u} + 2\mathbf{v} = (5,4,-9)$.

2 $|\mathbf{u}| = \sqrt{6}$, $\theta_1 = \theta_3 = 65°54'$, $\theta_2 = 144°44'$.

4 $\theta = 136°40'$

5 $(-\frac{9}{26},-\frac{3}{26},\frac{6}{13})$

7 $\frac{3}{7}\sqrt{14}$

9 $(x,y,z) = (3,-5,7) + t(-5,6,-3), -\infty < t < \infty$.

11 $(x,y,z) = (0,\frac{33}{43},-\frac{29}{43}) + t(1,-\frac{6}{43},\frac{17}{43}), -\infty < t < \infty$.

13 $14x - 13y - 18z = -19$

14 $(x,y,z) = (1,-3,4) + s(0,7,-4) + t(2,5,-1), -\infty < s < \infty, -\infty < t < \infty$.

15 $2x + 5y - z = 9$

EXERCISES 3.3

1 (*b*) If $\mathbf{v} \neq \mathbf{0}$ and $a \neq 1$, then $a\mathbf{v} \neq \mathbf{v}$.

5 The space of Example 3.3.4 is a proper subspace.

6 This is a proper subspace of the space of continuous functions.

7 If $a \neq 0$, then $a^{-1}(a\mathbf{u}) = \mathbf{u} = a^{-1}\mathbf{0} = \mathbf{0}$.

8 The origin, lines through the origin, or the whole space.

9 If $AX_1 = 0$, $AX_2 = 0$, then $A(aX_1 + bX_2) = 0$, for all a and b.

EXERCISES 3.4

5 Independent. $(1,2,3) = (1,0,1) + 2(0,1,1)$.

6 Dependent. $(1,2,3,4)$ cannot be expressed.

7 Independent. $(4,6,7,-5) = 2(1,0,1,0) - (0,2,-1,3) + 2(1,4,2,-1)$.

8 Dependent.

9 Independent.

12

$$W(0) = \begin{vmatrix} 1 & 1 & 1 & \cdots & 1 \\ 0 & 1 & 1 & \cdots & 1 \\ 0 & 0 & 2 & \cdots & 2 \\ \cdots & \cdots & \cdots & \cdots & \cdots \\ 0 & 0 & 0 & \cdots & k! \end{vmatrix} = 1!2!3!\ldots k!$$

Independent.

14

$$W(0) = \begin{vmatrix} 1 & 0 & 0 \\ 1 & 1 & 0 \\ 1 & 2 & 2 \end{vmatrix} = 2$$

Independent.

15 Independent.

16 $W'(x) \equiv 0$. f and g are independent.

EXERCISES 3.5

1 (*b*)
2 (*a*) 2 (*b*) 4 (*c*) 3 (*d*) 1
3 $(4,5,6) = \frac{7}{2}(1,1,1) - \frac{3}{2}(1,-1,1) + (2,0,3)$
4 $(4,-2,4,-2) = 4(1,0,1,0) - 2(0,1,0,1)$
5 (*a*),(*b*).
6 The functions $1,x,x^2,\ldots,x^k$ are independent for arbitrary k.
8 The converse is not true.
9 (1,1,1,1), (0,1,0,1), (1,0,2,0), (0,1,0,−1).

EXERCISES 3.6

1 $(\mathbf{u}\cdot\mathbf{v}) = 5 = (\mathbf{v}\cdot\mathbf{u}),\ (2\mathbf{u}\cdot\mathbf{v}) = 10,\ (\mathbf{u}\cdot 4\mathbf{u} + 3\mathbf{v}) = 71.$
2 $(f\cdot g) = 0, \|f\| = \sqrt{\pi}, \|g\| = \sqrt{\pi}.$
3 $(f\cdot g) = 0$
5 $\|\mathbf{u}+\mathbf{v}\|^2 - \|\mathbf{u}\|^2 - \|\mathbf{v}\|^2 = 2(\mathbf{u}\cdot\mathbf{v}) = 0$
6 $\|\mathbf{u}+\mathbf{v}\|^2 + \|\mathbf{u}-\mathbf{v}\|^2 = (\mathbf{u}+\mathbf{v}\cdot\mathbf{u}+\mathbf{v}) + (\mathbf{u}-\mathbf{v}\cdot\mathbf{u}-\mathbf{v}) = 2\|\mathbf{u}\|^2 + 2\|\mathbf{v}\|^2$
8 $\text{Re}(\mathbf{u}\cdot\mathbf{v}) = |(\mathbf{u}\cdot\mathbf{v})| = \|\mathbf{u}\|\,\|\mathbf{v}\|$
9 Apply the Cauchy inequality to $|f|^{1/2}$ and $|f|^{3/2}$.
11 This norm does not satisfy the parallelogram rule.

EXERCISES 3.7

1 $(1/\sqrt{2},0,1/\sqrt{2}),\ (0,-1,0),\ (-1/\sqrt{2},0,1/\sqrt{2}).$
3 $1/\sqrt{2},\ \sqrt{\tfrac{3}{2}}x,\ \sqrt{\tfrac{5}{2}}(\tfrac{3}{2}x^2 - \tfrac{1}{2}).$
4 $\delta_{ij} = (\mathbf{v}_i\cdot\mathbf{v}_j) = \sum_{k=1}^{n} a_{ki}a_{kj},\ \mathbf{u}_k = \sum_{k=1}^{n} a_{ki}\mathbf{v}_i.$
5 $\delta_{ij} = (\mathbf{v}_i\cdot\mathbf{v}_j) = \sum_{k=1}^{n} a_{ki}\bar{a}_{kj},\ \mathbf{u}_k = \sum_{k=1}^{n} \bar{a}_{ki}\mathbf{v}_i.$
6 $\sum_{i=1}^{n} c_i\mathbf{v}_i = 0 \Rightarrow AC = 0$ has only the trivial solution.
$\mathbf{u}_k = \sum_{k=1}^{n} b_{ik}\mathbf{v}_i$, where $B = A^{-1}$.
7 $(1/\sqrt{2},-1/\sqrt{2},0),\ (1/\sqrt{6},1/\sqrt{6},-2/\sqrt{6}),\ (1/\sqrt{3},1/\sqrt{3},1/\sqrt{3}).$
11 Let $\mathbf{u}_1, \mathbf{u}_2, \mathbf{u}_3$ be the basis of Exercise 7. Then projection of $\mathbf{v} = (1,2,3)$ on the given plane is $(\mathbf{v}\cdot\mathbf{u}_1)\mathbf{u}_1 + (\mathbf{v}\cdot\mathbf{u}_2)\mathbf{u}_2 = (-1,0,1).$
12 (2,3,2,3)

EXERCISES 3.8

4 Each coordinate of a Cauchy sequence is a Cauchy sequence of real numbers.
5 Each coordinate of a Cauchy sequence is a Cauchy sequence of complex numbers.
6 Each coordinate (with respect to some basis) of a Cauchy sequence is a Cauchy sequence of real numbers (or complex numbers).
8 If f_n is a Cauchy sequence, it converges uniformly to a continuous function on $\{x \mid a \le x \le b\}$.
9 $1, \sqrt{3}(2x - 1), \sqrt{5}(6x^2 - 6x + 1).$
12 $\int_0^{2\pi} \cos x \sin nx\, dx = 0,\ n = 1,2,3,\ldots.$

EXERCISES 4.2

1 $f(a\mathbf{u}_1 + b\mathbf{u}_2) = (ax_1 + bx_2, -ay_1 - by_2)$
$= a(x_1,-y_1) + b(x_2,-y_2)$
Null space $= \{\mathbf{0}\}$; range $= R^2$.

3 $f(a\mathbf{u}_1 + b\mathbf{u}_2) = (ax_1 + bx_2, ay_1 + by_2, -az_1 - bz_2)$
$= a(x_1,y_1,-z_1) + b(x_2,y_2,-z_2)$
Null space $= \{\mathbf{0}\}$; range $= R^3$.

5 Let $\mathbf{z} = (1/\sqrt{3},1/\sqrt{3},1/\sqrt{3})$. Then $f(\mathbf{u}) = (\mathbf{u}\cdot\mathbf{z})\mathbf{z}$.
$f(a\mathbf{u}_1 + b\mathbf{u}_2) = (a\mathbf{u}_1 + b\mathbf{u}_2\cdot\mathbf{z})\mathbf{z} = a(\mathbf{u}_1\cdot\mathbf{z})\mathbf{z} + b(\mathbf{u}_2\cdot\mathbf{z})\mathbf{z}$.
Null space $= \{(x,y,z)|x + y + z = 0\}$; range $= \{(x,y,z)|x = y = z\}$.

7
$$\begin{pmatrix} y_1 \\ y_2 \\ y_3 \end{pmatrix} = \begin{pmatrix} 1 & -1 & 2 & -1 \\ -1 & 2 & -3 & 1 \\ 1 & -3 & 4 & -1 \end{pmatrix} \begin{pmatrix} x_1 \\ x_2 \\ x_3 \\ x_4 \end{pmatrix}$$

Null space = subspace spanned by $(1,0,0,1)$, $(-1,1,1,0)$;
range $= \{(y_1,y_2,y_3)|y_1 + 2y_2 + y_3 = 0\}$.

9 $f(a\mathbf{u} + b\mathbf{v}) = au_1 + bv_1 + au_2 + bv_2 + \cdots + au_n + bu_n$
$= a(u_1 + u_2 + \cdots + u_n) + b(v_1 + v_2 + \cdots + v_n)$
Null space $= \{\mathbf{u}|u_1 + u_2 + \cdots + u_n = 0\}$; range $= C^1$.

11 (*a*) Null space $= \{f|f \equiv \text{constant}\}$; range $= \{f|f = \int_0^x g(t)\,dt + f_0, g \text{ continuous}\}$.
(*b*) Null space $= \{f|\int_a^b f(x)\,dx = 0\}$; range $= R^1$.

12 Let $a = b = 0$. $f(a\mathbf{u} + b\mathbf{v}) = f(\mathbf{0}) = \mathbf{0}$.

14 $f(x) = cx$, c = real constant.

16 $A\mathbf{u} = \mathbf{0}$ always has nontrivial solutions if $n > m$.

18 (1) $f^{-1}(\mathbf{v}) = (x, -y)$
(3) $f^{-1}(\mathbf{v}) = (x,y,-z)$
(4) $f^{-1}(\mathbf{v}) = (-x,-y,z)$
(6) $f^{-1}(\mathbf{v}) = (1/c)\mathbf{v}$, $c \neq 0$

20 $f(\mathbf{u}) = \frac{1}{2}(x + y, x + y)$, $g(\mathbf{u}) = (x,-y)$
$(f \circ g)(\mathbf{u}) = \frac{1}{2}(x + y, -x - y)$
$(g \circ f)(\mathbf{u}) = \frac{1}{2}(x - y, x - y)$
Composition is not commutative. Composition is associative.

EXERCISES 4.3

1 Zero matrix in either case.

3
$$\begin{pmatrix} 1 & 0 & 0 \\ 0 & 1 & 0 \\ 0 & 0 & 0 \end{pmatrix}$$

5 (*a*) $\begin{pmatrix} 1 & 0 \\ 0 & -1 \end{pmatrix}$ (*b*) $\begin{pmatrix} 0 & 1 \\ 1 & 0 \end{pmatrix}$

7 (*a*) $\begin{pmatrix} -1 & 0 & 0 \\ 0 & -1 & 0 \\ 0 & 0 & 1 \end{pmatrix}$ (*b*) $\begin{pmatrix} -1 & 0 & 0 \\ 0 & -1 & 0 \\ 0 & 0 & 1 \end{pmatrix}$

8 (*a*) $\begin{pmatrix} 1 & -1 & 2 & -1 \\ -1 & 2 & -3 & 1 \\ 1 & -3 & 4 & -1 \end{pmatrix}$

(*b*) Basis for null space: (1,0,0,1), (−1,1,1,0).
Basis for domain: (1,0,0,1), (−1,1,1,0), (1,0,1,−1), (1,1,0,−1).

(*c*) $\begin{pmatrix} 0 & 0 & 4 & 1 \\ 0 & 0 & -5 & 0 \\ 0 & 0 & 6 & -1 \end{pmatrix}$

10

(*a*) Invertible, $A^{-1} = \begin{pmatrix} \frac{3}{2} & \frac{3}{2} & -1 \\ \frac{1}{2} & -\frac{1}{2} & 0 \\ -1 & -1 & 1 \end{pmatrix}$

(*b*) Not invertible.

EXERCISES 4.4

1 (*a*) $\begin{pmatrix} 0 & 1 \\ 1 & 0 \end{pmatrix}$ (*b*) $\begin{pmatrix} \frac{1}{2} & \frac{1}{2} \\ -\frac{1}{2} & \frac{1}{2} \end{pmatrix}$

(*c*) $\begin{pmatrix} 1 & -1 \\ 1 & 1 \end{pmatrix}$ (*d*) $\begin{pmatrix} 1 & 0 \\ 0 & -1 \end{pmatrix}$

4 $f(1,-1,1) = (1,-1,1)$. Basis: (1,−1,1), (1,1,0), (1,−1,−2).

Representation: $\begin{pmatrix} 1 & 0 & 0 \\ 0 & -\frac{1}{2} & \frac{3}{2} \\ 0 & -\frac{1}{2} & -\frac{1}{2} \end{pmatrix}$

5 $B = PAP^{-1}$, $B^{-1} = PA^{-1}P^{-1}$.
$B^2 = (PAP^{-1})(PAP^{-1}) = PA^2P^{-1}$, etc.
$B^{-2} = (PA^{-1}P^{-1})(PA^{-1}P^{-1}) = PA^{-2}P^{-1}$, etc.

6 Clearly $\lim_{k\to\infty} (I + D + D^2 + \cdots + D^k) =$

$$\begin{pmatrix} (1-\lambda_1)^{-1} & 0 & 0 & \cdots & 0 \\ 0 & (1-\lambda_2)^{-1} & 0 & \cdots & 0 \\ 0 & 0 & (1-\lambda_3)^{-1} & \cdots & 0 \\ \cdots & \cdots & \cdots & \cdots & \cdots \\ 0 & 0 & 0 & \cdots & (1-\lambda_n)^{-1} \end{pmatrix}$$

EXERCISES 4.5

1 (*a*) $\lambda_1 = 1$, $\mathbf{u}_1 = a(1,-1)$
$\lambda_2 = 3$, $\mathbf{u}_2 = a(1,1)$

(*c*) $\lambda_1 = 1$, $\mathbf{u}_1 = a(1,0)$
$\lambda_2 = 2$, $\mathbf{u}_2 = a(1,1)$

(*d*) $\lambda_1 = 1 - i\sqrt{6}$, $\mathbf{u}_1 = a(i\sqrt{6}/2,1)$
$\lambda_2 = 1 + i\sqrt{6}$, $\mathbf{u}_2 = a(1, i\sqrt{6}/3)$

2 (*a*) $\lambda_1 = 2$, $\mathbf{u}_1 = a(1,-1,0)$
$\lambda_2 = 2$, $\mathbf{u}_2 = a(0,0,1)$
$\lambda_3 = 4$, $\mathbf{u}_3 = a(1,1,0)$

(*c*) $\lambda_1 = 4$, $\mathbf{u}_1 = a(1,-1,0)$
$\lambda_2 = 2\sqrt{2}$, $\mathbf{u}_2 = a(-1,3,2 - 2\sqrt{2})$
$\lambda_3 = -2\sqrt{2}$, $\mathbf{u}_3 = a(1,1,2 + 2\sqrt{2})$

(*e*) $\lambda_1 = 0$, $\mathbf{u}_1 = a(1,-4,7)$
$\lambda_2 = 6$, $\mathbf{u}_2 = a(1,2,1)$
$\lambda_3 = -11$, $\mathbf{u}_3 = a(-3,1,1)$

3 Relative to the basis (1,−1), (1,1), the representation is $\begin{pmatrix} 1 & 0 \\ 0 & -1 \end{pmatrix}$.

5 Relative to the basis (5,1,0), (0,1,−2), (2,0,1), the representation is $\begin{pmatrix} 0 & 0 & 0 \\ 0 & 2 & 0 \\ 0 & 0 & -1 \end{pmatrix}$.

6 If $AX = \lambda X, X \neq 0$, then $A^2X = \lambda AX = \lambda^2 X$, etc.

8 If $AX = \lambda X, \lambda \neq 0, X \neq 0$, then $A^{-1}(AX) = X = \lambda A^{-1}X$.

10 $p(A)\mathbf{X}_i = p(\lambda_i)\mathbf{X}_i = 0$, since $p(\lambda_i) = 0$. Let **X** be any vector. Then $\mathbf{X} = c_1\mathbf{X}_1 + c_2\mathbf{X}_2 + \cdots + c_n\mathbf{X}_n$ and $p(A)\mathbf{X} = 0$. Now successively let $\mathbf{X} = \mathbf{e}_1, \mathbf{e}_2, \mathbf{e}_3, \ldots, \mathbf{e}_n$.

12 Characteristic vectors corresponding to different characteristic values are independent.

14 $x = ae^t + be^{3t}$, $y = -ae^t + be^{3t}$, where a and b are arbitrary constants.

16 $(d^2x/dt^2) - 3(dx/dt) - 4x = (\lambda^2 - 3\lambda - 4)e^{\lambda t} = 0$

EXERCISES 4.6

1

(*a*) $P^{-1} = \begin{pmatrix} 1 & 0 & 1 \\ 0 & 1 & 0 \\ -1 & 0 & 1 \end{pmatrix} \qquad P = \begin{pmatrix} \frac{1}{2} & 0 & -\frac{1}{2} \\ 0 & 1 & 0 \\ \frac{1}{2} & 0 & \frac{1}{2} \end{pmatrix} \qquad PAP^{-1} = \begin{pmatrix} 2 & 0 & 0 \\ 0 & 2 & 0 \\ 0 & 0 & 4 \end{pmatrix}$

(*c*) $P^{-1} = \begin{pmatrix} 1 & 0 & -2 \\ 2 & 1 & 1 \\ 0 & 1 & -1 \end{pmatrix} \qquad P = \begin{pmatrix} \frac{1}{3} & \frac{1}{3} & -\frac{1}{3} \\ -\frac{1}{3} & \frac{1}{6} & \frac{5}{6} \\ -\frac{1}{3} & \frac{1}{6} & -\frac{1}{6} \end{pmatrix} \qquad PAP^{-1} = \begin{pmatrix} 0 & 0 & 0 \\ 0 & 0 & 0 \\ 0 & 0 & 6 \end{pmatrix}$

2 (*a*) $x_1 = -e^{2t} + 2e^{4t}, x_2 = 2e^{2t}, x_3 = e^{2t} + 2e^{4t}$.

(*c*) $x_1 = e^{6t}, x_2 = \frac{5}{2} - \frac{1}{2}e^{6t}, x_3 = \frac{5}{2} + \frac{1}{2}e^{6t}$.

3

(*a*) $P^{-1} = \begin{pmatrix} 1 & -1 \\ i & i \end{pmatrix} \qquad P = \begin{pmatrix} \frac{1}{2} & -i/2 \\ -\frac{1}{2} & -i/2 \end{pmatrix} \qquad PAP^{-1} = \begin{pmatrix} 0 & 0 \\ 0 & 2 \end{pmatrix}$

(*c*) $P^{-1} = \begin{pmatrix} 1 & 1 & -1 \\ 1 & -1 & 1 \\ 0 & i & 2i \end{pmatrix} \qquad P = \begin{pmatrix} \frac{1}{2} & \frac{1}{2} & 0 \\ \frac{1}{3} & -\frac{1}{3} & -i/3 \\ -\frac{1}{6} & \frac{1}{6} & -i/3 \end{pmatrix} \qquad PAP^{-1} = \begin{pmatrix} 2 & 0 & 0 \\ 0 & 0 & 0 \\ 0 & 0 & 6 \end{pmatrix}$

4 $x_1 = (i/2)e^{2t} + (\frac{1}{3} + \frac{1}{3}i) - (\frac{1}{3} - \frac{1}{6}i)e^{6t}$

$x_2 = (i/2)e^{2t} - (\frac{1}{3} + \frac{1}{3}i) + (\frac{1}{3} - \frac{1}{6}i)e^{6t}$

$x_3 = (\frac{1}{3}i - \frac{1}{3}) + (\frac{2}{3}i + \frac{1}{3})e^{6t}$

6 An ellipsoid $6(x')^2 + 9(y')^2 + 15(z')^2 = 1$. The axes are the lines through the origin and the points (1,1,1), (1,0,−1), (1,−2,1), respectively.

EXERCISES 4.7

3 $(A - \lambda I)X = P^{-1}(J - \lambda I)PX$

4

(*a*) $\begin{pmatrix} -2 & 0 & 0 \\ 0 & 4 & 1 \\ 0 & 0 & 4 \end{pmatrix} \qquad$ (*c*) $\begin{pmatrix} 1 & 0 & 0 \\ 0 & 1 & 1 \\ 0 & 0 & 1 \end{pmatrix}$

5 (*a*) $x_1 = -\frac{5}{2}e^{-2t} + \frac{7}{2}e^{4t} + 3te^{4t}$

$x_1 = 3e^{-2t} - \frac{1}{2}e^{4t} - 3te^{4t}$

$x_3 = \frac{5}{2}e^{-2t} + \frac{1}{2}e^{4t} + 3te^{4t}$

(*c*) $x_1 = e^t + 2te^t$

$x_2 = 2e^t - 2te^t$

$x_3 = 3e^t - 2te^t$

EXERCISES 5.2

2 $y(t) = (\dot{y}_0/\omega) \sin \omega t + y_0 \cos \omega t$

4 Since the differential equation is nonlinear the difference between two solutions does not satisfy the equation.

5 $(1/2\pi)\sqrt{g/l}$

6 $y(t) = y_0 \cos \omega t + (\dot{y}_0/\omega) \sin \omega t$ is a continuous function of y_0 and $\dot{y}_0$ for fixed t.

EXERCISES 5.3

1 (*a*) First order linear. (*c*) Second order linear.
(*e*) Third order linear. (*g*) Second order linear.

2 $F(x,y,u,u_x,u_y,u_{xx},u_{yy},\ldots,\partial^n u/\partial x^n, \partial^n u/\partial x^{n-1}\,\partial y,\ldots,\partial^n u/\partial x\,\partial y^{n-1}, \partial^n u/\partial y^n) = 0$
where at least one of the nth order partial derivatives appears.

3 $$f(x,y) + a_0(x,y)u + a_{10}(x,y)u_x + a_{01}(x,y)u_y + a_{20}(x,y)u_{xx} + a_{11}(x,y)u_{xy} + a_{02}(x,y)u_{yy} + \cdots + a_{n0}(x,y)\partial^n u/\partial x^n + a_{n-1,1}(x,y)(\partial^n u/\partial x^{n-1}\,\partial y) + \cdots + a_{1,n-1}(x,y)(\partial^n u/\partial x\,\partial y^{n-1}) + a_{0n}(x,y)(\partial^n u/\partial y^n) = 0$$
where at least one of $a_{n0}, a_{n-1,1},\ldots,a_{1,n-1}, a_{0n}$ is not identically zero.

4 (*a*) Second order linear. (*c*) Second order linear.
(*e*) First order nonlinear. (*g*) Second order nonlinear.

EXERCISES 5.4

1 (*a*) $y(t) = 1 + (y_0 - 1)\exp(-\tfrac{1}{2}t^2)$
(*b*) $y(t) = (t-1)(t+3)/2t$
(*c*) $y(t) = 1 + (y_0 - 1)e^{(1-e^t)}$
(*d*) $y(t) = e^t/(1+t)$
(*e*) $y(t) = \sin t - \cos t$

2 Approximately 3 hours 48 minutes.

3 Approximately 166,000 years ago.

4 .9 pounds per gallon.

5 mg/k, k = air resistance/unit velocity.

6 $\dot{A} = r(t)/y_1(t)$, $y_1(t) = \exp(-\int_0^t q(\tau)\,d\tau)$.

EXERCISES 5.5

1 (*a*) Separable: $\cos y = \cos t_0 \cos y_0 \sec t$.
(*c*) Bernoulli: $y^2 = 3t/(3ct^3 - 2)$, where c depends on initial conditions.
(*e*) Reducible to separable: $(y + t - 1)^3(y + 2t - 2)^{-2} = c$, where c depends on initial conditions. Note that $y = 1 - t$ and $y = 2 - 2t$ are also solutions.

2 $$\frac{\partial}{\partial y}(QMG(F)) = (QM)_y G(F) + QMG'(F)F_y = (QN)_t G(F) + QNG'(F)F_t = \frac{\partial}{\partial t}(QNG(F))$$

4 (*a*) Integrating factor is $1/t$: $yt + e^y - y_0t_0 - e^{y_0} = 0$.
(*c*) Integrating factor is $1/ty(y - 2t^2)$: $y = ty_0/t_0$. Note that $y = 2t^2$ is also a solution.

EXERCISES 5.6

2 $(mv_0/k) - (m^2g/k^2) \ln (1 + kv_0/mg)$

3 $\sqrt{50}$ miles per second.

4 (*a*) $\frac{1}{2}y^2 + x^2 = k^2$

(*c*) $(x + y)^3/(x - y) = k$ and $y = x$

5 $x^9 + 300(y - \sqrt{x^2 + y^2})^8 = 0$

7 $y^2 = 8x$

EXERCISES 5.7

1 $y = (1 - t)^{-1}$

EXERCISES 5.8

1 Take $b = 1, c = 1$; then $M = 4, a = \frac{1}{4}$.

2 $y(t) = 1 + \int_0^t \tau y^2\, d\tau$

$y_1 = 1 + \int_0^t \tau\, d\tau = 1 + (t^2/2)$

$y_2 = 1 + \int_0^t \tau[1 + (\tau^2/2)]^2\, d\tau = 1 + (t^2/2) + (t^4/4) + (t^6/24)$

3 $y = 2/(2 - t^2) = 1 + (t^2/2) + (t^4/4) + \cdots, |t| < \sqrt{2}$

5 $y \equiv 0, 0 \leq t \leq a.$

$y = [(t^2 - a^2)/4]^2, a \leq t.$

EXERCISES 6.2

1 $$W(0) = \begin{vmatrix} 1 & 1 \\ 1 & -1 \end{vmatrix} = -2$$

3 $$W(0) = \begin{vmatrix} 0 & 1 \\ \omega & 0 \end{vmatrix} = -\omega \neq 0$$

5 e^{2t}, e^{3t}.

7 $y = \frac{1}{4} + c_1 \sin 2t + c_2 \cos 2t$

9 $y = -1 + c_1 t + c_2 t^{-1}$

11 $y = \frac{1}{4} + \frac{3}{4} \cos 2t$

13 $y = -1 + \frac{3}{2}t + \frac{1}{2}t^{-1}$

14 $\dot{W}(t) = (d/dt)(y_1\dot{y}_2 - y_2\dot{y}_1) = y_1\ddot{y}_2 - y_2\ddot{y}_1 = -p(y_1\dot{y}_2 - y_2\dot{y}_1)$

$W(t) = W(t_0) \exp\left(-\int_{t_0}^t p(\tau)\, d\tau\right)$

16 $$\dot{W}(t) = \begin{vmatrix} y_1 & y_2 & \cdots & y_n \\ \dot{y}_1 & \dot{y}_2 & \cdots & \dot{y}_n \\ \cdots & \cdots & \cdots & \cdots \\ y_1^{(n)} & y_2^{(n)} & \cdots & y_n^{(n)} \end{vmatrix}$$

EXERCISES 6.3

1 $y = -\frac{1}{2}te^{-t} + c_1 e^t + c_2 e^{-t}$

3 $y = (1/\omega)t \sin \omega t + c_1 \sin \omega t + c_2 \cos \omega t$

5 $y = \frac{1}{3}t + \frac{7}{9} + c_1 e^{2t} + c_2 e^{3t}$

7 $y = \frac{2}{9}e^t + \frac{7}{36} + \frac{7}{12}e^{2t} - \frac{7}{18}te^{2t}$

9 $y = -\cos t \ln (\sec t + \tan t) + c_1 \sin t + c_2 \cos t$

EXERCISES 6.4

3 $q_1 = p_1 + r,\ q_2 = p_2 + rq_1,\ q_3 = p_3 + rq_2, \ldots$

6 $y = c_1e^t + c_2te^t + c_3t^2e^t + c_4t^3e^t + c_5e^{-2t} + c_6te^{-2t} + c_7t^2e^{-2t} + c_8t^3e^{-2t}$

7 (*b*) $y = c_1e^t + c_2e^{-t} + c_3 \sin t + c_4 \cos t$

(*d*) $y = c_1 + c_2e^t + c_3e^{-(1/2)t} \cos(\sqrt{3}t/2) + c_4e^{-(1/2)t} \sin(\sqrt{3}t/2)$

(*f*) $y = c_1e^t \sin t + c_2e^t \cos t + e_3te^t \sin t + c_4te^t \cos t$
$+ c_5e^{-t} \sin t + c_6e^{-t} \cos t + c_7te^{-t} \sin t + c_8te^{-t} \cos t$

EXERCISES 6.5

1 (*a*) D^3

(*c*) $(D^2 - 4D + 5)^2$

2 (*b*) $y = c_1e^{2t} + c_2e^{3t} + 2te^{3t} + \frac{1}{10} \cos t - \frac{1}{10} \sin t$

(*d*) $y = c_1 \cos 2t + c_2 \sin 2t + \frac{1}{8}t^2 \sin 2t - \frac{3}{16}t \cos 2t$

(*f*) $y = c_1e^t \sin 2t + c_2e^t \cos 2t + \frac{1}{16}te^t \sin 2t - \frac{1}{8}t^2e^t \cos 2t$

3 $y = e^t + \frac{1}{2}t^2e^t$

4 $y = 2 + \frac{1}{2} \sin t - \cos t + \frac{1}{2}te^t - e^t$

EXERCISES 6.6

1 Period $= \pi/4\sqrt{6}$ seconds; frequency $= 4\sqrt{6}/\pi$ hertz; amplitude = 1 inch.

2 Amplitude $= \frac{1}{12}[1 + \frac{12}{32}]^{1/2}$ ft; phase $= \tan^{-1}\sqrt{\frac{3}{8}} = 31°30'$.

4 $y = c_1 \cos \omega t + c_2 \sin \omega t + (f_0/2\omega m)t \sin \omega t$

5 $\tau = 4\pi\ 10^{-3}$ seconds.

6 $I = .141 \sin(120\pi t - \theta)$ amperes, $\theta = 57°22'$.

EXERCISES 6.7

1 $G(t,\tau) = 1 - e^{(\tau - t)}$

3 $G(t,\tau) = e^{2(t-\tau)} - e^{(t-\tau)}$

6 $y(t) = \frac{1}{3}\int_0^t (e^{3(t-\tau)} - 1)f(\tau)\,d\tau - \frac{2}{3}\int_0^t (e^{3(t-\tau)} - 1)y(\tau)\,d\tau$

10
$$G(x,\xi) = \frac{\sin \omega\xi \sin \omega(L - x)}{\omega \sin \omega L} \qquad 0 \le \xi \le x$$
$$= \frac{\sin \omega x \sin \omega(L - \xi)}{\omega \sin \omega L} \qquad x \le \xi \le L$$

12
$$G(x,\xi) = \frac{Q(\xi)[Q(1) - Q(x)]}{Q(1)P(\xi)} \qquad 0 \le \xi \le x$$
$$= \frac{Q(x)[Q(1) - Q(\xi)]}{Q(1)P(\xi)} \qquad x \le \xi \le 1$$

where $P(x) = \exp(-\int_0^x p(\xi)\,d\xi)$ and $Q(x) = \int_0^x P(\xi)\,d\xi$.

16 $y(x) = \int_0^1 G(x,\xi)f(\xi)\,d\xi - \int_0^1 G(x,\xi)q(\xi)y(\xi)\,d\xi$
where $G(x,\xi)$ is the Green's function of Exercise 12.

EXERCISES 7.2

2 $\mathscr{L}[\cos \omega t] = s/(s^2 + \omega^2)$

4 $\mathscr{L}[t^\alpha] = \Gamma(\alpha + 1)s^{-(\alpha+1)}$

6 $\mathscr{L}[t^n e^{at}] = n!(s - a)^{-(n+1)}$
8 $\mathscr{L}[e^{bt} \sin \omega t] = \omega/[(s - b)^2 + \omega^2]$
$\mathscr{L}[e^{bt} \cos \omega t] = (s - b)/[(s - b)^2 + \omega^2]$
10 $\omega/s^2(s^2 + \omega^2)$
11 $\mathscr{L}[f(t)]/s$
13 $[1 - \cos \omega t]/\omega^2$
14 $\dfrac{1}{a - b} e^{at} + \dfrac{1}{b - a} e^{bt}$

EXERCISES 7.3

4 $\mathscr{L}[t \sinh \omega t] = 2s\omega/(s^2 - \omega^2)^2$
$\mathscr{L}[t \cosh \omega t] = (s^2 + \omega^2)/(s^2 - \omega^2)^2$
6 $(e^{-as} - e^{-bs})/s$
8 $\dfrac{1}{s^2} - \dfrac{c}{s} \dfrac{e^{-sc}}{1 - e^{-sc}}$
10 $(D^2 - 2aD + a^2 + \omega^2)^2 y = 0, y(0) = 0, \dot{y}(0) = 0, \ddot{y}(0) = 2\omega, \dddot{y}(0) = 6\omega a.$

EXERCISES 7.4

4 $\frac{1}{4} - \frac{7}{40} e^{-t} - \frac{1}{12} e^{t} - \frac{15}{8} e^{3t} + \frac{111}{60} e^{4t}$
6 $\frac{7}{2} - 9te^{-t} - 4e^{-t} + \frac{3}{2}e^{-t} \cos t + 4e^{-t} \sin t$
8 $e^{-t} + te^{-t} - e^{-t} \cos t - \frac{3}{2}e^{-t} \sin t - \frac{1}{2}te^{-t} \sin t$
9 $u(t - 2)f(t - 2)$ where $f(t) = \frac{1}{16}e^{-t} + \frac{1}{16}te^{t} - \frac{1}{16}e^{t} + \frac{1}{32}e^{-t} \sin 2t.$

EXERCISES 7.5

1 (a) $y(t) = \frac{1}{10}e^{t} + \frac{1}{5} \sin 2t + (-\frac{3}{10} + \frac{4}{3}y_0 + \frac{1}{3}\dot{y}_0)e^{-t} + (\frac{1}{5} - \frac{1}{3}y_0 - \frac{1}{3}\dot{y}_0)e^{-4t}$
(c) $y(t) = -\frac{3}{26} \cos 3t - \frac{1}{13} \sin 3t + (\frac{3}{26} + y_0)e^{-t} \cos 2t$
$+ [\frac{9}{52} + (y_0/2) + (\dot{y}_0/2)]e^{-t} \sin 2t$

2 $y(t) = (2y_0 + \dot{y}_0)e^{-t} - (y_0 + \dot{y}_0)e^{-2t} + f(t)$
$- 2u(t - a)f(t - a) + u(t - 2a)f(t - 2a)$
where $f(t) = \frac{1}{2}t + e^{-t} - \frac{1}{4}e^{-2t} - \frac{3}{4}.$

3 $y(t) = f(t) + 2 \sum_{k=1}^{\infty} (-1)^k u(t - ka)f(t - ka)$
$f(t)$ as in Exercise 2.
4 $x(t) = (t/2)e^{-t} - (t/2)e^{-2t} + e^{-t}(1 + \frac{3}{2}x_0 - \frac{1}{2}y_0) + e^{-2t}(-1 - \frac{1}{2}x_0 + \frac{1}{2}y_0)$
$y(t) = (t/2)e^{-t} - \frac{3}{2}te^{-2t} + e^{-t}(1 + \frac{3}{2}x_0 - \frac{1}{2}y_0) + e^{-2t}(-1 - \frac{3}{2}x_0 + \frac{3}{2}y_0)$

5 $y(t) = \sum_{n=0}^{\infty} \dfrac{(-1)^n t^{3n}(3n - 2)(3n - 5)\ldots(4)(1)}{(3n)!}$

6 $y(t) = \sum_{n=0}^{\infty} \dfrac{(-1)^n t^{2n+1}}{2^{2n+1} n!(n + 1)!}$

EXERCISES 7.6

1 $X = -733$ ohms, $Z = 947.3$ ohms, $\theta = -50°47'$, $I = .1056 \sin (800\pi t - \theta)$.
3 Let $a = (1/RC) + (R/2H)$, $b = 1/HC$, $d = 1/2RHC^2$.

Then $E_{\text{out}} = \dfrac{E_0}{2} \dfrac{\omega^3(\omega^3 - b\omega)}{(d - a\omega^2)^2 + (b\omega - \omega^3)^2} \sin \omega t$

$+ \dfrac{E_0}{2} \dfrac{\omega^3(d - a\omega^2)}{(d - a\omega^2)^2 + (b\omega - \omega^3)^2} \cos \omega t$

4 $E_{\text{out}} = RME_0 \int_0^t \sin \omega\tau \, A(t - \tau)\, d\tau$

where $\mathscr{L}[A(t)] = \dfrac{s^3}{(H^2 - M^2)s^4 + 2RHs^3 + [R + (2H/C)]s^2 + (2R/C)s + (1/C^2)}$

EXERCISES 7.7

2 $B_n(t;t^3) = \dfrac{1}{n^3}[n(n-1)(n-2)t^3 + 3n(n-1)t^2 + nt]$

3 $B_n(t;f(t)) = \dfrac{1}{(b-a)^n} \sum_{k=0}^{\infty} \binom{n}{k} f\left[(b-a)\dfrac{k}{n} + a\right](t-a)^k(b-t)^{n-k}$

4 $B_n(t;|t|) = 2^{-n} \sum_{k=0}^{n} \binom{n}{k} \left|-1 + 2\dfrac{k}{n}\right| (1+t)^k(1-t)^{n-k}$

6 (*a*) $\tan^{-1}(\omega/s)$ (*b*) $\ln[(s-a)/s]$ (*c*) $\frac{1}{2}\ln[(s+\omega)/(s-\omega)]$

EXERCISES 8.2

1 (*a*) $R = \infty$ (*c*) $R = 1$ (*e*) $R = \infty$ (*g*) $R = 1$

2 $y_1 = \sum_{k=0}^{\infty} \dfrac{(3k-2)(3k-5)\ldots(4)(1)}{(3k)!} x^{3k} \qquad |x| < \infty$

$y_2 = \sum_{k=0}^{\infty} \dfrac{(3k-1)(3k-4)\ldots(2)(1)}{(3k+1)!} x^{3k+1} \qquad |x| < \infty$

3 $y_1 = 1 + \sum_{k=1}^{\infty} \dfrac{(-\lambda)(4-\lambda)(8-\lambda)\ldots(4k-4-\lambda)}{(2k)!} x^{2k} \qquad |x| < \infty$

$y_2 = x + \sum_{k=1}^{\infty} \dfrac{(2-\lambda)(6-\lambda)\ldots(4k-2-\lambda)}{(2k+1)!} x^{2k+1} \qquad |x| < \infty$

5 $y_1 = 1 + \sum_{k=1}^{\infty} \dfrac{(-\lambda)(4-\lambda)(16-\lambda)\ldots(4k^2-8k+4-\lambda)}{(2k)!} x^{2k} \qquad |x| < 1$

$y_2 = x + \sum_{k=1}^{\infty} \dfrac{(1-\lambda)(9-\lambda)\ldots(4k^2-4k+1-\lambda)}{(2k+1)!} x^{2k+1} \qquad |x| < 1$

7 $y = c_1 y_1 + c_2 y_2 + \frac{1}{3}x$

where $y_1 = xe^{-x^2/2}$, $y_2 = \sum_{k=0}^{\infty} [(-2)^k k! x^{2k}]/(2k)!$

EXERCISES 8.3

1 $m_1 = \nu$, $m_2 = -\nu$.

2 At $x = 1$, $m_1 = m_2 = 0$; at $x = -1$, $m_1 = m_2 = 0$.

3 $m_1 = m_2 = 0$

$y = 1 + \sum_{k=1}^{\infty} \dfrac{(-\lambda)(1-\lambda)(2-\lambda)\ldots(k-1-\lambda)}{(k!)^2} x^k \qquad |x| < \infty$

5 $m_1 = m_2 = 0$

$$y = \sum_{k=0}^{\infty} \frac{x^{2k}}{2^{2k}(k!)^2} \qquad |x| < \infty$$

6 $m_1 = \frac{1}{2}, m_2 = 0.$

$$y_1 = 1 - x - \tfrac{3}{5}x^2 - \tfrac{69}{125}x^3 - \cdots \qquad |x| < 1$$

$$y_2 = x^{1/2}\left(1 + \frac{x}{6} + \frac{13x^2}{120} + \frac{(13)(61)}{(120)(72)}x^3 + \cdots\right) \qquad |x| < 1$$

EXERCISES 8.4

1 Prove the identities $J_{1/2}(x) = \sqrt{2/\pi}\,(\sin x/\sqrt{x})$ and $J_{-1/2}(x) = \sqrt{2/\pi}\,(\cos x/\sqrt{x})$.

3 $t^2\ddot{w} + t\dot{w} + (t^2 - \frac{1}{9})w = 0$

4 $t^2\ddot{w} + t\dot{w} + (t^2 - \nu^2)w = 0 \qquad t = \alpha x^{\beta}$

6 $$y = J_n(x)\ln x - \sum_{k=0}^{n-1} \frac{(n-k-1)!x^{2k-n}}{2^{k-n+1}k!} + \frac{n(n+2)}{2^{n+3}(n+1)!}x^{2n+2} - \frac{n^2(n^2+4n+1)}{2^{n+7}(n+2)!}x^{2n+4} + \cdots$$

EXERCISES 8.5

1 $\lambda_n^2 = n^2\pi^2/L^2,\ n = 0,1,2,\ldots;\ y_n = \cos(n\pi x/L).$

2 $\lambda_n^2 = (2n+1)^2\pi^2/4L^2,\ n = 0,1,2,\ldots;\ y_n = \cos[(2n+1)\pi x/2L].$

5 $y_n = \sin \lambda_n x$, where the λ_n are all the positive solutions of $\tan\lambda = -\lambda$.

7 $y_n = J_1(\sqrt{\lambda_n}x)$, where the λ_n are all the positive solutions of $J_1(\sqrt{\lambda}) = 0$.

9 $y = (1 - \frac{1}{2}x)\cos x + B\sin x$, where $B = \frac{1}{2} + \frac{1}{2}\tan 1 + \sec 1$.

10 No solutions.

11 $y = x + B\sin \pi x$, where B is arbitrary.

EXERCISES 9.2

2 $\dot{u}_1 = u_2,\ \dot{u}_2 = e^t - 6u_1 - 5u_2.$

3 $\dot{u}_1 = u_2,\ \dot{u}_2 = -6t^{-1}u_1 - 6u_2;\ y = t^{-2} - t^{-3}, 0 < t < \infty.$

5 $y_1 = 2/(2 - t^2),\ y_2 \equiv 0.$

EXERCISES 9.3

4 $$\mathbf{Y} = c_1\begin{pmatrix}1\\-1\end{pmatrix} + c_2\begin{pmatrix}1\\1\end{pmatrix}e^t$$

5 $$\mathbf{Y} = \begin{pmatrix}1\\-1\end{pmatrix}$$

6 $$\mathbf{Y} = c_1\begin{pmatrix}1\\-1\end{pmatrix} + c_2\begin{pmatrix}1\\1\end{pmatrix}e^t + (\tfrac{1}{4}e^t + \tfrac{1}{8}t^2)\begin{pmatrix}1\\-1\end{pmatrix} + (\tfrac{1}{4}t + \tfrac{1}{4}te^{-t} + \tfrac{1}{4}e^{-t})\begin{pmatrix}1\\1\end{pmatrix}e^t$$

10 $$\mathbf{Y} = c_1\begin{pmatrix}1\\0\\0\end{pmatrix}e^t + c_2\begin{pmatrix}2\\1\\0\end{pmatrix}e^{2t} + c_3\begin{pmatrix}6t-3\\3t\\1\end{pmatrix}e^{2t}$$

11 $$\mathbf{Y} = c_1\left[\begin{pmatrix}1\\1\end{pmatrix}\cos t - \begin{pmatrix}1\\0\end{pmatrix}\sin t\right] + c_2\left[\begin{pmatrix}1\\1\end{pmatrix}\sin t + \begin{pmatrix}1\\0\end{pmatrix}\cos t\right]$$

EXERCISES 9.4

1 (*a*) $\mathbf{Y} = c_1\begin{pmatrix}1\\-1\end{pmatrix}e^t + c_2\begin{pmatrix}1\\1\end{pmatrix}e^{3t}$

(*c*) $\mathbf{Y} = c_1e^t\left[\begin{pmatrix}1\\0\end{pmatrix}\cos 2t - \begin{pmatrix}0\\2\end{pmatrix}\sin 2t\right] + c_2e^t\left[\begin{pmatrix}1\\0\end{pmatrix}\sin 2t + \begin{pmatrix}0\\2\end{pmatrix}\cos 2t\right]$

2

(*a*) $\mathbf{Y} = c_1\begin{pmatrix}0\\0\\1\end{pmatrix}e^{2t} + c_2\begin{pmatrix}1\\-1\\0\end{pmatrix}e^{2t} + c_3\begin{pmatrix}1\\1\\0\end{pmatrix}e^{4t}$

(*d*) $\mathbf{Y} = c_1\begin{pmatrix}-3\\4\\2\end{pmatrix}e^{-t} + c_2\begin{pmatrix}0\\1\\-1\end{pmatrix}e^{2t} + c_3\left[\begin{pmatrix}1\\1\\1\end{pmatrix}e^{2t} + \begin{pmatrix}0\\1\\-1\end{pmatrix}te^{2t}\right]$

3 $\mathbf{Y} = (c_1 + t - \frac{1}{4}\cos 2t)\begin{pmatrix}\cos t - \sin t\\ \cos t\end{pmatrix} + (c_2 - \frac{1}{2}t - \frac{1}{4}\sin 2t)\begin{pmatrix}\sin t + \cos t\\ \sin t\end{pmatrix}$

6 $y_1 = \dfrac{y_1(0) - y_2(0)}{2}e^t + \dfrac{y_1(0) + y_2(0)}{2}e^{3t}$

$y_2 = \dfrac{y_2(0) - y_1(0)}{2}e^t + \dfrac{y_1(0) + y_2(0)}{2}e^{3t}$

7 $y_1 = \dfrac{y_1(0) - y_2(0)}{2}e^t + \dfrac{y_1(0) + y_2(0)}{2}(e^t + te^t)$

$y_2 = \dfrac{y_2(0) - y_1(0)}{2}e^t + \dfrac{y_1(0) + y_2(0)}{2}(e^t - te^t)$

9 $\mathbf{Y} = c_1\begin{pmatrix}-3\\4\\2\end{pmatrix}e^{-t} + c_2\begin{pmatrix}0\\1\\-1\end{pmatrix}e^{2t} + c_3\left[\begin{pmatrix}1\\1\\0\end{pmatrix}e^{2t} + \begin{pmatrix}0\\1\\-1\end{pmatrix}te^{2t}\right]$

$+ \begin{pmatrix}-3\\4\\2\end{pmatrix}te^{-t} - \frac{2}{3}\begin{pmatrix}0\\1\\-1\end{pmatrix}e^{-t}$

where $c_1 = \dfrac{-y_1(0) + y_2(0) + y_3(0)}{9}$

$c_2 = \dfrac{-2y_1(0) + 2y_2(0) - 7y_3(0)}{9} + \frac{2}{3}$

$c_3 = \dfrac{2y_1(0) + y_2(0) + y_3(0)}{3}$

EXERCISES 9.5

1 (*b*) $y_1 = c_1e^t + c_2e^{-t} + c_3\cos t + c_4\sin t$
$y_2 = -c_1e^t - c_2e^{-t} + c_3\cos t + c_4\sin t$

(*d*) $y_1 = c_1e^{-t} + c_2e^{2t} + c_3e^{4t} - \frac{1}{8}$
$y_2 = 3c_1e^{-t} - \frac{1}{3}c_3e^{4t} + \frac{1}{2}t - \frac{3}{8}$

2 (*a*) $y_1 = c_1e^{-(1/3)t}\cos\dfrac{\sqrt{2}}{3}t + c_2e^{-(1/3)t}\sin\dfrac{\sqrt{2}}{3}t$

$y_2 = \left(c_1 + \dfrac{\sqrt{2}}{2}c_2\right)e^{-(1/3)t}\cos\dfrac{\sqrt{2}}{3}t + \left(c_2 - \dfrac{\sqrt{2}}{2}c_1\right)e^{-(1/3)t}\sin\dfrac{\sqrt{2}}{3}t$

(*c*) No solutions.

INDEX

A CATALOG OF SELECTED

DOVER BOOKS

IN SCIENCE AND MATHEMATICS

Astronomy

BURNHAM'S CELESTIAL HANDBOOK, Robert Burnham, Jr. Thorough guide to the stars beyond our solar system. Exhaustive treatment. Alphabetical by constellation: Andromeda to Cetus in Vol. 1; Chamaeleon to Orion in Vol. 2; and Pavo to Vulpecula in Vol. 3. Hundreds of illustrations. Index in Vol. 3. 2,000pp. 6⅛ x 9¼.
Vol. I: 0-486-23567-X
Vol. II: 0-486-23568-8
Vol. III: 0-486-23673-0

EXPLORING THE MOON THROUGH BINOCULARS AND SMALL TELESCOPES, Ernest H. Cherrington, Jr. Informative, profusely illustrated guide to locating and identifying craters, rills, seas, mountains, other lunar features. Newly revised and updated with special section of new photos. Over 100 photos and diagrams. 240pp. 8¼ x 11. 0-486-24491-1

THE EXTRATERRESTRIAL LIFE DEBATE, 1750–1900, Michael J. Crowe. First detailed, scholarly study in English of the many ideas that developed from 1750 to 1900 regarding the existence of intelligent extraterrestrial life. Examines ideas of Kant, Herschel, Voltaire, Percival Lowell, many other scientists and thinkers. 16 illustrations. 704pp. 5⅜ x 8½. 0-486-40675-X

THEORIES OF THE WORLD FROM ANTIQUITY TO THE COPERNICAN REVOLUTION, Michael J. Crowe. Newly revised edition of an accessible, enlightening book re-creates the change from an earth-centered to a sun-centered conception of the solar system. 242pp. 5⅜ x 8½. 0-486-41444-2

ARISTARCHUS OF SAMOS: The Ancient Copernicus, Sir Thomas Heath. Heath's history of astronomy ranges from Homer and Hesiod to Aristarchus and includes quotes from numerous thinkers, compilers, and scholasticists from Thales and Anaximander through Pythagoras, Plato, Aristotle, and Heraclides. 34 figures. 448pp. 5⅜ x 8½.
0-486-43886-4

A COMPLETE MANUAL OF AMATEUR ASTRONOMY: TOOLS AND TECHNIQUES FOR ASTRONOMICAL OBSERVATIONS, P. Clay Sherrod with Thomas L. Koed. Concise, highly readable book discusses: selecting, setting up and maintaining a telescope; amateur studies of the sun; lunar topography and occultations; observations of Mars, Jupiter, Saturn, the minor planets and the stars; an introduction to photoelectric photometry; more. 1981 ed. 124 figures. 25 halftones. 37 tables. 335pp. 6½ x 9¼. 0-486-42820-8

AMATEUR ASTRONOMER'S HANDBOOK, J. B. Sidgwick. Timeless, comprehensive coverage of telescopes, mirrors, lenses, mountings, telescope drives, micrometers, spectroscopes, more. 189 illustrations. 576pp. 5⅝ x 8¼. (Available in U.S. only.)
0-486-24034-7

STAR LORE: Myths, Legends, and Facts, William Tyler Olcott. Captivating retellings of the origins and histories of ancient star groups include Pegasus, Ursa Major, Pleiades, signs of the zodiac, and other constellations. "Classic."—Sky & Telescope. 58 illustrations. 544pp. 5⅜ x 8½. 0-486-43581-4

Chemistry

THE SCEPTICAL CHYMIST: THE CLASSIC 1661 TEXT, Robert Boyle. Boyle defines the term "element," asserting that all natural phenomena can be explained by the motion and organization of primary particles. 1911 ed. viii+232pp. $5^3/_8 \times 8^1/_2$. 0-486-42825-7

RADIOACTIVE SUBSTANCES, Marie Curie. Here is the celebrated scientist's doctoral thesis, the prelude to her receipt of the 1903 Nobel Prize. Curie discusses establishing atomic character of radioactivity found in compounds of uranium and thorium; extraction from pitchblende of polonium and radium; isolation of pure radium chloride; determination of atomic weight of radium; plus electric, photographic, luminous, heat, color effects of radioactivity. ii+94pp. $5^3/_8 \times 8^1/_2$. 0-486-42550-9

CHEMICAL MAGIC, Leonard A. Ford. Second Edition, Revised by E. Winston Grundmeier. Over 100 unusual stunts demonstrating cold fire, dust explosions, much more. Text explains scientific principles and stresses safety precautions. 128pp. $5^3/_8 \times 8^1/_2$. 0-486-67628-5

MOLECULAR THEORY OF CAPILLARITY, J. S. Rowlinson and B. Widom. History of surface phenomena offers critical and detailed examination and assessment of modern theories, focusing on statistical mechanics and application of results in mean-field approximation to model systems. 1989 edition. 352pp. $5^3/_8 \times 8^1/_2$. 0-486-42544-4

CHEMICAL AND CATALYTIC REACTION ENGINEERING, James J. Carberry. Designed to offer background for managing chemical reactions, this text examines behavior of chemical reactions and reactors; fluid-fluid and fluid-solid reaction systems; heterogeneous catalysis and catalytic kinetics; more. 1976 edition. 672pp. $6^1/_8 \times 9^1/_4$. 0-486-41736-0 $31.95

ELEMENTS OF CHEMISTRY, Antoine Lavoisier. Monumental classic by founder of modern chemistry in remarkable reprint of rare 1790 Kerr translation. A must for every student of chemistry or the history of science. 539pp. $5^3/_8 \times 8^1/_2$. 0-486-64624-6

MOLECULES AND RADIATION: An Introduction to Modern Molecular Spectroscopy. Second Edition, Jeffrey I. Steinfeld. This unified treatment introduces upper-level undergraduates and graduate students to the concepts and the methods of molecular spectroscopy and applications to quantum electronics, lasers, and related optical phenomena. 1985 edition. 512pp. $5^3/_8 \times 8^1/_2$. 0-486-44152-0

A SHORT HISTORY OF CHEMISTRY, J. R. Partington. Classic exposition explores origins of chemistry, alchemy, early medical chemistry, nature of atmosphere, theory of valency, laws and structure of atomic theory, much more. 428pp. $5^3/_8 \times 8^1/_2$. (Available in U.S. only.) 0-486-65977-1

GENERAL CHEMISTRY, Linus Pauling. Revised 3rd edition of classic first-year text by Nobel laureate. Atomic and molecular structure, quantum mechanics, statistical mechanics, thermodynamics correlated with descriptive chemistry. Problems. 992pp. $5^3/_8 \times 8^1/_2$. 0-486-65622-5

ELECTRON CORRELATION IN MOLECULES, S. Wilson. This text addresses one of theoretical chemistry's central problems. Topics include molecular electronic structure, independent electron models, electron correlation, the linked diagram theorem, and related topics. 1984 edition. 304pp. $5^3/_8 \times 8^1/_2$. 0-486-45879-2

Engineering

DE RE METALLICA, Georgius Agricola. The famous Hoover translation of greatest treatise on technological chemistry, engineering, geology, mining of early modern times (1556). All 289 original woodcuts. 638pp. $6\frac{3}{4}$ x 11. 0-486-60006-8

FUNDAMENTALS OF ASTRODYNAMICS, Roger Bate et al. Modern approach developed by U.S. Air Force Academy. Designed as a first course. Problems, exercises. Numerous illustrations. 455pp. $5\frac{3}{8}$ x $8\frac{1}{2}$. 0-486-60061-0

DYNAMICS OF FLUIDS IN POROUS MEDIA, Jacob Bear. For advanced students of ground water hydrology, soil mechanics and physics, drainage and irrigation engineering and more. 335 illustrations. Exercises, with answers. 784pp. $6\frac{1}{8}$ x $9\frac{1}{4}$. 0-486-65675-6

THEORY OF VISCOELASTICITY (SECOND EDITION), Richard M. Christensen. Complete consistent description of the linear theory of the viscoelastic behavior of materials. Problem-solving techniques discussed. 1982 edition. 29 figures. xiv+364pp. $6\frac{1}{8}$ x $9\frac{1}{4}$. 0-486-42880-X

MECHANICS, J. P. Den Hartog. A classic introductory text or refresher. Hundreds of applications and design problems illuminate fundamentals of trusses, loaded beams and cables, etc. 334 answered problems. 462pp. $5\frac{3}{8}$ x $8\frac{1}{2}$. 0-486-60754-2

MECHANICAL VIBRATIONS, J. P. Den Hartog. Classic textbook offers lucid explanations and illustrative models, applying theories of vibrations to a variety of practical industrial engineering problems. Numerous figures. 233 problems, solutions. Appendix. Index. Preface. 436pp. $5\frac{3}{8}$ x $8\frac{1}{2}$. 0-486-64785-4

STRENGTH OF MATERIALS, J. P. Den Hartog. Full, clear treatment of basic material (tension, torsion, bending, etc.) plus advanced material on engineering methods, applications. 350 answered problems. 323pp. $5\frac{3}{8}$ x $8\frac{1}{2}$. 0-486-60755-0

A HISTORY OF MECHANICS, René Dugas. Monumental study of mechanical principles from antiquity to quantum mechanics. Contributions of ancient Greeks, Galileo, Leonardo, Kepler, Lagrange, many others. 671pp. $5\frac{3}{8}$ x $8\frac{1}{2}$. 0-486-65632-2

STABILITY THEORY AND ITS APPLICATIONS TO STRUCTURAL MECHANICS, Clive L. Dym. Self-contained text focuses on Koiter postbuckling analyses, with mathematical notions of stability of motion. Basing minimum energy principles for static stability upon dynamic concepts of stability of motion, it develops asymptotic buckling and postbuckling analyses from potential energy considerations, with applications to columns, plates, and arches. 1974 ed. 208pp. $5\frac{3}{8}$ x $8\frac{1}{2}$. 0-486-42541-X

BASIC ELECTRICITY, U.S. Bureau of Naval Personnel. Originally a training course; best nontechnical coverage. Topics include batteries, circuits, conductors, AC and DC, inductance and capacitance, generators, motors, transformers, amplifiers, etc. Many questions with answers. 349 illustrations. 1969 edition. 448pp. $6\frac{1}{2}$ x $9\frac{1}{4}$. 0-486-20973-3

ROCKETS, Robert Goddard. Two of the most significant publications in the history of rocketry and jet propulsion: "A Method of Reaching Extreme Altitudes" (1919) and "Liquid Propellant Rocket Development" (1936). 128pp. $5^3/_8$ x $8^1/_2$. 0-486-42537-1

STATISTICAL MECHANICS: PRINCIPLES AND APPLICATIONS, Terrell L. Hill. Standard text covers fundamentals of statistical mechanics, applications to fluctuation theory, imperfect gases, distribution functions, more. 448pp. $5^3/_8$ x $8^1/_2$. 0-486-65390-0

ENGINEERING AND TECHNOLOGY 1650–1750: ILLUSTRATIONS AND TEXTS FROM ORIGINAL SOURCES, Martin Jensen. Highly readable text with more than 200 contemporary drawings and detailed engravings of engineering projects dealing with surveying, leveling, materials, hand tools, lifting equipment, transport and erection, piling, bailing, water supply, hydraulic engineering, and more. Among the specific projects outlined-transporting a 50-ton stone to the Louvre, erecting an obelisk, building timber locks, and dredging canals. 207pp. $8^3/_8$ x $11^1/_4$. 0-486-42232-1

THE VARIATIONAL PRINCIPLES OF MECHANICS, Cornelius Lanczos. Graduate level coverage of calculus of variations, equations of motion, relativistic mechanics, more. First inexpensive paperbound edition of classic treatise. Index. Bibliography. 418pp. $5^3/_8$ x $8^1/_2$. 0-486-65067-7

PROTECTION OF ELECTRONIC CIRCUITS FROM OVERVOLTAGES, Ronald B. Standler. Five-part treatment presents practical rules and strategies for circuits designed to protect electronic systems from damage by transient overvoltages. 1989 ed. xxiv+434pp. $6^1/_8$ x $9^1/_4$. 0-486-42552-5

ROTARY WING AERODYNAMICS, W. Z. Stepniewski. Clear, concise text covers aerodynamic phenomena of the rotor and offers guidelines for helicopter performance evaluation. Originally prepared for NASA. 537 figures. 640pp. $6^1/_8$ x $9^1/_4$. 0-486-64647-5

INTRODUCTION TO SPACE DYNAMICS, William Tyrrell Thomson. Comprehensive, classic introduction to space-flight engineering for advanced undergraduate and graduate students. Includes vector algebra, kinematics, transformation of coordinates. Bibliography. Index. 352pp. $5^3/_8$ x $8^1/_2$. 0-486-65113-4

HISTORY OF STRENGTH OF MATERIALS, Stephen P. Timoshenko. Excellent historical survey of the strength of materials with many references to the theories of elasticity and structure. 245 figures. 452pp. $5^3/_8$ x $8^1/_2$. 0-486-61187-6

ANALYTICAL FRACTURE MECHANICS, David J. Unger. Self-contained text supplements standard fracture mechanics texts by focusing on analytical methods for determining crack-tip stress and strain fields. 336pp. $6^1/_8$ x $9^1/_4$. 0-486-41737-9

STATISTICAL MECHANICS OF ELASTICITY, J. H. Weiner. Advanced, self-contained treatment illustrates general principles and elastic behavior of solids. Part 1, based on classical mechanics, studies thermoelastic behavior of crystalline and polymeric solids. Part 2, based on quantum mechanics, focuses on interatomic force laws, behavior of solids, and thermally activated processes. For students of physics and chemistry and for polymer physicists. 1983 ed. 96 figures. 496pp. $5^3/_8$ x $8^1/_2$. 0-486-42260-7

Mathematics

FUNCTIONAL ANALYSIS (Second Corrected Edition), George Bachman and Lawrence Narici. Excellent treatment of subject geared toward students with background in linear algebra, advanced calculus, physics and engineering. Text covers introduction to inner-product spaces, normed, metric spaces, and topological spaces; complete orthonormal sets, the Hahn-Banach Theorem and its consequences, and many other related subjects. 1966 ed. 544pp. $6^1/_8$ x $9^1/_4$. 0-486-40251-7

DIFFERENTIAL MANIFOLDS, Antoni A. Kosinski. Introductory text for advanced undergraduates and graduate students presents systematic study of the topological structure of smooth manifolds, starting with elements of theory and concluding with method of surgery. 1993 edition. 288pp. $5^3/_8$ x $8^1/_2$. 0-486-46244-7

VECTOR AND TENSOR ANALYSIS WITH APPLICATIONS, A. I. Borisenko and I. E. Tarapov. Concise introduction. Worked-out problems, solutions, exercises. 257pp. $5^5/_8$ x $8^1/_4$. 0-486-63833-2

AN INTRODUCTION TO ORDINARY DIFFERENTIAL EQUATIONS, Earl A. Coddington. A thorough and systematic first course in elementary differential equations for undergraduates in mathematics and science, with many exercises and problems (with answers). Index. 304pp. $5^3/_8$ x $8^1/_2$. 0-486-65942-9

FOURIER SERIES AND ORTHOGONAL FUNCTIONS, Harry F. Davis. An incisive text combining theory and practical example to introduce Fourier series, orthogonal functions and applications of the Fourier method to boundary-value problems. 570 exercises. Answers and notes. 416pp. $5^3/_8$ x $8^1/_2$. 0-486-65973-9

COMPUTABILITY AND UNSOLVABILITY, Martin Davis. Classic graduate-level introduction to theory of computability, usually referred to as theory of recurrent functions. New preface and appendix. 288pp. $5^3/_8$ x $8^1/_2$. 0-486-61471-9

AN INTRODUCTION TO MATHEMATICAL ANALYSIS, Robert A. Rankin. Dealing chiefly with functions of a single real variable, this text by a distinguished educator introduces limits, continuity, differentiability, integration, convergence of infinite series, double series, and infinite products. 1963 edition. 624pp. $5^3/_8$ x $8^1/_2$. 0-486-46251-X

METHODS OF NUMERICAL INTEGRATION (SECOND EDITION), Philip J. Davis and Philip Rabinowitz. Requiring only a background in calculus, this text covers approximate integration over finite and infinite intervals, error analysis, approximate integration in two or more dimensions, and automatic integration. 1984 edition. 624pp. $5^3/_8$ x $8^1/_2$. 0-486-45339-1

INTRODUCTION TO LINEAR ALGEBRA AND DIFFERENTIAL EQUATIONS, John W. Dettman. Excellent text covers complex numbers, determinants, orthonormal bases, Laplace transforms, much more. Exercises with solutions. Undergraduate level. 416pp. $5^3/_8$ x $8^1/_2$. 0-486-65191-6

RIEMANN'S ZETA FUNCTION, H. M. Edwards. Superb, high-level study of landmark 1859 publication entitled "On the Number of Primes Less Than a Given Magnitude" traces developments in mathematical theory that it inspired. xiv+315pp. $5^3/_8$ x $8^1/_2$. 0-486-41740-9

CALCULUS OF VARIATIONS WITH APPLICATIONS, George M. Ewing. Applications-oriented introduction to variational theory develops insight and promotes understanding of specialized books, research papers. Suitable for advanced undergraduate/graduate students as primary, supplementary text. 352pp. $5^3/_8 \times 8^1/_2$. 0-486-64856-7

MATHEMATICIAN'S DELIGHT, W. W. Sawyer. "Recommended with confidence" by *The Times Literary Supplement,* this lively survey was written by a renowned teacher. It starts with arithmetic and algebra, gradually proceeding to trigonometry and calculus. 1943 edition. 240pp. $5^3/_8 \times 8^1/_2$. 0-486-46240-4

ADVANCED EUCLIDEAN GEOMETRY, Roger A. Johnson. This classic text explores the geometry of the triangle and the circle, concentrating on extensions of Euclidean theory, and examining in detail many relatively recent theorems. 1929 edition. 336pp. $5^3/_8 \times 8^1/_2$. 0-486-46237-4

COUNTEREXAMPLES IN ANALYSIS, Bernard R. Gelbaum and John M. H. Olmsted. These counterexamples deal mostly with the part of analysis known as "real variables." The first half covers the real number system, and the second half encompasses higher dimensions. 1962 edition. xxiv+198pp. $5^3/_8 \times 8^1/_2$. 0-486-42875-3

CATASTROPHE THEORY FOR SCIENTISTS AND ENGINEERS, Robert Gilmore. Advanced-level treatment describes mathematics of theory grounded in the work of Poincaré, R. Thom, other mathematicians. Also important applications to problems in mathematics, physics, chemistry and engineering. 1981 edition. References. 28 tables. 397 black-and-white illustrations. xvii + 666pp. $6^1/_8 \times 9^1/_4$. 0-486-67539-4

COMPLEX VARIABLES: Second Edition, Robert B. Ash and W. P. Novinger. Suitable for advanced undergraduates and graduate students, this newly revised treatment covers Cauchy theorem and its applications, analytic functions, and the prime number theorem. Numerous problems and solutions. 2004 edition. 224pp. $6^1/_2 \times 9^1/_4$. 0-486-46250-1

NUMERICAL METHODS FOR SCIENTISTS AND ENGINEERS, Richard Hamming. Classic text stresses frequency approach in coverage of algorithms, polynomial approximation, Fourier approximation, exponential approximation, other topics. Revised and enlarged 2nd edition. 721pp. $5^3/_8 \times 8^1/_2$. 0-486-65241-6

INTRODUCTION TO NUMERICAL ANALYSIS (2nd Edition), F. B. Hildebrand. Classic, fundamental treatment covers computation, approximation, interpolation, numerical differentiation and integration, other topics. 150 new problems. 669pp. $5^3/_8 \times 8^1/_2$. 0-486-65363-3

MARKOV PROCESSES AND POTENTIAL THEORY, Robert M. Blumental and Ronald K. Getoor. This graduate-level text explores the relationship between Markov processes and potential theory in terms of excessive functions, multiplicative functionals and subprocesses, additive functionals and their potentials, and dual processes. 1968 edition. 320pp. $5^3/_8 \times 8^1/_2$. 0-486-46263-3

ABSTRACT SETS AND FINITE ORDINALS: An Introduction to the Study of Set Theory, G. B. Keene. This text unites logical and philosophical aspects of set theory in a manner intelligible to mathematicians without training in formal logic and to logicians without a mathematical background. 1961 edition. 112pp. $5^3/_8 \times 8^1/_2$. 0-486-46249-8

Math—Decision Theory, Statistics, Probability

INTRODUCTION TO PROBABILITY, John E. Freund. Featured topics include permutations and factorials, probabilities and odds, frequency interpretation, mathematical expectation, decision-making, postulates of probability, rule of elimination, much more. Exercises with some solutions. Summary. 1973 edition. 247pp. $5^{3}/_{8}$ x $8^{1}/_{2}$.
0-486-67549-1

STATISTICAL AND INDUCTIVE PROBABILITIES, Hugues Leblanc. This treatment addresses a decades-old dispute among probability theorists, asserting that both statistical and inductive probabilities may be treated as sentence-theoretic measurements, and that the latter qualify as estimates of the former. 1962 edition. 160pp. $5^{3}/_{8}$ x $8^{1}/_{2}$.
0-486-44980-7

APPLIED MULTIVARIATE ANALYSIS: Using Bayesian and Frequentist Methods of Inference, Second Edition, S. James Press. This two-part treatment deals with foundations as well as models and applications. Topics include continuous multivariate distributions; regression and analysis of variance; factor analysis and latent structure analysis; and structuring multivariate populations. 1982 edition. 692pp. $5^{3}/_{8}$ x $8^{1}/_{2}$. 0-486-44236-5

LINEAR PROGRAMMING AND ECONOMIC ANALYSIS, Robert Dorfman, Paul A. Samuelson and Robert M. Solow. First comprehensive treatment of linear programming in standard economic analysis. Game theory, modern welfare economics, Leontief input-output, more. 525pp. $5^{3}/_{8}$ x $8^{1}/_{2}$. 0-486-65491-5

PROBABILITY: AN INTRODUCTION, Samuel Goldberg. Excellent basic text covers set theory, probability theory for finite sample spaces, binomial theorem, much more. 360 problems. Bibliographies. 322pp. $5^{3}/_{8}$ x $8^{1}/_{2}$. 0-486-65252-1

GAMES AND DECISIONS: INTRODUCTION AND CRITICAL SURVEY, R. Duncan Luce and Howard Raiffa. Superb nontechnical introduction to game theory, primarily applied to social sciences. Utility theory, zero-sum games, n-person games, decision-making, much more. Bibliography. 509pp. $5^{3}/_{8}$ x $8^{1}/_{2}$. 0-486-65943-7

INTRODUCTION TO THE THEORY OF GAMES, J. C. C. McKinsey. This comprehensive overview of the mathematical theory of games illustrates applications to situations involving conflicts of interest, including economic, social, political, and military contexts. Appropriate for advanced undergraduate and graduate courses; advanced calculus a prerequisite. 1952 ed. x+372pp. $5^{3}/_{8}$ x $8^{1}/_{2}$. 0-486-42811-7

FIFTY CHALLENGING PROBLEMS IN PROBABILITY WITH SOLUTIONS, Frederick Mosteller. Remarkable puzzlers, graded in difficulty, illustrate elementary and advanced aspects of probability. Detailed solutions. 88pp. $5^{3}/_{8}$ x $8^{1}/_{2}$. 0-486-65355-2

PROBABILITY THEORY: A CONCISE COURSE, Y. A. Rozanov. Highly readable, self-contained introduction covers combination of events, dependent events, Bernoulli trials, etc. 148pp. $5^{5}/_{8}$ x $8^{1}/_{4}$. 0-486-63544-9

THE STATISTICAL ANALYSIS OF EXPERIMENTAL DATA, John Mandel. First half of book presents fundamental mathematical definitions, concepts and facts while remaining half deals with statistics primarily as an interpretive tool. Well-written text, numerous worked examples with step-by-step presentation. Includes 116 tables. 448pp. $5^{3}/_{8}$ x $8^{1}/_{2}$. 0-486-64666-1

Math—Geometry and Topology

ELEMENTARY CONCEPTS OF TOPOLOGY, Paul Alexandroff. Elegant, intuitive approach to topology from set-theoretic topology to Betti groups; how concepts of topology are useful in math and physics. 25 figures. 57pp. 5³/₈ x 8¹/₂. 0-486-60747-X

A LONG WAY FROM EUCLID, Constance Reid. Lively guide by a prominent historian focuses on the role of Euclid's Elements in subsequent mathematical developments. Elementary algebra and plane geometry are sole prerequisites. 80 drawings. 1963 edition. 304pp. 5³/₈ x 8¹/₂. 0-486-43613-6

EXPERIMENTS IN TOPOLOGY, Stephen Barr. Classic, lively explanation of one of the byways of mathematics. Klein bottles, Moebius strips, projective planes, map coloring, problem of the Koenigsberg bridges, much more, described with clarity and wit. 43 figures. 210pp. 5³/₈ x 8¹/₂. 0-486-25933-1

THE GEOMETRY OF RENÉ DESCARTES, René Descartes. The great work founded analytical geometry. Original French text, Descartes's own diagrams, together with definitive Smith-Latham translation. 244pp. 5³/₈ x 8¹/₂. 0-486-60068-8

EUCLIDEAN GEOMETRY AND TRANSFORMATIONS, Clayton W. Dodge. This introduction to Euclidean geometry emphasizes transformations, particularly isometries and similarities. Suitable for undergraduate courses, it includes numerous examples, many with detailed answers. 1972 ed. viii+296pp. 6¹/₈ x 9¹/₄. 0-486-43476-1

EXCURSIONS IN GEOMETRY, C. Stanley Ogilvy. A straightedge, compass, and a little thought are all that's needed to discover the intellectual excitement of geometry. Harmonic division and Apollonian circles, inversive geometry, hexlet, Golden Section, more. 132 illustrations. 192pp. 5³/₈ x 8¹/₂. 0-486-26530-7

THE THIRTEEN BOOKS OF EUCLID'S ELEMENTS, translated with introduction and commentary by Sir Thomas L. Heath. Definitive edition. Textual and linguistic notes, mathematical analysis. 2,500 years of critical commentary. Unabridged. 1,414pp. 5³/₈ x 8¹/₂. Three-vol. set.
Vol. I: 0-486-60088-2 Vol. II: 0-486-60089-0 Vol. III: 0-486-60090-4

SPACE AND GEOMETRY: IN THE LIGHT OF PHYSIOLOGICAL, PSYCHOLOGICAL AND PHYSICAL INQUIRY, Ernst Mach. Three essays by an eminent philosopher and scientist explore the nature, origin, and development of our concepts of space, with a distinctness and precision suitable for undergraduate students and other readers. 1906 ed. vi+148pp. 5³/₈ x 8¹/₂. 0-486-43909-7

GEOMETRY OF COMPLEX NUMBERS, Hans Schwerdtfeger. Illuminating, widely praised book on analytic geometry of circles, the Moebius transformation, and two-dimensional non-Euclidean geometries. 200pp. 5⁵/₈ x 8¹/₄. 0-486-63830-8

DIFFERENTIAL GEOMETRY, Heinrich W. Guggenheimer. Local differential geometry as an application of advanced calculus and linear algebra. Curvature, transformation groups, surfaces, more. Exercises. 62 figures. 378pp. 5³/₈ x 8¹/₂. 0-486-63433-7

History of Math

THE WORKS OF ARCHIMEDES, Archimedes (T. L. Heath, ed.). Topics include the famous problems of the ratio of the areas of a cylinder and an inscribed sphere; the measurement of a circle; the properties of conoids, spheroids, and spirals; and the quadrature of the parabola. Informative introduction. clxxxvi+326pp. $5^3/_8$ x $8^1/_2$. 0-486-42084-1

A SHORT ACCOUNT OF THE HISTORY OF MATHEMATICS, W. W. Rouse Ball. One of clearest, most authoritative surveys from the Egyptians and Phoenicians through 19th-century figures such as Grassman, Galois, Riemann. Fourth edition. 522pp. $5^3/_8$ x $8^1/_2$. 0-486-20630-0

THE HISTORY OF THE CALCULUS AND ITS CONCEPTUAL DEVELOPMENT, Carl B. Boyer. Origins in antiquity, medieval contributions, work of Newton, Leibniz, rigorous formulation. Treatment is verbal. 346pp. $5^3/_8$ x $8^1/_2$. 0-486-60509-4

THE HISTORICAL ROOTS OF ELEMENTARY MATHEMATICS, Lucas N. H. Bunt, Phillip S. Jones, and Jack D. Bedient. Fundamental underpinnings of modern arithmetic, algebra, geometry and number systems derived from ancient civilizations. 320pp. $5^3/_8$ x $8^1/_2$. 0-486-25563-8

THE HISTORY OF THE CALCULUS AND ITS CONCEPTUAL DEVELOPMENT, Carl B. Boyer. Fluent description of the development of both the integral and differential calculus—its early beginnings in antiquity, medieval contributions, and a consideration of Newton and Leibniz. 368pp. $5^3/_8$ x $8^1/_2$. 0-486-60509-4

GAMES, GODS & GAMBLING: A HISTORY OF PROBABILITY AND STATISTICAL IDEAS, F. N. David. Episodes from the lives of Galileo, Fermat, Pascal, and others illustrate this fascinating account of the roots of mathematics. Features thought-provoking references to classics, archaeology, biography, poetry. 1962 edition. 304pp. $5^3/_8$ x $8^1/_2$. (Available in U.S. only.) 0-486-40023-9

OF MEN AND NUMBERS: THE STORY OF THE GREAT MATHEMATICIANS, Jane Muir. Fascinating accounts of the lives and accomplishments of history's greatest mathematical minds—Pythagoras, Descartes, Euler, Pascal, Cantor, many more. Anecdotal, illuminating. 30 diagrams. Bibliography. 256pp. $5^3/_8$ x $8^1/_2$. 0-486-28973-7

HISTORY OF MATHEMATICS, David E. Smith. Nontechnical survey from ancient Greece and Orient to late 19th century; evolution of arithmetic, geometry, trigonometry, calculating devices, algebra, the calculus. 362 illustrations. 1,355pp. $5^3/_8$ x $8^1/_2$. Two-vol. set. Vol. I: 0-486-20429-4 Vol. II: 0-486-20430-8

A CONCISE HISTORY OF MATHEMATICS, Dirk J. Struik. The best brief history of mathematics. Stresses origins and covers every major figure from ancient Near East to 19th century. 41 illustrations. 195pp. $5^3/_8$ x $8^1/_2$. 0-486-60255-9

Physics

OPTICAL RESONANCE AND TWO-LEVEL ATOMS, L. Allen and J. H. Eberly. Clear, comprehensive introduction to basic principles behind all quantum optical resonance phenomena. 53 illustrations. Preface. Index. 256pp. 5³/₈ x 8¹/₂. 0-486-65533-4

QUANTUM THEORY, David Bohm. This advanced undergraduate-level text presents the quantum theory in terms of qualitative and imaginative concepts, followed by specific applications worked out in mathematical detail. Preface. Index. 655pp. 5³/₈ x 8¹/₂. 0-486-65969-0

ATOMIC PHYSICS (8th EDITION), Max Born. Nobel laureate's lucid treatment of kinetic theory of gases, elementary particles, nuclear atom, wave-corpuscles, atomic structure and spectral lines, much more. Over 40 appendices, bibliography. 495pp. 5³/₈ x 8¹/₂. 0-486-65984-4

A SOPHISTICATE'S PRIMER OF RELATIVITY, P. W. Bridgman. Geared toward readers already acquainted with special relativity, this book transcends the view of theory as a working tool to answer natural questions: What is a frame of reference? What is a "law of nature"? What is the role of the "observer"? Extensive treatment, written in terms accessible to those without a scientific background. 1983 ed. xlviii+172pp. 5³/₈ x 8¹/₂. 0-486-42549-5

AN INTRODUCTION TO HAMILTONIAN OPTICS, H. A. Buchdahl. Detailed account of the Hamiltonian treatment of aberration theory in geometrical optics. Many classes of optical systems defined in terms of the symmetries they possess. Problems with detailed solutions. 1970 edition. xv + 360pp. 5³/₈ x 8¹/₂. 0-486-67597-1

PRIMER OF QUANTUM MECHANICS, Marvin Chester. Introductory text examines the classical quantum bead on a track: its state and representations; operator eigenvalues; harmonic oscillator and bound bead in a symmetric force field; and bead in a spherical shell. Other topics include spin, matrices, and the structure of quantum mechanics; the simplest atom; indistinguishable particles; and stationary-state perturbation theory. 1992 ed. xiv+314pp. 6¹/₈ x 9¹/₄. 0-486-42878-8

LECTURES ON QUANTUM MECHANICS, Paul A. M. Dirac. Four concise, brilliant lectures on mathematical methods in quantum mechanics from Nobel Prize-winning quantum pioneer build on idea of visualizing quantum theory through the use of classical mechanics. 96pp. 5³/₈ x 8¹/₂. 0-486-41713-1

THIRTY YEARS THAT SHOOK PHYSICS: THE STORY OF QUANTUM THEORY, George Gamow. Lucid, accessible introduction to influential theory of energy and matter. Careful explanations of Dirac's anti-particles, Bohr's model of the atom, much more. 12 plates. Numerous drawings. 240pp. 5³/₈ x 8¹/₂. 0-486-24895-X

ELECTRONIC STRUCTURE AND THE PROPERTIES OF SOLIDS: THE PHYSICS OF THE CHEMICAL BOND, Walter A. Harrison. Innovative text offers basic understanding of the electronic structure of covalent and ionic solids, simple metals, transition metals and their compounds. Problems. 1980 edition. 582pp. 6¹/₈ x 9¹/₄. 0-486-66021-4